Nutritional Neurosciences

Series Editor
Mohamed Essa, Sultan Qaboos University
Qaboos, Oman

This book series aims to publish volumes focusing on both basic and clinical research in the field of nutritional neuroscience with a focus on delineating the effect of nutrition on brain function and behavior. The books will examine the role of different nutrients, food agents and supplements (both macro and micro) on brain health, neurodevelopment, neurochemistry, and behaviour. The books will examine the influence of diet, including phytochemicals, antioxidants, dietary supplements, food additives, and other nutrients on the physiology and metabolism of neurons, neurotransmitters and their receptors, cognition, behavior, and hormonal regulations.

The books will also cover the influence of nutrients and dietary supplements on the management of neurological disorders. It details the mechanism of action of phytonutrients on signaling pathways linked with protein folding, aggregation, and neuroinflammation. The books published in the series will be useful for neuroscientists, nutritionists, neurologists, psychiatrists, and those interested in preventive medicine.

Jai Malik • Subhash C. Mandal
Prerna Sarup • G. David Lin
Rachel W. Li
Editors

Diet and Alzheimer's Disease: Let Food be Our Medicine

Springer

Editors
Jai Malik
University Institute of Pharmaceutical Sciences
Panjab University
Chandigarh, India

Prerna Sarup
University Institute of Pharmaceutical Sciences
Panjab University
Chandigarh, India

Rachel W. Li
Department of Surgery
Australian National University Canberra
Canberra, ACT, Australia

Subhash C. Mandal
Department of Pharmaceutical Technology
Jadavpur University
Kolkata, West Bengal, India

G. David Lin
Research School of Chemistry
Australian National University Canberra
Canberra, ACT, Australia

ISSN 2730-6712 ISSN 2730-6720 (electronic)
Nutritional Neurosciences
ISBN 978-981-95-2735-9 ISBN 978-981-95-2736-6 (eBook)
https://doi.org/10.1007/978-981-95-2736-6

This Springer imprint is published by the registered company Springer Nature Singapore Pte Ltd.
The registered company address is: 152 Beach Road, #21-01/04 Gateway East, Singapore 189721, Singapore

The book is dedicated:
To the millions affected by Alzheimer's disease—may this book be a source of knowledge, comfort, and possibility.
To the caregivers and families who walk this path with courage and grace—your love is the light in the darkness.

To the scientific community who is working tirelessly in the field of Alzheimer's disease and those who are highlighting the role of diet and dietary components in various diseases.

Preface

Hippocrates once said, "Thy Food Thy Medicine"—Let your Food Be your Medicine; though this saying is centuries old, it is still relevant but rather its relevance has increased manyfold in recent past. Over the period of time, the importance of diet has been realized just for providing nutrients to the body. But, in the recent past its therapeutic aspects have garnered significant attention. It has been found that a good diet not only helps in mitigating the symptoms or slowing the progression of various diseases but also delaying their onset. Ample evidences, both preclinical and clinical, have proved the efficacy of diet or a dietary pattern (when followed for sufficient period of time) in not only slowing the progression but delaying the onset of diseases especially the degenerative disorders. With the advancement in the field of therapeutics, there has been significant increase in the life expectancy of man which brings along with it the burden of degenerative disorders. Among various degenerative disorders, neurodegenerative disorders (dementia, Alzheimer's disease, Parkinson's disease, Huntington's disease, etc.) have wide epidemiology, which not only affect the patients but also the caregivers or family members. Alzheimer's disease is a complex and multifaceted neurodegenerative disorder that affects millions of people worldwide. While significant progress has been made in understanding the disease's pathophysiology, effective prevention and treatment strategies remain elusive. However, emerging evidence suggests that dietary interventions may play a crucial role in mitigating the risk and progression of Alzheimer's disease. This book brings together cutting-edge research on the relationship between diet and Alzheimer's disease (AD), highlighting the potential benefits of various dietary patterns, foods, nutrients, and dietary compounds in preventing and managing the disease. It also provides the latest scientific insights into the connection between nutrition and Alzheimer's, shining a light on the protective potential of certain foods, nutrients, and natural compounds. Each chapter dives into a specific aspect of this relationship, offering both depth and practical implications. Chapter 1 gives an overview of the pathophysiology of the disease highlighting various hypothesis that play an important role in the pathogenesis of the disease. It also highlights current therapeutic interventions and other therapeutic strategies that form the basis for the drug discovery of AD. Chapter 2 introduces the concept of nutrition-based neuroprotection, laying the groundwork for innovative dietary and complementary therapeutic strategies with a special emphasis on the molecular basis of protective effect of dietary components against AD. Chapter 3 examines the role of inflammatory pathways in the pathogenesis of AD and how dietary anti-inflammatory nutrients may reduce Alzheimer's risk, focusing on their biochemical actions and protective mechanisms. Chapter 4 delves into the Mediterranean diet, exploring its core principles and strong evidence from both animal and human studies supporting its neuroprotective effects. Chapter 5 highlights a new aspect of the pathogenesis, the gut brain axis, and how gut microbiome plays an important role in neuroprotection. It emphasizes on the exciting role of probiotics and prebiotics in managing AD, and how a balanced gut microbiome can influence brain health and helps combat neuroinflammation. Chapter 6 highlights curcumin, the active compound in turmeric, and its promising role in preventing or slowing Alzheimer's through its antioxidant and anti-amyloid properties. Chapter 7 presents a detailed look at the second largest consumed beverage after water, i.e., tea and its active constituents—catechins, and unravels their molecular mechanisms and potential benefits for brain function. Chapter 8 offers an Ayurvedic perspective on diet and lifestyle, emphasizing mindfulness and holistic eating in managing cognitive health. Chapter 9 focuses on nuts—rich in healthy fats, antioxidants, and bioactive compounds—as a brain-boosting dietary addition and their role in managing AD. Chapter 10 discusses the balance of omega-3 and omega-6 fatty acids and their roles in neuroprotection and inflammation modulation. It also highlights the dietary sources of these important classes of fatty acids. Chapter 11 explores ginger, its active constituents, and its potent molecular mechanisms that support cognitive function and reduce neurodegeneration in AD. Chapter 12 reviews various carotenoids and their antioxidant-rich profiles, which show promise in protecting neural tissue and also delay the onset and mitigate the symptoms of AD. Chapter 13 discusses polyphenols as emerging modulators of Alzheimer's pathogenesis, with potential to intervene at various stages of the disease. Chapter 14 presents a thorough analysis of Maca whose roots are consumed for its nutritional and medicinal values. The chapter apprises the readers about its neuroprotective effects from both in vitro and in vivo studies in AD. Chapter 15 focuses on perhaps one of the most extensively studied phytomolecules, resveratrol. The chapter focuses on detailing its ability to support blood-brain barrier integrity and its molecular pathways to influence the onset, progression, and symptoms of AD. Chapter 16 explores the cognitive benefits of mushrooms and their unique neuroactive compounds that support brain health. Chapter 17 revisits an interesting and a new area of brain health through the gut-brain axis. It highlights how modulating gut health or gut microbiome population and diversity through diet may offer new frontiers in Alzheimer's care. Chapter 18 delves into the benefits of extra virgin olive oil, revealing how its unique properties may slow Alzheimer's onset and progression. Chapter 19 investigates the role of different vitamins in cognitive preservation, focusing on how deficiencies or supplementation can impact brain aging and can be helpful in AD. Finally, Chap. 20 concludes with an in-depth journey of one of the largest class of dietary compounds, polyphenols—tracing their path through the body as they target key Alzheimer's mechanisms, from improving mitochondrial health to safeguarding neural communication.

The demonstrated ability of the authors to present wide and complicated information with regard to the molecular mechanism of the dietary components and their role in managing AD makes this book a useful tool for the researchers working the field of diet and AD, academicians, dieticians, biochemists, students, food biochemists, clinicians, and nutritionists. Each chapter in this book has a list of references that can also help the readers for supplementary and detailed reading of the topics of their interest. In a nut shell, this book serves as a vital resource for understanding how nutrition can influence brain health and offers evidence-backed insights into how simple dietary choices may pave the way for effective Alzheimer's prevention and management strategies. We hope that our attempts to bring the widespread information on diet and its role in AD at a single platform will help in initiating more detailed mechanistic studies on the molecular/cellular pathways of dietary components and their effect on the disease biomarkers.

Chandigarh, India — Jai Malik
Kolkata, West Bengal, India — Subhash C. Mandal
Chandigarh, India — Prerna Sarup
Canberra, ACT, Australia — G. David Lin
Canberra, ACT, Australia — Rachel W. Li

Contents

Editors and Contributors

About the Editors

Jai Malik As an Associate Professor of Pharmacognosy at University Institute of Pharmaceutical Sciences, Panjab University, Chandigarh, Dr. Malik has a strong research focus on natural product standardization and their pharmacology, particularly on plant drugs having neuroprotective potential. He has published peer-reviewed publications in reputed journals, two books, and various book chapters. His work bridges traditional medicine and modern pharmacology. He has received accolades including the PD Sethi Memorial Award, best research paper awards at various national and international conferences for his contributions to phytochemical research, and fellowship by UNIDO. His ongoing studies explore plant-based therapeutics for neurological disorders, aligning with the global shift toward evidence-based herbal medicine.

Subash C. Mandal Dr. Subhash C. Mandal, PhD, is Professor at Division of Pharmacognosy, Department of Pharmaceutical Technology, Jadavpur University, Kolkata, India. He is the recipient of prestigious awards including Endeavour Research Award, Government of Australia; Distinguished Education & Research Award, AAiPS, United States; UGC Research Award, India; and Outstanding University Teachers' Award, Department of higher education Government of West Bengal, India. He has supervised more than 35 doctoral and 30 MPharm scholars and has more than 200 research publications, several patents, books, and book chapters to his credit.

Prerna Sarup Currently working as a faculty (ad hoc) at University Institute of Pharmaceutical Sciences, Panjab University, Chandigarh, Dr. Prerna is a recipient of 8th Global Excellence Teacher Award (2018) for excellent achievements in education field on National Education Day by ACFI and UGC Research Fellowships in Science for Meritorious Students (UGC-RFSMS) to pursue her doctoral degree from Panjab University. She has been associated with providing quality pharmacy education as an academician and researcher from past 20 years. She has presented her research at various national and international conferences, wherein she was conferred with Best Paper Award thrice. She has published various research and review articles, in prestigious journals, primarily in areas of natural product isolation, drug discovery, development of herbal formulations, and drug delivery.

G. David Lin Dr. G. David Lin is Adjunct Associate Professor at the Research School of Chemistry, Australian National University. His research focuses on the chemistry, pharmacology, and biotechnology of natural products. He has held academic positions in both Australia and the United States, and currently serves in the Australian public service. He established the Natural Products Research Laboratory at Southern Cross University and was a founding Senior Lecturer in Pharmacy at the University of Canberra.

Rachel W. Li Associate/Professor Rachel W. Li is a molecular pharmacologist and osteoimmunologist with interests in understanding the processes that control a "foreign body reaction or response" initiated by biomaterials implanted/natural compounds into bone or exposed to human cells. With her clinician background, she led several clinical trials in antiviral and anti-inflammatory drugs and successfully transferred an intellectual property to the pharmaceutical industry. She established the Trauma and Orthopaedic Research Unit Laboratory (TORU Lab) at the Australian National University, which pioneered basic orthopedic research in the ACT region. She is currently leading her team in clinical-based laboratory research focuses on osteoimmunology and biomaterials for medical implants. She has made great research contributions to the fields of osteoimmunology and medical education in the USA and Australia.

Contributors

Navidha Aggarwal M. M. College of Pharmacy, Maharishi Markandeshwar (Deemed to be University), Ambala, Haryana, India

Touqeer Ahmed Neurobiology Laboratory, Department of Biomedicine, Atta-ur-Rahman School of Applied Biosciences (ASAB), National University of Sciences and Technology (NUST), Islamabad, Pakistan

Md Yousuf Ansari Ibne Seena Pharmacy College, Shahabad, Hardoi, Uttar Pradesh, India

G. Babu Central Ayurveda Research Institute (CARI), Central Council for Research in Ayurvedic Sciences (CCRAS), Kolkata, West Bengal, India

Kajal Bagri Department of Pharmacology, Central University of Punjab, Bathinda, Punjab, India

Sayani Banerjee Department of Pharmacology and Toxicology, National institute of Pharmaceutical Education and Research, Kolkata, West Bengal, India

Sugato Banerjee Department of Pharmacology and Toxicology, National institute of Pharmaceutical Education and Research, Kolkata, West Bengal, India

Seema Bansal M. M. College of Pharmacy, Maharishi Markandeshwar (Deemed to be University), Ambala, Haryana, India

Saima Bashir Neurobiology Laboratory, Department of Biomedicine, Atta-ur-Rahman School of Applied Biosciences (ASAB), National University of Sciences and Technology (NUST), Islamabad, Pakistan

Divya Dhawal Bhandari University Institute of Pharmaceutical Sciences, Panjab University, Chandigarh, India

Mahendra Bishnoi National Agri-Food Biotechnology Institute, Sahibzada Ajit Singh Nagar, Punjab, India

Prarit Chandel Chitkara University School of Pharmacy, Chitkara University, Baddi, Himachal Pradesh, India

Sunayna Choudhary M. M. College of Pharmacy, Maharishi Markandeshwar (Deemed to be University), Ambala, Haryana, India

John Curtin School of Medical Research, The Australian National University, Canberra, ACT, Australia

Jyoti Dahiya Central Ayurveda Research Institute (CARI), Central Council for Research in Ayurvedic Sciences (CCRAS), Kolkata, West Bengal, India

Rahul Deshmukh Department of Pharmacology, Central University of Punjab, Bathinda, Punjab, India

Rakesh Kumar Dolai Dr. B. C. Roy College of Pharmacy and AHS, Durgapur, West Bengal, India

Meenakshi Garg Department of Food Technology, Bhaskaracharya College of Applied Science, New Delhi, Delhi, India

Ankita Ghosh Central Ayurveda Research Institute (CARI), Central Council for Research in Ayurvedic Sciences (CCRAS), Kolkata, West Bengal, India

Nilanjan Ghosh Department of Pharmaceutical Technology, Jadavpur University, Kolkata, West Bengal, India

Anwar-ul-Hassan Gilani Higher Education Commission, H-9, Islamabad, Pakistan

Gamze GÖGER Department of Pharmacognosy, Faculty of Pharmacy, Afyonkarahisar Health Sciences University, Afyonkarahisar, Türkiye

Gandhodi Gopi Department of Pharmaceutical Chemistry, Mahathi College of Pharmacy, Madanapalle, Andhra Pradesh, India

Kiranmai Gudimetla GITAM School of Pharmacy, GITAM University Hyderabad Campus, Rudraram, Telangana, India

Amit Kumar Halder Dr. B. C. Roy College of Pharmacy and AHS, Durgapur, West Bengal, India

Emani Sai Sri Jayanthi Department of Pharmacy Practice, Pulla Reddy Institute of Pharmacy, Jawaharlal Nehru Technological University, Hyderabad, Telangana, India

Nitin Jhangra Chitkara University School of Pharmacy, Chitkara University, Baddi, Himachal Pradesh, India

Dipesh Joshi Novotech (Australia) Pty Ltd., Sydney, NSW, Australia

Neha Kanojia Chitkara University School of Pharmacy, Chitkara University, Baddi, Himachal Pradesh, India

Songül Karakaya Department of Pharmaceutical Botany, Faculty of Pharmacy, Atatürk University, Erzurum, Türkiye

Navkiran Kaur Amity Institute of Biotechnology, Amity University, Noida, India

Areena Suhail Khan Neurobiology Laboratory, Department of Biomedicine, Atta-ur-Rahman School of Applied Biosciences (ASAB), National University of Sciences and Technology (NUST), Islamabad, Pakistan

Poonam Khare Department of Allied Health Sciences, Evergreen Valley College, San Jose, CA, USA

Ritu Khosla Department of Biomedical Science, Acharya Narendra Dev College, University of Delhi, New Delhi, Delhi, India

Aleyna Kopuz Department of Graphic Arts, Faculty of Fine Arts, Atatürk University, Erzurum, Türkiye

Shilpa Kumari Department of Pharmacology, Central University of Punjab, Bathinda, Punjab, India

Pradeep Kumar Department of Pharmacology and Toxicology, National institute of Pharmaceutical Education and Research, Kolkata, West Bengal, India

G. David Lin Research School of Chemistry, the Australian National University, Canberra, ACT, Australia

Rachel W. Li School of Medicine and Psychology, the Australian National University, Canberra, ACT, Australia

John Curtin School of Medical Research, the Australian National University, Canberra, ACT, Australia

Jai Malik University Institute of Pharmaceutical Sciences, Panjab University, Chandigarh, India

Subhash Chandra Mandal Department of Pharmaceutical Technology, Jadavpur University, Kolkata, West Bengal, India

Anupam K. Mangal Central Ayurveda Research Institute (CARI), Central Council for Research in Ayurvedic Sciences (CCRAS), Kolkata, West Bengal, India

Rani Mansuri School of Pharmaceutical Sciences, Apeejay Stya University, Gurugram, Haryana, India

Vinyas Mayasa GITAM School of Pharmacy, GITAM University Hyderabad Campus, Rudraram, Telangana, India

Soumya Mitra Dr. B. C. Roy College of Pharmacy and AHS, Durgapur, West Bengal, India

Department of Pharmaceutical Technology, Jadavpur University, Kolkata, West Bengal, India

Monika School of Pharmaceutical Sciences, Apeejay Stya University, Gurugram, Haryana, India

Poonam Negi Chitkara College of Pharmacy, Chitkara University, Rajpura, India

Vinod Kumar Nelson Department of Pharmaceutical Chemistry, Mahathi College of Pharmacy, Madanapalle, Andhra Pradesh, India

Nitish Nirala Department of Pharmacology and Toxicology, National institute of Pharmaceutical Education and Research, Kolkata, West Bengal, India

Prachi Patel Amity Institute of Biotechnology, Amity University, Noida, India

Arpit Raj School of Pharmaceutical Sciences, Apeejay Stya University, Gurugram, Haryana, India

Alison Rodger Research School of Chemistry, the Australian National University, Canberra, ACT, Australia

Sudipta Roy Central Ayurveda Research Institute (CARI), Central Council for Research in Ayurvedic Sciences (CCRAS), Kolkata, West Bengal, India

Monika Sachdeva Fatima College of Health Sciences, Alain, UAE

Nagehan SALTAN Department of Pharmaceutical Botany, Faculty of Pharmacy, Anadolu University, Eskişehir, Türkiye

Prerna Sarup University Institute of Pharmaceutical Sciences, Panjab University, Chandigarh, India

Pranshul Sethi Department of Pharmacology, School of Pharmaceutical Sciences, Shoolini University, Solan, Himachal Pradesh, India

Richa Shri Department of Pharmaceutical Sciences and Drug Research, Punjabi University, Patiala, Punjab, India

Nishtha Singh Department of Pharmacology, School of Pharmaceutical Sciences, Shoolini University, Solan, Himachal Pradesh, India

Paul N. Smith School of Medicine and Psychology, the Australian National University, Canberra, ACT, Australia

Komal Thapa Chitkara University School of Pharmacy, Chitkara University, Baddi, Himachal Pradesh, India

Komal Tyagi University Institute of Pharmaceutical Sciences, Panjab University, Chandigarh, India

Liliana Vargas-Murga Department of Chemical and Agricultural Engineering and Agrifood Technology, Polytechnic School, Universitat de Girona, Girona, Catalonia, Spain

Nitin Verma Chitkara University School of Pharmacy, Chitkara University, Baddi, Himachal Pradesh, India

Sabina Yasmin Department of Pharmaceutical Chemistry, College of Pharmacy, King Khalid University, Abha, Saudi Arabia

Hafize Yuca Department of Pharmacognosy, Faculty of Pharmacy, Atatürk University, Erzurum, Türkiye

Alzheimer's Disease: Pathophysiological and Therapeutical Insights

1

Komal Tyagi, Prerna Sarup, and Jai Malik

Abstract

Alzheimer's disease (AD) is the most common form of dementia that is considered as a multifactorial neurodegenerative disease which causes a gradual cognitive decline along with a significant socio-economic burden. In this chapter, a unitive approach is adopted to give an overview of AD pathophysiology, beginning with historical foundations and its classification into early-onset and late-onset variants, alongside familial and sporadic forms. It also critically examines both accepted and novel theories explaining AD pathogenesis-including the cholinergic deficit, amyloid-β cascade, tau hyperphosphorylation, neuroinflammation, oxidative stress, mitochondrial dysfunction, cholesterol dysregulation, vascular compromise, and aberrant cell cycle hypotheses. This chapter further evaluates the restrictions of the traditional treatment and explains the shift towards the disease-modifying therapeutic strategies, including monoclonal antibodies targeting amyloid (e.g., aducanumab, lecanemab, donanemab), tau inhibitors, and combination therapies. Also, the therapeutic strategies like phytotherapy, anti-oxidants, gut-microbiome, etc. are found to be as emerging approaches in the modern world.

Keywords

Neurodegenerative disorders · Amyloid · Tau · Neuroinflammation · Acetylcholinesterase inhibitors

K. Tyagi · P. Sarup · J. Malik (✉)
University Institute of Pharmaceutical Sciences, Panjab University, Chandigarh, India

J. Malik et al. (eds.), *Diet and Alzheimer's Disease: Let Food be Our Medicine*,
Nutritional Neurosciences, https://doi.org/10.1007/978-981-95-2736-6_1

1.1 Introduction

Neurodegenerative disorders (NDs) are chronic, progressive conditions characterized by the selective loss of neurons, primarily within the central nervous system (CNS), and in some cases, the peripheral nervous system (PNS) (Alkahtani et al. 2023). These disorders are more prevalent among individuals over the age of 65, although certain familial or early-onset forms may present much earlier in life (Hou et al. 2019). Dementia, Alzheimer's disease (AD), Parkinson's disease (PD), Amyotrophic lateral sclerosis (ALS), and stroke are some of the common NDs that are observed in elderly patients, and amongst these, dementia is the most common one. Philippe Pinel, who is considered as the "Father of modern psychiatry", gave the term dementia in nineteenth century for cognitive impairments that were converted into clinical illnesses and could be treated. But it was at the beginning of the twentieth century when a German psychiatrist named Alois Alzheimer first identified AD. Though initially he identified AD as a different type of dementia as he found some protein abnormalities (later identified as beta amyloid (Aβ) plaques and neurofibrillary tangles) in the brain of his patient, it was Kraepelin who gave the actual name to this disease as Alzheimer's Disease in honour of Dr. Alois Alzheimer (Beyreuther et al. 2006; Ferrari and Sorbi 2021; Reichman and Rose 2012).

Dementia is characterized by loss of memory, and AD is the most common cause of dementia. In addition to the impact on memory, AD also impacts the other cognitive skills (thinking, speech, decision making, etc.) that make the condition of the patient more appalling as he cannot perform routine tasks effectively. Such an inability then brings certain behavioural changes such as aggression and depression, in the patients. AD is sufficiently affecting 2% of the population in developed countries (Mattson 2004). AD contributes to almost 50–60% of all cases of dementia, making it the most prevalent type, and hence, the frequency of AD is expected to double during the next 20 years, and estimates predict that 81.1 million individuals will be affected by 2040 (Hardiman et al. 2011). A report from the European Union indicates that the number of deaths from dementias, including Alzheimer's disease, has been steadily increasing annually, starting with 198,320 deaths in 2011 to 316,989 fatalities in 2017, a 59.83% rise over these 6 years has been encountered. In 2020, approximately 6.2 million Americans aged 65 and older were living with AD, and this figure is expected to rise to 12.7 million by 2050 (Ciurea et al. 2023). Recently, a 19-year-old has been diagnosed by with AD making him the youngest patient (Jia et al. 2023). The disease has great social implications as it not only affects the patient but also the family members and caregivers. In addition to social implications, this dreadful disease also has economic implications in terms of treatment cost and loss of working hours of the family members to take care of their loved ones. Healthcare cost, which includes treatment expenses, caregivers, and loss of working hours, was USD 305 billion in 2020 and is expected to reach USD 1 trillion by 2050 (Wong 2020). Moreover, it is not only the socio-economic impact of this horrible disease, AD has also been identified as the fifth leading cause of death in adults above 65 years of age.

1.2 Classification of AD

Alzheimer's disease can be classified either on the basis of the age of onset into Early- and Late-onset AD or on the basis of its nature into familial- or sporadic-AD. Early onset AD occurs in people who are in their 30s or 40s or before age 60 and constitutes 1–6% of total AD cases, while the late onset AD (LOAD) occurs after the age of 60 or 65 years and contributes to more than 90% of total AD cases. Further, early-onset AD has four major characteristics/features which differentiate it from late-onset AD. First, it occurs mainly due to genetic mutations involving a direct autosomal dominant transmission. Such patients have a greater propensity for AD development and also suffer from some atypical clinical features (headaches, gait problems, seizures, etc.) (Gerritsen et al. 2021). Further, prevalence of comorbidity in patients with young-onset Alzheimer's disease is higher in comparison to the ones with late-onset (Gerritsen et al. 2016). Second, the progression of the disease in such patients is more aggressive (Wattmo and Wallin 2017) and have a greater risk of mortality than patients with late-onset AD (Chang et al. 2017). Third, early-onset AD patients suffer from disease for longer duration before they are diagnosed (Panegyres and Chen 2013). Fourth, patients with early-onset AD exhibit lesser semantic memory impairment and higher chances of attention, praxis, and visuospatial difficulties, and also higher psychosocial problems (Mendez 2017).

Further, familial AD mainly has a genetic origin and occurs due to a mutation in (amyloid precursor protein (APP) gene, presenilin1 (PSEN1) gene, and presenilin 2 (PSEN2) gene that is passed on to the next generation. On the contrary, sporadic AD occurs due to both genetic and environmental causes (Piaceri et al. 2013).

1.3 Pathophysiology of AD

Since its discovery, ample work has been done to understand the cause and its pathophysiology. Various hypotheses, namely cholinergic, amyloid cascade, tau hyperphosphorylation, mitochondrial dysfunction, genetic, etc., have been suggested over the period of time. All these factors cause extensive neuronal loss and bring various synaptic changes in the hippocampus, cerebral cortex, and other brain tissues programmed for cognitive and memory functions.

1.3.1 Cholinergic Hypothesis

Peter Davies and A.J.F. Maloney, in 1976, for the first time proposed the cholinergic hypothesis during their research on the activity of key enzymes involved in neurotransmitter production (Liu et al. 2019). The cholinergic hypothesis, for the first time, showed that cognitive functions like awareness, learning, attention, and memory are controlled by hippocampal cholinergic neurons, amygdala, frontal cortex, medial septum, and nucleus basalis (Sanabria-Castro et al. 2017). Acetylcholine (ACh), choline acetyltransferase (responsible for ACh synthesis), and acetylcholinesterase enzyme ('AChE', responsible for degradation of ACh) are major contributing factors that help in understanding the cholinergic hypothesis. Amongst the above-mentioned factors, ACh is essential for learning and memory, but as the acetylcholinesterase activity increases, choline acetyltransferase activity decreases, which leads to the loss in cholinergic activity (due to decreased ACh levels) that ultimately impairs memory, like that seen in aged individuals (Francis et al. 1999. Further, ACh is also capable of altering the expression of brain-derived neurotrophic factor (BDNF), which is essential for maintaining plasticity and survival of neurons in later age (Maurer and Williams 2017). These modifications in the cholinergic neuronal activity cause obstruction in ACh release, restriction in the expression of muscarinic and nicotinic receptors, along with the defective neurotrophic contribution and axonal transport. It has been analyzed in many neuroimaging studies that nicotinic receptors (specifically α4β2 subtype) are less in terms of quantity and density in AD patients as compared to the normal individuals, due to which the binding ability of α3, α4, and α7 subunits in the cortical and hippocampal region also decreases (Roberts et al. 2021). These disturbances in cholinergic function also lead to an increase in Aβ production, which further enhances the acetylcholinesterase activity to increase the ACh degradation (Yan and Feng 2004). Cholinergic functioning is also disrupted by the interplay of cholinergic and glutamatergic systems. In AD patients, hippocampal functioning of cholinergic neurons, amygdala, frontal cortex, etc., also gets impaired due to the abnormalities reported in the glutamatergic signaling.

Reduced cholinergic signalling is one of many downstream effects of the complex neuropathological changes in AD, reflecting neuronal loss and synaptic dysfunction rather than serving as a primary etiological factor. While therapies targeting cholinergic dysfunction, such as acetylcholinesterase inhibitors, cholinergic precursors, cholinergic receptor agonists, allosteric cholinergic receptor potentiators, NMDA receptor blockers, etc., provide symptomatic relief, they do not alter the underlying progression of AD (Thakur et al. 2018).

1.3.2 Amyloid (Aβ) Hypothesis

Despite ongoing debates and alternative theories, the amyloid hypothesis maintains its status as the most accepted model for AD pathogenesis. The presence of Aβ plaques is the mainstay in the pathogenesis of AD as they play a defining role in the disease progression (Paroni et al. 2019). The disorder is characterized by anomalies in the synthesis, secretion, and clearance of Aβ (Hardy and Higgins 1992). Aβ is made up of amyloid precursor protein (APP), in which a larger contribution comes from brain neurons, vascular and blood cells (including platelets), and astrocytes (to a minor extent) (Hampel et al. 2021). APP is a type I transmembrane glycoprotein which means it spans the membrane with a single transmembrane domain, and consists of an extracellular N-terminal domain and an intracellular C-terminal region (Müller and Zheng 2012). Cleavage of APP give rise to two fragments of protein, Aβ1–40 peptides, alongside the more neurotoxic Aβ1–42 variant.

The cleavage of APP is done via two ways, i.e., the amyloidogenic pathway and the non-amyloidogenic pathway. The amyloidogenic pathway, which is considered as the pathological pathway, produces Aβ and leads to AD. While the non-amyloidogenic pathway inhibits Aβ production naturally. The amyloidogenic pathway is characterized by a sequence of proteolytic events facilitated by β-secretase and γ-secretase. Initially, β-secretase cleaves the APP, yielding soluble β-APP (sAPPβ) and a membrane-tethered C-terminal fragment (CTFβ or C99). The C99 fragment (pathological fragment) then undergoes intramembrane cleavage by γ-secretase, resulting in the production of amyloid intracellular domain (AICD) and amyloid-beta (Aβ) peptides. This process ultimately gives rise to the Aβ40 and Aβ42 isoforms, with Aβ42 being particularly associated with neurotoxicity in Alzheimer's pathology (Azargoonjahromi 2024). The non-amyloidogenic pathway involves the cleavage of APP by α-secretase that resides in the plasma membrane within the Aβ region, producing soluble α-APP (sAPPα) and a membrane-bound C-terminal fragment, CTFα (or C83). The C83 fragment is subsequently processed by γ-secretase, leading to the release of non-toxic P3 peptides and the APP intracellular domain (AICD). This regulatory mechanism plays a crucial role in preserving the balance of APP metabolism and preventing the formation of toxic amyloidogenic products (Chen et al. 2024).

sAPPα formed during the non-amyloidogenic pathway is secreted into the extracellular milieu, where it contributes to the modulation of neuronal and synaptic dynamics and is regarded as neuroprotective (Sirisi et al. 2024). The γ-secretase enzyme is a multi-subunit (heteromeric) intramembrane protease complex comprising the following key components: Presenilin (PSEN1 or PSEN2), Nicastrin (NCT), Anterior Pharynx Defective 1 (APH-1), and Presenilin Enhancer 2 (PEN-2) (Ahmad et al. 2019b). Amongst these, the pathophysiology of AD and numerous other neuronal physiological processes, such as development, calcium homeostasis, and apoptosis, are profoundly implicated by PSEN1 and PSEN2 (Dong et al. 2012). It acts at the γ-site of APP, resulting in APP intracellular domain (AICD) and amyloid β peptide, leading to secretion, aggregation, and accumulation of Aβ in extracellular plaques owing to its low solubility.

The concentration of Aβ peptides in the brain can be balanced through a combination of physiological processes that regulate their production, clearance, and degradation (Evin and Weidemann 2002). P-glycoproteins and lipoprotein receptor-related protein-1 (LRP-1 receptors) facilitate the Aβ transportation via blood-brain barriers (BBB). With the ageing expression of LRP-1 receptor gets diminished, which causes accumulation of Aβ in the CNS, leading to fibril generation that produces senile plaques and induces neurotoxicity. This accumulation of Aβ further augments the tau hyperphosphorylation, making the condition more severe (McCormick et al. 2021) (Donahue et al. 2006). Apart from p-glycoprotein and LRP-1, Receptor for Advanced Glycation End Products (RAGE) also plays an important role in the regulation and clearance of Aβ. RAGE enhances the Aβ production by promoting the enzymatic activities of β- and γ-secretases (Cai et al. 2016). Since RAGE and LRP1 work in opposing directions to regulate Aβ transport across the BBB, their expression levels critically influence Aβ homeostasis. Under pathological conditions, upregulated RAGE expression and downregulated LRP1 expression lead to the progression of Aβ depositions (Chakraborty et al. 2021). It is these anomalies in the synthesis, secretion, and clearance of Aβ that lead to AD (Hardy and Higgins 1992).

Neprilysin (NEP) and insulin-degrading enzyme (IDE) are two of the key enzymes involved in the degradation of amyloid-beta (Aβ). NEP is a membrane-bound metalloprotease enzyme responsible for the breakdown of extracellular Aβ, whereas IDE is a zinc-metalloprotease that degrades both intracellular and extracellular Aβ. It has been found that in aged people, there has been downregulation of both NEP and IDE in cortical and hippocampal regions, which leads to inefficient clearance of Aβ and an increase in its accumulation (Sasaguri et al. 2021). Apart from all the enzymatic factors, mutation in certain genes also enhances the production of Aβ. The key genetic contributors are: APP, Presenilin 1 (PSEN1), Presenilin 2 (PSEN2) for familial AD, and Apolipoprotein E (APOE), Triggering Receptor Expressed on Myeloid Cells 2 (TREM2), and Clusterin genes for sporadic AD. Among all, mutations prominently in the APP gene on chromosome 21 result in overproduction or altered cleavage of Aβ, particularly the toxic Aβ42 peptide, promoting its aggregation into plaques. Since individuals with Down Syndrome have an extra copy of the APP gene on chromosome 21, they are more vulnerable or at high risk of AD, and many exhibit an early onset of AD (Sehar et al. 2022).

Despite enormous literature on Aβ being the mainstay of AD, recent reports have shown that the severity of the disease is not governed by Aβ (Kametani and Hasegawa 2018). Clinicopathological studies have shown a weak correlation between the Aβ levels and cognitive decline in patients aged above 85 years. The results have shown significant deposition of Aβ plaques without any clinical symptoms (Boccardi et al. 2017). Furthermore, most of the therapeutic interventions targeting Aβ have failed, yet still majority of drug discovery is targeted towards Aβ. Therefore, a debate whether Aβ should be considered as the mainstay of AD is on, and more detailed studies are warranted (Fig. 1.1).

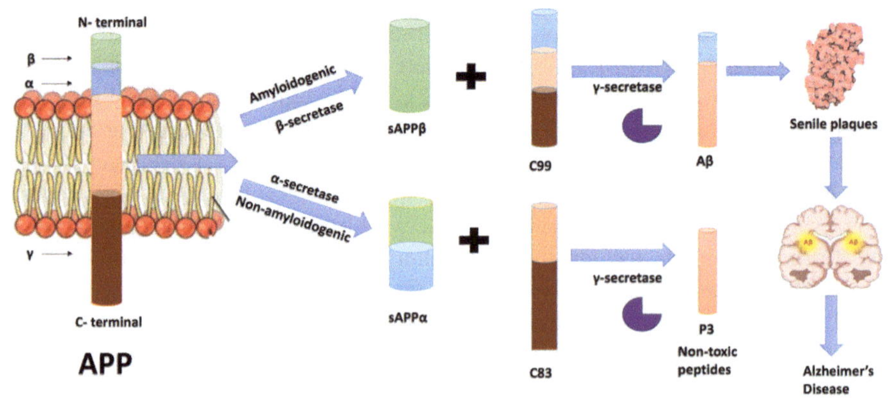

Fig. 1.1 Amyloid hypothesis and Alzheimer's disease

1.3.3 Tau Hyperphosphorylation Hypothesis

Gene encoding tau proteins is present on chromosome 17 and is also known as the microtubule-associated protein tau (MAPT)-gene. It has 15 exons and six isoforms, and the longest isoform of tau contains over 30 phosphorylated sites with 79 potential serine and threonine phosphate acceptor residues. Under normal physiological conditions, tau protein plays an important role in stabilizing the microtubules in the axonal region, regulating their assembly, and influencing the neuronal shape and growth (Guo et al. 2022). But under pathological conditions, normal tau gets N-glycosylated, thereby becoming a more favourable substrate for phosphorylation. It can undergo hyperphosphorylation by action of: *proline-directed protein kinases* [glycogen synthase kinase 3β (GSK3β); cyclin-dependent kinase 5 (CDK5), etc.]; *non-proline-directed kinases* (protein kinase A, B and C, Tau-tubulin kinases 1 and 2, etc.) and *protein kinases specifically for tyrosines* (FYN) (Rawat et al. 2022). Furthermore, once the tau gets glycosylated, it also becomes resistant to dephosphorylation by PP-2A. Once the tau filaments undergo hyperphosphorylation, they get dissociated from microtubules, thereby causing axonal degeneration and neuronal damage (Kanaan et al. 2013). The hyperphosphorylated filament also causes changes in structural conformity, and it starts getting converted into its oligomers, followed by the formation of helical filaments, which eventually form neurofibrillary tangles (NFTs). These NFTs then get accumulated in the pre- and post-synaptic regions, thereby causing synaptic dysfunction (plays a major role in AD progression), which can spread throughout the brain (Fig. 1.2). Compelling evidence has also shown that Aβ promotes the tau-induced AD pathology as:

1. Aβ causes dysregulation of kinases, especially glycogen synthase kinase 3-beta (GSK-3β) and CDK5 (cyclin-dependent kinase 5), leading to hyperphosphorylation of tau proteins. The hyperphosphorylated tau proteins form large helical filaments that further form aggregates, NFTs, which alter the structural integrity of microtubules that eventually leads to neuronal damage. Hyperphosphorylated tau also becomes resistant towards proteolytic degradation, which further increases the tau levels in the brain (Lu et al. 2019; Reddy 2013).
2. Accumulation of Aβ generates oxidative stress (OS), which further brings structural changes in tau proteins, thereby enhancing their aggregation in NFTs. Furthermore, Aβ-induced OS also impairs protein phosphatases like PP2A, thereby reducing tau dephosphorylation and clearance (Martin et al. 2013).
3. Aβ accumulation/plaques activate microglia and astrocytes, leading to a chronic neuro-inflammatory process. This significantly increases the pro-inflammatory cytokines (e.g., IL-1β, TNF-α), which further promote tau hyperphosphorylation and their aggregation in NFTs (Miao et al. 2023).

Intriguingly, it has been found that Aβ generally accumulates in brain areas like the neocortex, and then spreads to another region, finally entering the cerebellum. Unlike Aβ, tau pathology starts in the entorhinal cortex, followed by limbic regions, and then to the neo-cortex (van der Kant et al. 2020). This spatial paradox

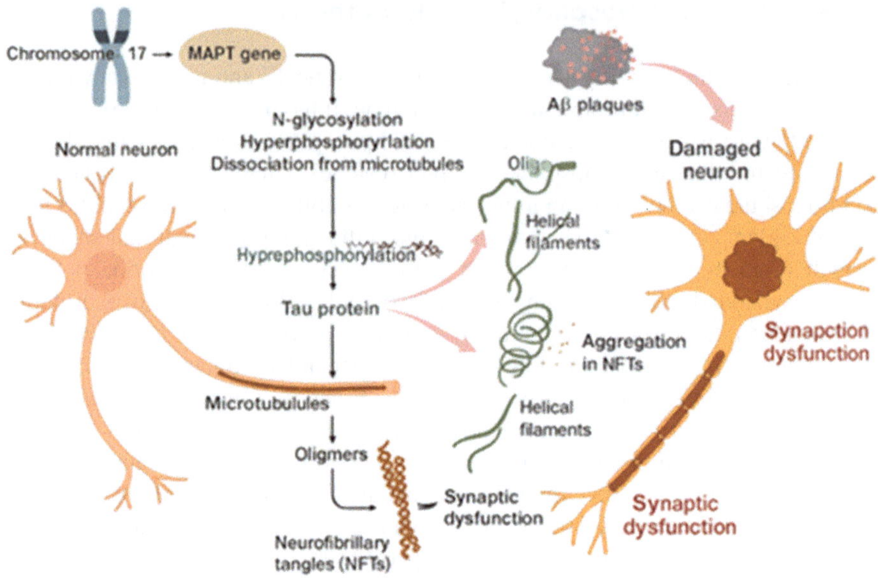

Fig. 1.2 Tau hyperphosphorylation and Alzheimer's disease

challenges the hypothesis that tauopathy succeeds the Aβ pathology. In addition to Aβ induced tau pathology, evidence has suggested that tau abnormalities can occur independent of Aβ. But one thing is common in both cases that hyperphosphorylation of tau precedes the NFT formation that leads to the progression of the disease.

1.3.4 Inflammatory Hypothesis

Inflammation acts as a double-edged sword, as acute inflammation is beneficial for clearing debris and maintaining homeostasis, while under chronic conditions, inflammation is detrimental. In AD, it is the chronic inflammatory process that becomes maladaptive and contributes to neurodegeneration (Darqui et al. 2021). According to the inflammatory hypothesis of AD, persistent neuroinflammation is a key factor in the development and progression of AD, and it is caused by Aβ plaques, NFTs, and other cellular damage, resulting in a prolonged immunological response that worsens neuronal injury and speeds up disease progression (Sun et al. 2022). Long-term use of nonsteroidal anti-inflammatory drugs (NSAIDs) has been associated with a reduced risk of developing Alzheimer's disease (AD) and delaying its onset. The protective effects of NSAIDs are believed to involve multiple mechanisms. One key pathway is the inhibition of cyclooxygenases (COX), enzymes that play a central role in the inflammatory process, thereby reducing neuroinflammation. Additionally, some NSAIDs can activate the peroxisome proliferator-activated receptor gamma (PPAR-γ), which helps regulate inflammatory responses and may directly target AD-related pathology. Experimental studies provide further support for these effects, showing that specific NSAIDs, such as ibuprofen, flurbiprofen, indomethacin, and sulindac, are capable of reducing Aβ levels (Gasparini et al. 2004).

Inflammatory response in neurons leads to activation of microglia, lymphocytes, and astrocytes, leading to the release of cytokines, chemokines, and the production of ROS (McGeer and McGeer 2013). Microglia, the resident immune cells of the central nervous system (CNS), play a dual role in Alzheimer's disease (AD), acting protectively in the early stages but becoming harmful as the disease progresses (Du et al. 2018) (Kaur et al. 2019). In the initial phases of AD, microglia contribute to reducing amyloid-beta (Aβ) plaque accumulation by clearing these toxic aggregates. However, as the disease advances, microglia are persistently exposed to Aβ and other stress signals, which push them into a chronic pro-inflammatory state, often referred to as the M1-like phenotype. In their activated state, microglia release pro-inflammatory cytokines, such as IL-1β and TNF-α, as well as reactive oxygen species (ROS), which together contribute to neuronal injury (Ahmad et al. 2019a). Further, continuous exposure of microglia to Aβ also decreases their ability to bind Aβ and clear them (Krabbe et al. 2013). Intriguingly, the capability of microglia to produce inflammatory cytokines remains unaffected which further aggravates the neuronal damage (Hickman et al. 2008). It has also been found that elevated levels of IL-1 have increased the production of APP, which further increases the Aβ load (Goldgaber et al. 1989). Elevated IL-1 levels also increase the production the IL-6 that activates CDK5, which eventually augments the tau hyperphosphorylation (Quintanilla et al. 2004).

The role of microglia in AD pathogenesis is also supported by triggering receptors expressed on myeloid cells-2 (TREM 2) expression (Kleinberger et al. 2014). The role of TERM2 in AD and other NDs has been extensively reviewed by various groups (Kulkarni et al. 2021) (Gratuze et al. 2018a, b). TREM 2 helps microglial activation and their phagocytic activity in inflammatory conditions but remains suppressed under pro-inflammatory conditions (Liu et al. 2020). Further, it is interesting to find that the main role of TREM 2 is to promote the microglia survival along with myeloids and help in their interaction with Aβ plaques (Jay et al. 2015). In vitro studies have shown that an increased TREM 2 expression stimulates the microglial-mediated phagocytosis/clearance of Aβ (Kleinberger et al. 2014; Melchior et al. 2010). Studies have also shown that any mutation in the TREM 2 gene, located on chromosome 6p21, significantly decreases its ability to recruit microglia and subsequent Aβ clearance, thereby increasing the Aβ load (Hickman and El Khoury 2014; Jin et al. 2015). TREM2 variant, R47H, has been found to increase the chances of LOAD by 2 to four-fold (Ruiz et al. 2014). In addition to their role in Aβ pathology, TREM 2 has also been found to play a role in tau-mediated AD pathogenesis. A positive correlation was found in the levels of TREM2 and hyperphosphorylated tau proteins in cerebrospinal fluid of AD patients (Piccio et al. 2016). Further, patients having the TREM2 R47H variant also showed higher levels of phosphorylated tau proteins (Lill et al. 2015).

The role of apolipoprotein E (ApoE) in AD pathogenesis has also been established. It has been found that ApoE also acts as a ligand of TREM 2 that stimulates the downstream inflammatory cascade (including activation of microglia) through intracellular adaptor DAP12 (DNAX-activation protein 12, also known as TYROBP) protein (Yeh et al. 2016; Bouchon et al. 2001). The APOE-ε4 allele is the strongest non-Mendelian genetic risk factor for Alzheimer's disease, with carriers more likely to develop cerebral amyloidosis at a younger age. Recent studies suggest APOE-ε4 also contributes to tau pathology and neurodegeneration independently of Aβ pathology (Riphagen et al. 2020).

1.3.5 Oxidative Stress Hypothesis

Oxidative stress (OS) is a condition in which there is an imbalance between the production (increases) and scavenging (decreases) of the free radicals. The brain is an energy hogging organ as it is one of the highest oxygen-consuming organs in the body. High oxygen consumption along with high lipidic nature, iron content, energy requirement, and low endogenous antioxidant defence make it highly vulnerable to OS. Various pathological conditions like PD, ALS, AD, and brain ischemia-reperfusion have been linked with OS (Markesbery 1997). Ample work has been done to understand the role of the OS in the pathogenesis of AD. OS is considered as an early event in AD as a significant increase in the free radical generation has been observed in patients, even before the development of Aβ plaques or NFTs, exhibiting mild cognitive impairment (MCI) (Castellani et al. 2002). The role of OS in AD can be understood through relation between: (1) OS and amyloid proteins; (2) OS and tau proteins (3) OS and neuroinflammation; (4) OS and mitochondrial dysfunction; (5) OS and heavy metals; (6) OS and gut microbiota and (7) OS and molecular pathways (Fig. 1.3).

The amyloid hypothesis is the mainstay in AD pathogenesis and studies have shown that Aβ in the brain causes OS, and OS also stimulates the production of Aβ. So, the question whether the Chicken came before the egg? still remain unanswered. But looking into the literature, one can surely say that OS and Aβ lead to a vicious cycle that aggravates the disease. OS directly upregulates the activity of β-secretase, inducing changes that favour the production of Aβ and its accumulation (Zhang et al. 2016a). OS boosts the amyloidogenic process by activating stress-activated protein kinases, such as c-Jun N-terminal kinase (JNK) and p38 mitogen-activated protein kinase (p38-MAPK), which further induce β-secretase expression. Furthermore, the kinases of the translation factor eIF2 must be activated for BACE1 mRNA translation to occur. These kinases phosphorylate eIF2 at its α subunit in response to harsh stimuli such as oxidative stress (Fanlo-Ucar et al. 2024). OS also upregulates the expression of γ-secretase, which in turn also regulates the increased expression of β-secretase induced by OS (Gratuze et al. 2018a, b); (Sayas and Ávila 2021); (Tamagno et al. 2012). This increased production of Aβ also increases the production of ROS, leading to OS and neuronal damage.

OS also increases the phosphorylation of tau proteins by upregulating the GSK3β and cyclin-dependent protein kinase-5 (CDK5) activity, which prevents the binding of tau proteins to microtubules, leading to increased NFTs and neuronal damage (Alavi Naini and Soussi-Yanicostas 2015). OS also inhibits the protein phosphatase

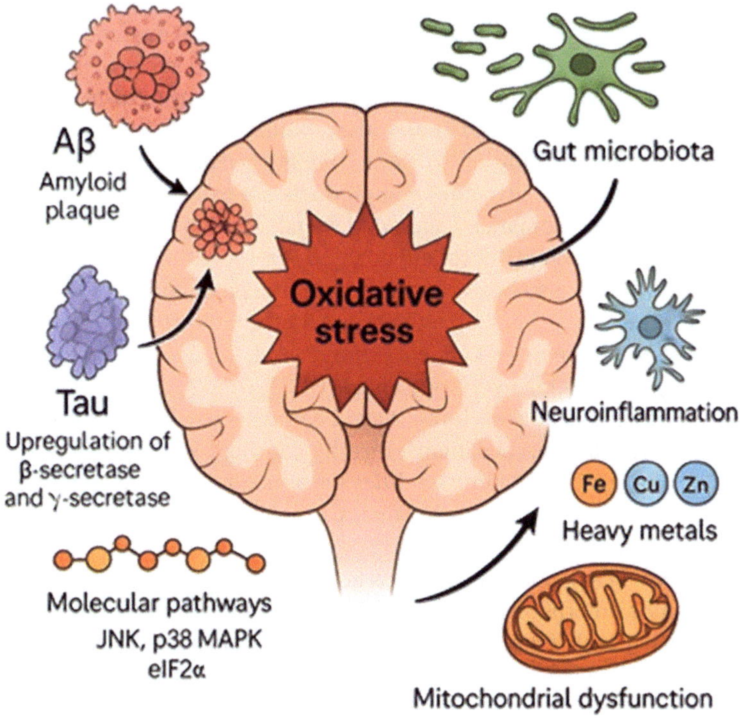

Fig. 1.3 Oxidative stress and Alzheimer's disease

2 (PPA2) activity, which further stimulates the ERK pathway of apoptosis in neurons.

OS also leads to mitochondrial dysfunction, thereby disturbing the energy dynamics of the cells. It leads to low energy metabolism, reduced expression of key enzymes such as α-ketoglutarate dehydrogenase, pyruvate dehydrogenase, and cytochrome oxidase, involved in phosphorylation (Gibson et al. 1998). For further details, see the section "Mitochondrial Cascade Hypothesis" below.

Activation of microglia and astrocytes is also induced by OS, which leads to the release of pro-inflammatory cytokines as a protective mechanism. The microglia and astrocytes are also activated by Aβ and NFTs, causing overburdening of the inflammatory process. While initially being a protective action against OS, Aβ and NFTs, constant exposure to the inflammatory cytokines (due to the vicious cycle of Aβ, NFTs, and OS) shifts this acute protective effect into a chronic and toxic inflammatory condition that further contributes to neuroinflammation and neurodegeneration. Such neuroinflammatory conditions further enhance the production of Aβ and NFTs. Furthermore, the activated microglia increased the production of capthesin B, which further stimulates the production of IL-1β, augmenting the inflammatory condition in neurons. In addition to this, capthesin B has also been found to degrade the mitochondrial transcription factor A (TFAM), a protein that helps in protecting mitochondrial DNA against age-related and other damages (Nakanishi 2020). This degradation of TFAM also contributes to the mitochondrial dysfunction and neuronal energy dysregulation. On the contrary, there are reports that have also highlighted the protective action of capthesin B in AD against Aβ-induced memory impairments (Embury et al. 2017) and also stimulates hippocampal neurogenesis (Moon et al. 2016). In addition to these, oxidative damage products from lipid, DNA, RNA oxidation, or glycation end products are extensively studied as neurodegeneration and AD biomarkers and are detectable in blood, plasma, serum, cerebrospinal fluid (CSF), etc. (Buccellato et al. 2021).

Heavy metals have also been correlated with the production of OS and progression of AD. Presence of Fe, Cu, and Zn in Aβ plaques correlates with Aβ protein distribution in the brain of AD patients (Fasae et al. 2021). Iron contributes to the Fenton reaction, producing hydroxyl radicals (•OH) from hydrogen peroxide (H_2O_2) (Z. Zhao 2019). Elevated iron levels have been detected in amyloid plaques and neurofibrillary tangles in AD patients, thereby causing localized oxidative damage (Wang et al. 2022a). Cu^{2+} binding modifies the structure of the Aβ peptide, facilitating interactions at its N-terminus. Further, Cu^{+2} also acts as a bridge between peptide molecules, thereby enhancing the stability of Aβ-Aβ complexes. Though it reduces β-sheets in Aβ monomers, it does not affect fibrillar Aβ; instead, it promotes the formation of amorphous aggregates. This was further supported by the TEM analysis, which revealed that Cu^{2+} increases the entanglement of striated Aβ ribbons (Cheng et al. 2013; Matheou et al. 2015). The Cu^{2+}–Aβ complex is more toxic than Aβ alone, indicating that Cu^{2+} amplifies Aβ's neurotoxic effects (Chen et al. 2023). Zinc stabilizes cellular structures and enzymes, but when accumulated at synaptic sites, it can encourage Aβ aggregation and impair mitochondrial function (Das et al. 2021). Thus, it is evident that the dysregulation of metal homeostasis exacerbates oxidative stress and neuronal injury. These processes underscore the interconnected roles of biometals and free radicals in promoting oxidative damage, which drives the neurodegenerative progression of AD.

1.3.6 Mitochondrial Cascade Hypothesis

Though the brain is an energy hogging organ, it lacks energy reserves and is largely dependent on the blood flow in its different parts and the energy produced by mitochondria. The mitochondrial cascade hypothesis firmly believes that the Aβ and or tau-hypothesis does not initiate the AD pathogenesis. Adlimoghaddam et al. showed that mitochondrial dysfunction occurs much before the Aβ and tau-induced pathogenic effects. The results of their fluorodeoxyglucose-PET imaging studies on the 3xTg-AD mouse evidently showed hypometabolism of glucose molecules and decreased mitochondrial complexes expression even in the brain areas without any AD pathology (Adlimoghaddam et al. 2019). According to this hypothesis, the brain requires a constant supply of energy from mitochondria to maintain the proper functioning of billions of neurons and 10^{14} synaptic junctions. The energy-producing capacity of mitochondria decreases with aging, and as it reaches a threshold, it adversely affects the function of neurons and cognitive capabilities. This also reduces the endurance of the brain towards various pathologies, and this is when the other indicators of the AD, i.e., Aβ and tau, kick in (Boccardi et al. 2017). The mitochondrial hypothesis also highlights the interaction between mitochondrial dysfunction and Aβ and tau production (Fig. 1.4). Altered mitochondrial functions cause enhanced expression and processing of APP to cause Aβ aggregation in the brain of AD patients (Dentoni et al. 2022). Furthermore, mitochondrial dysfunction also increases the production of mitochondrial C99 (direct precursor of Aβ) (Area-Gomez et al. 2018). Increased C99 levels further increase the ceramide production, which inhibits the mitochondrial respiratory process, thereby decreasing the ATP production. Reduction in mitochondrial capacity to generate ATP also increases the load of ROS, leading to OS that further deteriorates the condition (Bhatt et al. 2021). Likewise, various reports also showed that mitochondrial dysfunction leads to tau hyperphosphorylation and subsequent NFT formation (Guha et al. 2020; Melov et al. 2007; Samluk et al. 2022).

The mitochondrial cascade hypothesis also believes that the OS, high Ca^{+2} ion influx, and poor energy metabolism also trigger signalling pathways that alter the end products (Cascella and Cecchi 2021). Furthermore, during mitochondrial dysfunction, mitophagy (a process by which damaged mitochondria are removed from the cells to maintain cellular health) gets impaired and results in accumulation of damaged mitochondrial debris, thereby increasing the OS and neuroinflammation, leading to neuronal damage that further helps in the progression of AD (Pradeepkiran et al. 2024).

1.3.7 Vascular Hypothesis

Vascular causes are ranked second among the most common causes for cognitive decline. A plethora of evidence shows the role of cerebrovascular conditions in AD and cognitive decline (Arvanitakis et al. 2016; Iturria-Medina et al. 2016; Ruitenberg et al. 2005; Santos et al. 2017). It is also believed that as the age proceeds, cerebral microvascular (tiny blood vessels) pathology and hypoperfusion (reduction in blood flow) also contribute to the neurodegeneration process to cause cognitive decline (Fisher et al. 2022). This hypothesis incorporates the Critically Attained Threshold of Cerebral Hypoperfusion (CATCH) framework (de Montgolfier et al. 2020). Torre has reviewed the role of CATCH in AD (De la Torre 2000). According to him, chronic cerebral hypoperfusion—a sustained reduction in cerebral blood flow (CBF)—is a key initiating event that triggers the cascade of pathological changes leading to AD. This hypothesis states that when a critical point at which reduced blood flow becomes insufficient to meet the brain's metabolic demands is crossed, it leads to decreased production of ATP, followed by mitochondrial dysfunction, oxidative stress, and neurodegeneration (De la Torre 2021). This decrease in ATP production and mitochondrial dysfunction then leads to an increase in Aβ production and hyperphosphorylation of tau proteins. Furthermore, Torre also stressed the use of vascular risk factors in predicting AD (De La Torre 2018). The deposition of

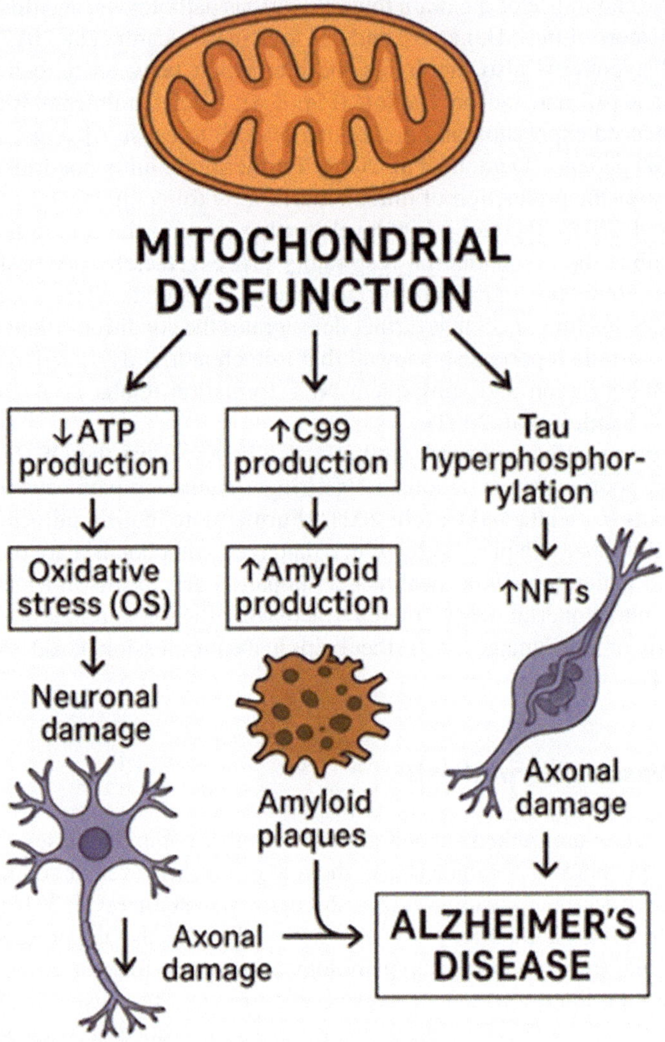

Fig. 1.4 Mitochondrial dysfunction pathway leading to Alzheimer' s disease

Aβ in the blood vessels leads to a condition called as cerebral amyloid angiopathy (CAA) that severely affects the occipital, parietal, frontal, and temporal lobes (Kalaria 2018). CAA precedes AD as CAA remains asymptomatic and AD develops decades after the first symptom appears, further supporting the Torre suggestion of using vascular events in predicting AD (Dubois et al. 2014). Thus, it is reasonable to say that cerebral hypoperfusion leads to a series of events starting with capillary collapse that cause reduced blood flow to the neurons, followed by hypoxic injury, tight junction breakdown, mitochondrial dysfunction, energy crisis, oxidative stress, inflammatory response, amyloid deposition and tau-hyperphosphorylation, and neurodegeneration (Eisenmenger et al. 2023).

1.3.8 Cholesterol Hypothesis

Cholesterol, a vital lipid in the central nervous system, plays an essential role in maintaining membrane integrity, supporting synaptic plasticity, and facilitating neuronal signalling (Hussain et al. 2019). However, emerging evidence suggests that disruptions in cholesterol metabolism and distribution are implicated in the pathogenesis of neurodegenerative diseases, particularly AD (Gliozzi et al. 2021). The cholesterol hypothesis posits that imbalance in brain cholesterol levels—through increased cholesterol levels, impaired cholesterol transport, and altered lipid raft dynamics—is central to the development and progression of AD. This hypothesis is supported by observations of elevated cholesterol levels in the brains of AD patients, even though the peripheral cholesterol is prohibited from entering the brain due to the BBB (Hu et al. 2023). High neuronal cholesterol levels affect the lipid rafts—highly specialized domains in the membrane, thereby promoting the conversion of APP to Aβ (Djelti et al. 2015; Wang et al. 2021). Furthermore, these lipid rafts are also specialized in different functions such as maintaining the cholinergic system, which is essential for memory, neuronal membrane stability, signal transduction, etc., for proper neuronal functioning (Feringa and Van der Kant 2021).

Barrett et al. have shown that there are specialized regions in amyloid precursor proteins for the binding of cholesterol. This bound cholesterol further accelerates the process of conversion of Aβ monomers to Aβ fibrils (Barrett et al. 2012). In addition to Aβ, cholesterol influences tau protein phosphorylation (that is Aβ-independent), leading to increased NFTs (van der Kant et al. 2020). This increased accumulation and subsequent hyperphosphorylation are largely due to the inhibition of normal proteasomal degradation of tau (van der Kant et al. 2019). Increased neuronal cholesterol also contributes to neuroinflammation that exacerbates neuronal damage and cognitive decline (Campos-Peña et al. 2022). Furthermore, altered cholesterol metabolism impacts the integrity of the blood-brain barrier (BBB), a critical component in maintaining brain homeostasis. In AD, cholesterol-related changes in the BBB's lipid composition contribute to its dysfunction, resulting in the leakage of toxic molecules and inflammatory mediators into the brain, which accelerates neurodegeneration (Fig. 1.5). Thus, cholesterol dysregulation serves as a significant factor in both the development and progression of Alzheimer's disease, impacting multiple pathways that contribute to brain injury and cognitive decline (Rappoport 2023).

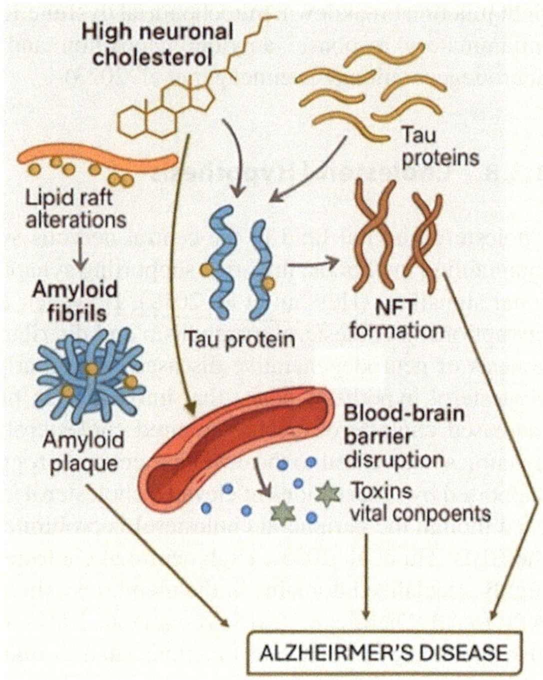

Fig. 1.5 Role of cholesterol in Alzheimer's disease pathophysiology

1.3.9 Insulin Signalling

Alzheimer's, being a metabolic disorder, has gained significant attention in the recent past with insulin-related events in the central core. Insulin plays a crucial role in regulating synapse formation, neuronal survival, neuroprotection, and synaptic plasticity through activation of the insulin receptor (IR) and the insulin-like growth factor (IGF) receptor (Pomytkin et al. 2018). So, impaired insulin signalling in the brain contributes to both neuronal dysfunction and neurodegeneration, leading to cognitive decline and dementia (Arvanitakis et al. 2020). Insulin is capable of increasing the glucose uptake to the plasma membrane through GLUTs (insulin-dependent glucose transporters), especially GLUT 4, which gets impaired in a diseased person (Sędzikowska and Szablewski 2021). Insulin and IGF dysfunction can be considered as a mechanism of pathogenesis of AD. It has been found that decreased insulin levels in the brain can disrupt the glucose metabolism in neurons, causing disturbed neuronal energetics and its function (Zhao et al. 2004). In diabetic conditions, more insulin is secreted from pancreas, which also crosses the BBB causing hyperinsulinemia. Under such conditions brain becomes insulin resistant, thereby disrupting even the normal insulin functions such as glucose metabolism, maintenance of synaptic plasticity, etc., leading to neuronal damage (Tian et al. 2023). Brain imaging studies of patients with insulin resistance have shown the axonal disruptions leading to cognitive impairments (Ryu et al. 2014). Furthermore, insulin resistance impairs the insulin receptor signalling, which further impairs

PI3K/AKT and MAPK pathways that downregulate an enzyme, insulin-degrading enzyme (IDE), responsible for insulin degradation (L. Zhao et al. 2004). IDE is also responsible for degrading Aβ proteins in brain cells. Thus, downregulation of IDE decreases the Aβ degradation and increases its accumulation. It has also been reported that during elevated insulin levels, the IDE affinity increases for insulin, which hinders the Aβ-degrading function of IDE, leading to an increase in Aβ levels in the brain, causing neuronal degradation (Albar et al. 2024; Qiu and Folstein 2006).

Insulin is also seen in tau aggregate formation through phosphorylation via an equilibrium between enzyme activities like kinase and phosphatase. Dysregulation of insulin homeostasis leads to diabetes mellitus (DM) which further increases the expression of GSK-3β and inhibits the phosphatase enzymes. GSK-3β is known for its role in upregulating the hyperphosphorylation of tau-proteins. Thus, one can say that insulin dysregulation also has an impact on tau hyperphosphorylation, leading to neuronal and cognitive damage (El Khoury et al. 2014). Planel et al. (2007) have also reviewed the role of insulin in tau-pathology. But it has also been found that temporary administration of insulin causes quick hyperphosphorylation, while prolonged administration gives just the reverse results, leading to a reduction in phosphorylation (Gratuze et al. 2018a, b).

1.3.10 Cell Cycle Hypothesis

According to the cell cycle hypothesis, disturbances in the cell cycle or mitogenic signalling, or both, occur in an AD patient (Liu 2022). In AD, neurons abnormally express cell cycle-related molecules (such as cell division cycle 2 (cdc2), cyclin-dependent kinase 4 (cdk4), and other members of the cdk family, etc.) and disrupt neuronal homeostasis, leading to failed cell cycle attempts, contributing to neurodegeneration through apoptosis, tau hyperphosphorylation, and other pathological changes (Advani and Kumar 2024). Another molecule GSK-3 is a multifunctional serine/threonine kinase that is critically involved in cell cycle regulation, glycogen metabolism, and gene transcription. In AD, GSK-3 hyperactivates the β-secretase (BACE1) that promotes Aβ production, and also drives neuroinflammation by activating microglia (Yu et al. 2023). These interconnected mechanisms contribute to synaptic dysfunction, neuronal death, and the progressive cognitive decline as seen in AD. Peroxisome proliferator-activated receptor gamma (PPARγ), a nuclear receptor and transcription factor, plays a protective role in AD. It decreases the expression and activity of β-secretase (BACE1), thereby reducing the cleavage of amyloid precursor protein (APP) into amyloidogenic Aβ fragments (Gu et al. 2023). Simultaneously, PPARγ enhances α-secretase activity, promoting the production of soluble APPα (sAPPα), a neuroprotective fragment that does not generate Aβ (Dar and Glazner 2020). Additionally, PPARγ activation shifts microglia toward an anti-inflammatory M2 phenotype, improving their capacity to phagocytose and degrade Aβ via liver kinase B1 (LKB1)-adenosine 5′-monophosphate-activated protein kinase (AMPK) signalling pathway (Cai et al. 2022). This is particularly significant as Aβ accumulation triggers the release of pro-inflammatory cytokines like IL-1β and TNF-α, which further exacerbate Aβ-related neurotoxicity (Kinney et al. 2018). Through these actions, PPARγ effectively mitigates Aβ production, aggregation, and inflammation, making it a promising therapeutic target in AD.

1.4 Alzheimer's Disease Therapeutics

Alzheimer's disease (AD) remains one of the most challenging neurodegenerative conditions, both in terms of clinical complexity and therapeutic response. Over the period of time, extensive research has been done to understand the pathophysiological aspects of AD. As a result, various hypotheses (discussed above) for its pathogenesis have been suggested, and most of the proposed therapeutic strategies revolve around these hypotheses. Despite decades of research, a definitive cure still remains elusive. However, the therapeutic landscape is evolving rapidly, driven by improved understanding of disease biology and novel drug development strategies. There has been a significant shift in the drug discovery paradigm for AD. As discussed ahead, the classical therapies dealt with symptoms only accompanied by serious side effects. Furthermore, these therapies did not modify the course/progression of the disease. Therefore, in the recent past, emphasis has been shifted towards the discovery of disease-modifying agents (Fig. 1.6).

This section critically delineates the AD therapeutics into: (I) therapeutic strategies—will provide a brief overview of the drug discovery approaches; and (II) current approved therapies. A better understanding of the AD pathophysiology has enabled the scientific community to suggest various therapeutic strategies for AD.

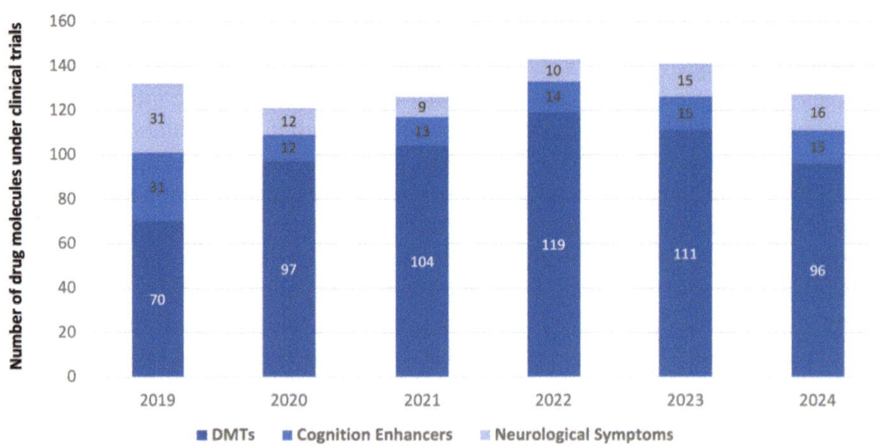

Fig. 1.6 Overview of the agents under clinical trials 2019–2024. *DMTs* disease modifying agents. (Cummings et al. 2019a, 2020, 2021, 2022, 2024)

1.4.1 Therapeutic Strategies

1.4.1.1 Acetylcholinesterase Inhibitors

The cholinergic hypothesis of AD suggests that cognitive decline in AD is primarily caused by the degeneration of cholinergic neurons in the basal forebrain, resulting in a deficit of acetylcholine (ACh), a neurotransmitter essential for memory and learning (Chen et al. 2022). This hypothesis has driven the development of therapies aimed at enhancing cholinergic signalling. One major therapeutic approach involves the use acetylcholinesterase inhibitors (AChEIs), which prevent the breakdown of acetylcholine, thereby increasing or maintaining its levels in the synaptic cleft and enhancing cholinergic neurotransmission. Approved drugs by the US-FDA in these categories include Donepezil, Rivastigmine, and Galantamine, while China gave the approval to an alkaloid Huperizine isolated from a Chinese moss *Huperzia serrata* (Family: Lycopodiaceae) for AD treatment (Jiang, 2024 #386).

Another approach involves targeting cholinergic receptors. Nicotinic acetylcholine receptor (nAChR) agonists stimulate these receptors, which play a key role in memory and attention, while muscarinic acetylcholine receptor (mAChR) agonists (especially M1 agonists) enhance cholinergic signalling by acting on muscarinic receptors (Akaike et al. 2018). These combined strategies aim to mitigate the cognitive deficits associated with AD by boosting cholinergic activity. During the animal studies, the results showed the failure of first-generation muscarinic agonists, which led to the need to invent second-generation M1 muscarinic agonists named as positive allosteric modulators (PAMs) (Monaco et al. 2024). AChEIs are a reliable option for symptomatic treatment, offering modest cognitive and functional benefits in early to moderate AD. However, their limitations, particularly the lack of disease-modifying effects, highlight the need for combination therapies or novel treatments targeting the underlying pathology of AD (Cummings et al. 2019b).

1.4.1.2 NMDA Receptor Antagonist

Memory formation also relies on NMDA receptors (NMDAR), which are ligand-gated ion channels (Babaei 2021). They are the key glutamate receptors for essential CNS and other cognitive functions. In general, Mg^{2+} regulates NMDAR, allowing controlled Ca^{2+} influx for physiological processes. Although overactivation of NMDAR due to glutamatergic neurotransmission leads to the extensive influx of Ca^{2+} causing oxidative stress, neuroinflammation, neuronal loss, and elevated p-Tau (Companys-Alemany et al. 2020). Hence, using antagonists to target NMDARs is a promising approach for treating neurological illnesses.

Memantine, the leading medication in this class, reduces pathogenic Ca^{2+} influx while maintaining neurotransmission. Despite its efficacy, the benefits are limited, prompting a hunt for stronger NMDAR antagonists in the case of mild to moderate AD (Świetlik et al. 2022). Pharmacokinetic and pharmacodynamic trials, as well as ongoing observational research, support the use of memantine as an adjunctive treatment for individuals with intermediate to advanced Alzheimer's disease who are also taking galantamine (Bertocchi et al. 2024). Furthermore, several NMDA receptor antagonists have failed in clinical studies due to unwanted extrapyramidal effects and pharmacokinetic issues (Egunlusi and Joubert 2024). Ifenprodil, a non-competitive NMDAR antagonist, selectively inhibits the GluN2B subunit, offering potential benefits in AD. Amyloid-beta (Aβ), a hallmark of AD, disrupts glutamatergic signalling by promoting excessive NMDAR activation, particularly GluN2B-containing receptors, leading to excitotoxicity, oxidative stress, and neuroinflammation. By blocking GluN2B, Ifenprodil reduces pathological Ca^{2+} influx, counteracting Aβ-induced toxicity and neuronal damage, also by dampening overactive extrasynaptic NMDAR signaling. Radiprodil (selective negative allosteric modulator of GluN2B) protects neurons and preserves cognitive function (Bertocchi et al. 2024). This mechanism distinguishes it from existing treatments like memantine, making it a promising candidate for improving NMDAR-targeted AD therapies. Another, non-competitive NMDA receptor antagonist, similar in profile to memantine, is Neramaxine which blocks excessive Ca^{2+} influx through overactivated NMDARs and enhances long-term memory in adult rats, as per the researchers (Zhang et al. 2016b).

1.4.1.3 Anti-Amyloid Agents

For the treatment of AD, anti-amyloid agents are one of the most tested regimes. The mechanism by which these agents work is the inhibition of Aβ production (by targeting β and γ secretases), an increase in their clearance, and avoiding the aggregation. Agents that inhibit the production of Aβ by targeting β and γ secretase have garnered special interest and various potential molecules are under clinical trials (Cummings et al. 2022, 2023). But again, these agents are also associated with some side effects. Drugs known for γ-secretase targeting can hamper Notch signalling, which is an essential signalling pathway for the normal biological functioning (Kumar et al. 2018). Similarly, some of the β-secretase targeting agents also cause side effects like blindness. An immunological approach for amyloid clearance has also been used (Tondo et al. 2024), and it has shown some promising results as three immunological drugs (discussed below) have been recently approved by the US-FDA for the treatment of mild AD.

1.4.1.4 Tau Inhibitors

Apart from anti-Aβ agents, tau inhibition has also attracted significant attention as one of the efficient strategies for therapeutic purposes. Methylene blue has been found as an effective anti-tau agent that acts by strongly decreasing insoluble Tau (Hochgräfe et al. 2015). Other than inhibitors, use of tau level reducers is also an acceptable approach to decrease the levels of tau in achieving the treatment of AD (e.g.; lithium chloride) (Congdon and Sigurdsson 2018). Another strategy against tau proteins is a combination of suppression of tau hyperphosphorylation and promoting proteosome fusion (clearance) (Hamano et al. 2021). Immunization is also a promising approach aimed at facilitating tau degradation in Alzheimer's disease (AD). This approach targets the accumulation and propagation of hyperphosphorylated tau, which forms neurofibrillary tangles, a central pathological hallmark in AD. It aims to stimulate the immune system to recognize and remove pathogenic tau species without disrupting normal tau function (Maphis et al. 2025).

1.4.1.5 Repurposing of Drugs

With a better understanding of the AD pathophysiology and the molecular pathways, drugs that are already used for some pathological conditions are now being repurposed for their beneficial role in AD. An anti-diabetic drug, liraglutide, has been combined with donepezil to modulate insulin resistance in the brain and cognitive outcomes. Liraglutide, a GLP-1 receptor agonist and originally used in type 2 diabetes, is being repurposed for AD due to its neuroprotective, anti-inflammatory, and insulin-sensitizing properties. A Phase 2 randomized, placebo-controlled study was conducted in the UK to assess liraglutide in individuals with early-stage AD. Results demonstrated that liraglutide may reduce cognitive decline by up to 18% (Carranza-Naval et al. 2021; Teixeira et al. 2025).

1.4.1.6 Phytotherapy

Since there is no synthetic molecule that works well for treating AD, plants, phytomolecules and phytoformulations have gained interest as possible treatment options. Given that, a lot of plants and phytomolecules exhibit their positive effects via multiple pathways like impacting Aβ production (inhibit β-secretase, activate γ-secretase), anti-inflammatory, mitigating oxidative stress, inhibiting acetylcholinesterase, affecting other molecular pathways, etc., and therefore, have gained significant attention in the recent past (Rajendran and Krishnan 2024). Malik (2017) has extensively reviewed the role of plants and phytomolecules in AD. Some of the plants/phytomolecules are mentioned in Table 1.1.

1.4.1.7 Antioxidants

Antioxidants are the substances that help in combating the elevated levels of oxidants in the body due to disturbances in the levels of oxidants and antioxidants. There are basically two types of antioxidant procedure followed up in our body based upon the mechanism of action, i.e., primary and secondary, in which the former helps in terminating the chain reactions and increasing ROS levels while latter helps in oxidizing the hydroperoxides into stable forms and regeneration of antioxidants so that the synergistic effects can be obtained (Pritam et al. 2022). If we talk about the intrinsic system of antioxidants present in our body, there is a natural system that implicates two types, *endogenous* (in body like glutathione peroxidase, catalase, glutathione reductase, superoxide dismutase) and *exogenous* (through diet), that protect and prevent our body from damage. There are certain antioxidant and nutraceuticals that help in neutralizing the free radicals and oxidative damage in the body, like molecular hydrogen that diffuses into mitochondria and nucleus to react with free radicals, generated due to β-amyloid, to simply reduce the hydroxyl and peroxynitrite levels in the body. Superoxide dismutases is a metalloenzymes that are primarily helpful in providing protection against ROS damage by reducing the superoxide anion free radical into molecular oxygen and hydrogen peroxide. Astaxanthin, a dietary antioxidant carotenoid molecule, has excellent antioxidant properties that help in reducing ROS and RNS levels. It is potentially capable of inhibiting the gene expression of mediators like TNF-α, PGE2, and IL-1 that are

Table 1.1 Some of the medicinal plants that have shown protective effects against AD

S. no	Medicinal plant/ constituent	Pharmacological activity	References
1	*Bacopa monniera* (Scrophulariaceae), Bacoside A and B	Upregulation of expression of the NMDA receptor subunit GluN2B, increases neuronal growth, antioxidant, and anti-amyloid	Holcomb et al. (2006), Limpeanchob et al. (2008), Rai et al. (2015), Ternchoocheep et al. (2012)
2	*Berberis species* (Berberidaceae), Berberine	AMPK pathway modulator, anti-amyloid, prevents tau hyperphosphorylation, increase amyloid autophagy	Akbar et al. (2021), Chen et al. (2020), Qin et al. (2020), Wang et al. (2018), Zhang et al. (2017)
3	*Convolvulus pluricaulis* (Convolvulaceae) *Scopoletin, Scopolin*	AChEI, reduced mRNA expression levels of aβ and tau	Bihaqi et al. (2012), Malik et al. (2016)
4	*Curcuma longa* (Zingiberaceae), curcumin	Anti-inflammatory, anti-tau, anti-amyloid	Ambegaokar et al. (2003), Bernardo et al. (2021), Hatami et al. (2019), Rane et al. (2017), Reddy et al. (2018), Zou et al. (2022)
5	*Ginkgo biloba* (Ginkgokaceae) *EGb 761*	Reduce neuroinflammation, anti-amyloid, antioxidant, and increase protein autophagy	Adebayo et al. (2022), Qin et al. (2018), Singh et al. (2019), Verma et al. (2020)
6	*Ginseng* (Araliaceae), Ginsenosides	Increase protein autophagy, decrease neuroinflammation, anti-amyloid, AChEI, BChEI, upregulate PI3K/Akt signalling pathway	Razgonova et al. (2019), Wang et al. (2022a, b)
7	*Withania somnifera* (Solanaceae) *Withanolides*	nAChR agonist, anti-tau, anti-amyloid, antioxidant, anti-inflammatory, improve neuronal growth	Dixit et al. (2021), Kurapati et al. (2013), Remya et al. (2016), Zhao et al. (2002)
8	*Resveratrol*	Antioxidant, anti-amyloid, anti-tau, anti-inflammatory	Ashrafizadeh et al. (2020), Ladiwala et al. (2010), Sun et al. (2019), Yao et al. (2015)
9	*Mori fructus* (Moraceae)	Upregulated Akt, BCL2, and reduced tau phosphorylation, apoptotic markers, ROS production, and caspase activity	Kim et al. (2015a)
10	*Ilex latifolia* (Aquifoliaceae)	Increased glutathione concentrations; inhibited lipid peroxidation, p-tau, and apoptosis	Kim et al. (2015b)
11	*Moringa oleifera* (Moringaceae)	Inhibited oxidative stress and tau hyperphosphorylation	Mahaman et al. (2018)
12	*Pterocarpus marsupium* (Fabaceae) *and Eugenia jambolana* (Myrtaceae)	Decreased Aβ1–42, total tau, p-tau levels; inhibited neuro-inflammatory changes; upregulated GLP-1 levels	Kosaraju et al. (2014)

mainly involved in the ageing process due to increased ROS in mitochondria. Also, it is able to diminish microglial stimulation, which makes AST a promising contributor in AD therapy (Singh et al. 2024).

1.4.1.8 Gut Microbiome

More and more recent studies have found a reciprocal relationship between gut microbiota and AD. The gut-brain axis, a sophisticated communication network between the central nervous system and the gastrointestinal tract, is the basis for this relationship. The hallmarks of AD, such as neuroinflammation, amyloid-β deposition, and cognitive impairment, may be made worse by dysbiosis, or an imbalance in the gut microbiota (Hasan et al. 2025). Sochocka et al. reviewed how gut microbiome dysbiosis causes neuroinflammation and plays a role in AD pathogenesis (Sochocka et al. 2019). The production of amyloid-like proteins by some bacteria may hasten the development of β-amyloid plaque. Change in the gut microbiome has been correlated with Aβ and tau-hyperphosphorylation and has been suggested as an indicator of AD (Ferreiro et al. 2023). In addition to this significant decrease in the microbiome population as well as diversity, it has also been reported in the AD patients in comparison to those of controls (Vogt et al. 2017). Furthermore, by producing reactive oxygen species (ROS), microbial imbalances exacerbate oxidative stress and cause brain damage. Together, these mechanisms upset brain balance and accelerate the development of AD. Microbes and their metabolites play an important role in the bidirectional gut-brain axis, wherein they stimulate the immune signalling in this process, leading to a neuroinflammatory process, and a positive modulation of gut microbiota may offer therapeutic potential for neuroinflammatory and age-related neurological disorders (Mou et al. 2022). Understanding the role of gut microbiome in the pathogenesis of AD has further emphasized the role of good dietary practices as a therapeutic strategy for AD.

1.4.2 Current Approved Treatments/Drugs

The pharmacological management of AD currently includes both symptomatic therapies and disease-modifying therapies (DMTs). The former primarily targets cholinergic and glutamatergic neurotransmission, while the latter focuses on modulating the progression of the disease.

1.4.2.1 Symptomatic Treatments

This class consists of AChEIs, NMDA receptor antagonist and their combination. These drugs do not alter disease progression but aim to temporarily improve cognitive functions, thereby providing only symptomatic relief. These include:

1.4.2.2 AChEIs

- Donepezil (Aricept): An acetylcholinesterase inhibitor for mild, moderate, and severe AD.
- Rivastigmine (Exelon): Offers both oral and transdermal formulations.
- Galantamine (Razadyne): Also enhances nicotinic receptor activity.

1.4.2.3 NMDA Receptor Inhibitor

- Memantine (Namenda): An NMDA receptor antagonist, often used in combination with cholinesterase inhibitors.

1.4.2.4 Combination of AChEI and NMDA Receptor Antagonist

- Memantine + Donepezil (Namzaric): The first FDA-approved fixed-dose combination (FDC), which combines cholinergic and glutamatergic modulation. It synergizes NMDA receptor antagonism with acetylcholinesterase inhibition and is widely prescribed for moderate-to-severe AD.

1.4.2.5 Disease-Modifying Therapies (Anti-Amyloid Agents)

Recent breakthroughs have introduced DMTs aimed at clearing or preventing aggregation of Aβ plaques, thereby delaying the progression of the disease:

- Aducanumab (Aduhelm, approved in June 2021): It is a monoclonal antibody that targets aggregated Aβ and removes them by dissolving/degrading them. The drug has been discontinued since January 2024.
- Lecanemab (Leqembi, approved in July 2023): It is a humanized IgG1 monoclonal antibody that targets protofibrils of Aβ and demonstrated a significant slowing of cognitive decline in early AD. But its accelerated approval has been controversial as brain swelling and bleeding have been reported in some patients.
- Donanemab (Kisunla, approved in July 2024): It targets N3pG-modified amyloid plaques, facilitating their clearance through Fcγ receptor-mediated phagocytosis (US-FDA n.d.).

1.5 Conclusion

Since Alzheimer's disease (AD) is a multidimensional, complex neurodegenerative illness, understanding and treating it requires an equally multifaceted approach. Although amyloid cascade and tau pathology have long dominated the conceptual landscape, it is now undeniable that AD results from a network of interconnected pathological mechanisms, such as neuroinflammation, oxidative stress, mitochondrial dysfunction, metabolic dysregulation, vascular insufficiency, and impaired proteostasis. A significant change in AD research and its drug discovery, from single-target models to integrative systems-based therapies, has been observed in the recent past. An important turning point in AD treatment is the recent acceleration of the development of disease-modifying treatments (DMTs), such as anti-amyloid monoclonal antibodies (e.g., Aducanumab, Lecanemab, Donanemab). However, they are not universally acceptable because of their low clinical efficacy (than expected), and adverse effects. There is growing agreement that effective treatment will probably necessitate multi-target intervention and customization to individual pathological signatures. This is further supported and evidenced by the pipeline's expanding diversity, which includes anti-amyloid agents, tau aggregation inhibitors, anti-inflammatory agents, mitochondrial stabilizers, neuroprotective phytochemicals, and combination therapies. Despite so many advancements in the understanding of pathophysiological aspects of AD and its drug discovery process, a safe and effective treatment is still missing. Apart from all these therapeutic interventions, various alternative therapies like exercise, yoga, proper and nutritious diet have also been suggested to delay the onset of the disease and to some extent mitigate some of the symptoms. In the end, defeating AD will necessitate a comprehensive approach that goes beyond conventional pharmacology to include systems biology techniques and lifestyle change. The discipline is currently in a better position than ever to achieve a paradigm change in the management of Alzheimer's disease—from palliation to prevention and reversal—thanks to developments in translational neuroscience, biomarker identification, and clinical trial design. Although there is still a long way to go.

References

Adebayo OG, Ben-Azu B, Ajayi AM, Wopara I, Aduema W, Kolawole TA et al (2022) Gingko biloba abrogate lead-induced neurodegeneration in mice hippocampus: involvement of NF-κB expression, myeloperoxidase activity and pro-inflammatory mediators. Biol Trace Elem Res 200(4):1736–1749. https://doi.org/10.1007/s12011-021-02790-3

Adlimoghaddam A, Snow WM, Greg S, Claudia P, Jelena D, Andrew LG et al (2019) Regional hypometabolism in the 3xTg mouse model of Alzheimer's disease. Neurobiol Dis 127:264–277

Advani D, Kumar P (2024) Uncovering cell cycle dysregulations and associated mechanisms in cancer and neurodegenerative disorders: a glimpse of Hope for repurposed drugs. Mol Neurobiol 61(11):8600–8630

Ahmad MH, Fatima M, Mondal AC (2019a) Influence of microglia and astrocyte activation in the neuroinflammatory pathogenesis of Alzheimer's disease: rational insights for the therapeutic approaches. J Clin Neurosci 59:6–11

Ahmad SS, Khan S, Kamal MA, Wasi U (2019b) The structure and function of α, β and γ-secretase as therapeutic target enzymes in the development of Alzheimer's disease: a review. CNS Neurol Disorders-Drug Targets 18(9):657–667

Akaike A, Shimohama S, Misu Y (2018) Nicotinic acetylcholine receptor signaling in neuroprotection

Akbar M, Shabbir A, Rehman K, Akash MSH, Shah MA (2021) Neuroprotective potential of berberine in modulating Alzheimer's disease via multiple signaling pathways. J Food Biochem 45(10):e13936

Alavi Naini SM, Soussi-Yanicostas N (2015) Tau hyperphosphorylation and oxidative stress, a critical vicious circle in neurodegenerative tauopathies. Oxid Med Cell Longev 2015(1):151979

Albar NY, Hassaballa H, Shikh H, Albar Y, Ibrahim AS, Mousa AH et al (2024) Retracted article: the interaction between insulin resistance and Alzheimer's disease: a review article. Postgrad Med 136(4):377–395

Alkahtani S, Al-Johani NS, Alarifi S (2023) Mechanistic insights, treatment paradigms, and clinical progress in neurological disorders: current and future prospects. Int J Mol Sci 24(2):1340

Ambegaokar SS, Wu L, Alamshahi K, Lau J, Jazayeri L, Chan S et al (2003) Curcumin inhibits dose-dependently and time-dependently neuroglial cell proliferation and growth. Neuroendocrinol Lett 24(6):469–469

Area-Gomez E, de Groof A, Bonilla E, Montesinos J, Tanji K, Boldogh I et al (2018) A key role for MAM in mediating mitochondrial dysfunction in Alzheimer disease. Cell Death Disease 9(3):335

Arvanitakis Z, Capuano AW, Leurgans SE, Bennett DA, Schneider JA (2016) Relation of cerebral vessel disease to Alzheimer's disease dementia and cognitive function in elderly people: a cross-sectional study. Lancet Neurol 15(9):934–943

Arvanitakis Z, Wang HY, Capuano AW, Khan A, Taïb B, Anokye-Danso F et al (2020) Brain insulin signaling, Alzheimer disease pathology, and cognitive function. Ann Neurol 88(3):513–525

Ashrafizadeh M, Zarrabi A, Najafi M, Samarghandian S, Mohammadinejad R, Ahn KS (2020) Resveratrol targeting tau proteins, amyloid-beta aggregations, and their adverse effects: an updated review. Phytother Res 34(11):2867–2888. https://doi.org/10.1002/ptr.6732

Azargoonjahromi A (2024) The duality of amyloid-β: its role in normal and Alzheimer's disease states. Mol Brain 17(1):44

Babaei P (2021) NMDA and AMPA receptors dysregulation in Alzheimer's disease. Eur J Pharmacol 908:174310

Barrett PJ, Song Y, Van Horn WD, Hustedt EJ, Schafer JM, Hadziselimovic A et al (2012) The amyloid precursor protein has a flexible transmembrane domain and binds cholesterol. Science 336(6085):1168–1171

Bernardo A, Plumitallo C, De Nuccio C, Visentin S, Minghetti L (2021) Curcumin promotes oligodendrocyte differentiation and their protection against TNF-α through the activation of the nuclear receptor PPAR-γ. Sci Rep 11(1):1–13

Bertocchi I, Cifarelli L, Oberto A, Eva CE, Sprengel R, Mirza NR, Muglia P (2024) Radiprodil, a selective GluN2B negative allosteric modulator, rescues audiogenic seizures in mice carrying the GluN2A (N615S) mutation. Br J Pharmacol 181(12):1886–1894

Beyreuther K, Haass C, Nitsch RM, Christen Y (2006) Alzheimer: 100 years and beyond. Springer

Bhatt S, Puli L, Patil CR (2021) Role of reactive oxygen species in the progression of Alzheimer's disease. Drug Discov Today 26(3):794–803

Bihaqi SW, Singh AP, Tiwari M (2012) Supplementation of convolvulus pluricaulis attenuates scopolamine-induced increased tau and amyloid precursor protein (AβPP) expression in rat brain. Indian J Pharmacol 44(5):593–598

Boccardi V, Comanducci C, Baroni M, Mecocci P (2017) Of energy and entropy: the ineluctable impact of aging in old age dementia. Int J Mol Sci 18(12):2672

Bouchon A, Hernández-Munain C, Cella M, Colonna M (2001) A DAP12-mediated pathway regulates expression of CC chemokine receptor 7 and maturation of human dendritic cells. J Exp Med 194(8):1111–1122

Buccellato FR, D'Anca M, Fenoglio C, Scarpini E, Galimberti D (2021) Role of oxidative damage in Alzheimer's disease and neurodegeneration: from pathogenic mechanisms to biomarker discovery. Antioxidants 10(9):1353

Cai Z, Liu N, Wang C, Qin B, Zhou Y, Xiao M et al (2016) Role of RAGE in Alzheimer's disease. Cell Mol Neurobiol 36:483–495

Carranza-Naval MJ, Vargas-Soria M, Hierro-Bujalance C, Baena-Nieto G, Garcia-Alloza M, Infante-Garcia C, Del Marco A (2021) Alzheimer's disease and diabetes: role of diet, microbiota and inflammation in preclinical models. Biomol 11(2):262

Cai Y, Liu J, Wang B, Sun M, Yang H (2022) Microglia in the neuroinflammatory pathogenesis of Alzheimer's disease and related therapeutic targets. Front Immunol 13:856376

Campos-Peña V, Pichardo-Rojas P, Sánchez-Barbosa T, Ortíz-Islas E, Rodríguez-Pérez CE, Montes P et al (2022) Amyloid β, lipid metabolism, basal cholinergic system, and therapeutics in Alzheimer's disease. Int J Mol Sci 23(20):12092

Cascella R, Cecchi C (2021) Calcium dyshomeostasis in Alzheimer's disease pathogenesis. Int J Mol Sci 22(9):4914

Castellani R, Hirai K, Aliev G, Drew KL, Nunomura A, Takeda A et al (2002) Role of mitochondrial dysfunction in Alzheimer's disease. J Neurosci Res 70(3):357–360

Chakraborty A, Sami SA, Marma KKS (2021) A comprehensive review on RAGE-facilitated pathological pathways connecting Alzheimer's disease, diabetes mellitus, and cardiovascular diseases. Egypt J Internal Med 33:1–10

Chang KJ, Hong CH, Lee KS, Kang DR, Lee JD, Choi SH et al (2017) Mortality risk after diagnosis of early-onset Alzheimer's disease versus late-onset Alzheimer's disease: a propensity score matching analysis. J Alzheimer's Dis 56(4):1341–1348
Chen M, Li L, Liu C, Song L (2020) Berberine attenuates Aβ-induced neuronal damage through regulating miR-188/NOS1 in Alzheimer's disease. Mol Cell Biochem 474:285–294
Chen Z-R, Huang J-B, Yang S-L, Hong F-F (2022) Role of cholinergic signaling in Alzheimer's disease. Molecules 27(6):1816
Chen L-L, Fan Y-G, Zhao L-X, Zhang Q, Wang Z-Y (2023) The metal ion hypothesis of Alzheimer's disease and the anti-neuroinflammatory effect of metal chelators. Bioorg Chem 131:106301
Chen J, Chen J-S, Li S, Zhang F, Deng J, Zeng L-H, Tan J (2024) Amyloid precursor protein: a regulatory hub in Alzheimer's disease. Aging disease 15(1):201
Cheng XR, Hau BY, Veloso AJ, Martic S, Kraatz H-B, Kerman K (2013) Surface plasmon resonance imaging of amyloid-β aggregation kinetics in the presence of epigallocatechin gallate and metals. Anal Chem 85(4):2049–2055
Ciurea VA, Covache-Busuioc R-A, Mohan AG, Costin HP, Voicu V (2023) Alzheimer's disease: 120 years of research and progress. J Med Life 16(2):173
Companys-Alemany J, Turcu AL, Bellver-Sanchis A, Loza MI, Brea JM, Canudas AM et al (2020) A novel NMDA receptor antagonist protects against cognitive decline presented by senescent mice. Pharmaceutics 12(3):284
Congdon EE, Sigurdsson EM (2018) Tau-targeting therapies for Alzheimer disease. Nat Rev Neurol 14(7):399–415
Cummings J, Lee G, Ritter A, Sabbagh M, Zhong K (2019a) Alzheimer's disease drug development pipeline: 2019. Alzheimer's Dement. Transl Res Clin Interv 5:272–293
Cummings JL, Tong G, Ballard C (2019b) Treatment combinations for Alzheimer's disease: current and future pharmacotherapy options. J Alzheimer's Dis 67(3):779–794
Cummings J, Lee G, Ritter A, Sabbagh M, Zhong K (2020) Alzheimer's disease drug development pipeline: 2020. Alzheimer's Dement. Transl Res Clin Interv 6(1):12050
Cummings J, Lee G, Zhong K, Fonseca J, Taghva K (2021) Alzheimer's disease drug development pipeline: 2021. Alzheimer's Dement. Transl Res Clin Interv 7(1):e12179
Cummings J, Lee G, Nahed P, Kambar MEZN, Zhong K, Fonseca J, Taghva K (2022) Alzheimer's disease drug development pipeline: 2022. Alzheimer's Dement. Transl Res Clin Interv 8(1):e12295
Cummings J, Zhou Y, Lee G, Zhong K, Fonseca J, Cheng F (2023) Alzheimer's disease drug development pipeline: 2023. Alzheimer's Dement. Transl Res Clin Interv 9(2):e12385
Cummings J, Zhou Y, Lee G, Zhong K, Fonseca J, Cheng F (2024). Alzheimer's disease drug development pipeline: 2024. Alzheimer's Dement. Transl Res Clin Interv 10(2):e12465
Dar NJ, Glazner GW (2020) Deciphering the neuroprotective and neurogenic potential of soluble amyloid precursor protein alpha (sAPPα). Cell Mol Life Sci 77:2315–2330
Darqui FS, Radonic LM, Beracochea VC, Hopp HE, López Bilbao M (2021) Peculiarities of the transformation of Asteraceae family species: the cases of sunflower and lettuce. Front Plant Sci 12:767459
Das N, Raymick J, Sarkar S (2021) Role of metals in Alzheimer's disease. Metab Brain Dis 36(7):1627–1639
De la Torre J (2000) Critically attained threshold of cerebral hypoperfusion: the CATCH hypothesis of Alzheimer's pathogenesis. Neurobiol Aging 21(2):331–342
De La Torre J (2018) The vascular hypothesis of Alzheimer's disease: a key to preclinical prediction of dementia using neuroimaging. J Alzheimer's Dis 63(1):35–52
De la Torre JC (2021) Deciphering Alzheimer's disease pathogenic pathway: role of chronic brain hypoperfusion on p-tau and mTOR. J Alzheimer's Dis 79(4):1381–1396
de Montgolfier O, Thorin-Trescases N, Thorin E (2020) Pathological continuum from the rise in pulse pressure to impaired neurovascular coupling and cognitive decline. Am J Hypertens 33(5):375–390

Dentoni G, Naia L, Portal B, Leal NS, Nilsson P, Lindskog M, Ankarcrona M (2022) Mitochondrial alterations in neurons derived from the murine app NL-F Knock-in model of Alzheimer's disease. J Alzheimer's Dis 90(2):565–583

Dixit H, Selvaa Kumar C, Dasgupta D, Gadewal N (2021) Molecular docking analysis of hyperphosphorylated tau protein with compounds derived from Bacopa monnieri and Withania somnifera. Bioinformation 17(9):798–804. https://doi.org/10.6026/97320630017798

Djelti F, Braudeau J, Hudry E, Dhenain M, Varin J, Bieche I et al (2015) CYP46A1 inhibition, brain cholesterol accumulation and neurodegeneration pave the way for Alzheimer's disease. Brain 138(8):2383–2398

Donahue JE, Flaherty SL, Johanson CE, Duncan JA, Silverberg GD, Miller MC et al (2006) RAGE, LRP-1, and amyloid-beta protein in Alzheimer's disease. Acta Neuropathol 112:405–415

Dong S, Duan Y, Hu Y, Zhao Z (2012) Advances in the pathogenesis of Alzheimer's disease: a re-evaluation of amyloid cascade hypothesis. Transl Neurodegener 1:1–12

Du X, Wang X, Geng M (2018) Alzheimer's disease hypothesis and related therapies. Transl Neurodegener 7:1–7

Dubois B, Feldman HH, Jacova C, Hampel H, Molinuevo JL, Blennow K et al (2014) Advancing research diagnostic criteria for Alzheimer's disease: the IWG-2 criteria. Lancet Neurol 13(6):614–629

Egunlusi AO, Joubert J (2024) NMDA receptor antagonists: emerging insights into molecular mechanisms and clinical applications in neurological disorders. Pharmaceuticals 17(5):639

Eisenmenger LB, Peret A, Famakin BM, Spahic A, Roberts GS, Bockholt JH et al (2023) Vascular contributions to Alzheimer's disease. Transl Res 254:41–53

El Khoury NB, Gratuze M, Papon M-A, Bretteville A, Planel E (2014) Insulin dysfunction and tau pathology. Front Cell Neurosci 8:22

Embury CM, Dyavarshetty B, Lu Y, Wiederin JL, Ciborowski P, Gendelman HE, Kiyota T (2017) Cathepsin B improves ß-amyloidosis and learning and memory in models of Alzheimer's disease. J Neuroimmune Pharmacol 12:340–352

Evin G, Weidemann A (2002) Biogenesis and metabolism of Alzheimer's disease Aβ amyloid peptides. Peptides 23(7):1285–1297

Fanlo-Ucar H, Picón-Pagès P, Herrera-Fernández V, Ill-Raga G, Muñoz FJ (2024) The dual role of amyloid Beta-peptide in oxidative stress and inflammation: unveiling their connections in Alzheimer's disease etiopathology. Antioxidants 13(10):1208

Fasae KD, Abolaji AO, Faloye TR, Odunsi AY, Oyetayo BO, Enya JI et al (2021) Metallobiology and therapeutic chelation of biometals (copper, zinc and iron) in Alzheimer's disease: limitations, and current and future perspectives. J Trace Elements Med Biol 67:126779

Feringa FM, Van der Kant R (2021) Cholesterol and Alzheimer's disease; from risk genes to pathological effects. Front Aging Neurosci 13:690372

Ferrari C, Sorbi S (2021) The complexity of Alzheimer's disease: an evolving puzzle. Physiol Rev 101(3):1047–1081

Ferreiro AL, Choi J, Ryou J, Newcomer EP, Thompson R, Bollinger RM et al (2023) Gut microbiome composition may be an indicator of preclinical Alzheimer's disease. Sci Transl Med 15(700):eabo2984

Fisher RA, Miners JS, Love S (2022) Pathological changes within the cerebral vasculature in Alzheimer's disease: new perspectives. Brain Pathol 32(6):e13061

Francis PT, Palmer AM, Snape M, Wilcock GK (1999) The cholinergic hypothesis of Alzheimer's disease: a review of progress. J Neurol Neurosur Psychiatry 66(2):137–147

Gasparini L, Ongini E, Wenk G (2004) Non-steroidal anti-inflammatory drugs (NSAIDs) in Alzheimer's disease: old and new mechanisms of action. J Neurochem 91(3):521–536

Gerritsen AA, Bakker C, Verhey FR, de Vugt ME, Melis RJ, Koopmans RT et al (2016) Prevalence of comorbidity in patients with young-onset Alzheimer disease compared with late-onset: a comparative cohort study. J Am Med Dir Assoc 17(4):318–323

Gerritsen AA, Bakker C, Bruls E, Verhey FR, Pijnenburg YA, Millenaar JK et al (2021) Psychotropic drug use in community-dwelling people with young-onset dementia: two-year course and determinants. Aging mental health 25(1):179–186

Gibson G, Sheu K-F, Blass J (1998) Abnormalities of mitochondrial enzymes in Alzheimer disease. J Neural Transm 105:855–870
Gliozzi M, Musolino V, Bosco F, Scicchitano M, Scarano F, Nucera S et al (2021) Cholesterol homeostasis: researching a dialogue between the brain and peripheral tissues. Pharmacol Res 163:105215
Goldgaber D, Harris HW, Hla T, Maciag T, Donnelly RJ, Jacobsen JS et al (1989) Interleukin 1 regulates synthesis of amyloid beta-protein precursor mRNA in human endothelial cells. Proc Natl Acad Sci 86(19):7606–7610
Gratuze M, Joly-Amado A, Vieau D, Buee L, Blum D (2018a) Mutual relationship between tau and central insulin signalling: consequences for AD and tauopathies? Neuroendocrinology 107(2):181–195
Gratuze M, Leyns CE, Holtzman DM (2018b) New insights into the role of TREM2 in Alzheimer's disease. Mol Neurodegener 13(1):66
Gu L, Ju Y, Hu M, Zheng M, Li Q, Zhang X (2023) Research progress of PPARγ regulation of cholesterol and inflammation in Alzheimer's disease. Metab Brain Dis 38(3):839–854
Guha S, Johnson GV, Nehrke K (2020) The crosstalk between pathological tau phosphorylation and mitochondrial dysfunction as a key to understanding and treating Alzheimer's disease. Mol Neurobiol 57(12):5103–5120
Guo Y, Li S, Zeng L-H, Tan J (2022) Tau-targeting therapy in Alzheimer's disease: critical advances and future opportunities. Ageing Neurodegener Dis 2(3)
Hamano T, Enomoto S, Shirafuji N, Ikawa M, Yamamura O, Yen S-H, Nakamoto Y (2021) Autophagy and tau protein. Int J Mol Sci 22(14):7475
Hampel H, Hardy J, Blennow K, Chen C, Perry G, Kim SH et al (2021) The amyloid-β pathway in Alzheimer's disease. Mol Psychiatry 26(10):5481–5503
Hardiman O, Doherty CP, Elamin M, Bede P (2011) Neurodegenerative disorders. Springer
Hardy JA, Higgins GA (1992) Alzheimer's disease: the amyloid cascade hypothesis. Science 256(5054):184–185
Hasan A, Scuderi SA, Capra AP, Giosa D, Bonomo A, Ardizzone A, Esposito E (2025) An updated and comprehensive review exploring the gut–brain axis in neurodegenerative disorders and neurotraumas: implications for therapeutic strategies. Brain Sci 15(6):654
Hatami M, Abdolahi M, Soveyd N, Djalali M, Togha M, Honarvar NM (2019) Molecular mechanisms of curcumin in neuroinflammatory disorders: a mini review of current evidences. Endocr Metab Immune Disord Drug Targets 19(3):247–258
Hickman SE, El Khoury J (2014) TREM2 and the neuroimmunology of Alzheimer's disease. Biochem Pharmacol 88(4):495–498
Hickman SE, Allison EK, El Khoury J (2008) Microglial dysfunction and defective β-amyloid clearance pathways in aging Alzheimer's disease mice. J Neurosci 28(33):8354–8360
Hochgräfe K, Sydow A, Matenia D, Cadinu D, Könen S, Petrova O et al (2015) Preventive methylene blue treatment preserves cognition in mice expressing full-length pro-aggregant human tau. Acta Neuropathol Commun 3:1–22
Holcomb LA, Dhanasekaran M, Hitt AR, Young KA, Riggs M, Manyam BV (2006) *Bacopa monniera* extract reduces amyloid levels in PSAPP mice. J Alzheimer's Dis 9(3):243–251
Hou Y, Dan X, Babbar M, Wei Y, Hasselbalch SG, Croteau DL, Bohr VA (2019) Ageing as a risk factor for neurodegenerative disease. Nat Rev Neurol 15(10):565–581
Hu Z-L, Yuan Y-Q, Tong Z, Liao M-Q, Yuan S-L, Jian Y et al (2023) Reexamining the causes and effects of cholesterol deposition in the brains of patients with Alzheimer's disease. Mol Neurobiol 60(12):6852–6868
Hussain G, Wang J, Rasul A, Anwar H, Imran A, Qasim M et al (2019) Role of cholesterol and sphingolipids in brain development and neurological diseases. Lipids Health Disease 18:1–12
Iturria-Medina Y, Sotero RC, Toussaint PJ, Mateos-Pérez JM, Evans AC (2016) Early role of vascular dysregulation on late-onset Alzheimer's disease based on multifactorial data-driven analysis. Nat Commun 7(1):11934

Jay TR, Miller CM, Cheng PJ, Graham LC, Bemiller S, Broihier ML et al (2015) TREM2 deficiency eliminates TREM2+ inflammatory macrophages and ameliorates pathology in Alzheimer's disease mouse models. J Exp Med 212(3):287–295

Jia J, Zhang Y, Shi Y, Yin X, Wang S, Li Y et al (2023) A 19-year-old adolescent with probable Alzheimer's disease. J Alzheimer's Dis 91(3):915–922

Jiang S, Gao BB, Ou YF, Zhao QS (2024) Lycopodium alkaloids from Huperzia serrata and their cholinesterase inhibitory activities. Phytochem 223:114114

Jin SC, Carrasquillo MM, Benitez BA, Skorupa T, Carrell D, Patel D et al (2015) TREM2 is associated with increased risk for Alzheimer's disease in African Americans. Mol Neurodegener 10:1–7

Kalaria RN (2018) The pathology and pathophysiology of vascular dementia. Neuropharmacology 134:226–239

Kametani F, Hasegawa M (2018) Reconsideration of amyloid hypothesis and tau hypothesis in Alzheimer's disease. Front Neurosci 12:328460

Kanaan NM, Pigino GF, Brady ST, Lazarov O, Binder LI, Morfini GA (2013) Axonal degeneration in Alzheimer's disease: when signaling abnormalities meet the axonal transport system. Exp Neurol 246:44–53

Kaur D, Sharma V, Deshmukh R (2019) Activation of microglia and astrocytes: a roadway to neuroinflammation and Alzheimer's disease. Inflammopharmacology 27:663–677

Kim HG, Park G, Lim S, Park H, Choi JG, Jeong HU et al (2015a) Mori Fructus improves cognitive and neuronal dysfunction induced by beta-amyloid toxicity through the GSK-3β pathway in vitro and in vivo. J Ethnopharmacol 171:196–204

Kim JY, Lee HK, Jang JY, Yoo JK, Seong YH (2015b) Ilex latifolia prevents amyloid β protein (25–35)-induced memory impairment by inhibiting apoptosis and tau phosphorylation in mice. J Med Food 18(12):1317–1326

Kinney JW, Bemiller SM, Murtishaw AS, Leisgang AM, Salazar AM, Lamb BT (2018) Inflammation as a central mechanism in Alzheimer's disease. Alzheimer's Dement. Transl Res Clin Interv 4:575–590

Kleinberger G, Yamanishi Y, Suárez-Calvet M, Czirr E, Lohmann E, Cuyvers E et al (2014) TREM2 mutations implicated in neurodegeneration impair cell surface transport and phagocytosis. Sci Transl Med 6(243):243ra286-243ra286

Kosaraju J, Madhunapantula SV, Chinni S, Khatwal RB, Dubala A, Nataraj SKM, Basavan D (2014) Dipeptidyl peptidase-4 inhibition by Pterocarpus marsupium and Eugenia jambolana ameliorates streptozotocin induced Alzheimer's disease. Behav Brain Res 267:55–65

Krabbe G, Halle A, Matyash V, Rinnenthal J, Eom G (2013) Functional impairment of microglia coincides with beta-amyloid deposition in mice

Kulkarni B, Kumar D, Cruz-Martins N, Sellamuthu S (2021) Role of TREM2 in Alzheimer's disease: a long road ahead. Mol Neurobiol 58:5239–5252

Kumar D, Ganeshpurkar A, Kumar D, Modi G, Gupta SK, Singh SK (2018) Secretase inhibitors for the treatment of Alzheimer's disease: long road ahead. Eur J Med Chem 148:436–452

Kurapati KR, Atluri VS, Samikkannu T, Nair MP (2013) Ashwagandha (*Withania somnifera*) reverses beta-amyloid1-42 induced toxicity in human neuronal cells: implications in HIV-associated neurocognitive disorders (HAND). PLoS One 8(10):e77624. https://doi.org/10.1371/journal.pone.0077624

Ladiwala ARA, Lin JC, Bale SS, Marcelino-Cruz AM, Bhattacharya M, Dordick JS, Tessier PM (2010) Resveratrol selectively remodels soluble oligomers and fibrils of amyloid Aβ into off-pathway conformers. J Biol Chem 285(31):24228–24237

Lill CM, Rengmark A, Pihlstrøm L, Fogh I, Shatunov A, Sleiman PM et al (2015) The role of TREM2 R47H as a risk factor for Alzheimer's disease, frontotemporal lobar degeneration, amyotrophic lateral sclerosis, and Parkinson's disease. Alzheimers Dement 11(12):1407–1416

Limpeanchob N, Jaipan S, Rattanakaruna S, Phrompittayarat W, Ingkaninan K (2008) Neuroprotective effect of Bacopa monnieri on beta-amyloid-induced cell death in primary cortical culture. J Ethnopharmacol 120(1):112–117

Liu R-M (2022) Aging, cellular senescence, and Alzheimer's disease. Int J Mol Sci 23(4):1989

Liu P-P, Xie Y, Meng X-Y, Kang J-S (2019) History and progress of hypotheses and clinical trials for Alzheimer's disease. Signal Transduct Target Ther 4(1):29

Liu W, Taso O, Wang R, Bayram S, Graham AC, Garcia-Reitboeck P et al (2020) Trem2 promotes anti-inflammatory responses in microglia and is suppressed under pro-inflammatory conditions. Hum Mol Genet 29(19):3224–3248

Lu T-T, Wan C, Yang W, Cai Z (2019) Role of Cdk5 in amyloid-beta pathology of Alzheimer's disease. Curr Alzheimer Res 16(13):1206–1215

Mahaman YAR, Huang F, Wu M, Wang Y, Wei Z, Bao J et al (2018) Moringa oleifera alleviates homocysteine-induced Alzheimer's disease-like pathology and cognitive impairments. J Alzheimer's Dis 63(3):1141–1159

Malik J (2017) Phytomedicine : a possible tool for Alzheimer's disease therapeutics. In: Farooqui T, Farooqui AA (eds) Neuroprotective effects of phytochemicals in neurological disorders, 1st edn. Wiley, Hoboken, pp 167–198

Malik J, Karan M, Vasisht K (2016) Attenuating effect of bioactive coumarins from *convolvulus pluricaulis* on scopolamine-induced amnesia in mice. Nat Prod Res 30(5):578–582

Maphis NM, Hulse J, Peabody J, Dadras S, Whelpley MJ, Kandath M et al (2025) Targeting of phosphorylated tau at threonine 181 by a Qβ virus-like particle vaccine is safe, highly immunogenic, and reduces disease severity in mice and rhesus macaques. Alzheimers Dement 21(3):e70101

Markesbery WR (1997) Oxidative stress hypothesis in Alzheimer's disease. Free Radical Biol Med 23(1):134–147

Martin L, Latypova X, Wilson CM, Magnaudeix A, Perrin M-L, Terro F (2013) Tau protein phosphatases in Alzheimer's disease: the leading role of PP2A. Ageing Res Rev 12(1):39–49

Matheou CJ, Younan ND, Viles JH (2015) Cu2+ accentuates distinct misfolding of Aβ (1–40) and Aβ (1–42) peptides, and potentiates membrane disruption. Biochem J 466(2):233–242

Mattson MP (2004) Pathways towards and away from Alzheimer's disease. Nature 430(7000):631–639

Maurer SV, Williams CL (2017) The cholinergic system modulates memory and hippocampal plasticity via its interactions with non-neuronal cells. Front Immunol 8:1489

McCormick JW, McCormick LA, Chen G, Vogel PD, Wise JG (2021) Transport of Alzheimer's associated amyloid-β catalyzed by P-glycoprotein. PLoS One 16(4):e0250371

McGeer PL, McGeer EG (2013) The amyloid cascade-inflammatory hypothesis of Alzheimer disease: implications for therapy. Acta Neuropathol 126:479–497

Melchior B, Garcia AE, Hsiung B-K, Lo KM, Doose JM, Thrash JC et al (2010) Dual induction of TREM2 and tolerance-related transcript, Tmem176b, in amyloid transgenic mice: implications for vaccine-based therapies for Alzheimer's disease. ASN Neuro 2(3):AN20100010

Melov S, Adlard PA, Morten K, Johnson F, Golden TR, Hinerfeld D et al (2007) Mitochondrial oxidative stress causes hyperphosphorylation of tau. PLoS One 2(6):e536

Mendez MF (2017) Early-onset Alzheimer's disease. Neurol Clin 35(2):263

Miao J, Ma H, Yang Y, Liao Y, Lin C, Zheng J et al (2023) Microglia in Alzheimer's disease: pathogenesis, mechanisms, and therapeutic potentials. Front Aging Neurosci 15:1201982

Monaco M, Trebesova H, Grilli M (2024) Muscarinic receptors and Alzheimer's disease: new perspectives and mechanisms. Curr Issues Mol Biol 46(7):6820–6835

Moon HY, Becke A, Berron D, Becker B, Sah N, Benoni G et al (2016) Running-induced systemic cathepsin B secretion is associated with memory function. Cell Metab 24(2):332–340

Mou Y, Du Y, Zhou L, Yue J, Hu X, Liu Y et al (2022) Gut microbiota interact with the brain through systemic chronic inflammation: implications on neuroinflammation, neurodegeneration, and aging. Front Immunol 13:796288

Müller UC, Zheng H (2012) Physiological functions of APP family proteins. Cold Spring Harb Perspect Med 2(2):a006288

Nakanishi H (2020) Microglial cathepsin B as a key driver of inflammatory brain diseases and brain aging. Neural Regen Res 15(1):25–29

Panegyres PK, Chen H-Y (2013) Differences between early and late onset Alzheimer's disease. Am J Neurodegener Dis 2(4):300

Paroni G, Bisceglia P, Seripa D (2019) Understanding the amyloid hypothesis in Alzheimer's disease. J Alzheimer's Dis 68(2):493–510

Piaceri I, Nacmias B, Sorbi S (2013) Genetics of familial and sporadic Alzheimer's disease. Front Biosci 5(1):167–177

Piccio L, Deming Y, Del-Águila JL, Ghezzi L, Holtzman DM, Fagan AM et al (2016) Cerebrospinal fluid soluble TREM2 is higher in Alzheimer disease and associated with mutation status. Acta Neuropathol 131:925–933

Planel E, Tatebayashi Y, Miyasaka T, Liu L, Wang L, Herman M et al (2007) Insulin dysfunction induces in vivo tau hyperphosphorylation through distinct mechanisms. J Neurosci 27(50):13635–13648

Pomytkin I, Costa-Nunes JP, Kasatkin V, Veniaminova E, Demchenko A, Lyundup A et al (2018) Insulin receptor in the brain: mechanisms of activation and the role in the CNS pathology and treatment. CNS Neurosci Ther 24(9):763–774

Pradeepkiran JA, Baig J, Seman A, Reddy PH (2024) Mitochondria in aging and Alzheimer's disease: focus on mitophagy. Neuroscientist 30(4):440–457

Pritam P, Deka R, Bhardwaj A, Srivastava R, Kumar D, Jha AK et al (2022) Antioxidants in Alzheimer's disease: current therapeutic significance and future prospects. Biology 11(2):212

Qin Y, Zhang Y, Tomic I, Hao W, Menger MD, Liu C et al (2018) Ginkgo biloba extract EGb 761 and its specific components elicit protective protein clearance through the autophagy-lysosomal pathway in tau-transgenic mice and cultured neurons. J Alzheimer's Dis 65(1):243–263

Qin S, Tang H, Li W, Gong Y, Li S, Huang J et al (2020) AMPK and its activator berberine in the treatment of neurodegenerative diseases. Curr Pharm Des 26(39):5054–5066

Qiu WQ, Folstein MF (2006) Insulin, insulin-degrading enzyme and amyloid-β peptide in Alzheimer's disease: review and hypothesis. Neurobiol Aging 27(2):190–198

Quintanilla RA, Orellana DI, González-Billault C, Maccioni RB (2004) Interleukin-6 induces Alzheimer-type phosphorylation of tau protein by deregulating the cdk5/p35 pathway. Exp Cell Res 295(1):245–257

Rai R, Singh HK, Prasad S (2015) A special extract of *Bacopa monnieri* (CDRI-08) restores learning and memory by upregulating expression of the NMDA receptor subunit GluN2B in the brain of scopolamine-induced amnesic mice. Evid Based Complement Alternat Med 2015:254303. https://doi.org/10.1155/2015/254303

Rajendran K, Krishnan UM (2024) Mechanistic insights and emerging therapeutic stratagems for Alzheimer's disease. Ageing Res Rev 102309:102309

Rane JS, Bhaumik P, Panda D (2017) Curcumin inhibits tau aggregation and disintegrates preformed tau filaments in vitro. J Alzheimer's Dis 60(3):999–1014

Rappoport A (2023) A lipid-raft theory of Alzheimer's disease. Annu Rev Biochem 94

Rawat P, Sehar U, Bisht J, Selman A, Culberson J, Reddy PH (2022) Phosphorylated tau in Alzheimer's disease and other tauopathies. Int J Mol Sci 23(21):12841

Razgonova MP, Veselov VV, Zakharenko AM, Golokhvast KS, Nosyrev AE, Cravotto G et al (2019) Panax ginseng components and the pathogenesis of Alzheimer's disease. Mol Med Rep 19(4):2975–2998

Reddy PH (2013) Amyloid beta-induced glycogen synthase kinase 3β phosphorylated VDAC1 in Alzheimer's disease: implications for synaptic dysfunction and neuronal damage. Biochim Biophys Acta Mol basis Dis 1832(12):1913–1921

Reddy PH, Manczak M, Yin X, Grady MC, Mitchell A, Tonk S et al (2018) Protective effects of Indian spice curcumin against amyloid-β in Alzheimer's disease. J Alzheimer's Dis 61(3):843–866

Reichman WE, Rose NS (2012) History and experience: the direction of Alzheimer's disease. Menopause 19(7):724–734

Remya C, Dileep K, Variayr E, Sadasivan C (2016) An in silico guided identification of nAChR agonists from *Withania somnifera*. Front Life Sci 9(3):201–213

Riphagen JM, Ramakers IH, Freeze WM, Pagen LH, Hanseeuw BJ, Verbeek MM et al (2020) Linking APOE-ε4, blood-brain barrier dysfunction, and inflammation to Alzheimer's pathology. Neurobiol Aging 85:96–103

Roberts JP, Stokoe SA, Sathler MF, Nichols RA, Kim S (2021) Selective coactivation of α7-and α4β2-nicotinic acetylcholine receptors reverses beta-amyloid–induced synaptic dysfunction. J Biol Chem 296:100402
Ruitenberg A, Den Heijer T, Bakker SL, Van Swieten JC, Koudstaal PJ, Hofman A, Breteler MM (2005) Cerebral hypoperfusion and clinical onset of dementia: the Rotterdam study. Ann Neurol 57(6):789–794
Ruiz A, Dols-Icardo O, Bullido MJ, Pastor P, Rodríguez-Rodríguez E, de Munain AL et al (2014) Assessing the role of the TREM2 p. R47H variant as a risk factor for Alzheimer's disease and frontotemporal dementia. Neurobiol Aging 35(2):444
Ryu SY, Coutu J-P, Rosas HD, Salat DH (2014) Effects of insulin resistance on white matter microstructure in middle-aged and older adults. Neurology 82(21):1862–1870
Samluk L, Ostapczuk P, Dziembowska M (2022) Long-term mitochondrial stress induces early steps of tau aggregation by increasing reactive oxygen species levels and affecting cellular proteostasis. Mol Biol Cell 33(8):ar67
Sanabria-Castro A, Alvarado-Echeverría I, Monge-Bonilla C (2017) Molecular pathogenesis of Alzheimer's disease: an update. Ann Neurosci 24(1):46–54
Santos CY, Snyder PJ, Wu W-C, Zhang M, Echeverria A, Alber J (2017) Pathophysiologic relationship between Alzheimer's disease, cerebrovascular disease, and cardiovascular risk: a review and synthesis. Alzheimer's Dement. Transl Res Clin Interv 7:69–87
Sasaguri H, Takamura R, Watamura N, Kakiya N, Ohshima T, Fujioka R, . . . Matsub Y (2021) Neprilysin-sensitive amyloidogenic Aβ versus IDE-sensitive soluble Aβ: a probable mechanistic cause for sporadic Alzheimer's disease. *bioRxiv*, 2021.2008. 2022.457281
Sayas CL, Ávila J (2021) GSK-3 and tau: a key duet in Alzheimer's disease. Cells 10(4):721
Sędzikowska A, Szablewski L (2021) Insulin and insulin resistance in Alzheimer's disease. Int J Mol Sci 22(18):9987
Sehar U, Rawat P, Reddy AP, Kopel J, Reddy PH (2022) Amyloid beta in aging and Alzheimer's disease. Int J Mol Sci 23(21):12924
Singh SK, Srivastav S, Castellani RJ, Plascencia-Villa G, Perry G (2019) Neuroprotective and antioxidant effect of Ginkgo biloba extract against AD and other neurological disorders. Neurotherapeutics 16(3):666–674
Singh S, Gupta G, Aran KR (2024) Emerging role of antioxidants in Alzheimer's disease: insight into physiological, pathological mechanisms and management. Pharm Sci Adv 2:100021
Sirisi S, Sánchez-Aced É, Belbin O, Lleó A (2024) APP dyshomeostasis in the pathogenesis of Alzheimer's disease: implications for current drug targets. Alzheimer's Res Ther 16(1):144
Sochocka M, Donskow-Łysoniewska K, Diniz BS, Kurpas D, Brzozowska E, Leszek J (2019) The gut microbiome alterations and inflammation-driven pathogenesis of Alzheimer's disease – a critical review. Mol Neurobiol 56:1841–1851
Sun X-Y, Dong Q-X, Zhu J, Sun X, Zhang L-F, Qiu M et al (2019) Resveratrol rescues tau-induced cognitive deficits and neuropathology in a mouse model of Tauopathy. Curr Alzheimer Res 16(8):710–722. https://doi.org/10.2174/1567205016666190801153751
Sun Y, Koyama Y, Shimada S (2022) Inflammation from peripheral organs to the brain: how does systemic inflammation cause neuroinflammation? Front Aging Neurosci 14:903455
Świetlik D, Kusiak A, Krasny M, Białowąs J (2022) The computer simulation of therapy with the NMDA antagonist in Excitotoxic neurodegeneration in an Alzheimer's disease-like pathology. J Clin Med 11(7):1858
Tamagno E, Guglielmotto M, Monteleone D, Tabaton M (2012) Amyloid-β production: major link between oxidative stress and BACE1. Neurotox Res 22:208–219
Teixeira LC, Luizon MR, Gomes KB (2025) Exploring the role of GLP-1 receptor agonists in Alzheimer's disease: a review of preclinical and clinical evidence. Receptor 4(1):2
Ternchoocheep K, Ingkaninan K, Yasothornsrikul S (2012) Tau protein attenuation ability of Bacopa monnieri exract on nerve growth factor-deprived PC12 cells in normal-serum and serum-free medium. Chiang Mai Med J 51(3):59–69
Thakur AK, Kamboj P, Goswami K, Ahuja K (2018) Pathophysiology and management of Alzheimer's disease: an overview. J Pharm Res 9(2):226–235

Tian Y, Jing G, Zhang M (2023) Insulin-degrading enzyme: roles and pathways in ameliorating cognitive impairment associated with Alzheimer's disease and diabetes. Ageing Res Rev 90:101999

Tondo G, De Marchi F, Bonardi F, Menegon F, Verrini G, Aprile D et al (2024) Novel therapeutic strategies in Alzheimer's disease: pitfalls and challenges of anti-amyloid therapies and beyond. J Clin Med 13(11):3098

US-FDA (n.d.) FDA approves treatment for adults with Alzheimer's disease. Retrieved from https://www.fda.gov/drugs/news-events-human-drugs/fda-approves-treatment-adults-alzheimers-disease

van der Kant R, Langness VF, Herrera CM, Williams DA, Fong LK, Leestemaker Y et al (2019) Cholesterol metabolism is a druggable axis that independently regulates tau and amyloid-β in iPSC-derived Alzheimer's disease neurons. Cell Stem Cell 24(3):363–375. e369

van der Kant R, Goldstein LS, Ossenkoppele R (2020) Amyloid-β-independent regulators of tau pathology in Alzheimer disease. Nat Rev Neurosci 21(1):21–35

Verma S, Sharma S, Ranawat P, Nehru B (2020) Modulatory effects of Ginkgo biloba against amyloid aggregation through induction of heat shock proteins in Aluminium induced neurotoxicity. Neurochem Res 45(2):465–490. https://doi.org/10.1007/s11064-019-02940-z

Vogt NM, Kerby RL, Dill-McFarland KA, Harding SJ, Merluzzi AP, Johnson SC et al (2017) Gut microbiome alterations in Alzheimer's disease. Sci Rep 7(1):13537

Wang S, He B, Hang W, Wu N, Xia L, Wang X et al (2018) Berberine alleviates tau hyperphosphorylation and axonopathy-associated with diabetic encephalopathy via restoring PI3K/Akt/GSK3β pathway. J Alzheimer's Dis 65(4):1385–1400

Wang H, Kulas JA, Wang C, Holtzman DM, Ferris HA, Hansen SB (2021) Regulation of beta-amyloid production in neurons by astrocyte-derived cholesterol. Proc Natl Acad Sci 118(33):e2102191118

Wang F, Wang J, Shen Y, Li H, Rausch W-D, Huang X (2022a) Iron dyshomeostasis and ferroptosis: a new Alzheimer's disease hypothesis? Front Aging Neurosci 14:830569

Wang N, Yang J, Chen R, Liu Y, Liu S, Pan Y et al (2022b) Ginsenoside Rg1 ameliorates Alzheimer's disease pathology via restoring mitophagy. J Ginseng Res. https://doi.org/10.1016/j.jgr.2022.12.001

Wattmo C, Wallin ÅK (2017) Early-versus late-onset Alzheimer's disease in clinical practice: cognitive and global outcomes over 3 years. Alzheimer's Res Ther 9:1–13

Wong W (2020) Economic burden of Alzheimer disease and managed care considerations. Am J Manag Care 26(8 Suppl):S177–S183

Yan Z, Feng J (2004) Alzheimer's disease: interactions between cholinergic functions and β-amyloid. Curr Alzheimer Res 1(4):241–248

Yao Y, Li J, Niu Y, Yu JQ, Yan L, Miao ZH et al (2015) Resveratrol inhibits oligomeric Aβ-induced microglial activation via NADPH oxidase. Mol Med Rep 12(4):6133–6139

Yeh FL, Wang Y, Tom I, Gonzalez LC, Sheng M (2016) TREM2 binds to apolipoproteins, including APOE and CLU/APOJ, and thereby facilitates uptake of amyloid-beta by microglia. Neuron 91(2):328–340

Yu H, Xiong M, Zhang Z (2023) The role of glycogen synthase kinase 3 beta in neurodegenerative diseases. Front Mol Neurosci 16:1209703

Zhang J-S, Zhou S-F, Wang Q, Guo J-N, Liang H-M, Deng J-B, He W-Y (2016a) Gastrodin suppresses BACE1 expression under oxidative stress condition via inhibition of the PKR/eIF2α pathway in Alzheimer's disease. Neuroscience 325:1–9

Zhang Y, Li P, Feng J, Wu M (2016b) Dysfunction of NMDA receptors in Alzheimer's disease. Neurol Sci 37:1039–1047

Zhang H, Zhao C, Cao G, Guo L, Zhang S, Liang Y et al (2017) Berberine modulates amyloid-β peptide generation by activating AMP-activated protein kinase. Neuropharmacology 125:408–417

Zhao Z (2019) Iron and oxidizing species in oxidative stress and Alzheimer's disease. Aging Medicine 2(2):82–87

Zhao J, Nakamura N, Hattori M, Kuboyama T, Tohda C, Komatsu K (2002) Withanolide derivatives from the roots of *Withania somnifera* and their neurite outgrowth activities. Chem Pharm Bull 50(6):760–765. https://doi.org/10.1248/cpb.50.760

Zhao L, Teter B, Morihara T, Lim GP, Ambegaokar SS, Ubeda OJ et al (2004) Insulin-degrading enzyme as a downstream target of insulin receptor signaling cascade: implications for Alzheimer's disease intervention. J Neurosci 24(49):11120–11126

Zou Y, Qi B, Tan J, Sun Y, Gong Y, Zhang Q (2022) Mechanistic insight into the disruption of tau R3–R4 protofibrils by curcumin and epinephrine: an all-atom molecular dynamics study. Phys Chem Chem Phys 24(34):20454–20465

Molecular Basis for the Protective Effect of Dietary Components Against Alzheimer's Disease

2

Kajal Bagri, Shilpa Kumari, Mahendra Bishnoi, and Rahul Deshmukh

Abstract

Amyloid-beta (Aβ) plaques, tau neurofibrillary tangles, oxidative stress, and neuroinflammation are the hallmarks of Alzheimer's disease (AD), a progressive neurodegenerative illness. Emerging evidence suggests that dietary components can play a protective role in AD by targeting key pathological mechanisms. This chapter explores the molecular basis of the neuroprotective effects of various dietary bioactives, including polyphenols, omega-3 fatty acids, vitamins, minerals, trace elements, and other phytochemicals. These compounds exert beneficial effects through multiple pathways, such as reducing Aβ aggregation, modulating tau phosphorylation, mitigating oxidative stress, and suppressing neuroinflammation. Additionally, dietary components influence gut microbiota composition and epigenetic modifications, further contributing to cognitive resilience. Understanding the molecular interactions between diet and AD pathology may pave the way for novel dietary interventions and therapeutic strategies. This chapter provides a comprehensive overview of current findings and highlights future directions for research on nutrition-based neuroprotection against AD.

Keywords

Alzheimer's disease · Dietary components · Polyphenols · Omega-3 fatty acids · Vitamins · Minerals · Trace elements

The authors Kajal Bagri and Shilpa Kumari have contributed equally in the preparation of this manuscript and should be considered as first authors.

K. Bagri · S. Kumari · R. Deshmukh (✉)
Department of Pharmacology, Central University of Punjab, Bathinda, Punjab, India
e-mail: rahul.deshmukh@cup.edu.in

M. Bishnoi
National Agri-Food Biotechnology Institute, Sahibzada Ajit Singh Nagar, Punjab, India

J. Malik et al. (eds.), *Diet and Alzheimer's Disease: Let Food be Our Medicine*, Nutritional Neurosciences, https://doi.org/10.1007/978-981-95-2736-6_2

Abbreviations

Aβ	Amyloid-beta
AD	Alzheimer's disease
NEP	Neprilysin
IDE	Insulin-degrading enzyme
LTP	long-term potentiation
CDK5	Cyclin-dependent kinase-5
GSK3β	Glycogen synthase kinase-3β
DYRK	Dual-specificity tyrosine-phosphorylation-regulated kinase
NFTs	Neurofibrillary tangles
ROS	Reactive oxygen species
TLRs	Toll-like receptors
IL	Interleukin
TNFα	Tumor necrosis factor-alpha
BACE1	Beta-site amyloid precursor protein cleaving enzyme 1
PTEN	Phosphatase and tensin homologue
PKB	Protein kinase B
CAT	Catalase
SOD	Superoxide dismutase
SIRT1	Sirtuin 1
NF-κB	Nuclear factor-kappa B
COX-2	Cyclooxygenase-2
EGCG	Epigallocatechin Gallate
ERK	Extracellular signal-regulated kinases
MAPK	Mitogen activated protein kinase
JNK	c-Jun N-terminal kinases
Keap1	Kelch Like ECH associated protein 1
Nrf2	Nuclear factor erythroid 2–related factor 2
ARE	Antioxidant Response Element
DHA	Docosahexaenoic Acid
EPA	Eicosapentaenoic Acid
PUFA	Polyunsaturated fatty acids
NMDA	N-methyl-D-aspartate
TBARS	Thiobarbituric acid reactive substances
PD	Parkinson's disease
APP	Amyloid precursor protein
TH	Tyrosine hydroxylase
WD	Wilson's disease
KD	Ketogenic diet
KB	Ketone bodies
Ac	Acetoacetate
ETC	Eelectron transport chain
mPTP	Mitochondrial permeability transition pore
mitoKATP	Mitochondrial ATP-sensitive potassium channels

UCP	Uncoupling proteins
4- HNE	4-hydroxynonenal
β-OHB	β-hydroxybutyrate

2.1 Introduction

Alzheimer's disease (AD) is a debilitating neurological illness that causes memory loss, increasing cognitive decline, and a reduction in the quality of life for those who have it. The prevalence of AD is increasing as the world's population ages, making it one of the most urgent public health issues. Despite significant advances in the understanding of AD's underlying mechanisms, treatment options remain limited, primarily focusing on alleviating symptoms rather than halting or reversing disease progression (Kumari and Deshmukh 2021). However, emerging research highlights the role of dietary components in potentially preventing or slowing the onset of Alzheimer's, offering a promising complementary approach to traditional treatments. The molecular mechanisms underlying Alzheimer's disease are complex, involving oxidative stress, inflammation, and the accumulation of amyloid-beta plaques and tau tangles in the brain. These processes disrupt neuronal function, leading to cognitive decline (Deshmukh et al. 2016; Gale et al. 2018). A growing body of evidence suggests that various dietary components can influence these mechanisms and offer neuroprotective effects. Among the most widely studied dietary factors are polyphenols, omega-3 fatty acids, vitamins, and specialized diets such as the ketogenic diet (Dhakal et al. 2019). These compounds have been shown to impact key pathways involved in Alzheimer's, including reducing inflammation, enhancing antioxidant defenses, and improving neuronal function.

Polyphenols, including flavonoids and curcumin, have garnered significant attention for their antioxidant and anti-inflammatory properties (Khan and Mukhtar 2007). These compounds are found in common dietary sources such as fruits, vegetables, tea, and spices. They can influence brain health by reducing oxidative damage, which plays a key role in AD pathology, as well as by modulating the accumulation of amyloid-beta and tau proteins, which are central to AD's progression. Similarly, omega-3 fatty acids, particularly docosahexaenoic acid (DHA), are crucial for maintaining brain structure and function (Kidd 2007; Pu et al. 2013). They are found in fatty fish and some plant sources and have been shown to support neuronal health by reducing inflammation and supporting synaptic plasticity, which may help delay cognitive decline in Alzheimer's patients.

In addition to individual compounds, certain dietary patterns, such as the ketogenic diet, Minerals have gained attention for their potential in managing neurodegenerative diseases like AD. The ketogenic diet, which is high in fats and low in carbohydrates, shifts the brain's energy source from glucose to ketones, providing neuroprotective benefits (Sridharan and Lee 2022). This diet has been linked to improved mitochondrial function, reduced oxidative stress, and decreased inflammation, all of which are important in mitigating the neuronal damage seen in AD.

Vitamins such as C and E also play a significant role in maintaining cognitive function. These vitamins contribute to important processes such as neurotransmitter synthesis, DNA repair, and the reduction of homocysteine levels, which have been associated with an increased risk of AD. Specifically, the strong antioxidant qualities of vitamin E aid in preventing oxidative damage to brain cells, while vitamin C helps regulate homocysteine metabolism, potentially lowering the risk of neurodegenerative processes (Mohajeri et al. 2015).

Overall, a growing body of research suggests that a diet rich in these beneficial compounds may support brain health and reduce the risk of Alzheimer's disease. By targeting the underlying mechanisms of oxidative stress, inflammation, and protein aggregation, dietary components such as polyphenols, omega-3 fatty acids, vitamins, and ketogenic diets may offer promising strategies for preventing or delaying the onset of AD. While more research is needed to understand the full extent of their impact, incorporating these dietary factors into daily life could provide a valuable tool in the fight against Alzheimer's disease.

2.2 Alzheimer's Disease Pathophysiology

Alzheimer's disease (AD) is a progressive neurodegenerative disorder characterized by cognitive decline, memory loss, and behavioral changes. The underlying pathophysiology of AD involves multiple interconnected mechanisms, including the accumulation of amyloid-beta (Aβ) plaques, hyperphosphorylation of tau proteins leading to neurofibrillary tangles (NFTs), oxidative stress, neuroinflammation, and mitochondrial dysfunction, all of which contribute to neuronal damage and synaptic dysfunction.

2.2.1 Amyloid-Beta (Aβ) Accumulation and Plaque Formation

Aβ accumulation is a hallmark of AD and contributes to the major pathologies of AD. Aβ is formed from the amyloid precursor protein (APP), a trans membrane protein that undergoes cleavage by α-, β-, and γ-secretases (Rahman 2009). Cleavage by β- and γ-secretases generates Aβ peptides, primarily Aβ1-42, which have a high tendency to aggregate into insoluble amyloid plaques. These plaques disrupt neuronal communication, impair synaptic plasticity, and induce neurotoxicity, ultimately leading to neuronal death. Presenilin-1 and presenilin-2, components of the γ-secretase complex, regulate Aβ production. Impaired clearance of Aβ, due to dysfunction in enzymes such as neprilysin (NEP) and insulin-degrading enzyme (IDE), further promotes its accumulation. Aβ plaques are also linked to long-term potentiation (LTP) impairment, which is critical for memory formation (Baranello et al. 2015).

2.2.2 Tau Hyperphosphorylation and Neurofibrillary Tangles

The tau hypothesis suggests that the abnormal phosphorylation of tau, a microtubule-associated protein, disrupts intracellular transport and contributes to AD pathology (Mietelska-Porowska et al. 2014). Normally, tau stabilizes microtubules, which are essential for axonal transport. However, kinases such as cyclin-dependent kinase-5 (CDK-5), glycogen synthase kinase-3β (GSK-3β), and dual-specificity tyrosine-phosphorylation-regulated kinase (DYRK) induce hyperphosphorylation of tau, separate it from the microtubules, and aggregate it into neurofibrillary tangles (NFTs). These tangles accumulate within neurons, impairing intracellular transport, disrupting synaptic function, and triggering neuronal apoptosis. The severity of tau pathology correlates with similar cognitive deficits as in patients of Alzheimer's disease (Raskin et al. 2015).

2.2.3 Oxidative Stress and Neuroinflammation

Oxidative stress imparts a significant role in AD pathogenesis. Elevated levels of reactive oxygen species (ROS) cause lipid peroxidation, DNA damage, and mitochondrial dysfunction. Aβ itself promotes oxidative stress by disrupting mitochondrial function and increasing free radical production (Yeung et al. 2021).

Chronic neuroinflammation, via activated microglia and astrocytes, further exacerbates neuronal damage. When Aβ binds to toll-like receptors (TLRs) on microglia, it triggers the release of pro-inflammatory cytokines such as interleukin-1 (IL-1), tumor necrosis factor-alpha (TNF-α), and interleukin-6 (IL-6). This inflammatory cascade leads to neuronal dysfunction and synaptic loss, accelerating AD progression (Cai et al. 2014a).

2.2.4 Neuronal Apoptosis and Mitochondrial Dysfunction

Mitochondrial dysfunction is a key feature of AD, leading to energy deficits and cellular apoptosis. Aβ impairs mitochondrial respiration by inhibiting electron transport chain complexes, leading to ATP depletion and increased oxidative stress. This dysfunction, combined with inflammation and tau pathology, triggers apoptosis through the activation of caspases and other cell death pathways (Azam et al. 2019). Alzheimer's disease is a multifactorial disorder with complex pathophysiological mechanisms, including Aβ accumulation, tau hyperphosphorylation, oxidative stress, neuroinflammation, and mitochondrial dysfunction. These processes collectively contribute to synaptic failure, neuronal loss, and cognitive decline. Understanding these mechanisms is needed for developing therapeutic strategies to slow down AD progression (Jembrek et al. 2013).

2.3 Molecular Mechanisms of Dietary Protection

AD is an age-related brain disorder and the most common neurogenerative disease more prevalent among the elder population. Only a small percentage of people (2–5% of all cases) with hereditary mutations in key genes appear to be diagnosed with AD before the age of 65. According to research, just one-third of the changes in our brains as we age may be attributed to hereditary risk factors (Sinai et al. 2017). Lifestyle choices and environmental conditions may influence the other two-thirds of non-genetic factors. In other words, while we cannot ignore the fact that our parents, our genes, and aging are risk factors for AD but we do have some control over other elements like our lifestyle, eating habits, environmental factors, etc.

Fruits and vegetables are considered healthy from ancient time as even they have considered to be antagonists of aging process and age-associated disorders (Song and Zhang 2023). Numerous epidemiological studies have supported these conclusions, and many specialist—including oncologists, cardiologists, angiologists, neurologists, and diabetologists—agrees that when they advise eating at least five servings of fruit and vegetables every day the results observed in patients are different. Diet is more than a food but a lifestyle that needs to be maintained and has a significant impact on the prevention of many illnesses as well as the risk of developing diabetes, heart disease, cancer, and other conditions that we are most afraid of. The same goes for the prevention of AD. Here in this section, we will discuss the role of dietary components and how they protect the pathological alteration in the development of AD, and what will be the impact of these dietary components at the molecular level in dementia (Fig. 2.1).

2.3.1 Polyphenols

A broad and prevalent class of phytochemicals known for their potent bioactive properties are polyphenols. Apples, berries, chocolate, herbs, red wine, turmeric, onions, and tea are among the fruits, vegetables, and drinks that contain them most. The most abundant polyphenols found in the human diet are flavonoid (Daikwo and Ogah 2024). Numerous natural polyphenols, including flavonoids, appear to have the ability to modulate the neuropathological alterations and cognitive impairment linked to AD through a number of mechanisms, such as the regulation of oxidation and inflammation as well as of Aβ metabolism, catabolism, and oligomerization (Niu et al. 2024). Diets high in polyphenols, like the Mediterranean diet, have long been thought by epidemiology to reduce cognitive impairment associated with aging, and certain studies show a lower chance of dementia (Dhakal et al. 2019). Additionally, eating flavonoids, a broad family of dietary polyphenol chemicals, slows down cognitive aging. Some of the polyphenols are discussed below, along with their molecular targets involved in AD pathology.

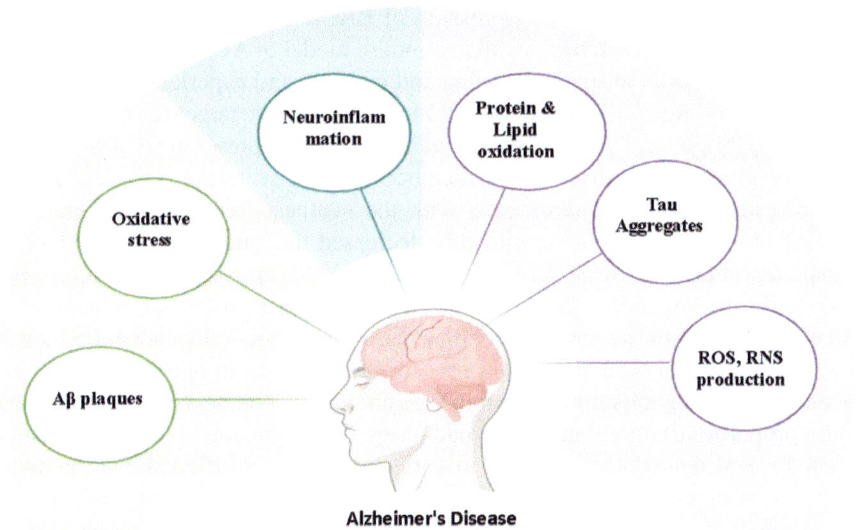

Fig. 2.1 Polyphenols as natural modulators of AD pathways: targeting amyloid-beta plaques, tau tangles, oxidative stress and neuroinflammation for neuroprotection

2.3.1.1 Curcumin (Turmeric)

In 1976, curcumin's ability to scavenge oxygen-free radicals was first demonstrated. Turmeric is made from the rhizome of *Curcuma longa,* a native of South and Southeast Asia and a member of the ginger family (Rathaur et al. 2012). In China and India, turmeric has long been used in traditional medicine and as a curry spice in culinary preparations. Curcuminoids, a commercially available combination of curcumin (75–80%), demethoxycurcumin (15–20%), and bisdemethoxycurcumin (3–5%), are responsible for the majority of turmeric's medicinal advantages (Fuloria et al. 2022). Numerous studies have documented curcumin's advantageous qualities, such as its anti-inflammatory, antioxidant, and anticancer effects, etc. Cognitive decline and behavioral abnormalities are hallmarks of AD (Joshi et al. 2022). The primary histological characteristics seen in AD brains are neurofibrillary tangles, plaques of senility, and widespread neuronal death. Tau is the primary component of neurofibrillary tangles, while Aβ is the primary component of senile plaques (Dementia 2016). Curcumin is a strong anti-inflammatory, rich antioxidant flavonoid that has been well explored in vitro as well as in vivo models of AD, and also in other neurodegenerative disorders (Mehla et al. 2010; Radbakhsh et al. 2021; Fuloria et al. 2022).

The pathological hallmark of AD and the role of curcumin on rectifying the alternations in AD are discussed one by one. β-secretase (BACE1) and Aβ aggregation are the key pathological targets AD. β-secretase and β-site APP-cleaving enzyme 1 (BACE-1) cleave APP, which is a crucial step in the production of Aβ. Curcumin has been proven that it decrease the expression of BACE1 (Vitória et al. 2022; Wu et al. 2022). By downregulating the expression of BACE1, intragastric curcumin treatment decreased the generation of Aβ in a mouse model of AD. Rats given curcumin showed better results in spatial learning and memory and experienced less synapse degeneration (Tang and Taghibiglou 2017). The enzymatic target for the creation of Aβ is GSK-3 (kinase-3), which also includes the protein presenilin-1 (PS-1), a substrate for glycogen synthase and a member of the γ-secretase complex. Both -γ secretase and GSK-3 are associated with the synthesis of Aβ (Sayas and Ávila 2021). Curcumin treatment significantly decreased the amount of Aβ produced in human neuroblastoma SHSY5Y cells (Huang et al. 2014). Additionally, there was a dose- and time-dependent drop in PS-1 and GSK-3 protein levels, indicating that curcumin reduced the generation of Aβ via inhibiting GSK-3-dependent PS1 activation. Curcumin has been shown to prevent fibrillar Aβ from aggregating and also encourage its disaggregation in both in vivo and in vitro models (Yang et al. 2005). These properties of curcumin on the pathology of AD are due to the mechanism of its chemical structure. In an in vitro investigation, it was concluded that the hydrophobicity of curcumin or the contacts between its keto or enol rings and the aromatic rings of Aβ dimers were destabilized the attraction necessary for the development of beta-sheets in Aβ plaques (Thapa et al. 2016). Furthermore, curcumin can destabilize beta-sheets because of its ability to interact with polar regions of the Aβ peptide through the polar hydroxyl groups on its two aromatic rings. This binding feature is made possible by the rigid linker that connects the two aromatic rings.

GSK-3β and reducing tau hyperphosphorylation are the pathological triggers of AD. The pathophysiology of AD depends heavily on hyperphosphorylated tau and its aggregation into neurofibrillary tangles (Ferrer et al. 2005). Several investigations have demonstrated that curcumin can stop tau hyperphosphorylation and neurotoxicity. GSK-3 is an enzyme that adds phosphate groups to serine and threonine amino acid residues and regulates tau phosphorylation. Curcumin inhibited tau hyperphosphorylation via the GSK-3 pathway, phosphatase and tensin homologue (PTEN), and protein kinase B (Akt) (Iqbal and Grundke-Iqbal 2008). Therefore, GSK-3 inhibition may shield cells against tau-induced neurotoxicity, and curcumin has been found to be one such inhibitor.

2.3.1.2 Resveratrol (Red Wine, Grapes)

Resveratrol, a polyphenol belonging to the stilbene family, was first isolated from *Veratrum grandiflorum* in 1940 by Takaoka. It is found in various plants, including grapes (*Vitis* spp.), peanuts (*Arachis hypogaea*), and blueberries (*Vaccinium* spp.), where it serves as a natural defence mechanism against environmental stressors (Georgieva et al. 2015). Structurally, it exists in two isoforms, trans- and cis-resveratrol, with the trans form being more stable and biologically active (Salehi et al. 2018). The neuroprotective potential of resveratrol is attributed to its ability to modulate multiple cellular pathways). By scavenging reactive oxygen species (ROS) and boosting endogenous antioxidant defense mechanisms like catalase (CAT) and superoxide dismutase (SOD), it functions as a potent antioxidant, additionally, it stimulates Sirtuin 1 (SIRT1), a deacetylase that is dependent on NAD+ and controls stress tolerance and cellular metabolism (Truong et al. 2018). Resveratrol could protect neurons in AD by inhibiting the apoptotic effects of p53 through SIRT1 upregulation. It is also seen that Resveratrol induced SIRT1 suppresses NF-KB signaling activation in microglia and astrocytes, thereby safeguarding AD neurons against Aβ-induced damage (Bartra et al. 2024).

In addition to its antioxidant role, resveratrol exerts strong anti-inflammatory effects by inhibiting nuclear factor-kappa B (NF-κB) signaling and reducing pro-inflammatory cytokines such as tumor necrosis factor-alpha (TNF-α), interleukin-6 (IL-6), and cyclooxygenase-2 (COX-2)(Batista-Jorge et al. 2024). These actions help mitigate the chronic neuroinflammation associated with AD. Furthermore, resveratrol promotes Aβ clearance through proteasomal and autophagic pathways while inhibiting β-secretase (BACE1), limiting Aβ production (Bai et al. 2021). It also prevents tau hyperphosphorylation, stabilizing microtubules and preserving neuronal integrity.

Since oxidative stress contributes significantly to the development of AD, it is an important target for treatment. Strong antioxidant resveratrol scavenges free radicals and stabilizes radical intermediates to prevent oxidative damage. To counteract reactive oxygen species (ROS), it activates SIRT1, which increases antioxidant enzymes such as catalase and superoxide dismutase (SOD) (Islam et al. 2022). Additionally, resveratrol improves mitochondrial function by lowering excessive ROS generation. It also upregulates antioxidant genes by activating the Nrf2 pathway. Despite being less potent as a direct antioxidant, it is a powerful long-term neuroprotective agent due to its capacity to alter gene expression. Together, these effects lessen amyloid-beta toxicity, preserve neurons, and enhance cognitive function in AD (Sarubbo et al. 2017).

2.3.1.3 Epigallocatechin Gallate (EGCG, Green Tea)

The risk of AD may be increased by a variety of lifestyle variables, low educational attainment, obesity, diabetes, hypertension, smoking, "unhealthy diet," and physical inactivity, in addition to a hereditary predisposition (Skýbová et al. 2021). According to estimates, dementia incidence can be reduced by at least 30% by removing modifiable lifestyle factors (Dominguez et al. 2021). This emphasizes how important lifestyle changes may be for the primary prevention of AD.

People have been drinking green tea for thousands of years. Several assertions on the advantages of consuming it were made and looked into. Given that green tea is becoming more and more popular in Western culture and that millions of people consume it daily worldwide, it is important to comprehend its impacts on the human brain (Khan and Mukhtar 2007). Steamed *Camellia sinensis* leaves are used to make green tea, which consist of variety of compounds, the most significant of which are catechins (30–42% of the weight of the solid extract), of which epigallocatechin gallate (EGCG) is the most abundant then caffeine (3–4%) and theanine (4–6%) (Cioanca et al. 2024). According to various reports available on EGCC, It has antioxidant and anti-inflammatory properties. According to a number of animal studies, EGCG can cross the blood-brain barrier (Cai et al. 2014b). According to studies on regular, daily use of at least 100 ml of green tea, the main long-term advantages are enhanced mental functions, a calmer mental state, and a decreased chance of dementia. Up to a daily intake of 500 ml, the effects seem to grow dose-dependently (Williams et al. 2020).

The most abundant catechin in green tea is EGCG, an ester of gallic acid, which makes up as much as 65% of the total catechin content and exhibits the highest biological activity, including antioxidative and neuroprotective properties documented in various reports (Afzal et al. 2022). Based on all of the available data, we may conclude that oxidative stress and AD have a complex relationship that suggests a positive feedback loop between the two. As such, oxidative stress and the aforementioned pathways could be targets for our attempts to reduce AD pathology. Antioxidants like green tea catechins may therefore present intriguing additions to our medicinal toolbox. EGCG has demonstrated the ability to function as an antioxidative system, neutralizing ROS, RNS, and other free radicals, including carbon and lipidic radicals, peroxyl, nitric oxide, and 1,1-diphenyl-3-picrylhydrazine derivatives (Afzal et al. 2022). EGCG increased the activity of endogenous antioxidants, i.e., SOD, glutathione-S-transferase, and catalase, etc., which proved to be enough to reduce oxidative stress in AD.

By controlling cell signaling, proliferation, differentiation, and death, kinases belonging to the PKC family contribute to the brain signaling network, which directly affects cancer, synaptic function, behavior, and cognition. Furthermore, PKC has been connected to synaptic plasticity, which facilitates the processes of long-term potentiation (LTP). PKC activation has been suggested to promote LTP and cognitive enhancement in the context of AD by assisting in the reduction of the amyloid load. Furthermore, the primary kinase implicated in tau phosphorylation, glycogen synthase kinase 3 (GSK3), is inhibited by PKC activation, offering yet another connection between AD pathogenesis and PKC pathways (Lordén et al. 2022). It has been demonstrated that EGCG contributes to PKC pathways by directly activating PKC through rapid phosphorylation, which supports the positive, neuroprotective effects in AD. Studies conducted in vivo on C57BL mice given EGCG at a dose of 2 mg/kg/day demonstrated that this catechin can also cause a rapid translocation of PKC, preventing PKC depletion and reversing the rise in the apoptotic protein Bax in neurons (Valverde-Salazar et al. 2023). A prospective target to treat AD pathogenesis, EGCG has also been shown to synergistically cause a rapid proteasomal degradation of BCL2 associated agonist of cell death (BAD) protein I, another proapoptotic protein, via PKC activation.

The fact that multifactorial pathogenic processes like AD and other neurodegenerative illnesses involve altered MAPK signaling should not be surprising. Three categories are used to classify MAPK: extracellular signal-regulated kinases (ERK), which are made up of the isoforms ERK1, ERK2, and ERK5; p38 kinases (Munoz and Ammit 2010), which have the isoforms α, β, γ, and δ; and c-Jun N-terminal kinases (JNK), which have three distinct isoforms called JNK1, JNK2, and JNK3. Both microglial activation and Aβ-induced oxidative stress have been shown to mediate MAPK p38 signaling in AD which not only triggers apoptosis but also functions as a tau protein kinase further exacerbating the condition. By activating

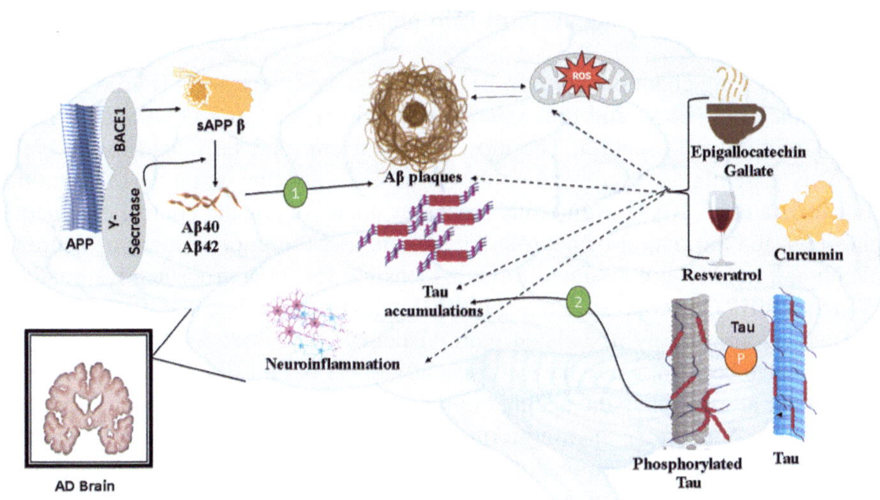

Fig. 2.2 Therapeutic potential of dietary modulation: Curcumin, EGCG, and Resveratrol Targeting Alzheimer's Disease Pathology through Amyloid-beta, Tau, and Neuroinflammatory Regulation

the Keap1/Nrf2/ARE pathway and upregulating ERK1/2 in response to oxidative stress, EGCG promotes cell survival and has neuroprotective effects (Ali et al. 2023; Valverde-Salazar et al. 2023). On the other hand, EGCG suppresses the ERK, p38, and JNK pathways, which inhibits the mitochondrial dysfunction and changed dynamics. EGCG protect tau protein hyperphosphorylation and its aggregation (Fig. 2.2).

2.3.2 Omega-3 Fatty Acids

Fish are high in polyunsaturated fats such as PUFA and MUFA. It has a high level of antioxidants. Fatty acids surround one-fifth of the dry component of the brain, where 20% fatty acid is in the form of DHA, which has anti-inflammatory properties protecting against AD (Keaskin 2013). Omega-3 fatty acids, particularly DHA (Docosahexaenoic Acid) and EPA (Eicosapentaenoic Acid), are essential for brain health and function. A healthy lifestyle is closely linked to brain function, memory, and emotional equilibrium, all of which are maintained by omega-3 fatty acids, especially DHA and EPA (Kidd 2007).

Due to our inability to generate them, omega-3 fatty acids are thought to be necessary for human health and must be received from diet. A diet high in omega-3 fatty acid, which can be obtained via fatty fish, seeds, or supplements, can benefit mental health by enhancing mood, lowering stress, and improving cognitive performance (Tur et al. 2012). A more vibrant and balanced life can be promoted by regularly including these vital nutrients in your routine, along with other healthful practices like exercise and sleep, to help prevent cognitive decline, anxiety, and depression.

The very long chain omega-3 (n-3) polyunsaturated fatty acids (PUFA), eicosapentaenoic acid (EPA) and docosahexaenoic acid (DHA), are essential for metabolic structure and function. The most prevalent omega-3 fatty acid in the brain, DHA is essential for preserving the structure and functionality of neurones (Echeverría et al. 2017). Numerous processes, including as the control of synaptic plasticity, the mitigation of neuroinflammation, and the improvement of neuronal membrane fluidity, are assumed to be responsible for its neuroprotective qualities (Vauzour 2012). Decreases in omega-3 PUFA levels in the brain and other peripheral tissues are closely associated with Alzheimer's disease. Alzheimer's disease patients have been shown to have consistently lower brain DHA levels.

Cellular alterations in the ageing brain could be the cause of memory and learning decline. A decrease in long-term potentiation (LTP) is one such alteration. Numerous studies have demonstrated that giving elderly rats more DHA or EPA in their diet overcomes age-related deficiencies in LTP. Changes in the expression of the glutamate receptor may mediate these effects (Horrocks and Farooqui 2004). Particularly N-methyl-D-aspartate (NMDA) glutamate receptors have an important role in synaptic mechanisms of learning and memory, such as LTP. Evidence supports a specific role for the NR2B subunit in learning and memory (Wang and Peng 2016). DHA greatly prevented the decline in learning capacity and lessened the oxidative stress that the amyloid peptide infusion caused in the hippocampus and cerebral cortex. By a number of unique and connected processes, omega-3 PUFAs affect amyloidogenic processing: DHA makes it easier for α-secretase and APP to interact, resulting in non-toxic fragments and preventing Aβ production. (Hashimoto et al. 2009). Being strong antioxidant, Omega-3 Fatty Acids supports mitochondrial functions by reducing ROS, apoptotic markers, and able to enhance the BDNF thus supports neurodegeneration (Azizi 2015).

2.3.3 Antioxidants and Vitamins

It is becoming clear that, Oxidative stress is suggested to play a major role in the pathogenesis of AD. The pathophysiology of AD is thought to be significantly influenced by oxidative stress (Zhao et al. 2005). Patients with AD have shown significant lipid peroxidation in the temporal cortex and mitochondrial and nuclear DNA oxidation in the parietal cortex. Furthermore, peroxidation end products, including heme oxygenase 1, superoxide dismutase 1, and peroxynitrite, were found in the brains of AD patients, particularly in senile plaques and neurofibrillary tangles (Wang et al. 2005). Furthermore, neurones and astrocytes in the hippocampus and cerebral cortex overexpress heme oxygenase 1, a cellular stress protein that is expressed in the brain and other tissues in response to oxidative stressors.

One of the most significant antioxidants that reduces damage to cerebral tissue from free radicals and harmful chain reactions is vitamin C. By giving up two of its electrons, vitamin C functions as an antioxidant, preventing the oxidation of other substances. Vitamin C is a well-known antioxidant that shields the body from oxidative stress and is a necessary ingredient for a variety of metabolic processes in all animals and plants (Doseděl et al. 2021). Despite eating a healthy diet, the study found that patients with AD had reduced plasma levels of vitamin C. In the CA1 region of the rat brain's hippocampal region, oral vitamin C decreased proinflammatory cytokines and oxidative stress brought on by injections of Aβ peptide. Exposure to β-Amyloid causes reactive oxygen species to be produced, which causes cultured neurons to undergo apoptosis and die (Monacelli et al. 2017). Vitamin C may shield cultured SH-SY5Y neuroblastoma cells from apoptotic cell death, whereas vitamin C pretreatment stopped both apoptosis and β-amyloid-induced cell death. Additionally, vitamin C-pretreated cells also showed a decrease in basal rates of endogenous β-amyloid generation (Heo et al. 2013). Furthermore, vitamin C has been reported to decrease acetylcholinesterase activity in mice. It has also been shown that vitamin C influences acetylcholine release from synaptic vesicles. In rats with sepsis-induced cognitive impairment, a high dosage of intraperitoneal vitamin C injection increased anti-inflammatory molecules including superoxide dismutase and IL-10 while decreasing oxidative stress, neuroinflammation, and BBB disruption in the hippocampus (Kuo and Yoshida 1980).

Vitamin E is an essential micronutrient for humans because it maintains the integrity of cell membranes. Depending on age, different countries recommend varying daily dietary intakes of vitamin E, ranging from 3 to 15 mg (Boothby and Doering 2005). Tocopherols and tocotrienols are found in small amounts in plant foods with low lipid content, such as fruit and vegetables, but in significant amounts in seeds and edible oils, such as those from almond, peanut, olive, palm, canola, corn, and soybean (Stuchlík and Zak 2002). Since vitamin E is the most significant lipophilic radical scavenger in vivo, its primary function is its antioxidant activity. Vitamin E scavenges free radicals mainly by hydrogen atom transfer reaction obtaining a non-radical product and a vitamin E radical, that may react with another radical to give a stable product, attack lipids, or react with a reducing agent such as vitamin C or ubiquinol to regenerate vitamin E (Wolf et al. 1998). Since α-tocopherol is the predominant form of vitamin E in tissues, it is the most researched. Interestingly, studies conducted in vitro have indicated that vitamin E may be able to lessen the oxidative damage caused by Aβ. Specifically, it was shown that vitamin E, possibly by scavenging Aβ-induced free radicals, may stop Aβ-induced protein oxidation, ROS generation, and neurotoxicity in primary rat embryonic hippocampus neuronal culture (George et al. 2000). A rise in lipid peroxidation, 4-hydroxynonenal (HNE), H2O2, thiobarbituric acid reactive substances (TBARS) generation, and apoptosis further showed that Aβ peptides induced oxidative stress (Fig. 2.3).

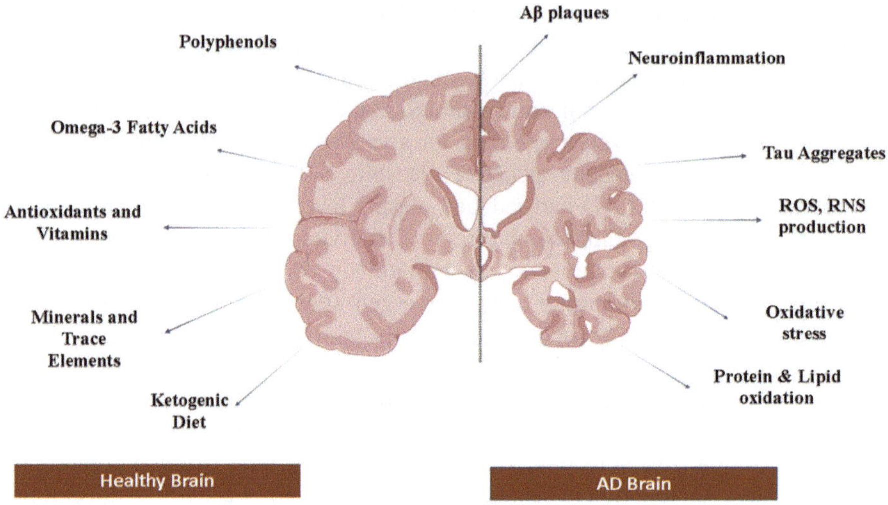

Fig. 2.3 Comparison of healthy and Alzheimer's brain, showing how dietary components reduce oxidative stress, inflammation, tau and amyloid buildup

2.4 The Trace Elements and Minerals

A balanced diet is essential for CNS neurotransmission and plasticity, but proper nutrient delivery, metabolism, and signaling are crucial. Vitamins and minerals facilitate nutrient utilization and also showed protection against oxidative damage (Scholefield et al. 2020). For example, vitamin B supports amino acid breakdown and carbohydrate metabolism, while vitamin C acts as a redox catalyst, scavenging free radicals. Trace elements like zinc, copper, iron, and manganese are vital for neurotransmission and metabolism, but their imbalance contributes to neurodegenerative diseases. Dysregulated copper and zinc levels are correlated to tau and amyloid-beta hypothesis in Alzheimer's disease (AD), while iron and manganese disruptions contribute to Parkinson's disease (PD). Toxic metals like lead and cadmium impair CNS health, exacerbating neuroinflammation, and aluminum is associated with AD neurofibrillary tangles (Long et al. 2019).

Studies show altered vitamin and mineral levels in AD patients, suggesting a role in disease pathogenesis. Chronic inflammation and oxidative stress, driven by overactive glial cells and a compromised blood-brain barrier, promote amyloid deposition and neuronal death. Micronutrients with anti-inflammatory and antioxidant properties may help slow AD progression by enhancing synaptic plasticity, reducing excitotoxicity, and modulating key pathways like CaMKII and BDNF-TrkB. Thus we need further studies to investigate their potential for therapeutic use (Savelieff et al. 2013).

Zinc Many molecular and metabolic mechanisms involved in cell viability depend on zinc. Its average concentration in the brain is between 10 and 15 μg/g of tissue, with variations in concentration seen in different brain regions, including those in mood regulation, learning, memory, and cognition. Zn2+ homeostasis disturbances may be linked to a number of central nervous system illnesses, including depression, anxiety, Parkinson's disease, and Alzheimer's disease. Amyloid precursor protein (APP) synthesis is aided by certain Zn2+ linked to certain transcription factors, including NF-κB and the tumor protein P53 (p53). Zn2+ can control the polymerization of Aβ by interacting with it (Awadalla et al. 2021). Low levels of Zn2+ have a protective effect and counteract β-amyloid's neurotoxic effects. Conversely, fibrillar Aβ aggregation increases with higher ion binding levels. Additionally, an increasing amount of data indicates that high Zn2+ levels in advanced AD may result in an excessively phosphorylated tau protein. Researchers have shown that regular zinc intake is linked to a slower rate of deterioration in patients with cognitive impairment and diagnosed AD, as well as a lower risk of cognitive decline. In another study, Zn abundance significantly decreased motor functions and coordination, as well as the levels of tyrosine hydroxylase (TH)-positive neuronal cells and the tyrosine hydroxylase protein (Awadalla et al. 2021).

Iron When compared to other parts of the brain, the concentrations of Fe are up to three times higher in the areas primarily engaged in the supervision of motor neuron activity (Long et al. 2019). It is claimed that the iron concentrations in the structures that form grey matter are higher than those in the structures that form white matter. The caudate nucleus, substantia nigra, and globus pallidus have the highest iron contents. This increased iron content implies that these structures are vulnerable to the disturbances brought on by movement disorders as a result of iron level variations. Iron is essential for maintaining healthy brain function because it plays a part in oxidative metabolism, myelin and neurotransmitter formation, and other processes. Its roles include DNA replication, myelin and neurotransmitter production, oxygen transport, glucose metabolism, and electron transport. The total amount of iron in the parts of the brain that are vulnerable to degenerative diseases such as Huntington's disease, Parkinson's disease, and Alzheimer's disease increases with age. An overabundance of reactive iron enters cells such as neurons, microglia, and astrocytes when iron homeostasis is not properly regulated. However, iron accumulation is very low in more than 95% of AD and PD cases, and the neurotoxic mechanism most likely involves a combination of iron and other toxins. Cognitive capacities are negatively impacted by higher brain iron levels in maturity. However, there is a persistent link between poorer memory function and elevated Fe levels in the thalamus, hippocampus, and caudate nuclei. Conversely, a stronger correlation was seen between decreased general cognitive function and higher putamen iron levels. Although iron is a redox-active metal, it is unknown how it contributes to the pathophysiology of AD. Fe inhibits Aβ aggregation and neuritic plaque formation, according to in vitro research. It can bind with hyperphosphorylated tau, causing it to aggregate and form NFTs, according to additional in vitro research. Fe3+ can be reduced to Fe2+ to reverse this process. To understand the role of metals in the pathophysiology of AD, metal chelators are now being developed (Drouin-Ouellet and Cicchetti 2012; Long et al. 2019).

Copper Copper is essential for brain metabolism, playing a role in enzymes like SOD1, ceruloplasmin, and cytochrome c oxidase. It is stored in the liver and regulated by Cu-dependent proteins. Excess copper can cause oxidative stress and lead to cuproptosis, a copper-dependent cell death mechanism. Wilson's disease (WD) is linked to copper overload due to ATPase gene mutations, leading to organ damage and neurological symptoms. In Alzheimer's disease (AD), copper interacts with tau and Aβ, promoting neurotoxicity. High Cu levels activate microglia and impair Aβ clearance, contributing to neurodegeneration. While excess copper is harmful, some studies suggest its potential therapeutic role by redistributing extracellular Cu2+ back into cells.

Aluminium Aluminum, a prevalent metallic element, is widely used in industrial and consumer goods. It is absorbed through food, cookware, pharmaceuticals, and vaccines but has limited solubility at physiological pH. Excess exposure, such as through dialysis or implanted materials, leads to toxicity. Aluminum disrupts metal balance, increases oxidative stress, and inhibits antioxidant enzymes, causing cell death. It crosses the blood–brain barrier via transferrin, accumulating in brain regions linked to neurodegenerative disorders like Alzheimer's disease, Parkinson's disease, and ALS. Aluminum impairs neurotransmission, reduces neurotrophic factors, and promotes inflammation. Its role in tau protein aggregation is associated with cognitive decline, highlighting its neurotoxic effects.

2.5 Energy Metabolism and Ketogenic Diet

The ketogenic diet (KD) is a high-fat, low-carb diet that shifts energy metabolism from glucose to ketone bodies (KBs) like β-hydroxybutyrate (β-OHB). This metabolic state, known as ketosis, enhances mitochondrial function, reduces oxidative stress, and lowers neurotoxic Aβ levels, potentially benefiting neurodegenerative diseases like Alzheimer's. KD improves mitochondrial respiration, decreases ROS production, and modulates key cellular pathways, including SIRT1 and Nrf2, which enhance antioxidant defense. Originally designed for epilepsy, KD is now explored for treating Alzheimer's by reducing inflammation, enhancing synaptic function, and improving memory. Its neuroprotective effects make it a promising therapeutic approach (Kayode et al. 2020).

2.5.1 Ketone Bodies as an Alternative Brain Fuel

Although the adult brain constitutes only 2% of body weight, it takes up to 20–23% of total energy, primarily relying on glucose. During fasting or low carbohydrate intake, ketone bodies (KBs) being the primary other energy source due to carbohydrate depletion and increased fatty acid availability. While acetone is rapidly expelled, β-hydroxybutyrate (βHB) and acetoacetate (AcAc) may work as alternatives for glucose in the brain. Plasma KB levels typically range from 100 to 250 mM, rising to 1 mM after prolonged exercise or fasting, and up to 6–8 mM during extended fasting by not causing acidosis. In diabetic ketoacidosis, levels can exceed 20 mM. The liver releases upto 300 g of KBs daily, contributing 5–20% of total energy expenditure depending on nutritional status (Scholl-Bürgi et al. 2015; Kayode et al. 2020).

In Alzheimer's disease (AD), impaired brain glucose uptake and metabolism lead to neuronal energy deprivation, potentially contributing to disease progression. Studies show reduced cerebral blood flow and oxygen use correlate with worsening dementia. Lower regional glucose metabolism in the posterior cingulate, temporal, parietal, and prefrontal cortex is linked to cognitive decline. Decreased brain glucose metabolism is an early feature of AD, possibly resulting from synapse loss and neuronal death. Notably, this metabolic decline is observed in preclinical stages, suggesting brain hypometabolism may increase AD risk before clinical symptoms appear (Target 2016).

2.5.2 Mechanisms Behind the Ketogenic Diet's Positive Effects

Initially, AD treatment focused on preventing histological damage (amyloid plaques, neurofibrillary tangles). When this approach failed, research shifted toward metabolic dysfunction, mitochondrial homeostasis, and neuronal signaling, leading to "metabolic therapy."

Ketogenic diets (KDs) offer neuroprotection through several mechanisms: increasing ATP production, reducing ROS via mitochondrial complex I, stabilizing synaptic activity, and modulating neurotransmitter metabolism. KD activates energy-sensing pathways (AMPK, mTOR, PPAR) and enhances antioxidant defenses through Nrf2 activation and NAD+/NADH regulation (Paoli et al. 2019). Elevated NAD+/NADH supports mitochondrial biogenesis and cellular respiration while improving electron transport chain (ETC) efficiency via uncoupling proteins (UCP) (Huang et al. 2018).

KD-mediated SIRT1 activation promotes mitochondrial biogenesis, balancing mitophagy and renewal. SIRT1 deacetylates PCG-1α, important factor for mitochondrial DNA replication and biogenesis. Additionally, KD prevents mitochondrial permeability transition pore (mPTP) opening, reducing apoptosis and enhancing ETC function by activating mitochondrial ATP-sensitive potassium channels (mitoKATP) (Paoli 2014).

2.6 Conclusion

To sum up, the increasing amount of evidence indicates that dietary elements such as EGCG, resveratrol, curcumin, omega-3 fatty acids, minerals, vitamins C and E, and the ketogenic diet have great potential to prevent AD. By focussing on the three main factors that cause AD oxidative stress, inflammation, and amyloid-beta accumulation these nutrients provide a multifaceted defence for brain health at the molecular level. While EGCG aids in the clearance of amyloid plaques, resveratrol and curcumin provide potent antioxidant and anti-inflammatory properties. Vitamins C and E preserve the brain from oxidative stress, whereas omega-3 fatty acids support neurones. The ketogenic diet improves overall brain function by giving the brain an alternate energy source, and minerals like magnesium and zinc help protect neurones. Together, these dietary components paint an exciting picture of potential ways to prevent or slow down the progression of AD. While more research is needed to fully understand their long-term impact, the evidence is compelling: a brain-boosting diet could be key to preserving cognitive function and offering new hope in the fight against Alzheimer's.

References

Afzal O et al (2022) Green tea catechins attenuate neurodegenerative diseases and cognitive deficits. Molecules 27(21):7604

Ali M et al (2023) Unravelling of molecular biomarkers in synaptic plasticity of Alzheimer's disease: critical role of the restoration of neuronal circuits. Ageing Res Rev 91:102069

Awadalla A et al (2021) Effect of zinc oxide nanoparticles and ferulic acid on renal ischemia/reperfusion injury: possible underlying mechanisms. Biomed Pharmacother 140:111686

Azam S et al (2019) Regulation of toll-like receptor (TLR) signaling pathway by polyphenols in the treatment of age-linked neurodegenerative diseases: focus on TLR4 signaling. Front Immunol 10:1000

Azizi H (2015) Effects of DHA and EGCG on the alpha-secretase mediated processing of APP. University of Guelph

Bai L et al (2021) Attenuation of Pb-induced Aβ generation and autophagic dysfunction via activation of SIRT1: neuroprotective properties of resveratrol. Ecotoxicol Environ Saf 222:112511

Baranello RJ et al (2015) Amyloid-beta protein clearance and degradation (ABCD) pathways and their role in Alzheimer's disease. Curr Alzheimer Res 12(1):32–46. https://doi.org/10.2174/1567205012666141218140953

Bartra C et al (2024) Resveratrol activates antioxidant protective mechanisms in cellular models of Alzheimer's disease inflammation. Antioxidants 13(2):177

Batista-Jorge GC et al (2024) Resveratrol effects on metabolic syndrome features: a systematic review and meta-analysis. Endocrine 5(2):225–243

Boothby LA, Doering PL (2005) Vitamin C and vitamin E for Alzheimer's disease. Ann Pharmacother 39(12):2073–2080

Cai Z, Hussain MD, Yan L-J (2014a) Microglia, neuroinflammation, and beta-amyloid protein in Alzheimer's disease. Int J Neurosci 124(5):307–321

Cai J et al (2014b) Epigallocatechin gallate (EGCG) attenuates infrasound-induced neuronal impairment by inhibiting microglia-mediated inflammation. J Nutr Biochem 25(7):716–725

Cioanca O et al (2024) Extraction and purification of catechins from tea leaves: an overview of methods, advantages, and disadvantages. Separations 11(6):171

Daikwo MA, Ogah CF (2024) Food sources of bioactive phytochemicals. In: Plant food phytochemicals and bioactive compounds in nutrition and health. CRC Press, pp 15–30

Dementia PM (2016) Dementia 2. Gerontologist 56(Suppl_3):536–536. https://doi.org/10.1093/geront/gnw162.2164

Deshmukh R et al (2016) Caffeic acid attenuates oxidative stress, learning and memory deficit in intra-cerebroventricular streptozotocin induced experimental dementia in rats. Biomed Pharmacother 81:56–62. https://doi.org/10.1016/j.biopha.2016.03.017

Dhakal S et al (2019) Dietary polyphenols: a multifactorial strategy to target Alzheimer's disease. Int J Mol Sci 20(20):5090

Dominguez LJ et al (2021) Nutrition, physical activity, and other lifestyle factors in the prevention of cognitive decline and dementia. Nutrients 13(11):4080

Doseděl M et al (2021) Vitamin C – sources, physiological role, kinetics, deficiency, use, toxicity, and determination. Nutrients 13(2):615

Drouin-Ouellet J, Cicchetti F (2012) Inflammation and neurodegeneration: the story "retolled". Trends Pharmacol Sci 33(10):542–551. https://doi.org/10.1016/j.tips.2012.07.002

Echeverría F et al (2017) Docosahexaenoic acid (DHA), a fundamental fatty acid for the brain: new dietary sources. Prostaglandins Leukot Essent Fat Acids 124:1–10

Ferrer I et al (2005) Current advances on different kinases involved in tau phosphorylation, and implications in Alzheimer's disease and tauopathies. Curr Alzheimer Res 2(1):3–18

Fuloria S et al (2022) A comprehensive review on the therapeutic potential of Curcuma longa Linn. In relation to its major active constituent curcumin. Front Pharmacol 13:820806

Gale SA, Acar D, Daffner KR (2018) Dementia. Am J Med 131(10):1161–1169. https://doi.org/10.1016/j.amjmed.2018.01.022

George S et al (2000) Development of a fish in vitro cell culture model to investigate oxidative stress and its modulation by dietary vitamin E. Mar Environ Res 50(1–5):541–544

Georgieva M et al (2015) Resveratrol: recommending its wide increase-a premature decision or not? A review. Int J Pharm Sci Res 6(9):3641

Hashimoto M et al (2009) The protective effect of dietary eicosapentaenoic acid against impairment of spatial cognition learning ability in rats infused with amyloid β (1–40). J Nutr Biochem 20(12):965–973

Heo J-H et al (2013) The possible role of antioxidant vitamin C in Alzheimer's disease treatment and prevention. Am J Alzheimers Dis Other Dement 28(2):120–125

Horrocks LA, Farooqui AA (2004) Docosahexaenoic acid in the diet: its importance in maintenance and restoration of neural membrane function. Prostaglandins Leukot Essent Fat Acids 70(4):361–372

Huang H-C et al (2014) Curcumin attenuates amyloid-β-induced tau hyperphosphorylation in human neuroblastoma SH-SY5Y cells involving PTEN/Akt/GSK-3β signaling pathway. J Recept Signal Transduct 34(1):26–37

Huang Q et al (2018) An 8-week, low carbohydrate, high fat, ketogenic diet enhanced exhaustive exercise capacity in mice part 2: effect on fatigue recovery, post-exercise biomarkers and antioxidation capacity. Nutrients 10(10):1339

Iqbal K, Grundke-Iqbal I (2008) Alzheimer neurofibrillary degeneration: significance, etiopathogenesis, therapeutics and prevention. J Cell Mol Med 12(1):38–55

Islam F et al (2022) Resveratrol and neuroprotection: an insight into prospective therapeutic approaches against Alzheimer's disease from bench to bedside. Mol Neurobiol 59(7):4384–4404

Jembrek M et al (2013) Hyperphosphorylation of tau by GSK-3β in Alzheimer's disease: the interaction of Aβ and sphingolipid mediators as a therapeutic target. Transl Neurosci 4(4):466–476

Joshi P et al (2022) Ameliorating potential of curcumin and its analogue in central nervous system disorders and related conditions: a review of molecular pathways. Phytother Res 36(8):3143–3180

Kayode TO et al (2020) Ketogenic diet: a nutritional remedy for some metabolic disorders. J Educ Health Sport 10(8):180–188

Keaskin L (2013) Assessing the dietary intake of omega 3 fatty acids and examining the effect of a 6 month supplementation of EPA on blood lipids and inflammatory cytokines in healthy Irish adults. University of Limerick

Khan N, Mukhtar H (2007) Tea polyphenols for health promotion. Life Sci 81(7):519–533

Kidd PM (2007) Omega-3 DHA and EPA for cognition, behavior, and mood: clinical findings and structural-functional synergies with cell membrane phospholipids. Altern Med Rev 12(3):207

Kumari S, Deshmukh R (2021) β-Lactam antibiotics to tame down molecular pathways of Alzheimer's disease. Eur J Pharmacol 895:173877

Kuo C-H, Yoshida H (1980) Ascorbic acid, an endogenous factor required for acetylcholine release from the synaptic vesicles. Jpn J Pharmacol 30(4):481–492

Long JM et al (2019) Novel upregulation of amyloid-β precursor protein (APP) by microRNA-346 via targeting of APP mRNA 5′-untranslated region: implications in Alzheimer's disease. Mol Psychiatry 24(3):345–363. https://doi.org/10.1038/s41380-018-0266-3
Lordén G et al (2022) Enhanced activity of Alzheimer disease-associated variant of protein kinase Cα drives cognitive decline in a mouse model. Nat Commun 13(1):7200
Mehla J et al (2010) Protective effect of curcumin against seizures and cognitive impairment in a pentylenetetrazole-kindled epileptic rat model. Life Sci 87(19–22):596–603
Mietelska-Porowska A et al (2014) Tau protein modifications and interactions: their role in function and dysfunction. Int J Mol Sci 15(3):4671–4713. https://doi.org/10.3390/ijms15034671
Mohajeri MH, Troesch B, Weber P (2015) Inadequate supply of vitamins and DHA in the elderly: implications for brain aging and Alzheimer-type dementia. Nutrition 31(2):261–275
Monacelli F et al (2017) Vitamin C, aging and Alzheimer's disease. Nutrients 9(7):670
Munoz L, Ammit AJ (2010) Targeting p38 MAPK pathway for the treatment of Alzheimer's disease. Neuropharmacology 58(3):561–568
Niu C, Dong M, Niu Y (2024) Natural polyphenol: their pathogenesis-targeting therapeutic potential in Alzheimer's disease. Eur J Med Chem 269:116359
Paoli A (2014) Ketogenic diet for obesity: friend or foe? Int J Environ Res Public Health 11(2):2092–2107
Paoli A et al (2019) Ketogenic diet and microbiota: friends or enemies? Genes 10(7):534
Pu H et al (2013) Omega-3 polyunsaturated fatty acid supplementation improves neurologic recovery and attenuates white matter injury after experimental traumatic brain injury. J Cereb Blood Flow Metab 33(9):1474–1484
Radbakhsh S et al (2021) Curcumin: a small molecule with big functionality against amyloid aggregation in neurodegenerative diseases and type 2 diabetes. Biofactors 47(4):570–586
Rahman A (2009) The role of adenosine in Alzheimers disease. Curr Neuropharmacol 7(3):207–216. https://doi.org/10.2174/157015909789152119
Raskin J et al (2015) Neurobiology of Alzheimer's disease: integrated molecular, physiological, anatomical, biomarker, and cognitive dimensions. Curr Alzheimer Res 12(8):712–722. https://doi.org/10.2174/1567205012666150701103107
Rathaur P et al (2012) Turmeric: the golden spice of life. Int J Pharm Sci Res 3(7):1987
Salehi B et al (2018) Resveratrol: a double-edged sword in health benefits. Biomedicine 6(3):91
Sarubbo F et al (2017) Effects of resveratrol and other polyphenols on the most common brain age-related diseases. Curr Med Chem 24(38):4245–4266
Savelieff MG et al (2013) Untangling amyloid-β, tau, and metals in Alzheimer's disease. ACS Chem Biol 8(5):856–865
Sayas CL, Ávila J (2021) GSK-3 and tau: a key duet in Alzheimer's disease. Cells 10(4):721
Scholefield M et al (2020) Evidence that levels of nine essential metals in post-mortem human-Alzheimer's-brain and ex vivo rat-brain tissues are unaffected by differences in post-mortem delay, age, disease staging, and brain bank location. Metallomics *12:952*
Scholl-Bürgi S et al (2015) Ketogenic diets in patients with inherited metabolic disorders. J Inherit Metab Dis 38(4):765–773
Sinai A et al (2017) Predictors of age of diagnosis and survival of Alzheimer's disease in down syndrome. J Alzheimer's Dis 61(2):717–728
Skýbová D et al (2021) Risk of chronic diseases limiting longevity and healthy aging by lifestyle and socio-economic factors during the life-course–a narrative review. Med Pr 72(5):535–548
Song L, Zhang S (2023) Anti-aging activity and modes of action of compounds from natural food sources. Biomolecules 13(11):1600
Sridharan B, Lee M-J (2022) Ketogenic diet: a promising neuroprotective composition for managing Alzheimer's diseases and its pathological mechanisms. Curr Mol Med 22(7):640–656
Stuchlík M, Zak S (2002) Vegetable lipids as components of functional foods. Biomed Pap Med Fac Univ Palacky Olomouc 146(2):3–10
Tang M, Taghibiglou C (2017) The mechanisms of action of curcumin in Alzheimer's disease. J Alzheimer's Dis 58(4):1003–1016

Target ANM (2016) Lactate Dehydrogenase. In: Ketogenic diet and metabolic therapies: expanded roles in health and disease, p 281

Thapa A, Jett SD, Chi EY (2016) Curcumin attenuates amyloid-β aggregate toxicity and modulates amyloid-β aggregation pathway. ACS Chem Neurosci 7(1):56–68

Truong V, Jun M, Jeong W (2018) Role of resveratrol in regulation of cellular defense systems against oxidative stress. Biofactors 44(1):36–49

Tur JA et al (2012) Dietary sources of omega 3 fatty acids: public health risks and benefits. Br J Nutr 107(S2):S23–S52

Valverde-Salazar V, Ruiz-Gabarre D, García-Escudero V (2023) Alzheimer's disease and green tea: epigallocatechin-3-gallate as a modulator of inflammation and oxidative stress. Antioxidants 12(7):1460

Vauzour D (2012) Dietary polyphenols as modulators of brain functions: biological actions and molecular mechanisms underpinning their beneficial effects. Oxidative Med Cell Longev 2012(1):914273

Vitória JJM, Trigo D, da Cruz e Silva OAB (2022) Revisiting APP secretases: an overview on the holistic effects of retinoic acid receptor stimulation in APP processing. Cell Mol Life Sci 79(2):101

Wang H, Peng R-Y (2016) Basic roles of key molecules connected with NMDAR signaling pathway on regulating learning and memory and synaptic plasticity. Mil Med Res 3:1–7

Wang J et al (2005) Increased oxidative damage in nuclear and mitochondrial DNA in Alzheimer's disease. J Neurochem 93(4):953–962

Williams JL et al (2020) The effects of green tea amino acid L-theanine consumption on the ability to manage stress and anxiety levels: a systematic review. Plant Foods Hum Nutr 75(1):12–23

Wolf R, Wolf D, Ruocco V (1998) Vitamin E: the radical protector. J Eur Acad Dermatol Venereol 10(2):103–117

Wu X et al (2022) A network pharmacology approach to identify the mechanisms and molecular targets of curcumin against Alzheimer disease. Medicine 101(34):e30194

Yang F et al (2005) Curcumin inhibits formation of amyloid β oligomers and fibrils, binds plaques, and reduces amyloid in vivo. J Biol Chem 280(7):5892–5901

Yeung AWK et al (2021) Reactive oxygen species and their impact in neurodegenerative diseases: literature landscape analysis. Antioxid Redox Signal 34(5):402–420. https://doi.org/10.1089/ars.2019.7952

Zhao ZC, Li F, Maiese K (2005) Oxidative stress in the brain: novel cellular targets that govern survival during neurodegenerative disease. Prog Neurobiol 75(3):207–246. https://doi.org/10.1016/j.pneurobio.2005.02.004

Anti-inflammatory Potential of Dietary Components and Benefits in Alzheimer's Disease

3

G. David Lin, Alison Rodger, Jai Malik, Paul N. Smith, and Rachel W. Li

Abstract

This chapter highlights the critical role of neuroinflammation in the onset and progression of Alzheimer's disease (AD). Despite decades of research and the development of pharmacological treatments since the 1980s, a definitive cure for AD remains elusive. Current pharmacotherapies primarily aim to alleviate symptoms rather than halt or reverse the disease process. Consequently, strategies focused on prevention and delaying disease onset have become increasingly important.

Among these strategies, anti-inflammatory dietary intervention holds significant promise due to its accessibility and daily relevance. Emerging research emphasizes the influence of the gut–brain axis, as well as antioxidant and anti-inflammatory pathways in linking diet to neuroinflammation and cognitive health. This chapter explores how the rich source of anti-inflammatory and antioxidant components in diets may serve as an adjunctive approach in modifying AD.

G. D. Lin (✉) · A. Rodger
Research School of Chemistry, the Australian National University, Canberra, ACT, Australia
e-mail: david.lin@anu.edu.au

J. Malik
University Institute of Pharmaceutical Sciences, Panjab University, Chandigarh, India

P. N. Smith
School of Medicine and Psychology, the Australian National University, Canberra, ACT, Australia

R. W. Li
School of Medicine and Psychology, the Australian National University, Canberra, ACT, Australia

John Curtin School of Medical Research, the Australian National University, Canberra, ACT, Australia

J. Malik et al. (eds.), *Diet and Alzheimer's Disease: Let Food be Our Medicine*, Nutritional Neurosciences, https://doi.org/10.1007/978-981-95-2736-6_3

The dietary modification interventions can be achieved by (1) increasing intake of anti-inflammatory dietary components via polyphenol-rich foods, omega-3 fatty acids, and essential vitamins; (2) adopting anti-inflammatory diets such as the Mediterranean or ketogenic diets; (3) modulating the gut microbiota through probiotics, prebiotics, and dietary fibre; (4) reducing processed foods and excessive sugar intake to minimize systemic inflammation.

The take home message is that incorporating anti-inflammatory foods into the daily diet offers an accessible and practical strategy to reduce neuroinflammation, support cognitive function, and potentially prevent or delay the onset of Alzheimer's disease.

Keywords

Alzheimer's disease · Neuroinflammation · Anti-inflammation food · Dietary component

Abbreviations

AD	Alzheimer's disease
COX	Cyclooxygenase
IL	Interleukin
CD	Cluster of differentiation
NLRP3	Nucleotide-Binding Oligomerization Domain (NOD)-Like Receptor (NLR) Pyrin Domain-Containing 3
NFκB	Nuclear factor kappa-light-chain-enhancer of activated B cells
NSAIDs	Non-steroidal anti-inflammatory drugs
NRF2	Nuclear factor erythroid 2-related factor 2
TNF	Tumour necrosis factor

3.1 Introduction

3.1.1 Alzheimer's Disease and Dementia

Alzheimer's disease (AD) is a devastating neurodegenerative disease for patients, their families, and society. The disease manifests as reduced cognitive function and memory loss, difficulty with language, disorientation with regard to time and place, and numerous neuropsychiatric symptoms, including mood changes and behavioural disturbances (Cummings 2004). These manifestations are similar to dementia, which is a set of symptoms (or a clincal symptom, Budson and Solomon 2022a) caused by various brain disorders, including AD, vascular dementia, Lewy body dementia, frontotemporal dementia, and mixed dementia. So, dementia itself is not a disease but is primarily caused by AD. Both dimentia and AD are different but manifest with similar symptoms (Budson and Solamon 2022a).

The burden of managing AD for all concerned will grow dramatically over the next 30–40 years as the population ages and the number of cases increases four- to five-fold, increasing the total cost of care to a trillion dollar per year (Budson and Solomon 2022b). In 2024, there were 6.9 million Americans aged 65 and older living with AD and by 2050, this number is expected to reach 16 million, with about 1 million new cases diagnosed annually (Alzheimer's Association 2024). In Australia and India, over 400,000 and 4 million people are living with dementia, respectively. The likelihood of developing AD in women is twice that of men and this difference is only partially understood (Moutinho 2025).

Early identification of Alzheimer's—ideally during the early memory and cognitive impairment stage—offers the best opportunity to slow disease progression and preserve cognitive function. This enables timely interventions, including pharmacological treatments, lifestyle modifications, including anti-inflammatory diets as described in this chapter, and cognitive therapies, all of which may be more effective before significant neuronal damage occurs and a diagnosis such as AD becomes clear.

Alzheimer's disease is multi-faceted and this is evident from our evolving understanding of AD from the cholinergic hypothesis in the 1970s (Terry and Buccafusco 2003) to the amyloid cascade in the 1990s and the current inflammation concept (Bolós et al. 2017).

3.1.2 Diagnosis and Treatment

Early recognition of signs and early diagnosis empowers patients and their families to plan for future care, access clinical trials, and adopt preventive strategies such as diet, exercise, and cardiovascular health, that target modifiable risk factors. A preventative and proactive approach to detection and treatment not only improves quality of life but also reduces the long-term burden on caregivers and healthcare systems.

Alzheimer's disease is a multi-faceted neurodegenerative disease with major pathology hallmarks including (1) amyloid-β (Aβ) plaques which are associated with neuroinflammation and disrupting neuron-to-neuron communication; (2) neurofibrillary tangles that lead to microtubule disintegration, impairing nutrient transport and causing cell death; (3) neuroinflammation that contributes to neuronal damage and progression of pathology; (4) neuronal loss and brain atrophy with shrinking brain tissue; and (5) cognitive decline and behavioural changes that progress from mild cognitive impairment to severe dementia (Serrano-Pozo et al. 2011). Advancements in diagnosis based on these hallmarks have been made and include biomarker analysis (e.g., amyloid-beta and tau proteins), neuroimaging, and genetic screening. These tools are helping clinicians detect AD earlier than previously. In addition, comprehensive medical and neurobehavioral assessment in clinical practice enables an appropriate and differential diagnosis of dementia (Budson and Solomon 2022a).

The differential diagnosis of dementia can be aided by artificial intelligence (AI). An AI model was developed by integrating diverse data sources, including demographics, medical history, neuropsychological assessments, and neuroimaging (Xue et al. 2024). Analysing 51,269 participants across nine datasets, the model identified 10 distinct dementia aetiologies with high accuracy, achieving an area under the receiver operating characteristic curve (AUROC) of 0.94 for distinguishing normal cognition, mild cognitive impairment, and dementia, and 0.96 for differentiating dementia subtypes. It demonstrated effectiveness in identifying mixed dementia cases (AUROC = 0.78) and improved neurologist diagnoses by 26.25%. Aligning with biomarker evidence and post-mortem findings, the model shows promise as a clinical screening tool, though further validation is required.

Current pharmaceutical intervention for AD includes (1) acetylcholinesterase inhibitors (tacrine hydrochloride, donepezil hydrochloride, galantamine, rivastigmine tartrate to increase acetylcholine in the brain) clinically used for mild to moderate AD (Suh and Checler 2002); (2) NMDA receptor antagonists (memantine), which regulate glutamate activity to protect neurons from moderate to severe AD (Budson and Solomon 2022c); and (3) recently FDA-approved anti-amyloid monoclonal antibodies that target amyloid-beta plaques and slow the progression of early AD and/or mild cognitive impairment. The FDA approval of aducanumab-avwa (Dunn 2021) and lecanemab-irmb (US Food and Drug Administration 2023) represents the new disease-modifying anti-amyloid-beta therapies against AD (Zhao et al. 2023).

3.1.3 Adjunctive Dietary Intervention

Despite the current progress in pharmaceutical treatment for AD, the therapeutic approach is far from satisfactory to arrest the progressive deterioration of brain functions largely due to a 'single drug single target' approach to the multifaceted AD affecting cognition, behaviour, and neural function. Many phytochemicals can counteract several pathogenic processes, such as the Aβ aggregation, neuroinflammation, and oxidative stress (Piccialli et al. 2022). Complex phytochemicals in the diet and generated during digestion and interaction with the gut microbiome may act synergistically on multiple targets of AD, despite each ingredient's action on any given target may be small (Lin and Li 2018).

In addition, David and co-workers communicated in Nature that the short-term consumption of diets composed entirely of animal or plant products alters microbial community structure and even overwhelms inter-individual differences in microbial gene expression (David et al. 2014). Altered gut microbiota affects gut barrier (e.g., leaking gut) and immune function, causes inflammation, and impacts neurological health via the gut–brain axis (Kearns 2024). There has been growing interest in understanding how dietary choices and components like polyphenols modulate systemic inflammation, neuroinflammation, and cognitive health through the detailed investigation of the gut–brain axis (McGrattan et al. 2019). The ongoing research has offered insights into potential dietary interventions for enhancing cognitive function, mitigating dementia and managing AD, and called to address modifiable risk factors as a promising strategy to prevent AD (Kearns 2024).

This chapter explores what is known about the role of neuroinflammation in the development of AD, along with the influence of the gut–brain axis linking diet to brain health. Following a brief account of inflammation and oxidative stress pathways that dietary components may target, the chapter highlights how anti-inflammatory dietary components including polyphenols, omega-3 fatty acids, and vitamins may help reducing neuroinflammation and benefit AD. In addition, other dietary interventions such as modulating gut microbiota with probiotics, prebiotics, and fibre; adopting anti-inflammatory diets (e.g., Mediterranean, ketogenic); and reducing processed foods and sugars are briefly discussed here and can be found in other chapters in this book.

3.2 Diet, Gut, and Neuroinflammation

3.2.1 Diet, Gut Microbiota, and Gut Leaking

Gut microbiota is a community of microorganisms living in the gastrointestinal tract. It includes bacteria, archaea, viruses, fungi, and protozoa, which are acquired from foods, the environment and feeding after birth.

It is not surprising that the foods and drinks that we consume every day have a big impact on our gut microbiota and digestion. David and co-workers (David et al. 2014) reported in Nature that the animal-based diet led to an increase in bile-tolerant microbes such as *Alistipes*, *Bilophila*, and *Bacteroides*, while reducing the presence of *Firmicutes* species involved in breaking down plant-derived polysaccharides, including *Roseburia*, *Eubacterium rectale*, and *Ruminococcus bromii*. The microbial changes resemble those observed between carnivorous and herbivorous animals, highlighting a balance between protein and carbohydrate fermentation (Muegge et al. 2011). Microorganisms from the consumed food—bacteria, fungi, and viruses—temporarily inhabited the gut. An association between dietary fat, bile acids, and the development of bacteria as a possible cause of inflammatory bowel disease is supported by increases in the quantity and activity of *Bilophila wadsworthia* on an animal-based diet (Devkota et al. 2012). Interestingly, colonization in a commensal of the human microbiota is established by the Toll-Like Receptor 2 (TLRP2) pathway (Round et al. 2011).

The role of gut permeability (gut leaking) in neurological disorders has been recently reviewed (Kearns 2024). Gut permeability refers to the intestinal lining's ability to selectively allow essential nutrients into the bloodstream while blocking harmful substances. Various factors, including poor diet, stress, infections, and genetic predispositions, can compromise gut integrity, leading to increased permeability. This condition facilitates the translocation of toxins and bacteria into systemic circulation, triggering widespread inflammation that impacts neurological health via the gut–brain axis. The gut–brain axis is a complex communication network between the gut and the central nervous system, which will be discussed in detail in the next section.

Tryptophan, through the kynurenine pathway, its main metabolic pathway, plays an important role in dysbiosis. Dysbiosis is an imbalance in the gut microbiota, which can lead to increased intestinal permeability and heightened systemic inflammation, both of which intensify neuroinflammation—a critical contributor to neurological diseases. When this pathway becomes disrupted during inflammation, it results in the buildup of harmful metabolites such as quinolinic acid, which are toxic to neurons and contribute to the development and progression of neurological conditions (Kearns 2024).

The gamma delta (γδ) T cells play a role in the gut–brain axis, which was impacted by the intestinal flora and meningeal IL-17$^+$ γδ T cells in the CNS Diseases (Wo et al. 2020), but the impact on AD remains unexplored. Intestinal dysbiosis causes an alteration in dendritic cell activity that alters the immune balance in the small intestine. This leads to an increase in regulatory T cells and a reduction in IL-17-positive γδ T cells. The γδT cells are abundant at barrier sites such as the gut, where up to 20% of intraepithelial lymphocytes in the human colon express the γδT cell receptors (Mayassi and Jabri 2018). Commensal gut bacteria may affect the host's immune system and then influence disease processes in several organs, for example, in ischemic stroke, where intestinal gamma delta T cells found in mucosal tissues can affect the outcome of ischemic stroke (Benakis et al. 2016). Benakis' group observed that antibiotic-induced alterations in the intestinal flora reduced ischemic brain injury in mice.

3.2.2 Neuroinflammation Via the Gut–Brain Axis

The idea of communication between the gut and brain dates back to early studies in neurogastroenterology in the 1960s–1980s, especially through the work of Michael Gershon, often called the "father of neurogastroenterology". His landmark book, "The Second Brain" (1998), described how the enteric nervous system operates independently and in concert with the brain, laying groundwork for the gut–brain axis concept. One of the earliest cited papers using the term is "Principles and clinical implications of the brain–gut–enteric microbiota axis" (Rhee et al. 2009).

Disruptions in the gut–brain axis are increasingly recognized as a contributing factor to neurodegenerative diseases, including AD (Kowalski and Mulak 2019). Imbalances in gut microbiota composition can compromise the integrity of the intestinal barrier, leading to heightened immune activation and systemic inflammation. This inflammatory cascade may weaken the blood-brain barrier (BBB), fostering neuroinflammation, neuronal damage, and ultimately, neurodegeneration.

Amyloid-β (Aβ) is not only a pathological hallmark of AD but also functions as an antimicrobial peptide within the innate immune system. When it is dysregulated, Aβ can exert toxic effects, contributing to disease progression. A critical aspect of this process involves bacterial amyloids, which, through molecular mimicry, may promote misfolding of endogenous proteins and prime microglial cells, exacerbating neuroinflammation. The propagation of Aβ pathology can occur at multiple levels within the gut–brain axis, including direct neuron-to-neuron transmission, blood-brain-barrier penetration, or interaction with support cells such as astrocytes, fibroblasts, microglia, and immune cells (Kowalski and Mulak 2019).

Whilst there are many types of mediators and varied kinds of inflammation, they can be generally understood from changes at cellular levels (cells responding to tissue damage or foreign material-invaded infection) and molecular levels (signalling inducers and mediators) (Lin and Li 2018). In the example of neuroinflammation, the cellular players are the microglial and astrocytic activation and cytokine release, and the molecule players are the nucleotide-binding oligomerization domain(NOD)-like receptor (NLR) pyrin domain-containing 3 (NLRP3) inflammasome in inflammation and the nuclear factor erythroid 2-related factor 2 (NRF2) transcription factor in oxidative stress. The NLRP3 inflammasome is presented below this section, and NRF2 transcript factor in Sect 3.3.

3.2.2.1 Microglial and Astrocytic Activation and Cytokine Release

The CNS's resident phagocytes, known as microglia, are widely dispersed throughout the brain. Inflammatory mediators are released as a result of misfolded and aggregated proteins attaching to pattern recognition receptors on microglia and astroglia, causing an innate immune response that contributes to the course and severity of the disease (Heneka et al. 2015). Receptors that identify danger-associated molecular patterns (DAMPs) or pathogen-associated molecular patterns (PAMPs) mediate the detection of pathogenic triggers. IL-1β, IL-6, and TNF-α are among the pro-inflammatory cytokines that are produced and elevated over time by the buildup of activated microglia in damaged areas during ageing and especially in pathological situations (Akiyama et al. 2000). A circle of neuroinflammatory processes, such as amyloidosis, neuronal death, decreased brain volume, and neurodegeneration, can be sustained by the production of these chemicals (Lee et al. 2010).

Microglia in AD use a variety of surface receptors, including SCARA1 (Scavenger Receptor Class A Member 1), CD36, CD14, α6β1 integrin, CD47, and toll-like receptors (TLR2, TLR4, TLR6, and TLR9), to identify and bind to soluble amyloid β (Aβ) oligomers and Aβ fibrils. It is thought that this interaction plays a part in the inflammatory response that the disease causes in the brain (Heneka et al. 2015). The accumulation of activated microglia in the damaged regions under aging and particularly under pathological conditions promotes sustained production and elevation of pro-inflammatory cytokines, including interleukin-1β, interleukin-6, and tumour necrosis factor-α (Akiyama et al. 2000). Production of these molecules can perpetuate a cycle of neuroinflammatory processes, including amyloidosis, neuronal death, reduced brain volume, and neurodegeneration (Lee et al. 2010).

Astrocytes contribute to cognitive function and are essential for maintaining synaptic transmission. Similar to microglia, astrocytes intensify the neuroinflammatory response and release cytokines, nitric oxide, and other potentially deadly chemicals after being exposed to Aβ (Sofroniew and Vinters 2010).

Cytokines are a collective name of lymphokines (cytokines secreted by lymphocytes), monokines (cytokines secreted by monocytes), interleukins (cytokines secreted by one leukocyte and acting on other leukocytes), chemokines (cytokines with chemotactic activities), etc. Cytokines are regulators of host responses to infection, trauma, inflammation and immune responses and can be broadly classified into proinflammatory and anti-inflammatory groups.

Proinflammatory cytokines, predominantly produced by activated macrophages, tend to make a disease or a condition worse. They are involved in the up-regulation of inflammatory reactions and include interleukin 1β (IL-1β), IL-6, and tumour necrosis factor-alpha (TNFα) (Dinarello 2000). By regulating the pro-inflammatory cytokine response, anti-inflammatory cytokines, such as IL-1 receptor antagonist, IL-4, IL-10, IL-11, and IL-13, reduce inflammation and encourage healing (Opal and DePalo 2000) In particular, IL-10 is a potent anti-inflammatory cytokine down-regulates pro-inflammatory cytokine receptors.

3.2.2.2 NOD-Like Receptor (NLR) Pyrin Domain-Containing 3 (NLRP3) Inflammasome

As an important immunological response, inflammation keeps tissue homeostasis protected under harmful circumstances and allows survival after infection or injury. But this protection comes at a cost as some tissues may lose their functionality and may contribute to the pathophysiology of a disease due to disturbed homeostasis (Medzhitov 2010). The combined effects of leukocyte and phagocyte migration, altered blood vessel wall permeability, and changes in microcirculation constitute an early inflammatory reaction (Baumert 1995). Occasionally, the inflammatory process itself may be more harmful to the body than the initial noxious stimuli. For instance, the host may not benefit from the inflammatory responses that take place in conditions like AD and rheumatoid arthritis. Clearly, not all inflammation is under full control by the body's defence mechanism and sometime the combined effect may often be excessive and uncontrolled, contributing to the development and progression of various chronic inflammatory diseases across the whole span of life (Furman et al. 2019). Inflammation covers a host of pathophysiological events described as acute or chronic, organ-specific, reversible or irreversible; but fundamentally, inflammation in neurogenerative diseases like AD and autoimmune conditions like rheumatoid arthritis is a state of chronic inflammation, which is among the most significant causes of death worldwide (Pahwa et al. 2023).

In various inflammatory and autoimmune diseases, nucleotide-binding oligomerization domain (NOD)-like receptor (NLR) family members are commonly involved. The NLR family plays diverse roles in pathogen sensing, inflammasome activation and immune regulation. Nucleotide-binding leucine-rich repeat (LRR) domain is located at the C-terminus (the tail end) of the NLRP3 protein. This domain, which consists of repeating structural motifs rich in the hydrophobic amino acid leucine, is also known as (NOD)-like receptors, belonging to a family of cytosolic pattern recognition receptors (PPRs) that are capable of detecting a broad range of sterile and pathogenic stimuli (Sundaram et al. 2024).

The NLRP is the subfamily of the NLR family containing the NLR pyrin domain-containing members (NLRPs), which are central to inflammasome formation, acting as sensors. Upon their activation, the NLRPs form a complex called the inflammasome, resulting in the activation of caspase-1, maturation and release of pro-inflammatory cytokines including IL-1β and IL-18, induction of pyroptosis (a type of inflammatory cell death), and coexistence of AD and chronic pain (Moradi and Mokhtari 2025). In the NLRP subfamily, the NLR pyrin domain-containing 3 (NLRP3) is the most studied inflammasome in the brain, as it is linked to AD and multiple sclerosis. The NLRP3 inflammasome pathway is activated by ATP, uric acid, amyloid-beta, and danger-associated molecular patterns (DAMPs), leading to excessive IL-1β production, chronic neuroinflammation, and neuronal damage (Kaur et al. 2025).

The NLRP3 inflammasome as an intracellular sensor is a major driver of AD pathology. In AD amyloid-beta (Aβ) plaques, tau tangles (abnormal intracellular tau protein), reactive oxygen species (ROS), mitochondrial dysfunction, dying neurons and debris (as part of DAMPs), and pathogen-associated molecular patterns (PAMPs) are all triggers for the activation of the NLRP3 Inflammasome. Once activated, NLRP3 assembles with apoptosis-associated speck-like (ASC) adaptor protein and pro-caspase-1, leading to caspase-1 activation, production of inflammatory cytokines (IL-1β and IL-18), and an inflammatory form of programmed cell death (pyroptosis) (Sharma et al. 2023). Researchers are actively developing NLRP3 inhibitors to reduce neuroinflammation in AD, including: (1). MCC950, a selective NLRP3 inhibitor that prevents IL-1β release, (2). DHA (Docosahexaenoic acid, an omega-3 fatty acids), and (3). dietary compounds that can modulate NLRP3 activity like polyphenols (curcumin, resveratrol, quercetin), which show anti-inflammatory effects by suppressing NLRP3 activation, as discussed in a selective review and virtual screening analysis—natural product inhibitors of the NLRP3 Inflammasome (El-Sayed et al. 2022).

3.2.2.3 Chronic Inflammation and Systemic Inflammation

The NLR pyrin domain-containing 3 (NLRP3) inflammasome plays a crucial role in maintaining a chronic inflammation, as discussed above. Chronic information is different from systemic information as compared in Table 3.1. Often and over time, especially if untreated, chronic inflammation can become systemic inflammation.

According to genome-wide research, several genes that raise the risk of sporadic AD also encode the elements that control the inflammatory response and glial clearance of misfolded proteins (Heneka et al. 2015). This provides strong evidence for a causative role of neuroinflammation in AD, with several mutations in microglial genes, for example, the triggering receptor expressed on myeloid cells 2 (TREM2) and CD33, being independently associated with elevated AD risk.

Epidemiological studies have shown that there is a delay in the onset and slow progression of AD on long-term use of non-steroidal anti-inflammatory drugs (NSAIDs), indicating a strong link between neuroinflammation and the pathophysiology of AD. (Lee et al. 2010).

Table 3.1 Chronic inflammation versus systemic inflammation

Types	Chronic inflammation	Systemic inflammation
Definition	Long-term, *persistent inflammation* that can last for months or years	An *inflammatory response that affects the entire body*, often through the bloodstream
Location	Can be *localized* (e.g., joints in rheumatoid arthritis) or *systemic*	More *widespread* than localized inflammation
Cause	Often triggered by ongoing stressors like: Persistent infections Autoimmune conditions Environmental toxins Poor diet/lifestyle	Severe infections (like sepsis) Autoimmune diseases (e.g., lupus) Chronic inflammatory states
Effect	Slowly damages tissues and organs, contributes to aging, and is linked to diseases like: Alzheimer's disease Type 2 diabetes Cancer Cardiovascular disease	Can lead to multi-organ dysfunction or amplify chronic disease processes. It is often measured by biomarkers like CRP (C-reactive protein) or IL-6 in blood tests

3.2.3 Managing Gut Dysbiosis and Neuroinflammation in AD

A growing body of research underscores the role of gut dysbiosis and gut–brain interactions in neurodegeneration, particularly in the aging population, where dietary habits and chronic inflammation further contribute to AD pathogenesis. Given this link, strategies targeting the gut microbiota—such as dietary modifications and probiotic supplementation—hold promise as potential preventive or therapeutic interventions for AD (Kowalski and Mulak 2019).

Management of AD through the gut–brain axis focuses on how gut microbiota, metabolites, and inflammation influence neurodegeneration. Therefore, strategies may include the following.

1. Modulating Gut Microbiota
    - Probiotics & prebiotics: Improve cognitive function by increasing beneficial bacteria (e.g., *Lactobacillus* and *Bifidobacterium*).
    - Dietary Fibre: Supports gut health and reduces systemic inflammation.
2. Anti-Inflammatory Diets
    - Mediterranean & ketogenic diets: Reduce neuroinflammation via short-chain fatty acids (SCFAs) and polyphenols.
    - Polyphenol-Rich Foods: Berries, green tea, and turmeric help modulate gut bacteria and reduce amyloid-beta toxicity.

3. Gut Metabolites and Brain Health
    - Short-chain fatty acids (SCFAs): Produced by gut bacteria, enhance neuroprotection and regulate brain inflammation.
    - Tryptophan & serotonin pathways: Gut bacteria influence neurotransmitter balance, affecting mood and cognitive function.
4. Reducing Gut Permeability
    - Avoid processed foods and excess sugar: Helps prevent "leaky gut," which can lead to systemic inflammation and AD progression.
    - Omega-3 fatty acids: Found in fish and flaxseeds, strengthen the gut barrier and support brain function.

3.3 Modifying Inflammation and Oxidative Stress via Diet

3.3.1 Inflammatory Pathways

Inflammation is a phenomenon well known to us since ancient civilizations with the five signs identified as redness, swelling, heat, pain, and the loss of function by Virchow, the founder of modern cellular pathology, in his classic work *Cellular Pathology* (*Die Cellularpathologie in ihrer Begründung auf physiologische und pathologische Gewebelehre*) published in 1858. The loss of function may induce an adaptive response, a so-called para-inflammation leading to the chronic inflammatory conditions associated with modern human diseases (Medzhitov 2008). For example, numerous human studies have demonstrated that the consumption of ultra-processed foods may cause chronic inflammation and has been associated with cognitive decline (a measure of function) in older adults with type 2 diabetes (Weinstein et al. 2021), US older adults (Cardoso et al. 2022), Brazilian older adults (Silva et al. 2022), adolescent students (Dos Santos et al. 2024), Black and White adults in the United States from 2003 to 2007 (Bhave et al. 2024), and 104 adults (age 50–65 years) with subjective cognitive decline (Tarkar et al. 2024).

A generic inflammatory pathway consists of inducers ➔ sensors ➔ mediators ➔ effectors, for example, lipopolysaccharide from foreign microorganisms ➔ Toll-like receptor 4 (TLR4) ➔ TNF-α, IL-6 and PGE2 ➔ endothelial cells, hepatocytes, leukocytes, the hypothalamus, etc. (Medzhitov 2008).

Traditionally, inflammation is initiated by classic triggers such as infection and tissue injury, which activate a well-coordinated immune response involving the recruitment of leukocytes and plasma proteins to the affected site. This acute inflammatory response is vital for eliminating pathogens, repairing tissue damage, and restoring homeostasis. However, inflammation is not solely a reaction to external threats—it also serves as an adaptive mechanism to maintain tissue integrity and function under various stress conditions. Tissues can experience a distinct, lower-grade immune response termed para-inflammation, which primarily relies on tissue-resident macrophages and other local immune components to modulate immune activity (Medzhitov 2008). Para-inflammation is an intermediate state that helps tissues adapt to environmental stressors, aging, and metabolic imbalances. However, prolonged para-inflammation can contribute to the development of chronic inflammatory diseases, including neurodegenerative disorders, cardiovascular disease, and metabolic syndromes. As modern lifestyles increasingly expose individuals to persistent stressors such as poor diet, pollution, and sedentary behaviour, para-inflammation may underlie many prevalent human diseases, highlighting the need for a deeper understanding of lifestyles, including dietary intervention on inflammation and long-term health.

Progress has been made in understanding inflammation at the cellular and molecular levels in the cases of infection and tissue injury. Both molecular (e.g., inducers, sensors, mediators) and cellular (e.g., effector cells) events are involved in the acute inflammatory response. Inflammation is exceedingly complex and equally fascinating, e.g., inflammatory mediators generated from various pathways play critical roles in the healing processes; however, they can cause damage if their activity becomes excessive or prolonged (Medzhitov 2008).

The hunt for the inflammatory mediators, with a focus on identifying compounds that interact with molecules in exudation, cellular adhesion, enhanced capillary permeability, and local vasodilation, has been very successful. Based on their biochemical properties, inflammatory mediators can be grouped as below (Medzhitov 2008).

- Vasoactive amines (e.g., histamine and serotonin).
- Vasoactive peptides (e.g., substance P and bradykinin).
- Complement fragments C3a, C4a and C5a (anaphylatoxins).
- Lipid mediators (e.g., eicosanoids and platelet-activating factor PAF).
- Cytokines (e.g., TNF-α, IL-1, IL-6, IL-8, INF-γ, and eotaxin).
- Proteolytic enzymes (e.g., cathepsins and matrix metalloproteinases).
- Reactive species (reactive oxygen species and nitric oxide).

There are also anti-inflammatory mediators, including IL-10, TGF-β, lipoxins, and resolvins.

These mediators coordinate immune responses, leading to the amplification or suppression of inflammation. Nevertheless, it is unclear how these mediators react to inflammatory inducers, either sequentially or hierarchically. But it is clear that the inflammatory mediators specifically affect the functional states of the cells and tissues. The cells and tissues are the effectors of an inflammatory response, including neutrophils, macrophages, mast cells, dendritic cells, T cells, B cells, endothelial cells and hepatocytes. These effector cells may generate effector molecules, including reactive oxygen species, nitric oxide, antibodies, enzymes and proteases, and many other mediators.

This section will focus on inflammatory mediators involved in the following pathways.

- Arachidonic acid pathways.
- Nuclear transcription factor kB (NF-kB) pathways.

- Inducible nitric oxide pathways.

3.3.1.1 Arachidonic Acid Pathways

Eicosanoids are strong mediators of inflammatory pathways and are generated from the arachidonic acid (AA) pathway. In this pathway, potent inflammatory mediators, mainly prostaglandins (PGs; for example PG-E_2) and, leukotrienes are produced by the metabolic action of cyclooxygenases (COX1 and COX2) and lipoxygenases on AA, respectively, followed by further actions of synthases (Lin and Li 2018). The mediators produce effects on increased blood vessel permeability, blood vessel dilatation, granulocyte chemotaxis, cartilage degradation, and B- and T-lymphocyte proliferation (Smith 1992). Furthermore, prostaglandins and leukotrienes are essential and responsible for the activation of inflammatory cells (such as neutrophils and macrophages) and the release of other mediators (such as the secretion of serotonin in platelets and histamine in mast cells and basophils). Prostaglandins and prostacyclins (produced by prostacyclin synthase from some prostaglandins) are the two important groups of eicosanoids that can produce pro-inflammatory and anti-inflammatory actions, respectively. Eicosanoids are extensively explored inflammatory mediators as every cell in the body can generate them from AA, abundantly present in the phospholipid membranes (Khanapure et al. 2007). Dietary components and phytochemicals can target these pathways and mediators to exhibit their protective effects, for example, in arachidonic acid pathways, targeting mediators of inflammatory and anaphylactic reactions (prostaglandins), mediators of vasodilation, anti-inflammation (prostacyclins), mediators of vasoconstriction (thromboxanes), mediators of asthmatic and allergic reactions (leukotrienes) and anti-inflammatory mediators to dampen inflammation (lipoxins, which are produced by the conversion of AA by 15-lipoxygenase).

3.3.1.2 Nuclear Factor Kappa-Light-Chain-Enhancer of Activated B Cells (NF-kB) Pathways

The NF-κB within macrophages plays important roles in the pathogenesis of various diseases largely by controlling the expression of pro-inflammatory genes, including cytokines, chemokines and adhesion molecules during the regulation of local and systemic inflammation, inducible COX-2 and nitric oxide synthase (Barnes and Karin 1997). It is part of the NLRP3 inflammasome discussed in Sect. 3.2. Nuclear transcription factor-κB is maintained in an inactive state in the cytoplasm by binding to the inhibitor of κB (IκB). Phosphorylation of IκB by the inhibitor of κB kinase (IKK), via the canonical NF-κB pathway, results in subsequent ubiquitination of IκB and its proteasomal degradation. The degradation of IκB liberates NF-κB, which moves to the nucleus and activates the transcription of its target genes.

Many natural products such as curcumin, resveratrol, epigallocatechin gallate, quercetin, capsaicin, and silybin act like IKK, thereby preventing IkB degradation and eventually inhibiting NF-kB activation (Bremner and Heinrich 2002). Food components in our diets are rich in these natural products.

3.3.1.3 Inducible Nitric Oxide (iNO) Pathways

Nitric oxide (NO) is a simple gas that was discovered in the late 1980s to be a mediator of macrophage-induced tumour cytotoxicity, vessel dilation, and neurotransmission. Since then, it has exhibited various important physiological activities such as smooth muscle relaxation, lysis of tumour cells and destruction of microorganisms, among many others (Lowenstein 1992; Moncada and Higgs 1993). L-arginine (via its guanidino nitrogen leads to synthesis of NO with the help of three isoforms of NO synthases (NOS): (i) a neural form (nNOS); (ii) an endothelial form (eNOS); and (iii) an inducible form (iNOS). The three forms can be differentiated in term of the response to the stimulation of cytokines or endo-toxins. In response to inflammatory stimuli, both iNOS and COX-2 are induced, and the production of NO further stimulates the formation of COX-2. This leads to the production of both NO and PGE_2 and significant cross-talks between PGE_2/NO and COX2/iNOS (Weinberg 2000). It is obvious that inhibiting both iNOS and COX2 would provide a better anti-inflammatory action than a single inhibition. The multiple components in the diet may partially act in this way.

It is well understood that cytokines frequently work by activating transcriptional factors such as NF-kB and protein kinases such as PKC, which in turn regulate target gene expression to maintain the inflammatory state, such as the activation of iNOS and COX-2. Cytokines form a network, allowing for functional redundancy in the inflammatory and immune systems. In many cases, individual cytokines have many biological actions, and various cytokines can have the same activity, necessitating more than a single network malfunction to significantly alter the outcome of cytokine-mediated inflammatory processes (Haddad 2002).

Non-steroidal anti-inflammatory drugs (NSAIDs), such as aspirin, ibuprofen, indomethacin, naproxen, piroxicam, diclofenac, and methyl salicylate, are one of three groups of anti-inflammatory drugs (the other two are glucocorticoids and biologics) that target inflammatory mediators (Rainsford 2007). These medications inhibit eicosanoid synthesis, specifically prostaglandins. Many food components that are physically similar to NSAIDs can interact with the inflammation mediators in a direct or indirect manner.

3.3.2 Oxidative Stress and Nuclear Factor Erythroid 2-Related Factor 2 (NRF2) Pathways

Under stress conditions, reactive oxygen species (ROS) and reactive nitrogen species (RNS), proteinase 3, cathepsin G and elastase are released from neutrophils with attempts to kill the invading agents. Collateral and unavoidable damage occurs to host tissues as well because the toxic contents from the granules of activated neutrophils do not discriminate between microbial and host targets (Nathan 2006). Oxidative stress can be understood by comparison with inflammation. The comparison of their pathways is summarized in Table 3.2.

The nuclear factor erythroid 2-related factor 2 (NRF2) pathway is important in activating antioxidant genes (Saha et al. 2020). The NRF2 pathway is human body's

Table 3.2 Oxidative stress pathways vs. inflammation pathways

	Oxidative stress (OS) pathways	Inflammation pathways
Trigger	Imbalance between ROS production and antioxidant defence	Injury, infection, or immune activation
Key molecules	ROS (e.g., superoxide, hydrogen peroxide)	Cytokines (e.g., TNF-α, IL-6, IL-1β)
	Antioxidants (e.g., glutathione, SOD)	Immune cells (e.g., macrophages, microglia)
	Nrf2, Keap1, mitochondrial signalling	NF-κB, JAK/STAT, MAPK
Primary role	Cellular signalling	Defend against pathogens
	Redox homeostasis	Clear damaged cells
	Detoxification and repair	Initiate tissue repair
When pathological	Persistent ROS→oxidative damage to DNA, lipids, proteins	Chronic activation → tissue damage, fibrosis, autoimmunity
When protective	NRF2-ARE pathway activates antioxidant genes	Resolution phase involves resolvins, IL-10, TGF-β to stop inflammation
When uncontrolled	Mitochondrial dysfunction	Chronic inflammation
	Apoptosis/necrosis	Autoimmune diseases
	Aging, cancer, neuro-degeneration	Atherosclerosis, Alzheimer's, IBD

master switch for protecting itself against stress, toxins, and inflammation. It is a transcription factor—a protein that enters the cellular nucleus and activates genes that help defend against damage. When activated, NRF2 turns on hundreds of protective genes to boost antioxidant production (like glutathione), detoxify harmful substances (through enzymes like NAD(P)H: quinone oxidoreductase 1 (NQO1) and heme oxygenase-1 (HO-1)), reduce inflammation and protect cells from oxidative stress. Under normal (non-stressful) conditions, NRF2 is kept in check by a gatekeeper protein called KEAP1 (Kelch-like ECH-associated protein 1), which marks it for destruction. Under oxidative stress or exposure to plant compounds (like sulforaphane or curcumin), NRF2 breaks free, goes to the cell nucleus, and activates survival genes (Saha et al. 2020).

Some natural activators of NRF2 are sulforaphane (in broccoli sprouts), curcumin (turmeric), green tea (EGCG), resveratrol (red grapes, wine), flavonoids, terpenoids, and carotenoids (Moratilla-Rivera et al. 2023).

3.3.3 Dietary Modifications

The old saying "you are what you eat" is supported by arecent study, which has revealed a positive link between diet with higher inflammatory potential and cognitive decline among elderly individuals in America (Zhang et al. 2024).

Our understanding of the dietary components that can act on inducers, mediators, and effectors of anti-inflammatory pathways also supports and explains some observations of human studies, such as a cross-sectional study on association between dietary inflammatory index and cognitive impairment (Zhang et al. 2024), a meta-analysis of the association between the traditional Mediterranean diet and dementia and Alzheimer disease (Nucci et al. 2024), and a review on the emerging healthy diets, DASH (Dietary Approach to Stop Hypertension) and MIND (Mediterranean-DASH diet intervention for neurodegenerative delay), for the maintenance of cognitive performance (Pistollato et al. 2018).

Dietary modification via gut microbiota will affect the inflammatory and oxidative stress pathways as discussed above.

Diet can modify both inflammation and ROS pathways as both are interconnected and interact with each other in a vicious cycle:

- ROS→Inflammation: excessive ROS activates NF-κB, triggering pro-inflammatory gene expression (e.g., TNF-α, IL-1β). ROS also oxidizes proteins and lipids, making them look like "foreign bodies," leading to immune activation.
- Inflammation → ROS: activated immune cells (e.g., neutrophils, macrophages) produce ROS as part of the inflammatory response. This creates a feedback loop: more inflammation → more ROS→more damage → more inflammation.

In the brain, microglial activation produces ROS and cytokines, breaking the blood-brain barrier and damaging neurons. Antioxidants such as vitamin C, E, and polyphenols, and anti-inflammatory components such as omega-3 s, curcumin, and fibre target both ROS and anti-inflammatory pathways. NRF-2 activators such as sulforaphane and resveratrol suppress oxidative stress and indirectly reduce inflammation. Emerging research increasingly highlights the role of diet as a modulator of inflammation and cognitive health. Several studies have explored how dietary patterns and specific nutrients influence inflammatory processes and cognitive functions. Diet is known to modulate inflammation as detailed in a review (Calder et al. 2017), which highlights the ability of the nutrients in modifying the age-related inflammatory processes. Another review (Pistollato et al. 2018) focuses on the clinical studies on the nutritional patterns associated with the maintenance of neurocognitive functions and the risk of dementia and AD. Other reviews and human studies from 2018 are summarized in Table 3.3. These publications underscore the growing interest in understanding how dietary choices influence inflammation and cognitive health, offering insights into potential dietary interventions for enhancing cognitive function and mitigating age-related cognitive decline and AD.

3.4 Anti-inflammatory Dietary Components for Alzheimer's

3.4.1 Overview of Dietary Components

Throughout recorded history in all cultures, foods as medicines and remedies have been practised all over the world and have provided a rich story for healthcare.

The growing awareness of the connection between diet and health has led to increased consumer interest in nutraceuticals, particularly those found in fruits and

Table 3.3 Selected dietary pattern studies and reviews

Title	Key findings	References
Diet and inflammation in cognitive ageing and Alzheimer's disease	This review discusses how nutrition influences the immune system and neuroinflammatory processes associated with AD, highlighting the impact of dietary patterns on cognitive ageing	McGrattan et al. (2019)
Dietary patterns, inflammatory biomarkers, and cognition in older adults: An analysis of three population-based cohorts	This research investigates the effect of inflammatory dietary patterns on cognitive function in three population-based cohorts	Ortega et al. (2024, 2025)
Modulation of gut microbiota through dietary intervention in neuroinflammation and cognitive function	The paper reviews how Western dietary patterns affect gut microbiota, leading to neuroinflammation and cognitive decline, and discusses dietary interventions that may mitigate these effects	Ayten and Bilici (2024)
The gut microbiome, immune modulation, and cognitive decline: Insights on the gut–brain axis	This review clarifies the complex interactions among gut microbiota, immune regulation, and cognitive disorders, evaluating their potential as therapeutic targets	Zhang et al. (2025)

vegetables. Berries are especially valued for their rich content of both nutritional and non-nutritional bioactive compounds, including flavonoids, anthocyanins, phenolic acids, stilbenes, tannins, vitamins, minerals and essential oils (Nile and Park 2014). These compounds have demonstrated strong antioxidant, anti-inflammatory, anticancer, antimicrobial and neuroprotective effects in both laboratory and animal studies.

In this section, the common anti-inflammatory dietary components are overviewed under six groups (1) flavonoids of polyphenols, which include flavones, flavonols, flavanones, anthocyanins, catechins and isoflavones; (2) non-flavonoids of polyphenols, which include phenolic acids, stilbenes, curcuminoids, gingerols, and tannins; (3) ommega-3 fatty acids; (4) vitamins; (5) probiotics and prebiotics; and (6) others.

3.4.2 Flavonoids of Polyphenols

Polyphenols, also called phenolics or phenolic compounds, are ubiquitous in plants, a major group of plant secondary metabolites, and an essential part of our vegetables and foods. They are so-called phenolics because they all contain at least an aromatic ring attached to at least a hydroxy group.

Polyphenols can be broadly classed into flavonoids (biosynthetically a phenylpropanoid type), which contain a unique feature of (A + B) two rings substituted by a third phenol-type ring C, and non-flavonoids, which lack the unique feature of A, B and C rings but have a general phenol-type structure.

Polyphenols are found to interfere with many mediators or pathways, including arachidonic acid metabolites and nitric oxide (Calixto et al. 2003), cytokines

(Calixto et al. 2004) and NF-κB signalling pathways (Seeram et al. 2005), NLRP3 inflammasome, oxidative stress and NRF2 pathways as have been discussed above.

Polyphenols from dietary foods and medicinal plants include simple phenolic acids (hydroxybenzoic acids (e.g., salicylic acid) and hydroxycinnamic acids), flavonoids (flavones, flavonols, flavanones, flavanols, and anthocyanidins), tannins, stilbenes, curcuminoids, coumarins, lignans, quinones and xanthones. Flavonoids are a major group of secondary metabolites from plants that have been recognised for potential health benefits a long time ago (Cody et al. 1986). Flavonoids, the most abundant polyphenols, are biosynthesised in plants through the amino acids phenylalanine and tyrosine in the shikimic acid pathway (Barros et al. 2016) and form a class of phytochemicals called phenylpropanoids.

Flavonoids derived from plants play a significant role in preventing neuroinflammation through multiple pathways, such as downregulation of transcription factor activity, e.g., NF-$_{K}$B, inhibition of pro-inflammatory cell signalling pathways, etc. (Rendeiro et al. 2015). Flangan et al. highlighted their beneficial roles against neuroinflammation in a comprehensive review of flavonoids and age-related cognitive decline and neurodegeneration (Flanagan et al. 2018).

Based on their structures, major flavonoids are classified as flavones, flavonols, flavanones, anthocyanins, flavans, and isoflavones in Table 3.4.

Quercetin is a well-known flavonoid, as it gained much attention as an active principle of vegetables and fruits against inflammation and inflammation-associated disorders such as cancer, cardiovascular disease, and neurodegeneration (Ossola et al. 2009). For example, *Allium cepa L.* (onion) is rich in quercetin and organosulfur compounds, and the in vitro bioaccessibility and anti-inflammatory effects of a quercetin-rich extract, focusing on different onion parts and varieties, have been investigated. Its "Golden" variety's tunic extract, having the highest quercetin content, was selected. Using a design of experiments approach, optimal extraction conditions were identified as 40% ethanol and 1 g tunic/20 mL solvent. HPLC analysis revealed 14 flavonols, including 10 quercetin derivatives. Gastro-duodenal digestion increased some quercetin derivatives, indicating good bioaccessibility. The extract significantly reduced PGE_2 levels in J774 cells but had no effect on IL-1β, TNF-α, or nitric oxide. These findings support the potential of optimized *A. cepa* tunic extract as a source of bioavailable anti-inflammatory compounds (Ullah et al. 2022).

Anthocyanins are a diverse group of phytochemicals that give rise to wonderful colours in foods and have potent antioxidant and anti-inflammatory action that benefits the management of acute and chronic conditions, including Alzheimer's disease. Anthocyanin sources include (1) berries: blueberries, cherries, raspberries, blackberries; (2) purple sweet potatoes, containing high content of peonidin and cyanidin; (3) grapes, especially with red, purple and black varieties for table grapes and for wine production, which are major sources of dietary anthocyanins; (4) plums, particularly with dark skinned varieties; (5) eggplant, the purple skin containing a unique anthocyanin called nasunin; and (6) red cabbage. Berries are especially valued for their rich content of both nutritional and non-nutritional bioactive

Table 3.4 Six types of flavonoids based on their structures

Flavonoid type	Key structure	Typical example
Flavones	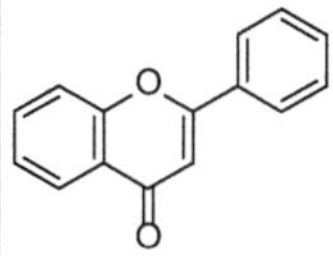	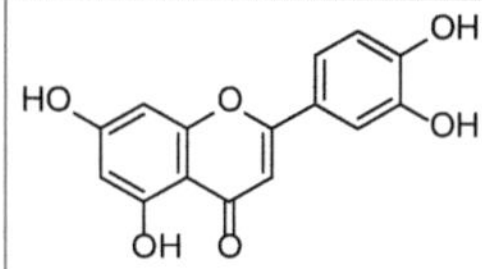 luteoline
Flavonols	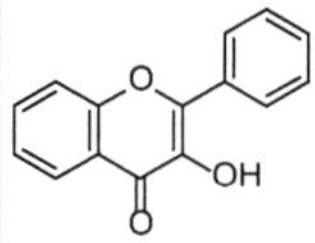	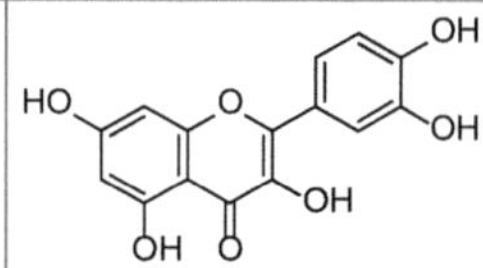 quercetin
Flavanones	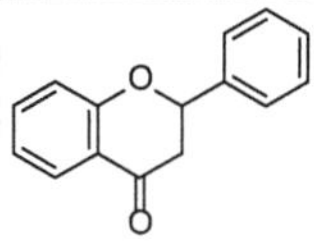	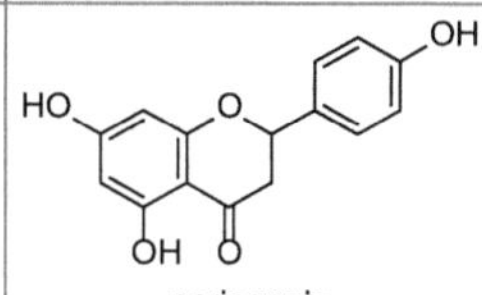 naringenin
Anthocyanidins and anthocyanins	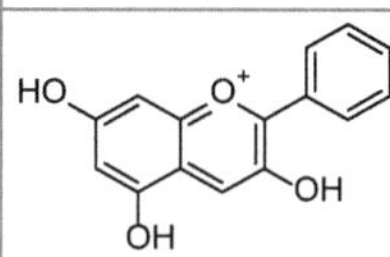	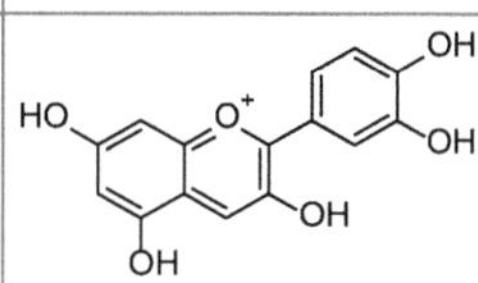 cyanidin
Flavans	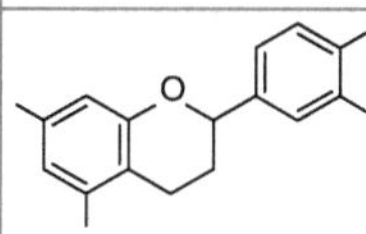	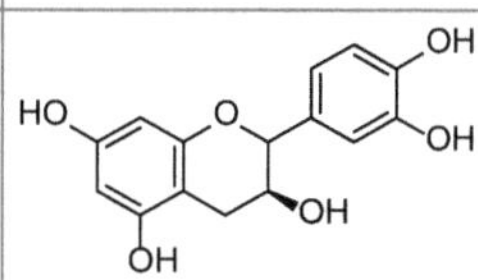 (+)catechin
Isoflavones	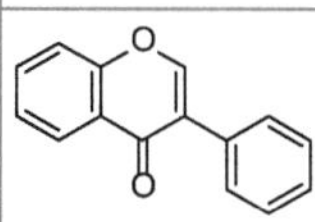	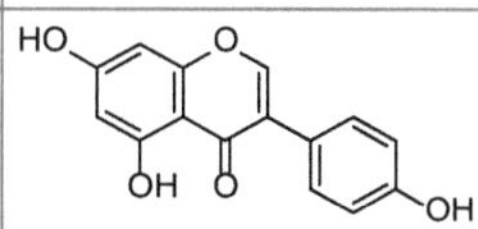 genistein

compounds, including flavonoids, anthocyanins, phenolic acids, stilbenes, tannins, vitamins, minerals, and essential oils (Nile and Park 2014).

Some phytochemcials (–)-epigallocatechin-3-gallate (EGCG), curcumin, trans-resveratrol, quercetin purified from food components, and racemosic acid isolated from Ficus Racemosa (Li et al. 2004) are unique molecules, with each possessing bioactivities against multiple molecular targets, including cyclooxygenases, lipoxygenases, microsomal PGE2 synthase-1, NF-κB signalling, and inducible nitric oxide (Lin and Li 2018).

3.4.3 Non-flavonoid Polyphenols

Non-flavonoid polyphenols include phenolic acids, stilbenes, curcuminoids, gingerols, tannins, and lignans. Stilbenes are best represented by resveratrol, which has gained much attention as an active principle of grapes and red wines against inflammation and inflammation-associated disorders. Resveratrol suppresses the inflammatory effect by inhibiting tumour necrosis factor receptor-associated factor 6 (TRAF6) expression and also by attenuating toll-like receptor 4 (TLR4)-mediated NF-κB activation (Jakus et al. 2013). Huang et al. have extensively reviewed its different mechanisms of action against neuroinflammation and its beneficial role in neurodegenerative disorders (Huang et al. 2021).

Curcumin, the bioactive compound found in turmeric, is classified as a diarylheptanoid and a polyphenol belonging to the larger group of curcuminoids, which are phenolic pigments responsible for turmeric's yellow colour. Curcumin is a molecule with multifaceted activities such as anticancer, antiproliferative, pro-apoptotic, antimetastatic, antiangiogenic, and anti-inflammatory as shown in various preclinical and clinical studies (Esatbeyoglu et al. 2012). The anti-inflammatory potential of curcumin is by virtue of its ability to modulate various signalling pathways, including NF-κB, TNF, IL-6, IL-1, COX-2, and 5-LOX, in addition to the activation of caspases and downregulation of antiapoptotic gene products (Hatami et al. 2019). 728 curcumin analogues were reported from 1815 to mid-2009 and reviewed by Agrawal and Mishra (2010) for various pharmacological activities in various models and various cell lines. Some are shown to have antioxidant, anti-HIV, antimutagenic, antiangiogenic, antimalarial, antitubercular, antiandrogenic, and COX inhibitory activities. A few analogues have shown very potent results and may be considered as clinical candidates for the development of future therapeutics. Curcumin itself contains two structural isomers (tautomers): enol and keto forms (shown below), which are in equilibrium and inter-convertible.

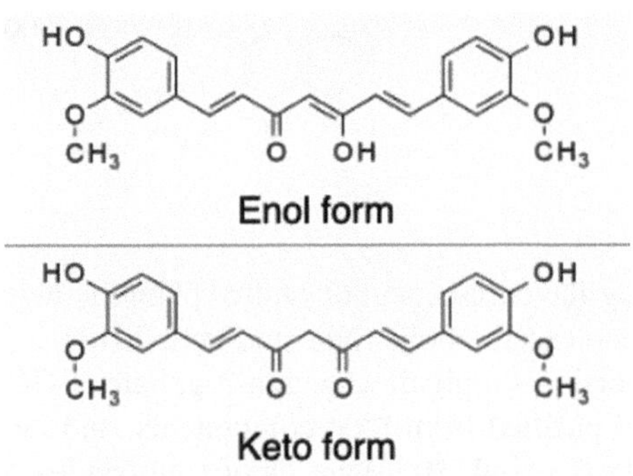

Curcumin attenuated Aβ42-induced IL-1β, IL-6, and TNF-α formation in microglia. Curcumin exhibited an inhibitory effect on phosphorylation of extracellular signal-regulated kinases 1 and 2 (ERK1/2) and p38 mitogen-activated protein kinase (MAPK) in Aβ42-activated microglia, and reduced inflammatory cytokine production (Shi et al. 2015). Curcumin also inhibited the IKK activity, thereby inhibiting the nuclear translocation of the NF-κB signalling pathway.

Gingerols are a class of phenolic compounds, specifically beta-hydroxy ketones, found in ginger (*Zingiber officinale*) and related to its pungent taste and various biological activities, including anti-inflammatory and antioxidant effects. The structure of gingerol itself is shown below. Gingerols are potent inhibitors against a number of prostaglandin biosynthesizing enzymes, including 5-lipoxygenase (5-LO) and leukotriene synthetases.

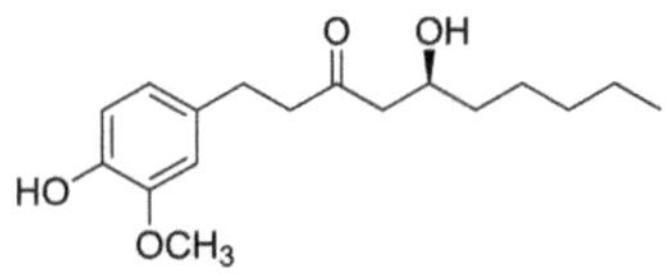

Tannins are a class of naturally occurring polyphenolic compounds found in many plants. They have astringent properties, meaning they can bind to proteins and other organic compounds, which often results in a dry or puckering sensation in the mouth (like when drinking strong tea or red wine). Health benefits of tannins include antioxidant and anti-inflammatory action, among others. The beneficial role of tannins has also been suggested for neurodegenerative disorders (Hussain et al. 2019). There are two types: hydrolysable and condensed tannins. Hydrolysable tannins can be broken down by hydrolysis with acid, base, or enzymes into smaller phenolic acids, as in Table 3.5.

Condensed tannins, or proanthocyanidins, are polymers of flavan-3-ol units and do not hydrolyze easily. Examples are given in Table 3.6.

Tannic acid has shown beneficial effects against various neurodegenerative disorders like AD, PD (Parkinson's disease), etc. Tannic acid has shown neuroprotection against lipopolysaccharide, STZ, and cerebral hypoperfusion-induced neuroinflammation by modulating NF-κB and NRF2 pathways (Sehati et al. 2021; Wu et al. 2019).

Lignans are a group of plant polyphenols formed from the dimerization of phenylpropanoid units. Some are found directly in plants, while others are converted by gut bacteria into bioactive forms called enterolignans (like enterolactone and enterodiol), which are believed to be responsible for many of lignans' health effects (Rodríguez-García et al. 2019). Important plant lignan compounds (precursors) are presented in Table 3.7.

Matairesinol (structure below) is a plant lignan, a type of phenylpropanoid, found in various seeds (Sicilia et al. 2003), grains, and soybeans, that is a precursor to mammalian lignans and has been shown to have estrogenic activity and potential anticancer properties.

Table 3.5 Hydrolysable tannins

Compound	Structure type	Found in
Gallic acid	Simple phenolic acid	Tea, oak bark
Ellagic acid	Dimer of gallic acid	Pomegranates, berries
Tannic acid	Poly-galloyl glucose	Tea, wine, tree bark
Gallotannins	Gallic acid + glucose core	Witch hazel, oak
Ellagitannins	Ellagic acid precursors	Berries, walnuts

Table 3.6 Condensed tannins

Compound	Structure type	Found in
Catechin	Flavan-3-ol monomer	Green tea, cocoa
Epicatechin	Flavan-3-ol isomer	Chocolate, berries
Procyanidin B1–B4	Dimers of catechin/epicatechin	Grape seeds, apples
Prodelphinidins	Related to gallocatechin	Berries, legumes

Table 3.7 Key plant lignan compounds (precursors) (Burgberger et al. 2025)

Compound	Found in	Notes
Matairesinol	Rye, oats, and cruciferous vegetables	Precursor to enterolactone
Pinoresinol	Olive oil, sesame seeds, tea	Also found in wine and vegetables
Lariciresinol	Flaxseed, lentils, garlic, carrots	Converted into enterodiol and enterolactone
Secoisolariciresinol	Flaxseeds, sesame seeds, sunflower seeds	One of the most abundant dietary lignans

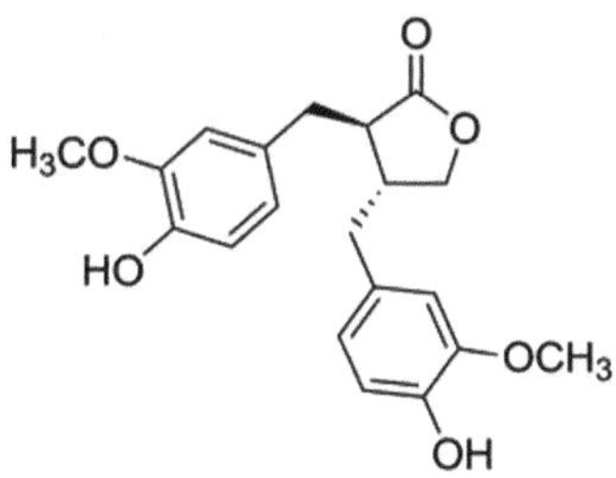

Enterodiol and enterolactone are two key metabolites of lignana in human gut (enterolignans). Enterodiol is metabolized from secoisolariciresinol or lariciresinol and has mild estrogenic and antioxidant activity. Enterolactone is metabolized from matairesinol, or lariciresinol, and has strong phytoestrogen, anti-cancer, cardioprotective properties. Lignans have also been found effective in neurodegenerative disorders by mitigating neuroinflammation (Li et al. 2023). The role of lignans in AD through different mechanisms has also been reviewed by (Han et al. 2022)

3.4.4 Omega-3 Fatty Acids

Long-chain omega-3 polyunsaturated fatty acids (PUFAs) are more abundant in the phhospholipids of inflammatory cells as there is an increase in their intake. The incorporation of eicosapentaenoic acid (EPA) and docosahexaenoic acid (DHA) into human inflammatory cells occurs at the expense of arachidonic acid, leaving less substrate available for the production of traditional inflammatory eicosanoids such as prostaglandin E_2. Omega-3 PUFAs may affect inflammation and inflammatory processes via altering eicosanoid synthesis, but they also have non-eicosanoid-mediated effects on cell signalling and gene expression. Therefore, EPA and DHA are thought to have anti-inflammatory properties (Calder et al. 2017).

Resolvins as specialized pro-resolving mediators (SPMs) are synthesized from omega-3 PUFAs, particularly EPA and DHA, which are commonly found in fatty fish (like salmon, mackerel) and certain algae. There are two series of resolvins: E- and D-series. In the E-series of resolvins, EPA is acted upon by 5-lipoxygenase (5-LOX) and 15-lipoxygenase (15-LOX) and is converted into resolvin E1 (RvE1) and other E-series of resolvins. While in D-series of resolvins, DHA is acted upon by 15-LOX and 5-LOX and other lipoxygenases and is converted into resolvin D1 (RvD1), RvD2, RvD3, and others. Resolvins are not produced by saturated fatty acids, which produce an inflammatory action mainly by activating TLR4 signalling in the hypothalamus (Milanski et al. 2009). Anti-inflammatory resolvins are important to promote resolution of inflammation without suppressing the immune system, reduce chronic inflammation and tissue damage, and help in conditions like neuroinflammation in AD, rheumatoid arthritis, and cardiovascular disease.

Regular consumption of fish (salmon, mackerel, sardines) will supply a good source of DHA and EPA. Other natural sources of omega-3 fatty acids are algal oils, chia seeds, flaxseeds and walnuts (a source of alpha-linolenic acid, which can be converted to EPA and DHA, though less efficiently).

3.4.5 Vitamins

Vitamin D is anti-inflammatory and supports the immune system by regulating immune cells, cytokine production, and oxidative stress. Food sources of vitamin D include egg yolks, salmon, mackerel, and sardines. Menéndez and Manucha (2024) and Kouba et al. (2023) have comprehensively reviewed the neuroprotective properties of vitamin D against neuroinflammation and AD, respectively. The mechanisms of action of vitamin D in neuroinflammation include the five important areas:

1. *Regulation of immune cells:* Vitamin D receptors (VDR) are present on microglial cells. The binding of vitamin D to the VDR helps in regulating microglial activation. Overactive microglia are a major source of pro-inflammatory cytokines in the brain. Vitamin D reduces the activation of microglia, preventing them from releasing excessive pro-inflammatory cytokines (like TNF-α, IL-1β, and IL-6).
2. *Modulation of cytokine production:* Vitamin D helps balance the levels of pro- and anti-inflammatory cytokines. It reduces the production of pro-inflammatory cytokines (such as IL-6, TNF-α, and IL-1β) but increases the production of anti-inflammatory cytokines (such as IL-10 and TGF-β), which help shift the immune response toward resolution of inflammation.
3. *Neuroprotective effects:* Vitamin D may promote the production of neurotrophic factors (like BDNF), which promote neuronal health and function. It also protects neurons from oxidative stress and apoptosis (cell death), both of which are key components of neuroinflammation.
4. *Regulation of the blood-brain barrier (BBB):* Vitamin D helps maintain the integrity and strength of the BBB, which prevents the leakage of immune cells

and inflammatory mediators into the brain, thus reducing the potential for chronic brain inflammation.

5. *Reduction of ROS:* By promoting the antioxidant defence systems (e.g., glutathione), vitamin D can reduce oxidative stress, which is often a trigger for neuroinflammation.

Vitamin E refers to a group of fat-soluble compounds, but the most biologically active form is α-tocopherol. It is well-known for its antioxidant role in protecting cells from oxidative damage, but it also plays a critical role in modulating inflammation, particularly in the brain (Jabeen et al. 2023).

Vitamin E acts as a potent antioxidant, helping to neutralize reactive oxygen species (ROS) and free radicals, which are often produced during inflammation. Vitamin E can reduce the activation of microglia, preventing them from producing excessive inflammatory molecules, such as pro-inflammatory cytokines (like TNF-α and IL-1β). Vitamin E has also been shown to decrease the production of inflammatory mediators, such as prostaglandins (like PGE_2) and leukotrienes (Betti et al. 2011). Vitamin E is primarily found in nuts, seeds and vegetable oils, including almonds, hazelnuts, sunflower seeds, oils from sunflower, wheat germ, safflower, spinach, kale, and avocados.

The B-group of vitamins are essential for nerve function, energy metabolism, and the regulation of inflammation in the brain. They often work synergistically in the body, especially B6 (pyridoxine), B12 (cobalamin), and folate (B9), which are involved in the methylation cycle. Vitamin B6 is crucial for amino acid metabolism, neurotransmitter synthesis, and myelin (the protective sheath around nerve cells) formation. It has shown neuroprotective action by attenuating neuroinflammation in BV2 (a microglial cell line that mimics immune responses in the brain) and SH-SY5Y (a neuronal cell line derived from human neuroblastoma that can be differentiated into functional neurons) (Nguyen et al. 2025). Vitamin B12 is involved in DNA synthesis, red blood cell formation, and nervous system health. It also plays a crucial role in methylation, which is vital for controlling the expression of certain genes related to inflammation. A deficiency in B12 can lead to neurological damage, cognitive decline, and elevated homocysteine levels, which contribute to neuroinflammation and oxidative stress. Vitamin B12 helps reduce homocysteine levels, which are associated with increased risk of Alzheimer's and vascular diseases (both of which are linked to chronic neuroinflammation). B12 supplementation has been studied for its potential to reduce neuroinflammation and slow the progression of neurodegenerative diseases like Alzheimer's and Parkinson's (Cassiano et al. 2023). Folate (B9) has been shown to regulate inflammatory cytokine production and folate supplementation has been studied for its potential to reduce neuroinflammation in Alzheimer's disease (Shaikh and Roy 2023). Folate also supports the production of serotonin, which has mood-regulating effects and may help prevent neuroinflammation associated with depression. Food Sources of B6 are poultry, fish, potatoes, bananas, fortified cereals; of B12 are meat, fish, and dairy products; and of folate are leafy green vegetables (spinach and kale), legumes and citrus fruits.

3.4.6 Probiotics and Prebiotics

Aging also leads to alteration in the gut microbiota diversity as there is an increase in the pathogenic bacterial population, which further triggers inflammatory pathways. This dysbiosis leads to chronic inflammation and a decrease in bacterial metabolites such as short-chain fatty acids, which play an important role in increasing the expression of neurotrophic factors (Romo-Araiza and Ibarra 2020). Probiotics and Prebiotics can modulate gut microbiota, which may promote the production of short-chain fatty acids (SCFA) for beneficial bacteria in the gut and thus reduce neuroinflammation in the gut–brain axis (Shikh et al. 2023). A detailed study on the effect of regular usage (for more than 6 months) of *Lactobacillus delbrueckii* subsp. *bulgaricus* 8481 on elderly subjects (aged >65 years) has showed that there was a significant increase in the number of recent thymus emigrant CD_{31}^{+} T-cells and a decrease in the number of CD_8^{+} CD_{28}^{null} T-cells, along with the prevention of cytomegalovirus reactivation (Moro-García et al. 2013). The study indicates that probiotic consumption could counteract some hallmarks of immunosenescence related to T-cell immunity. In addition, in the probiotic group decreased serum IL-8 and increased beta-defensin 2 levels were also observed, but there was no effect on serum IL-6 or TNF-α concentrations, indicating a non-generalized effect on low grade inflammation (LGI) in elderly subjects. Although these results indicate that probiotics could have an effect on both LGI and immunosenescence, whether these changes translate to clinical benefit is not known. Another study showed that *Mycobacterium vaccae* NCTC 11659 mitigated the neuroinflammatory process induced by stress and anxiety in mice by regulating T cell (Treg)-dependent mechanism (Reber et al. 2016). These Tregs downregulate the proinflammatory processes by generating anti-inflammatory cytokines (IL-10, TGF-β, IL-35, and IL-4).

Breast milk is considered a good prebiotic as it contains prebiotics such as human milk oligosaccharides (HMOs) that are indigestible to babies but are food for beneficial gut bacteria. The HMOs act as prebiotics, especially promoting the growth of *Bifidobacteria* in the infant gut. There are 200 types of HMOs and they can mimic binding sites on gut cells, trapping pathogens like *E. coli* or *Campylobacter*, so they cannot cause infection. The HMOs help maintain the intestinal lining, reducing "leaky gut" and systemic inflammation, potentially supporting brain and cognitive development. For more details, please see a recent review (Urashima et al. 2025).

3.4.7 Others

Carotenoids are natural pigments found in plants. They give many fruits and vegetables their red, orange, and yellow colours. There are over 600 types, but the most well-known include beta-carotene (in carrots, sweet potatoes), lutein and zeaxanthin (in leafy greens such as spinach and kale), lycopene (in tomatoes and watermelon), and astaxanthin (in salmon and shrimp). Carotenoids act as antioxidants to neutralize ROS, protect cells from oxidative stress, strengthen immune function, and promote healthy skin and eyes. Carotenoids such as β-carotene, lycopene, astaxanthin, lutein, and zeaxanthin have been shown in various in vitro and in vivo studies to activate NRF2, upregulate NRF2 target genes (like HO-1, NQO1), reduce oxidative stress via NRF2-mediated antioxidant defence and inhibit proinflammatory signalling pathways (like NF-κB, MAPK) (Moratilla-Rivera et al. 2023).

Phytosterols, glucosinolates, nuts, oilseeds, fruits, vegetables, cereals and legumes (Albuquerque et al. 2020) are also contributing to anti-inflammatory and antioxidant action in the diet. Bioactive compounds found in legumes that have demonstrated an anti-inflammatory effect in non-clinical studies have been compiled (Juarez-Chairez et al. 2022), and they include peptides, polyphenols and saponins.

Aromatic spices are used in the cooking of a meal worldwide, for example, ginger used in Chinese cooking and Turmeric in the Indian cooking. These spices contain rich, super anti-inflammatory compounds beneficial for AD. So are edible mushrooms and fermented foods like yogurt and kimchi.

3.5 Limitations and Future Directions

Our insights into the mechanistic actions of diet on neuroinflammation are mainly obtained from animal studies. When they are translated to humans, the effect may be different due to many factors, including bioavailability and metabolism. The past two decades of research into the pathogenesis of AD have been driven largely by the amyloid hypothesis, the neuroinflammation that is associated with AD has been assumed to be merely a response to pathophysiological events. However, new data from preclinical and clinical studies have established that immune system-mediated actions in fact contribute to and drive AD pathogenesis. These insights have suggested both novel and well-defined potential therapeutic targets for AD, including microglia and several cytokines. In addition, as inflammation in AD primarily concerns the innate immune system—unlike in "typical" neuroinflammatory diseases such as multiple sclerosis and encephalitides—the concept of neuroinflammation in AD may need refinement (Heppner et al. 2015).

Future directions on dietary components and benefits in AD research may be focused on nutrigenomics and precision nutrition, synergistic effects of dietary components and integrating diet with conventional AD therapies.

3.6 Conclusion

In summary, this chapter has highlighted the critical role of neuroinflammation in the onset and progression of Alzheimer's disease (AD). Despite decades of research and the development of pharmacological treatments since the 1980s, a definitive cure for AD remains elusive. Current pharmacotherapies primarily aim to alleviate symptoms such as memory loss rather than halt or reverse the disease process. Consequently, strategies focused on prevention and delaying disease onset have become increasingly important.

Among these strategies, anti-inflammatory dietary intervention holds significant promise due to its accessibility and daily relevance. Emerging research emphasizes the influence of the gut–brain axis, as well as antioxidant and anti-inflammatory pathways, in linking diet to neuroinflammation and cognitive health. This chapter has explored how modifying diet with rich sources of anti-inflammatory and antioxidant components may serve as an adjunctive approach in managing AD.

The dietary modification interventions may include:

- Increasing intake of anti-inflammatory dietary components, polyphenol-rich foods, omega-3 fatty acids (from fish and flaxseeds), and essential vitamins (D, E, B6, B12, and folate).
- Adopting anti-inflammatory diets such as the Mediterranean or ketogenic diets.
- Modulating the gut microbiota through probiotics, prebiotics, and dietary fibre.
- Reducing processed foods and excessive sugar intake to minimize systemic inflammation.

Looking ahead, future research may benefit from exploring integrated therapeutic approaches—combining dietary strategies with conventional pharmacological agents, including acetylcholinesterase inhibitors (e.g., donepezil, galantamine), N-methyl-D-aspartate (NMDA) receptor antagonists (e.g., memantine, Budson and Solomon 2022c), and anti-amyloid monoclonal antibodies (e.g., lecanemab-avwa, donanemab-irmb). Such a multifaceted approach could offer a more effective and personalized pathway for the management of Alzheimer's disease.

Modifying risk factors, particularly targeting neuroinflammation, immune mechanisms, gut microbiota, diets, and anti-inflammatory dietary components presented in this chapter, could lead to future preventive and therapeutic strategies for Alzheimer's disease.

Incorporating anti-inflammatory foods into the daily diet offers an accessible and practical strategy to reduce neuroinflammation, support cognitive function, and potentially prevent or delay the onset of Alzheimer's disease.

References

Agrawal DK, Mishra PK (2010) Curcumin and its analogues: potential anticancer agents. Med Res Rev 30:818–860

Akiyama H, Barger S, Barnum S, Bradt B, Bauer J, Cole GM et al (2000) Inflammation and Alzheimer's disease. Neurobiol Aging 21(3):383–421. https://doi.org/10.1016/S0197-4580(00)00124-X

Albuquerque TG, Nunes MA, Bessada SMF, Costa HS, Oliveira MBPP (2020) Chapter 14 – biologically active and health promoting food components of nuts, oilseeds, fruits, vegetables, cereals, and legumes. In: Pico Y (ed) Chemical analysis of food, 2nd edn. Academic Press, pp 609–656

Alzheimer's Association (2024) Alzheimer's disease facts and figures (Special report mapping a better future for dementia care navigation). Retrieved from https://www.alz.org/alzheimers-dementia/facts-figures?utm_source=chatgpt.com

Ayten Ş, Bilici S (2024) Modulation of gut microbiota through dietary intervention in Neuroinflammation and Alzheimer's and Parkinson's diseases. Curr Nutr Rep 13(2):82–96. https://doi.org/10.1007/s13668-024-00539-7

Barnes PJ, Karin M (1997) Nuclear factor-κB – a pivotal transcription factor in chronic inflammatory diseases. N Engl J Med 336:1066–1071

Barros J, Serrani-Yarce JC, Chen F, Baxter D, Venables BJ, Dixon RA (2016) Role of bifunctional ammonia-lyase in grass cell wall biosynthesis. Nat Plants 2(6):16050. https://doi.org/10.1038/nplants.2016.50

Baumert PW (1995) Acute inflammation after injury. Postgrad Med 97:35–49

Benakis C, Brea D, Caballero S, Faraco G, Moore J, Murphy M et al (2016) Commensal microbiota affects ischemic stroke outcome by regulating intestinal gamma delta T cells. Nat Med 22(5):516–523. https://doi.org/10.1038/nm.4068

Betti M, Minelli A, Ambrogini P, Ciuffoli S, Viola V, Galli F et al (2011) Dietary supplementation with α-tocopherol reduces neuroinflammation and neuronal degeneration in the rat brain after kainic acid-induced status epilepticus. Free Radic Res 45(10):1136–1142

Bhave VM, Oladele CR, Ament Z, Kijpaisalratana N, Jones AC, Couch CA et al (2024) Associations between ultra-processed food consumption and adverse brain health outcomes. Neurology 102(11):e209432. https://doi.org/10.1212/WNL.0000000000209432

Bolós M, Perea Juan R, Avila J (2017) Alzheimer's disease as an inflammatory disease. Biomol Concepts 8(1):37–43. https://doi.org/10.1515/bmc-2016-0029

Bremner P, Heinrich M (2002) Natural products as targeted modulators of the nuclear factor-kappaB pathway. J Pharm Pharmacol 54:453–472

Budson AE, Solomon PR (2022a) Chapter 3 – subjective cognitive decline, mild cognitive impairment, and dementia. In: Memory loss, Alzheimer's disease, and dementia, 3rd edn. Elsevier Inc., pp 38–47

Budson AE, Solomon PR (2022b) Chapter 23 – future treatments of memory loss, Alzheimer's disease, and dementia. In: Memory loss, Alzheimer's disease, and dementia, 3rd edn. Elsevier Inc., pp 237–253

Budson AE, Solomon PR (2022c) Chapter 20 – Memantine. In: Memory loss, Alzheimer's disease, and dementia, 3rd edn. Elsevier Inc., pp 217–223

Burgberger M, Mierziak J, Augustyniak B, Wojtasik W, Kulma A (2025) The Power of Lignans: Plant Compounds with Multifaceted Health-Promoting Effects. Metabolites 15(9):589. https://doi.org/10.3390/metabo15090589

Calder PC, Bosco N, Bourdet-Sicard R, Capuron L, Delzenne N, Dore J et al (2017) Health relevance of the modification of low grade inflammation in ageing (inflammageing) and the role of nutrition. Ageing Res Rev 40:95–119. https://doi.org/10.1016/j.arr.2017.09.001

Calixto JB, Otuki MF, Santos ARS (2003) Anti-inflammatory compounds of plant origin. Part I. Action on arachidonic acid pathway, nitric oxide and nuclear factor κ B (NF-κB). Planta Med 69:973–983

Calixto JB, Campos MM, Otuki MF, Santos ARS (2004) Anti-inflammatory compounds of plant origin. Part II. Modulation of pro-inflammatory cytokines, chemokines and adhesion molecules. Planta Med 70:93–103

Cardoso R et al (2022) Association between ultra-processed food consumption and cognitive performance in US older adults: a cross-sectional analysis of the NHANES 2011–2014. Eur J Nutr 61(8):3975–3985. https://doi.org/10.1007/s00394-022-02911-1

Cassiano LMG, da Silva Oliveira M, Coimbra RS (2023) Vitamin B12 as a neuroprotectant in neuroinflammation. In: Vitamins and minerals in neurological disorders. Academic Press, pp 399–416

Cody V, Middleton E, Harborne JB (1986) Plant flavonoids and medicine: biochemical, pharmacological and structure-activity relationships. In: Proceedings of a symposium held in Buffalo, New York, July 22–26, 1985, Volume 1. The University of Michigan: Liss

Cummings JL (2004) Alzheimer's disease. N Engl J Med 351(1):56–67

David LA, Maurice CF, Carmody RN, Gootenberg DB, Button JE, Wolfe BE et al (2014) Diet rapidly and reproducibly alters the human gut microbiome. Nature 505(7484):559–563. https://doi.org/10.1038/nature12820

Devkota S, Wang Y, Musch MW, Leone V, Fehlner-Peach H, Nadimpalli A et al (2012) Dietary-fat-induced taurocholic acid promotes pathobiont expansion and colitis in Il10−/− mice. Nature (London) 487(7405):104–108. https://doi.org/10.1038/nature11225

Dinarello CA (2000) Proinflammatory cytokines. Chest 118:503–508

Dos Santos JVL, de Melo ISV, Costa CACB, de Almeida LC, Silva DR, Ferro DC et al (2024) Association between ultra-processed food consumption and cognitive performance among adolescent students from underdeveloped cities in Brazil: a cross-sectional study. Int J Public Health 69:1607658. https://doi.org/10.3389/ijph.2024.1607658

Dunn WH (2021) BLA accelerated approval: we have approved your BLA for Aduhelm (aducanumab-avwa)

El-Sayed S, Freeman S, Bryce RA (2022) A selective review and virtual screening analysis of natural product inhibitors of the NLRP3 Inflammasome. Molecules (Basel, Switzerland) 27(19):6213. https://doi.org/10.3390/molecules27196213

Esatbeyoglu T, Huebbe P, Ernst IMA, Chin D, Wagner AE, Rimbach G (2012) Curcumin – from molecule to biological function. Angew Chem Int Ed 51:5308–5332

Flanagan E, Müller M, Hornberger M, Vauzour D (2018) Impact of flavonoids on cellular and molecular mechanisms underlying age-related cognitive decline and neurodegeneration. Curr Nutr Rep 7(2):49–57. https://doi.org/10.1007/s13668-018-0226-1

Furman D, Campisi J, Verdin E, Carrera-Bastos P, Targ S, Franceschi C et al (2019) Chronic inflammation in the etiology of disease across the life span. Nat Med 25(12):1822–1832. https://doi.org/10.1038/s41591-019-0675-0

Haddad JJ (2002) Cytokines and related receptor-mediated signaling pathways. Biochem Biophys Res Commun 297:700–713

Han N, Wen Y, Liu Z, Zhai J, Li S, Yin J (2022) Advances in the roles and mechanisms of lignans against Alzheimer's disease. Front Pharmacol 13

Heneka MT, Carson MJ, El Khoury J, Landreth GE, Brosseron F, Feinstein DL et al (2015) Neuroinflammation in Alzheimer's disease. Lancet Neurol 14(4):388–405. https://doi.org/10.1016/S1474-4422(15)70016-5

Hatami M, Abdolahi M, Soveyd N, Djalali M, Togha M, Honarvar NM (2019) Molecular mechanisms of curcumin in neuroinflammatory disorders: a mini review of current evidences. Endocr Metab Immune Disord Drug Targets 19(3):247–258

Heppner FL, Ransohoff RM, Becher B (2015) Immune attack: the role of inflammation in Alzheimer disease. Nat Rev Neurosci 16(6):358–372. https://doi.org/10.1038/nrn3880

Huang J, Huang N, Xu S, Luo Y, Li Y, Jin H, Yu C, Shi J, Jin F (2021) Signaling mechanisms underlying inhibition of neuroinflammation by resveratrol in neurodegenerative diseases. J Nutr Biochem 88:108552

Hussain G, Huang J, Rasul A, Anwar H, Imran A, Maqbool J, Razzaq A, Aziz N, Makhdoom EUH, Konuk M, Sun T (2019) Putative roles of plant-derived tannins in neurodegenerative and neuropsychiatry disorders: an updated review. Molecules (Basel, Switzerland) 24(12):2213

Jabeen K, Rehman K, Akash MSH, Nadeem A, Mir TM (2023) Neuroprotective and cardiometabolic role of vitamin E: alleviating neuroinflammation and metabolic disturbance induced by AlCl3 in rat models. Biomedicine 11(9):2453

Jakus PB, Kalman N, Antus C, Radnai B, Tucsek Z, Gallyas F, Sumegi B, Veres B (2013) TRAF6 is functional in inhibition of TLR4-mediated NF-κB activation by resveratrol. J Nutr Biochem 24(5):819–823

Juarez-Chairez MF, Meza-Marquez OG, Marquez-Flores YK, Jimenez-Martinez C (2022) Potential anti-inflammatory effects of legumes: a review. Br J Nutr 128(11):2158–2169. https://doi.org/10.1017/S0007114522000137

Kaur B, Biby S, Namme JN, More S, Xu Y, Zhang S (2025) Biological and therapeutic significance of targeting NLRP3 inflammasome in the brain and the current efforts to develop brain-penetrant inhibitors. Adv Pharmacol 1990(102):103–157. https://doi.org/10.1016/bs.apha.2024.10.004

Kearns R (2024) Gut-brain Axis and Neuroinflammation: the role of gut permeability and the kynurenine pathway in neurological disorders. Cell Mol Neurobiol 44(1):64. https://doi.org/10.1007/s10571-024-01496-z

Khanapure SP, Garvey DS, Janero DR, Letts LG (2007) Eicosanoids in inflammation: biosynthesis, pharmacology, and therapeutic Frontiers. Curr Top Med Chem 7(3):311–340. https://doi.org/10.2174/156802607779941314

Kouba BR, Camargo A, Rodrigues ALS (2023) Neuroinflammation in Alzheimer's disease: potential beneficial effects of vitamin D. Metab Brain Dis 38(3):819–829

Kowalski K, Mulak A (2019) Brain-gut-microbiota Axis in Alzheimer's disease. J Neurogastroenterol Motil 25(1):48–60. https://doi.org/10.5056/jnm18087

Lee Y-J, Han SB, Nam S-Y, Oh K-W, Hong JT (2010) Inflammation and Alzheimer's disease. Arch Pharm Res 33(10):1539–1556. https://doi.org/10.1007/s12272-010-1006-7

Li RW, Leach DN, Myers SP, Lin GD, Leach GJ, Waterman PG (2004) A new anti-inflammatory glucoside from *Ficus racemosa* L. Planta Med 70:421–426

Li Z, Zheng Y, Liu K, Liang Y, Lu J, Li Q, Zhao B, Liu X, Li X (2023) Lignans as multi-targeted natural products in neurodegenerative diseases and depression: recent perspectives. Phytother Res 37(12): 5599–5621

Lin GD, Li RW (2018) Natural products targeting inflammation processes and multiple mediators. In: Mandal SC, Mandal VK and Konishi T (eds) Natural products & drug discovery: an integrated approach. Elsevier, pp 277–308

Lowenstein CJ (1992) Nitric oxide, a novel biologic messenger. Cell 70:705–707

Mayassi T, Jabri B (2018) Human intraepithelial lymphocytes. Mucosal Immunol 11(5):1281–1289. https://doi.org/10.1038/s41385-018-0016-5

McGrattan AM, McGuinness B, McKinley MC, Kee F, Passmore P, Woodside JV, McEvoy CT (2019) Diet and inflammation in cognitive ageing and Alzheimer's disease. Curr Nutr Rep 8(2):53–65. https://doi.org/10.1007/s13668-019-0271-4

Medzhitov R (2008) Origin and physiological roles of inflammation. Nature 454:428–435

Medzhitov R (2010) Inflammation 2010: new adventures of an old flame. Cell 140(6):771–776. https://doi.org/10.1016/j.cell.2010.03.006

Menéndez SG, Manucha W (2024) Vitamin D as a modulator of neuroinflammation: implications for brain health. Curr Pharm Des 30(5):323–332

Milanski M, Degasperi G, Coope A, Morari J, Denis R, Cintra DE et al (2009) Saturated fatty acids produce an inflammatory response predominantly through the activation of TLR4 signaling in hypothalamus: implications for the pathogenesis of obesity. J Neurosci 29(2):359–370

Moncada S, Higgs A (1993) The L-arginine-nitric oxide pathway. N Engl J Med 329:2002–2012

Moradi F, Mokhtari T (2025) Role of NLRP3 Inflammasome in chronic pain and Alzheimer's disease – a review. J Biochem Mol Toxicol 39(2):e70071. https://doi.org/10.1002/jbt.70071

Moratilla-Rivera I, Sanchez M, Valdes-Gonzalez JA, Gomez-Serranillos MP (2023) Natural products as modulators of Nrf2 signaling pathway in neuroprotection. Int J Mol Sci 24(4). https://doi.org/10.3390/ijms24043748

Moro-García MA, Alonso-Arias R, Baltadjieva M, Fernández Benítez C, Fernández Barrial MA, Díaz Ruisánchez E et al (2013) Oral supplementation with lactobacillus delbrueckii subsp. bulgaricus 8481 enhances systemic immunity in elderly subjects. Age 35(4):1311–1326. https://doi.org/10.1007/s11357-012-9434-6

Moutinho S (2025) Women twice as likely to develop Alzheimer's disease as men – but scientists do not know why. Nat Med 31(3):704–707. https://doi.org/10.1038/s41591-025-03564-3

Muegge BD, Kuczynski J, Knights D, Clemente JC, González A, Fontana L et al (2011) Diet drives convergence in gut microbiome functions across mammalian phylogeny and within humans. Science 332(6032):970–974. https://doi.org/10.1126/science.1198719

Nathan C (2006) Neutrophils and immunity: challenges and opportunities. Nat Rev Immunol 6(3):173–182. https://doi.org/10.1038/nri1785

Nguyen HD, Jo WH et al (2025) The anti-inflammatory effects of vitamin B6 on neuroinflammation and neuronal damage caused by 1, 2-diacetylbenzene in BV2 microglial and sH-SY5Y cells. Immunopharmacol Immunotoxicol 47(3):273–286

Nile SH, Park SW (2014) Edible berries: bioactive components and their effect on human health. Nutrition 30(2):134–144. https://doi.org/10.1016/j.nut.2013.04.007

Nucci D, Sommariva A, Degoni LM, Gallo G, Mancarella M, Natarelli F et al (2024) Association between Mediterranean diet and dementia and Alzheimer disease: a systematic review with meta-analysis. Aging Clin Exp Res 36(1):77. https://doi.org/10.1007/s40520-024-02718-6

Opal SM, DePalo VA (2000) Anti-inflammatory cytokines. Chest 117:1162–1172

Ortega N, Schütte L, de Crom TOE, Voortman T, Okereke OI, Vinceti M et al (2024) Dietary patterns, inflammatory biomarkers and cognition in older adults: an analysis of three population-based cohorts. Clin Nutr (Edinburgh, Scotland) 43(10):2336–2343. https://doi.org/10.1016/j.clnu.2024.08.027

Ortega N, Schütte L, de Crom TOE, Voortman T, Okereke OI, Vinceti M et al (2025) Corrigendum to "dietary patterns, inflammatory biomarkers and cognition in older adults: an analysis of three population-based cohorts". Clin Nutr (Edinburgh, Scotland) 45:134–135. https://doi.org/10.1016/j.clnu.2025.01.013

Ossola B, Kääriäinen TM, Männistö PT (2009) The multiple faces of quercetin in neuroprotection. Expert Opin Drug Saf 8(4):397–409. https://doi.org/10.1517/14740330903026944

Pahwa R, Goyal A, Jialal I (2023) Chronic inflammation. In *StatPearls [Internet]*. Treasure Island (FL): StatPearls Publishing

Piccialli I, Tedeschi V, Caputo L, D'Errico S, Ciccone R, De Feo V et al (2022) Exploring the therapeutic potential of phytochemicals in Alzheimer's disease: focus on polyphenols and monoterpenes. Front Pharmacol 13:876614. https://doi.org/10.3389/fphar.2022.876614

Pistollato F, Iglesias R, Ruiz R, Aparicio S, Crespo J, Lopez L et al (2018) Nutritional patterns associated with the maintenance of neurocognitive functions and the risk of dementia and Alzheimer's disease: a focus on human studies. Pharmacol Res 131:32–43

Rainsford KD (2007) Anti-inflammatory drugs in the 21st century. Subcell Biochem 42:3–27

Reber SO, Siebler PH, Donner NC, Morton JT, Smith DG, Kopelman JM, Lowe KR, Wheeler KJ, Fox JH, Hassell JE Jr, Greenwood BN, Jansch C, Lechner A, Schmidt D, Uschold-Schmidt N, Fuchsl AM, Langgartner D, Walker FR, Hale MW, Lopez Perez G, Van Treuren W, Gonzalez A, Halweg-Edwards AL, Fleshner M, Raison CL, Rook GA, Peddada SD, Knight R, Lowry CA (2016) Immunization with a heat-killed preparation of the environmental bacterium *mycobacterium vaccae* promotes stress resilience in mice. Proc Natl Acad Sci USA 113:E3130–E3139

Rendeiro C, Rhodes JS, Spencer JPE (2015) The mechanisms of action of flavonoids in the brain: direct versus indirect effects. Neurochem Int 89:126–139. https://doi.org/10.1016/j.neuint.2015.08.002

Rhee SH, Pothoulakis C, Mayer EA (2009) Principles and clinical implications of the brain–gut–enteric microbiota axis. Nat Rev Gastroenterol Hepatol 6(5):306–314. https://doi.org/10.1038/nrgastro.2009.35

Rodríguez-García C, Sánchez-Quesada C, Toledo E, Delgado-Rodríguez M, Gaforio JJ (2019) Naturally lignan-rich foods: a dietary tool for health promotion? Molecules (Basel, Switzerland) 24(5):917

Romo-Araiza A, Ibarra A (2020) Prebiotics and probiotics as potential therapy for cognitive impairment. Med Hypotheses 134:109410

Round JL, Lee SM, Li J, Tran G, Jabri B, Chatila TA, Mazmanian SK (2011) The toll-like receptor 2 pathway establishes colonization by a commensal of the human microbiota. Science 332(6032):974–977. https://doi.org/10.1126/science.1206095

Saha S, Buttari B, Panieri E, Profumo E, Saso L (2020) An overview of Nrf2 signaling pathway and its role in inflammation. Molecules 25(22). https://doi.org/10.3390/molecules25225474

Seeram NP, Ichikawa H, Shishodia S, Aggarwal BB (2005) Preventive and therapeutic effects of plant polyphenols through suppression of nuclear factor-kappa B. In: Hiramatsu TYM, Packer L (eds) Molecular interventions in lifestyle-related diseases. CRC Press LLC, Boca Raton, pp 243–275

Sehati F, Ahmadi I, Farivar N, Ranjbaran M, Sadat-Shirazi MS, Nabavizadeh F, Mahla Shavakandi S, Ashabi G (2021) Tannic acid protects aged brain against cerebral hypoperfusion via modulation of Nrf2 and inflammatory pathways. Neurosci Lett 765:136263

Serrano-Pozo A, Frosch MP, Masliah E, Hyman BT (2011) Neuropathological alterations in Alzheimer disease. Cold Spring Harb Perspect Med 1(1):a006189–a006189. https://doi.org/10.1101/cshperspect.a006189

Shaikh A, Roy H (2023) Folate deprivation induced neuroinflammation impairs cognition. Neurosci Lett 807:137264

Sharma B, Satija G, Madan A, Garg M, Alam MM, Shaquiquzzaman M et al (2023) Role of NLRP3 Inflammasome and its inhibitors as emerging therapeutic drug candidate for Alzheimer's disease: a review of mechanism of activation, regulation, and inhibition. Inflammation 46(1):56–87. https://doi.org/10.1007/s10753-022-01730-0

Shi X, Zheng Z, Li J, Xiao Z, Qi W, Zhang A et al (2015) Curcumin inhibits aβ-induced microglial inflammatory responses in vitro: involvement of ERK1/2 and p38 signaling pathways. Neurosci Lett 594:105–110. https://doi.org/10.1016/j.neulet.2015.03.045

Shikh EV, Nikolaeva NB, Molchanova NB, Elizarova EV (2023) Correction of gut dysbiosis as a promising direction in the prevention of neuroinflammation and cognitive impairment. Vopr Pitan 92(6):107–119

Silva G, Freiria C, Brito T, Borim FAS, Corona LP (2022) Association of cognitive status and consumption of unprocessed and ultra-processed foods in brazilian older adults. Innov Aging 6(Supplement_1):822–822. https://doi.org/10.1093/geroni/igac059.2957

Sicilia T, Niemeyer HB, Honig DM, Metzler M (2003) Identification and stereochemical characterization of lignans in flaxseed and pumpkin seeds. J Agric Food Chem 51:1188

Sofroniew MV, Vinters HV (2010) Astrocytes: biology and pathology. Acta Neuropathol 119(1):7–35. https://doi.org/10.1007/s00401-009-0619-8

Smith WL (1992) Prostanoid biosynthesis and mechanisms of action, Am J Physiol 263:181e191

Suh Y, Checler F (2002) Amyloid precursor protein, presenilins, and alpha-Synuclein: molecular pathogenesis and pharmacological applications in Alzheimer's disease. Pharmacol Rev 54:469–525

Sundaram B, Tweedell RE, Prasanth Kumar S, Kanneganti TD (2024) The NLR family of innate immune and cell death sensors. Immunity 57(4):674–699. https://doi.org/10.1016/j.immuni.2024.03.012

Tarkar A, Couch S, Summer S, Krikorian R (2024) Association between ultra-processed food intake and cognitive performance in adults with subjective cognitive decline. J Acad Nutr Diet 124(10):A33–A33. https://doi.org/10.1016/j.jand.2024.07.030

Terry AV, Buccafusco JJ (2003) The cholinergic hypothesis of age and Alzheimer's disease-related cognitive deficits: recent challenges and their implications for novel drug development. J Pharmacol Exp Ther 306(3):821–827. https://doi.org/10.1124/jpet.102.041616

Ullah H, Minno AD, Santarcangelo C, Tantipongpiradet A, Dacrema M, Matteo RD et al (2022) In vitro bioaccessibility and anti-inflammatory activity of a chemically characterized Allium cepa L. extract rich in quercetin derivatives optimized by the Design of Experiments. Molecules 27(24). https://doi.org/10.3390/molecules27249065

Urashima T, Ajisaka K, Ujihara T, Nakazaki E (2025) Recent advances in the science of human milk oligosaccharides. BBA Adv 7:100136. https://doi.org/10.1016/j.bbadva.2024.100136

US Food and Drug Administration (2023) FDA converts Novel Alzheimer's disease treatment to traditional approval: converted Leqembi (lecanemab-irmb), indicated to treat adult patients with Alzheimer's Disease [Press release]

Weinberg JB (2000) Nitric oxide synthase 2 and cyclooxygenase 2 interactions in inflammation. Immunol Res 22:319–341

Weinstein G, Vered S, Ivancovsky-Wajcman D, Zelber-Sagi S, Ravona-Springer R, Heymann A, Beeri MS (2021) Consumption of ultra-processed food and cognitive decline among older adults with type-2 diabetes. Alzheimers Dement 17(S10):n/a. https://doi.org/10.1002/alz.055110

Wo J, Zhang F, Li Z, Sun C, Zhang W, Sun G (2020) The role of Gamma-Delta T cells in diseases of the central nervous system. Front Immunol 11:580304. https://doi.org/10.3389/fimmu.2020.580304

Wu Y, Zhong L, Yu Z, Qi J (2019) Anti-neuroinflammatory effects of tannic acid against lipopolysaccharide-induced BV2 microglial cells via inhibition of NF-κB activation. Drug Dev Res 80(2):262–268

Xue C, Kowshik SS, Lteif D, Puducheri S, Jasodanand VH, Zhou OT et al (2024) AI-based differential diagnosis of dementia etiologies on multimodal data. Nat Med 30(10):2977–2989. https://doi.org/10.1038/s41591-024-03118-z

Zhang Y, Peng Y, Deng W, Xiang Q, Zhang W, Liu M (2024) Association between dietary inflammatory index and cognitive impairment among American elderly: a cross-sectional study. Front Aging Neurosci 16:1371873–1371873. https://doi.org/10.3389/fnagi.2024.1371873

Zhang R, Ding N, Feng X, Liao W (2025) The gut microbiome, immune modulation, and cognitive decline: insights on the gut-brain axis. Front Immunol 16:1529958. https://doi.org/10.3389/fimmu.2025.1529958

Zhao Z, Liu Y, Ruan S, Hu Y (2023) Current anti-amyloid-β therapy for Alzheimer's disease treatment: from clinical research to Nanomedicine. Int J Nanomedicine 18:7825–7845. https://doi.org/10.2147/IJN.S444115

Role of Mediterranean Diet in Alzheimer's Disease: Pre-clinical and Clinical Evidence

4

Rani Mansuri, Monika, Arpit Raj, Sabina Yasmin, and Md Yousuf Ansari

Abstract

The Mediterranean Diet (MD) has a profound effect on Alzheimer's disease, with cultural adaptations aiding in sustained adherence over time. It centres around olive oil, legumes, whole grains, fruits, vegetables, moderate consumption of fish and dairy, and limited intake of meat. Nutritional intervention studies reveal that the MD can lower serum lipid levels, oxidative stress, inflammation, and platelet aggregation, while also influencing hormones and growth factors associated with cancer progression. Conversely, the Western Diet (WD), which is characterized by high levels of fats, refined carbohydrates, and red meat, along with low consumption of fish, fruits, and plant-based foods, adversely affects the immune system and contributes to dysbiosis—an imbalance in gut structure and function leading to inflammation. Clinical research confirms the pivotal role of diet in shaping gut microbiota and overall health. The WD is linked to negative health outcomes, such as obesity and various non-communicable diseases (NCDs), including cancer, cardiovascular diseases, and type 2 diabetes. Chronic oxidative stress, amplified by elevated reactive oxygen species (ROS), is closely associated with degenerative conditions like Alzheimer's disease, Parkinson's disease, and other inflammatory and cardiovascular disorders. This dynamic is bidirectional; gut microbiota not only processes nutrients but is also significantly influenced by dietary choices. Observational studies further underscore the MD's

R. Mansuri · Monika · A. Raj
School of Pharmaceutical Sciences, Apeejay Stya University, Gurugram, Haryana, India

S. Yasmin
Department of Pharmaceutical Chemistry, College of Pharmacy, King Khalid University, Abha, Saudi Arabia
e-mail: sahussain@kku.edu.sa

M. Y. Ansari (✉)
Ibne Seena Pharmacy College, Shahabad, Hardoi, Uttar Pradesh, India

J. Malik et al. (eds.), *Diet and Alzheimer's Disease: Let Food be Our Medicine*, Nutritional Neurosciences, https://doi.org/10.1007/978-981-95-2736-6_4

protective benefits against global health challenges, including cardiovascular disease (CVD), recurrent cardiac events, CVD-related mortality, metabolic syndrome, abdominal obesity, and gestational weight gain.

Keywords

Mediterranean diet · Alzheimer disease · Clinical evidence · Anti-atherogenic · Western diet

Abbreviations

MD	Mediterranean diet
WD	Western diet
NCDs	Non-communicable diseases
ROS	Reactive oxygen species
CVD	Cardiovascular disease
PBC	Perceived behavioural control
SWOT	Strengths, Weaknesses, Opportunities, and Threats
NAFLD	Non-alcoholic fatty liver disease
UPFDs	Ultra-processed food and drinks
AD	Alzheimer's disease
APP	Amyloid Precursor Protein
OLGs	Oligodendroglia
DASH	Dietary Approaches to Stop Hypertension
RCTs	Randomized controlled trials
MUFAs	Monounsaturated fatty acids

4.1 Introduction

Recent dietary intervention studies have explored the barriers and facilitators to adopting a Mediterranean diet (MD) pattern; however, these interventions lacked cultural adaptation (Guasch-Ferré and Willett 2021). Adapting the MD culturally can improve its long-term adherence at a population level. This method retains the diet's essential components, including high intake of olive oil, legumes, cereals, fruits, and vegetables, moderate consumption of fish, dairy products like yogurt and cheese, and wine, while limiting meat intake (Keys et al. 1986). Incorporating locally available foods allows the MD to be modified into a culturally acceptable dietary model, addressing nutritional inadequacies in non-Mediterranean populations with suitable substitutions. The nutritional impacts of these modifications, however, require evaluation. In Australia, few MD models have been designed based on traditional principles within a multi-ethnic context (Ros 2010). Nutrient profiles from previous MD interventions were utilized to identify crucial food components and create a 2-week meal plan. The barriers to adopting the MD in Australia were examined using a Strengths, Weaknesses, Opportunities, and Threats (SWOT) framework, alongside strategies aimed at overcoming these challenges. This model has been proposed for clinical trials in patients with Non-alcoholic Fatty Liver Disease (NAFLD) and secondary coronary heart disease (CHD) prevention, with initial data from currently available intervention studies in Australia emphasizing its potential to tackle environmental and health concerns. However, significant variability among studies that claim to test the MD pattern makes it hard to assess adherence and efficacy, and understanding whether this variability affects health outcomes is vital (Serra-Majem et al. 2019).

Key strategies for implementing MD interventions in non-Mediterranean settings, such as Australia, include dietician-led efforts providing written materials, food hampers, consistent interaction with healthcare providers, and cooking classes (Hidalgo-Mora et al. 2020). However, these resource-intensive approaches may not be feasible for widespread application. To achieve population-wide adoption, a multi-sectorial framework engaging the food industry, retailers, policymakers, regulatory authorities, and health professionals is essential. This framework should integrate a socio-ecological model to ensure the MD is cost-effective, accessible, and culturally adapted (Amezaga et al. 2008; Hidalgo-Mora et al. 2020).

In Australia, an online survey grounded in the Theory of Planned Behaviour examined barriers and enablers to adhering to the Mediterranean Diet (MD). These factors were grouped into attitudes (suitability, taste, restrictions, food waste), social norms (food culture), and perceived behavioural control (PBC) (motivation, affordability, time, access, knowledge, cooking skills, natural conditions, etc.) (Amezaga et al. 2008). Among these, PBC was identified as the most impactful determinant of adherence intention, highlighting perceived health benefits and improved dietary quality as key advantages, while noting adherence challenges. Future strategies should focus on enhancing self-efficacy to minimize perceived obstacles and encourage adherence (Amezaga et al. 2008). Although MD transferability research primarily emphasizes the UK and Australia, adaptations have been successfully implemented elsewhere. For instance, del Campo et al. customized the MD for low-income Hispanic American groups to mitigate cardiovascular disease (CVD) risk, resulting in improved dietary practices and self-reported engagement (Maggini et al. 2009).

4.1.1 Comparative Studies: Mediterranean Versus Western Diets

The MD is celebrated for its remarkable health benefits. It features high-fibre complex carbohydrates derived from cereals, legumes, vegetables, and fruits, alongside polyunsaturated fatty acids known for their anti-atherogenic and anti-inflammatory effects, found in olive oil and nuts. Furthermore, it includes bioactive compounds like flavonoids, phytosterols, terpenes, and polyphenols, which exhibit potent antioxidative properties (García-Montero et al. 2021). This diet also provides a balanced array of micronutrients essential for preventing malnutrition and immunodeficiency's (Stolfi et al. 2022). Nutrient-rich foods in the MD help reduce cardiovascular risk factors by mitigating inflammation caused by nutrient-poor, high-calorie diets. Adherence to the MD promotes gut health, as shown by the PREDIMED study, which found that better adherence correlated with higher concentrations of beneficial bacteria and SCFAs (short-chain fatty acids) (Wu et al. 2019). The MD counteracts the pathological microbiota status induced by Western diets, leading to decreased dysbiosis and LPS levels, improved gut barrier function, and better permeability whereas also boosting immunity (Pecora et al. 2020; Yasmin et al. 2025). Nutritional intervention trials have documented reductions in serum lipid levels, oxidative stress, inflammation, and platelet aggregation, along with modulation of hormones and growth factors involved in cancer pathogenesis and microbial metabolism (Aguilera 2019). Overall, the MD promotes proper host metabolism function and is significant for managing various pathologies (Barrea et al. 2021).

In contrast, the Western Diet (WD) is associated with negative health outcomes, driving the obesity epidemic and contributing to various non-communicable diseases (NCDs) such as cancer, cardiovascular diseases, osteoporosis, autoimmune disorders, and type 2 diabetes (Esposito et al. 2017; Garcia-Mantrana et al. 2018; Tosti et al. 2018). The WD is characterized by excessive consumption of unhealthy fats, refined grains, sugar, salt, and alcohol, along with minimal intake of fruits and vegetables (Serra-Majem et al. 2019). This dietary pattern adversely impacts gut microbiota and the immune system, promoting inflammation and impaired gut health (Gombart et al. 2020). Ultra-processed foods and drinks (UPFDs), a defining feature of the WD, are highly profitable, hyper-palatable, and convenient. These products frequently contain non-traditional ingredients, such as hydrogenated oils and additives, and are linked to increased risks of morbidity and mortality (Krznaric et al. 2019). The WD is rich in harmful components, including added sugars, carbohydrates, and saturated fats, but deficient in essential micronutrients like vitamins and trace elements (Christ et al. 2019). Children are particularly susceptible to the detrimental effects of UPFDs, experiencing gut dysbiosis and immune system alterations. Hence, education, awareness, and public health initiatives are crucial to mitigate the harmful consequences of the WD (Shively et al. 2019).

4.1.2 Core Features of the Mediterranean Diet

The MD is a celebrated model of healthy eating, renowned for its health benefits and rich culinary tradition (Statovci et al. 2017). This diet emphasizes a high consumption of fruits, vegetables, whole grains, legumes, nuts, and seeds, making plant-based foods a central component as shown in Fig. 4.1. Olive oil, particularly extra-virgin, is the primary source of fat, contributing beneficial monounsaturated fats and polyphenols (Statovci et al. 2017). Whole grains, including wheat, barley, and oats, provide essential nutrients and fibre, supporting both cardiovascular and

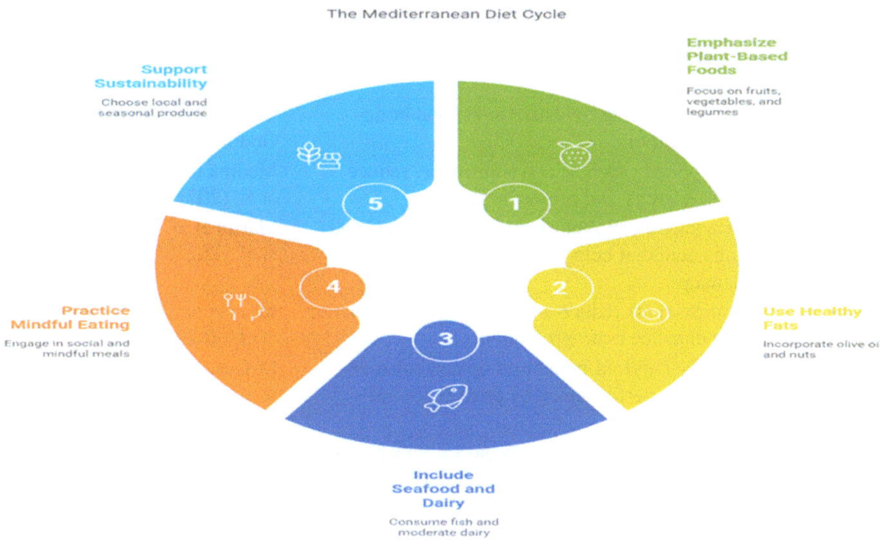

Fig. 4.1 Core feature of Mediterranean diet, it highlights key components such as high plant-based food, limited red meat and sweets, healthy fats, and sustainable lifestyle factors

metabolic health. Moderate intake of fish and seafood delivers omega-3 fatty acids, which have anti-inflammatory effects and promote heart health. Dairy products, particularly yogurt and cheese, are also consumed in moderation, offering calcium and probiotics. Eggs are also included, providing high-quality proteins (Kopp 2019). The MD allows for limited intake of red meat and sweets, promoting lean proteins like poultry and plant-based sources. Red wine, consumed in moderation and usually with meals, adds a unique element, providing antioxidants like resveratrol. Beyond specific food groups, the MD encompasses lifestyle factors such as regular physical activity, social meals, and mindful eating, which contribute to its overall health benefits. Traditional food preparation methods, focusing on freshness and minimal processing, enhance the nutritional quality and sustainability of the diet. Embracing seasonal, locally sourced produce, the MD supports environmental sustainability and biodiversity. This holistic approach not only supports physical health but also fosters social and cultural connections, embodying a way of life that promotes well-being and longevity (Kopp 2019). Combining these elements, the MD is a comprehensive and adaptable dietary pattern that can be applied globally for improved health outcomes and environmental benefits (as shown in Fig. 4.1 and summarized in Table 4.1).

Table 4.1 Impact of MD on cognitive function

MD components	Impact on cognitive function in Alzheimer's disease	References
Legumes	Aid in reducing oxidative stress and brain inflammation	García-Montero et al. (2021), Godos et al. (2024)
Red wine	Compounds such as resveratrol may reduce neurotoxicity and protect brain cells	Medina (2009), Valero-Vello et al. (2021)
Vegetables	Provide essential nutrients like vitamin C and folic acid that help protect neurons from damage	Martín et al. (2011), Zhou et al. (2021)
Olive oil	Antioxidants reduce amyloid-beta accumulation and enhance neuroprotection	Khan et al. (2019), Schepici et al. (2020)
Fruits	Reduce amyloid-beta build up and enhance cognitive performance by providing vital nutrients	Di Rosa et al. (2020), Salis et al. (2020)
Fish	Rich in vitamin D and omega-3 s, they protect neurons from oxidative damage and improve brain function	Kozlowska and Szostak-Wegierek (2014), Mattioli et al. (2020)

4.2 Overview of Alzheimer's Disease (AD)

AD is the neurodegenerative disorders responsible for over 80% of dementia cases among the elderly age groups that lies above ≥60 years. Worldwide (Araya Quintanilla et al. 2020). It is often diagnosed either by the cognitive score that is marked by their cognitive abilities, functionality, behaviour and learning capacity, while some relate to nutritional value like the MeDi score. At the microscopic level, AD is defined by three key neuropathological features: the deposition of extracellular Aβ plaques, the formation of intracellular NFTs, and progressive degeneration of neurons. Among them, the two former causes are the biomarkers of AD and the latter is considered as a co-morbid disease (Avallone et al. 2019).

This disease is initially identified by these two factors, named NFTs (Neurofibrillary Tangles) and abnormal deposition of plaques. This was diagnosed by the great German Psychiatrist Dr. Alois Alzheimer in 1906 while performing an autopsy of a patient affected by dementia (Bivona et al. 2019). From the diagnostic results these are considered as hallmarks of Alzheimer's disease (AD), but they are not exclusive be considered the main cause of AD, because some other studies show that ageing is also a factor of AD and other neurodegenerative disorders (Sousa and Guimarães 2015). At the diagnostic level, it was identified in AD, plaques and NFTs are concentrated in specific brain regions. This deposition of Aβ plaques is expected to promote these neurodegenerative diseases, and this hypothesis is named as "amyloid hypothesis" (Weih et al. 2007).

The amyloid-based hypothesis shows that the amyloid precursor protein is typically cleaved by α-secretase (Pasinetti and Eberstein 2008). However, in ADs, abnormal processing by γ-secretase and β-secretase disrupts the balance between the production and clearance of amyloid-beta (Aβ) peptides (Scheltens 2009). This imbalance causes Aβ peptides to aggregate into soluble oligomers, which eventually form fibrils with an insoluble β-sheet structure and are deposited as diffuse senile plaques (Ramesh et al. 2010). Under normal conditions, Aβ peptides are small and soluble, produced by the cleavage of APP by α-, β-, and γ-secretases. The disruption in Aβ production and clearance leads to the formation of various toxic oligomeric species, including protofibrils, fibrils, and lastly leads to plaque formation, depending on the degree of oligomerization (as shown in Fig. 4.1) (Gillette-Guyonnet et al. 2013). But the exact cause of Aβ formation is still under medical surveillance, as according to medical reports, there are many other factors that also lead to aggregation of plaques like peptide's sequence, concentration, and stability conditions (Vassallo and Scerri 2013).

According to a recent research study, a new variant of AD gets discovered whose primary cause is marked by cleavages of γ-secretase, especially APP (Amyloid Precursor Protein) at 42 no. amino acid peptide named Aβ42. Clinically, it was investigated that this peptide is primarily insoluble and mostly prone to aggregate at specified location in the brain, which were formerly characterized as beta-amyloid plaques. In contrast, α-secretase plays a protective role by cleaving APP at a site that prevents the formation of Aβ peptides.

Clinically, it was stated that Aβ-42 oligomers are produced through positive collaborative activities of neurons and their associated astrocytes (Armstrong 2019). These combined responses result in oxidative damage, enhancing tau hyperphosphorylation, and causing toxicity to synapses and mitochondria (Arriagada et al. 1992; Doruk et al. 2010). The role of Aβ42 senile plaques, which typically appear during the later stages of AD, is also significant. Usually, these plaques attract microglia, whose activation triggers the release of pro-inflammatory cytokines such as IL-1β, TNF-α, and IFN-γ. These cytokines further stimulate nearby astrocytes and neurons to produce additional Aβ-42 oligomers, perpetuating their production and spread (Bell and Ball 1990).

Oligodendroglia (OLGs), which are part of the neuron–astrocyte complex, are also affected by Aβ oligomers, leading to their destruction (Heyman et al. 1983). Aβ oligomers play a key role in neuronal and vascular degeneration. This degeneration is often marked by oxidative stress, to which OLGs are mainly vulnerable due to their small Glutathione (GSH) levels and elevated iron content, reducing their ability to neutralize oxygen radicals (as shown in Table 4.2) (Lahiri and Maloney 2010). Furthermore, Aβ42 oligomers have an increased capacity to damage cholesterol-rich membranes, such as those in OLGs and Myelin, which is also a contributing factor to progression of AD pathology (Xu et al. 2014).

4.2.1 Lifestyle and Nutritional Risk Factors

With the evolving age groups and persistent a pedigree forms of family pedigree are considered as well-established risk factors for any dreadful condition like AD. Pharmacologically, these former factors are irreversible and have historically fostered a sense of hopelessness regarding AD prevention. However, recent research that involves a clinical study and with obtained observational data indicates that the

Table 4.2 Associated risk factors that leads to AD progression

Factor	Description
Genetic factor	This is often marked by mutations in genes like PSEN I& II and APP. Associated with early-onset AD. The presence of the APOE ε4 gene variant significantly increases in onset of AD
Age	Age ≥ 65 are more prone to AD
Smoking	Elevation in inflammatory action and oxidative stress leads to alteration in neuronal function
Cardiovascular health	Severe vascular conditions like elevation in cholesterol level in blood, uneven fat deposition that leads to elevation in BP, which creates pressure on normal cardiac function that leads to alteration in neuronal function
Family history	Having a past history of AD is more likely to manifest AD in upcoming generations
Head trauma	Severe or repeated head injuries can cause lasting damage, which elevates the risk of AD
Physical inactivity	Irregularity in physical exercise is also considered as a key factor of neuronal dysfunction because it creates an ischemic condition
Diet	Diets lower in antioxidants, vitamins, and omega-3 fatty acids are also linked factors for neuronal dysfunctions.
Mental inactivity	Less involvement in IQ-based activities like mentally stimulating activities, reading, or solving aptitude problems, is also considered a risk of AD
Low education levels	Lowering in cognitive response is a key indication of AD
Exposure to toxins	Excessive exposure to heavy metals, like Al (aluminium), Pb (lead), etc., or host allergens present in the surrounding also a leading cause of neuronal damage
Chronic infections	Clinically, it was shown that viral infections like herpes simplex virus (HSV-1) are also a leading cause of neuronal dysfunction
Amyloid plaques	Formation of the altered beta-amyloid protein in the brain creates plaques that disrupt positive interactions between neurons
Neurofibrillary tangles	Abnormality in the tau protein leads to the formation of tangles within neurons, impairing their function and survival
Inflammation	Excessive inflammation in brain causes a neuronal dysfunction
Sex	In comparison to male individuals, females are more prone to AD because of excessive hormonal changes during their menopause phase
Depression	Long term untreated depression is also a leading cause of AD
Oxidative stress	Excessive free radicals also damages the neuronal cells
Chronic infections	Clinically, it was shown that viral infections like HSV-1 (herpes simplex virus) are also a leading cause of neuronal dysfunction

environment and day-to-day lifecycle factors also influence the risk of AD. So, adopting a healthy lifestyle could potentially lower the risk of AD, along with overall health, by reducing the likelihood of other age-related diseases (Armstrong 2013).

With many clinical evidence and evolving time, indicates that traditional cardiometabolic risk factors, which include hypertension, diabetes, etc., are linked to the progression of cognitive decline and AD. On the other hand, practices such as calorie restriction, consuming antioxidant-rich foods, and following specific dietary patterns possess a positive result in slowing down the progression of metabolic and neurodegenerative disorders (Arsene et al. 2011).

Table 4.3 Nutritional factor and impact of AD

Nutritional factor	Description	Impact on AD
Deficiency of Vit. D	Present in fatty fish, fortified foods, and sunlight	Inflammation and risk of cognitive impairment.
Deficiency of complex Vitamin B	Found in B1, B12, B9, B6	Homocysteine levels, alteration in glucose metabolism
Deficiency of Vit. E	Present in nuts, vegetable oils, seeds	Antioxidants protection Oxidative stress
Deficiency of omega-3 FA	Found in walnuts, and flaxseeds	Neuronal membrane integrity, inflammation
Deficiency of high saturated and trans fats	Found in fried items, fatty meats	Oxidative stress, enhances amyloid plaque formation
Deficiency of antioxidants	Found in Vitamin C, E, vegetables, nuts	Oxidative stress
Imbalance in iron, copper, and zinc	Found in shellfish, red meat, fortified food	Oxidative stress, amyloid plaque aggregation
Deficiency of magnesium	Found in nuts, leafy vegetables	Impairs synaptic plasticity and memory function
Excessive sugar carbs	Found in snacks, sodas and refined grains	Insulin resistances Amyloid Plaques
Deficiency of protein	Found in legumes, dairy products, lean meats	Impair neurotransmitter synthesis
Imbalance in gut microbiomes	Influenced by low-fibres diets and foods	Inflammation and disrupts brain health

By adopting a MD or the DASH's diet that comprises fruits, lean proteins, vegetables, whole grains, and healthy fats, has been linked to a lower risk of developing AD (as shown in Table 4.3) (Gentleman et al. 1993).

4.3 Mechanisms Linking MD to Brain Health

4.3.1 Oxidative Stress and Neuroinflammation Modulation

Mitochondria, often called the cell's powerhouse, depend on oxygen to produce energy, with 1–5% of this oxygen converting into reactive oxygen species (ROS)—harmful by-products that make mitochondria a primary source of intracellular ROS. These reactive species can damage DNA, lipids, proteins, and carbohydrates. Free radicals are heightened by lifestyle factors such as stress, unhealthy diets, smoking, alcohol intake, radiation, infections, drugs, and pollution (Findeis 2007). When excess free radicals overwhelm enzymes and antioxidants, oxidative stress occurs. Glutamate, a neurotransmitter, contributes to oxidative stress in brain cells, likely via ionotropic receptor activation (Stein et al. 2014). Chronic oxidative stress, harder to neutralize due to elevated ROS levels, is linked to degenerative diseases like Alzheimer's disease (AD), Parkinson's disease, cancers, atherosclerosis, diabetes mellitus, inflammatory diseases, and cardiovascular diseases (Larsson 1963). Amyloid beta-protein (Aβ) induces free radical generation in neuronal cells,

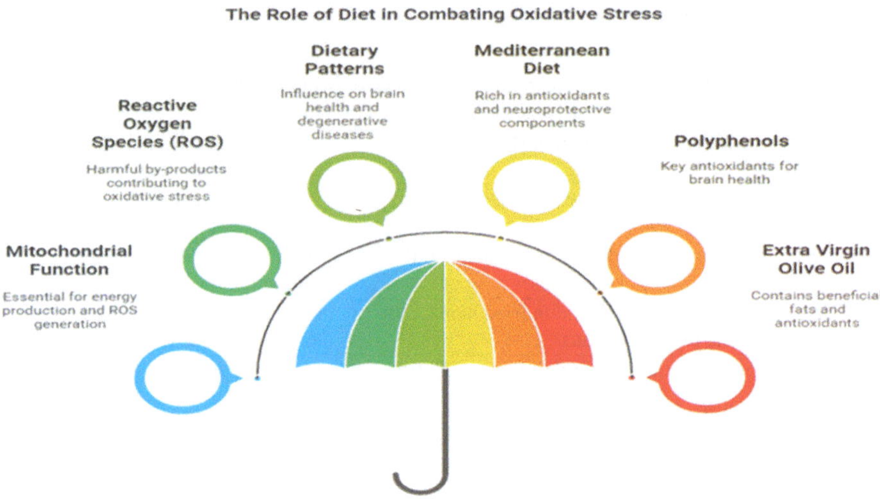

Fig. 4.2 Role of diet in combating oxidative stress: diet plays a crucial role in combating oxidative stress by providing antioxidants that neutralize harmful free radicals in the body

causing oxidative stress and cell death, with beta-amyloid plaques as a hallmark of AD (Lassale et al. 2019). Dietary patterns influence oxidative stress, brain inflammation, and are linked to depression (see Fig. 4.2) (Wojtunik-Kulesza et al. 2016). Systemic inflammation affects emotion regulation through neurotransmitters like serotonin, dopamine, and noradrenaline (Bhat et al. 2015). Studies show individuals with depression have higher oxidative stress markers and lower antioxidant markers than healthy controls (Gilgun-Sherki et al. 2001; Gomez-Pinilla and Nguyen 2012). The Mediterranean-style diet, emphasizing plant-based foods, healthy fats, and moderate animal products, is renowned for its health benefits (Aridi et al. 2017). Rooted in the traditional practices of Mediterranean regions, it encourages preparing food, sharing meals, consuming local produce, and socializing during meals (Cheignon et al. 2018). Scientific studies support the MD's role in promoting cardiovascular health, reducing inflammation, and improving cognitive function (Marx et al. 2017). Its neuroprotective effects stem from its wealth of antioxidants, minerals, and vitamins. Fruits and vegetables, central to this diet, offer vitamins C and E, along with polyphenols, effective scavengers of free radicals that cause neuronal damage. These antioxidants collectively provide robust protection against oxidative stress, a major factor in neurodegeneration. Polyphenols, crucial for brain health, mitigate oxidative stress and inflammation while modulating brain-gut communication (Picone et al. 2024). Extra virgin olive oil, a staple in the MD, is rich in polyphenols, tocopherols (vitamin E), and carotenoids, offering anti-inflammatory and cardiovascular-protective effects (Nishi et al. 2021). Olive oil contains monounsaturated fats, mainly oleic acid, associated with improved cognitive function and anti-inflammatory benefits. Its antioxidants and healthy fats contribute to neuroprotective properties, potentially reducing age-related cognitive decline. Daily intake of diverse fruits and vegetables—tomatoes, leafy greens, berries, and peppers—supports redox homeostasis(Silva and Pogačnik 2017). These foods supply vitamins C and E, flavonoids, and carotenoids, delivering a variety of antioxidants (Silva and Pogačnik 2017). Additionally, fruits, nuts, vegetables, and red wine provide antioxidant and anti-inflammatory properties, defending against free radicals that lead to neurodegenerative diseases (Hong et al. 2022).

4.3.2 Impact of Beta-Amyloid and Tau Pathology

The MeDi diet is known for its beneficial health consequences, like maintaining healthy brain that remarkably decreases neuronal dysfunction. As it comprises of Bucket of numerous healthy vegetables, fruits, legumes, whole grains, essential fish meats, and healthy fats like olive oil, as they are also marked by its limiting factors regarding saturated fats, dairy, and red meat (Westfall and Pasinetti 2019).

Recent clinical studies show that having a regular routine following an MDcan provoked to help in preventing the formation of AD biomarkers, named Amyloid and Tau proteins in the brain. As Amyloid forms plaques and Tau creates tangles. This diet may also protect against brain shrinkage that usually slows down memory, thinking problems over time and thus proved to improve the cognitive score of AD patients (Westfall and Pasinetti 2019).

It is clinically proven that the MeDi Diet has a potent ability to maintain a healthy brain. As its key ingredient mainly focuses on healthy fats, good antioxidant properties, and efficacious anti-inflammatory properties, which shield the brain from the host harmful effects (as summarized in Table 4.4).

Many other relevant research studies also suggested the possible link with Neurological problems in terms of therapeutic benefits in biological mechanisms. As clinical studies suggest that oxidative stress damages neuronal health that is key marker of AD progression (Grubić Kezele and Ćurko-Cofek 2022). The MeDi is an antioxidant, such as complex phenols, vitamins C's and E's, and carotenoids that helps in reducing the markers of oxidative stress like isoprostanes (Sakurai et al. 2021). This antioxidant effect could partly explain the diet's association with a reduced risk of Alzheimer's (as depicted in Fig. 4.3).

Additionally, vascularized health may play a role in this connection. The MeDidiet shows the lower risk of vascular problems, which are known to increase the likelihood of AD (Trichopoulou et al. 2003). While some studies show that vascular diseases do not have a mediator in the diet-Alzheimer's link, more precise methods, like imaging to detect subtle vascular changes, should be explored in future research (Praticò 2002).

4.3.3 Role of Microbiota

Gut Microbiota belongs to the diverse community of micro-organisms, including Bacteria, Fungi, Protozoans, Viruses, etc., that have niches in the gastrointestinal tract (Singh et al. 2002; Sánchez-Moreno et al. 2006). Usually, these microbes exist in a symbiotic relationship with the host's intestine. Although the brain was once

Table 4.4 Impact of MD on tau protein in AD treatment

Components of Mediterranean diet	Mechanism	Impact on amyloid protein	Impact on tau protein
Omega-3 F.A., flavonoids and polyphenols	Shows Anti-Inflammatory Effects	Neuroinflammation	Pathways causing inflammation that trigger tau kinase
Fasting-like patterns, polyphenols, caloric moderation	Autophagy promotion	Clearances of amyloid-beta plaques	Excretes excess unused tau protein
Whole grains, nuts, olive oils, fatty fish	Promotes healthy heart	Brain perfusion rate by amyloid-beta deposition	Brain health by tau protein
Vitamin C&E, polyphenols	Shows down fall in oxidative stress	Antioxidants and amyloid-beta Aggregation	Preserve neuronal health from oxidative stress.
Vitamin's, polyphenols, and omega-3 F.A.	Improved Mitochondrial dysfunction	Oxidative damage	Tau dysfunction
Omega-3 F.A., berries, tea	Modulation of an altered genetic character	Improve gene expression	Alters the tau production cycle in a controlled manner
Legumes, whole grains	Insulin sensitivity improvement	Insulin resistance	Tau hyperphosphorylation

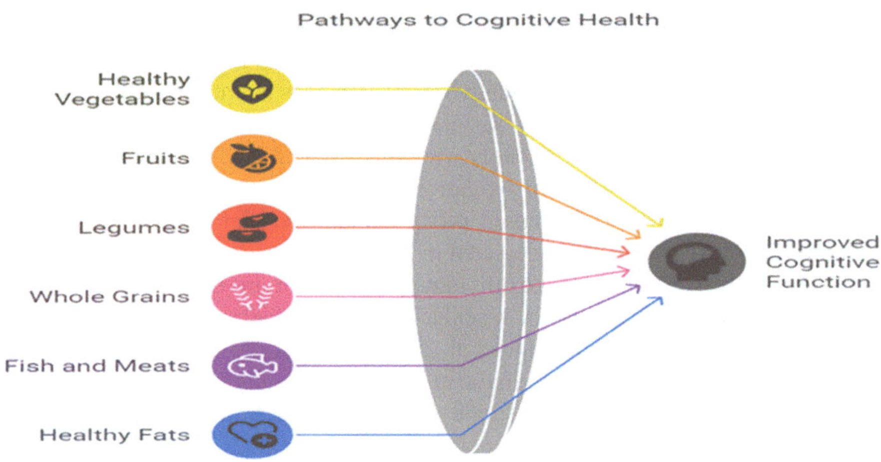

Fig. 4.3 Impact of MD in cognitive health during AD, Mediterranean Diet (MeDi) linked with cognitive health and a reduced risk of Alzheimer's Disease (AD)

thought to be completely isolated from the microbiotic surroundings (Breteler 2000; Cryan et al. 2019). Current research study establishes the two-way communication pathway between Gut microbiota and Neuronal systems popularly named gut–brain axis. This interaction involves multiple systems, like ENS, CNS, ANS and all neuronal-hormonal balance axes's (Rhee et al. 2009).The connection between the gut and neurons is important for many physiological and pathological functions that include controlling insulin sensitivity, the appetite centre, glucose and fat metabolism, and the body's reaction against stress (Kowalski and Mulak 2019).

The link between diet and gut microbiota is a mutual interaction. While gut microbiota processes nutrients from food, dietary choices significantly influence the components of the intestinal microbial community. The metabolic functions of gut microbes largely depend on the availability of indigestible cellulose and Proteins that reach the Intestine. Studies on faecal samples from animals and humans have shown that dietary changes can alter the composition of the gut microbiota. In healthy individuals, a well-balanced diet promotes the development of a beneficial microbial community, where different bacterial species coexist in a state of mutual regulation and equilibrium (Stilling et al. 2014).

Several clinical studies reveal the link between diet and gut microbiota has a potential impact on health. As these include higher fats, refined carbohydrates and red meat, but on the other side, lower in fish, fruits, and plant-based foods, can negatively impact the immune system that causing dysbiosis. Dysbiosis refers to an imbalance in structure and function of the gut, which can lead to severe inflammation (Kaur et al. 2021). This imbalance often arises from an increased presence of pro-inflammatory microorganisms and a decrease in those that support immune tolerance. It shows the rapid improvement responses by shifting the diet towards the MDshows therapeutic effects. Research suggests that diet can alter up to 60% of its overall composition (Heck and Handa 2019). Pharmacology it was shows that in the gut, the dominant microbial groups belong to the phyla Bacteroidetes, Firmicutes, and Actinobacteria (Flint et al. 2012).

The human gut microbiota plays a potential role in the development of neurological disorders characterized by amyloid accumulation, like AD, although this evidence is still under medical surveillance (Kaoutari et al. 2013; Mojgani et al. 2025). Because any changes in gut microbiota composition, along with increased gut permeability, may lead to triggering systemic neuronal inflammation, and impairments in critical brain regions such as the hippocampus and cerebellum (as seen in Fig. 4.4). Additionally, these alterations may contribute to the onset of insulin resistance, a condition closely associated with the progression of AD (Zhang, Zhang et al. 2010). Another review study compiled the data and emphasized the role of ACE inhibitors in managing hypertension to preserve cognitive function (Yasmin et al. 2025).

Current research highlights that the MD positively impacts gut microbiota in geriatric patients by promoting a healthier composition and reducing frailty (Mafra and Fouque 2015). Increased levels of beneficial microbial taxa, such as *Roseburia* and *Faecalibacterium,* have been primarily linked with better cognitive function and lower levels of pro-inflammatory markers. These effects may be linked to higher production of SCFAs. The MD is also rich source of polyphenols, which emerging evidence suggests can help to prevent or slow AD progression by reducing plaque formation and protecting the integrity of the blood-brain barrier (as summarized in Table 4.5) (Schwartz and Boles 2013).

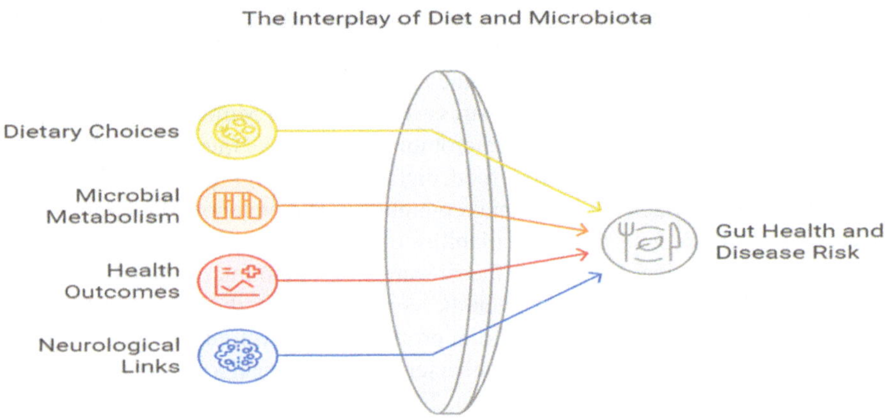

Fig. 4.4 Integration of MD and gut microbiota, MD plays a key role in shaping gut microbiota, promoting a diverse and healthy microbial environment

Table 4.5 Role of gut microbiota in AD treatment by using MD

Aim	Dietary intervention	Result	References
Linkage between MD and AD	Medi or MIND	Shows the significant and beneficial role of MD	Daulatzai (2014)
Linkage between MD and neurodegenerative problems	MIND	Enhancement in cognition power	Tomasik et al. (2015)
Impact of KD on patients with AD	KD or KS	Progression in acute and chronic type cognition	Zhang et al. (2008)
Linkage between MD and prolong-cognitive disorders	Medi	Shows beneficiary role of MD against cognitive problems	Bartochowski et al. (2020)
Impact of nutrition on learning memory	Different dietary patterns	It improves cognitive function	Kheirouri and Alizadeh (2022)
Effects of diets on CNS	Modified Atkins diet	It Significantly improves memory power	Grammatikopoulou et al. (2020)
Impact of KD on neuron metabolism and neurological diseases	KD or KS	Decline in AD symptoms	Limongi et al. (2020)
Linkage between learning memory and diet in AD	Omega-3, coconut oil, ginseng's and inositol's	At various disease omega-3 fatty acids proves to show positive results. This shows a positive response against cognition	Gutierrez et al. (2021)
Linkage between meat consumption and cognitive functions	Meat	NIL	Christensen et al. (2021)
Effects of supplementation's on cognitive impairment	Omega-3	Positive effects shown against cognitive problems	Jensen et al. (2020)
Linkage between grape polyphenols and AD	Grape polyphenols	NIL	Moreira et al. (2020)

4.4 Preclinical Studies

4.4.1 Pre-clinical Studies: Insights from Animal Models

4.4.1.1 Experimental Evidence in Rodent Models

Preclinical research on rodent models offers crucial information on regard to the health advantages of a Mediterranean diet. Studies on rats and mice have shown that it improves cardiovascular health, cognitive function, and metabolic health. According to studies, animals' intakes aMD that are higher in fruits, vegetables, nuts, and olive oil have better lipid profiles, less inflammation, and increased insulin sensitivity. Furthermore, studies indicate that by improving brain plasticity and lowering oxidative stress, this diet guards against neurodegenerative illnesses (Zhang et al. 2020).

4.4.2 Molecular Pathways Affected by the Mediterranean Diet

Numerous biological pathways that links to lifespan and illness prevention are influenced by the Mediterranean diet. By upregulating anti-inflammatory mediators and downregulating pro-inflammatory cytokines that regulates the inflammatory responses (Gkotzamanis and Panagiotakos 2020). By encouraging the activation of PPARs, which aid in controlling fatty acid oxidation and energy balance, the diet also influences lipid metabolism. Moreover, polyphenols found in MD staples like red wine and olive oil stimulate signalling pathways that includes sirtuins and AMPK, it plays a vital role in managing the cellular energy balance and ageing. These molecular processes that support the protective benefits of nutrition against neurodegeneration, cardiovascular disease, and metabolic disorders (El Gaamouch et al. 2021).

4.5 Clinical Evidence: Human Studies and Observations

The MD has garnered widespread attention for its remarkable health benefits, supported by extensive research and observational studies. It is frequently associated with reduced risks of chronic illnesses and an increase in life expectancy (Muralidharan et al. 2021). Researchers have consistently highlighted the diet's potential in preventing and managing Alzheimer disease, cardiovascular diseases, diabetes, certain cancers, and cognitive decline. While the MD originated in Mediterranean regions, its adaptability across different cultural and geographical contexts makes it a promising dietary model for global health promotion (Muralidharan et al. 2021). The evidence supporting the MD extends beyond physical health, as studies also reveal its neuroprotective effects and role in enhancing cognitive function. Despite the compelling evidence, limitations in current research underscore the need for long-term, methodologically robust studies to further validate its efficacy and broader implications. In this section, we explore epidemiological studies, cognitive outcomes, and the limitations of current clinical evidence to gain a comprehensive understanding of the Mediterranean diet's impact on human health (Muralidharan et al. 2021).

4.5.1 Epidemiological Studies and Longitudinal Research

Epidemiological studies have consistently highlighted the MD for its association with reduced chronic disease incidence and increased life expectancy(Estruch and Ros 2020) Observational research has linked the MD to improved outcomes in Alzheimer disease, coronary type II diabetes, heart disease, hypertension, embolisms, osteoporosis, and certain cancers, including breast, stomach, colorectal, and prostate cancers (Sánchez-Sánchez et al. 2020; Ranjan et al. 2021). In Denmark, the Euronut Survey on Nutrition and the Elderly studied the impact of the MD Score (MDS) on the survival rates of 202 elderly participants over a 6-year period. Using dietary diaries and food frequency questionnaires, researchers found that a one-point increase in the MDS correlated with a 21% reduction in overall mortality (Papadimitriou et al. 2023). In Australia, a cohort study involving 141 Anglo-Celts and 189 Greco-Australians in Melbourne aimed to replicate findings from rural Greek populations in an urban Australian setting. The study assessed dietary habits through validated frequency questionnaires and found that participants with higher MDS scores (≥4) had longer survival rates. Specifically, a one-point increase in the MDS was associated with a 17% reduction in general mortality (Georgoulis et al. 2022). Notably, 81% of Greco-Australians and 28% of Anglo-Celts adhered to four or more MD characteristics. Chinese researchers also investigated the MD's relevance to their population, noting similarities between Chinese dietary habits and the MD, such as high vegetable and fruit intake and low meat consumption (Bach et al. 2006). A study involving 1010 Chinese individuals across diverse regions, including Hong Kong, rural southern China, Sydney, and San Francisco, revealed that most participants achieved higher MDS scores than Greek populations. Middle-aged individuals (35–54 years), women, and rural residents demonstrated greater adherence to the MD, reflecting its adaptability across cultural contexts (Bach et al. 2006). These investigations emphasize the Mediterranean diet's ability to foster health and extend lifespan, showcasing its relevance even in non-Mediterranean settings. Ongoing studies are crucial to reinforce these conclusions and delve deeper into the wider implications of this dietary approach, as depicted in Table 4.6.

4.5.2 Cognitive Outcomes and Disease Progression

The MD is associated with significant health benefits, including reduced risks of cognitive decline and dementia. Studies in Greece have shown that adherence to MD correlates with better memory and lower dementia risk, with components like fish and whole cereals specifically linked to improved cognition (Osler and Schroll 1997). However, some studies did not find this connection due to factors like small sample sizes, unique

Table 4.6 Clinical study of MD

Study	Cohort details	Assessment method	Findings	References
Danish Study	202 elderly Danes, 6-year follow-up	3-day dietary diary and food frequency questionnaire	One-point increase in MDS predicted a 21% reduction in global mortality. Vegetables and legumes were combined into one component	Papadimitriou et al. (2023)
Australian Study	141 Anglo-Celts and 189 Greco-Australians in Melbourne	Extensive validated frequency questionnaires	High MDS scores (≥4) were associated with longer survival. 81% of Greco-Australians and 28% of Anglo-Celts adhered to 4 or more MD characteristics. 17% mortality reduction	Georgoulis et al. (2022)
Chinese Study	1010 participants across Hong Kong, rural Pan Yu (China), Sydney, and San Francisco	Adapted MD Score (MDS)	Majority had diets similar to the Mediterranean pattern, with higher scores in middle-aged individuals (35–54), women, and rural populations. Higher scores than Greek populations studied	Bach et al. (2006)

population traits, or inadequate methods for assessing MD adherence. Cohort studies and meta-analyses generally support MD's protective role against cognitive decline, as evidenced by Spanish studies showing better cognitive function among adherents. Conversely, an Australian trial failed to replicate these results, possibly due to shorter intervention periods, smaller samples, or a less traditional MD definition. MD is believed to enhance cognition through cardiovascular and cerebrovascular benefits, including fewer brain infarcts and reduced white matter hyper intensity volumes (Kouris-Blazos et al. 1999). Its antioxidant and anti-inflammatory properties may also counteract neuronal damage by reducing amyloid peptide accumulation. While fish consumption and non-refined cereals have shown consistent benefits, the combined effects of MD components appear to be more influential than individual foods. Although limitations like dietary reporting errors exist, MD shows promise as a population-level strategy for dementia prevention and cognitive health enhancement, warranting further robust trials and longitudinal studies (Trichopoulou et al. 1995).

4.5.3 Limitations of Current Clinical Evidence

The MD is widely recognized for its health benefits, but several limitations exist in the current evidence (Anastasiou et al. 2017). Many studies are observational, making it difficult to establish cause-and-effect relationships, while randomized controlled trials

(RCTs) face challenges like inconsistent adherence. Findings often lack generalizability due to population variability, as most studies focus on Mediterranean populations with distinct genetic, cultural, and lifestyle factors, raising questions about universal applicability (Charisis et al. 2021). Measuring adherence is also complex, as scoring systems fail to consider nuances like food quality, preparation methods, and cultural contexts. While short-term health benefits are well-documented, long-term studies confirming sustained outcomes are limited. Confounding factors like physical activity, social interactions, and environmental conditions associated with the Mediterranean lifestyle make it challenging to isolate the diet's specific effects (Charisis et al. 2021). Additionally, certain components, such as herbs, spices, fermented products, and cooking methods, remain underexplored. Despite strong evidence supporting its benefits, more robust, long-term, and diverse research is needed to better understand the Mediterranean diet's global impact and refine dietary guidelines (Martínez-Lapiscina et al. 2013).

4.6 Challenges in Translating Evidence to Practice

4.6.1 Barriers to Adoption of Mediterranean Diet

The Mediterranean Diet (MD) embodies the traditional eating habits of the Mediterranean region. It features a daily intake of fruits, vegetables, unrefined whole grains, and pulses, with significant amounts of monounsaturated fatty acids (MUFAs) primarily derived from olive oil (Finicelli et al. 2022). The diet also includes moderate consumption of fish and alcohol (predominantly red wine), low-to-moderate intake of dairy products such as yogurt and cheese, and minimal consumption of poultry and red meats. Observational research highlights the MD's protective effects against global health concerns like cardiovascular disease (CVD), recurrent cardiac events, CVD-related mortality, metabolic syndrome, abdominal obesity, and gestational weight gain. Randomized trials, such as the PREDIMED study, further affirm its advantages, demonstrating that a nut-enriched MD reduces CVD risk and improves secondary health outcomes. Additionally, greater adherence during pregnancy is linked to reduced offspring cardiometabolic risks, while poor adherence is associated with increased blood pressure and preeclampsia risks. (Godos et al. 2024). Despite the MD recognized advantages, challenges to adoption persist outside the Mediterranean region, with declining adherence even within Mediterranean and southern European populations (Hoffman and Gerber 2013). Studies have pinpointed barriers to following this dietary pattern, yet a detailed account is needed. The present work aims to consolidate research on barriers or facilitators affecting MD adoption and adherence, offering evidence-based insights into influencing factors. To date, no systematic review has focused on this specific purpose (Tsofliou et al. 2021).

4.6.2 Cultural and Socioeconomic Considerations

Recent dietary intervention studies have highlighted barriers and facilitators to adopting a MD, stressing the importance of cultural adaptation for sustained adherence while retaining key nutritional elements like olive oil, legumes, cereals, fruits, vegetables, fish, and moderate dairy, and wine intake (Spadafranca et al. 2018). Locally sourced substitutes, such as rapeseed oil in place of olive oil, can help maintain nutritional value and accessibility. Globally, MD adaptations have been explored (Romaguera et al. 2009). For instance, George et al. developed a culturally adapted MD model in Australia's multi-ethnic context, using a SWOT framework to identify barriers and implementing strategies aimed at patients with NAFLD and CHD prevention. While MD interventions address environmental and health concerns, variability in methodologies complicates evaluations of efficacy and adherence. Key strategies include dietitian-led interventions, written resources, recipes, and cooking classes (Martínez-González et al. 2017). Scalability requires a multi-sectorial approach involving policymakers, the food industry, and regulatory bodies, ensuring the diet remains culturally appropriate and affordable (Woodside et al. 2022; Martínez-González et al. 2017). In Australia, barriers and enablers were assessed through the Theory of Planned Behaviour, with PBC emerging as the most influential factor (Hoffman and Gerber 2014; McEvoy et al. 2022). Enhancing self-efficacy in future interventions could help mitigate barriers. In the USA, del Campo et al. trialed MD-style interventions with low-income Hispanic American women, achieving high engagement and improved dietary habits. These studies emphasize the value of culturally adapted MD interventions in diverse populations (George et al. 2018; Estrada Del Campo et al. 2019).

4.7 Future Prospective MD in AD

In order to alleviate perceived obstacles and encourage commitment to development and customised nutrition, it focusses on increasing self-efficacy. Through circulatory and cerebrovascular effects, like lowering of brain infarcts and the amount of white matter hyperintensities that aid in AD recovery, the MD is believed to improve cognitive performance. The MD region's traditionally having eating habits are reflected in their comparative diet, which includes a daily intake of fruits and vegetables, a high intake of unrefined whole grains and pulses, and significant intakes of monounsaturated fatty acids, mostly from olive oil. It also includes consuming little to moderate amounts of dairy products, moderate amounts of fish and alcohol (like red wines), and little to no poultry and red meat. Observational research indicated the MD's preventive benefits against global health problems, such as obesity in the abdomen and cardiovascular disease.

4.8 Conclusion

The MD plays a significant role in addressing AD and culturally adapting it can encourage long-term adherence at the population level. This approach maintains the diet's core components, which include high amounts of Olive oil's, Legume's, Whole grain's, Fruit's and Vegetable's, along with reasonable consumption of Fish, dairy (such as cheese and yogurt), and wine, while limiting meat intake. Nutritional intervention trials have documented reductions in serum lipid levels, oxidative stress, inflammation, platelet aggregation, as well as modulation of hormones and growth factors linked to cancer pathogenesis and microbial metabolism. Personal belief and confidence (PBC) emerged as the most influential factor affecting adherence intentions. It highlighted the perceived health benefits and improved diet quality as major advantages, while also noting that difficulties in adhering to the diet can be a challenge.

In contrast, the WD is associated with adverse health outcomes, contributing to the obesity crisis and various NCDs such as cancer, cardiovascular diseases, osteoporosis, autoimmune conditions, and type 2 diabetes. Chronic oxidative stress, which becomes harder to manage due to elevated levels of ROS, is linked to degenerative diseases like AD, PD, various cancers, etc. Clinical evidence reveals that the connection between diet and gut microbiota can significantly impact health. The WD typically includes higher levels of fats, refined carbohydrates, and red meat while being lower in fish, fruits, and plant-based foods, which can negatively affect the immune system and lead to dysbiosis. Dysbiosis refers to an imbalance in the gut's structure and function, which can result in severe inflammation. The link involving diet and gut microbiota is mutual; although the microbiota of the gut breaks down nutrients from Food, Dietary decisions also have a big impact on the makeup of the Intestinal microbial melieu. The availability of indigestible proteins and carbohydrates that reach the colon is crucial for the metabolic processes of gut microorganisms. Due to various reasons, such as limited sample numbers, distinct demographic characteristics, or insufficient techniques for evaluating adherence to the MD several studies have not discovered a substantial correlation between diet and gut health. The preventive role of the MD against cognitive decline is largely supported by cohort studies and meta-analyses; Spanish research show that followers have higher cognitive function. Conversely, an Australian trial failed to replicate these results, possibly due to shorter intervention periods, smaller sample sizes, or a less traditional definition of the MD.

Acknowledgements Authors are thankful to the Chairman, IBNE Seena Pharmacy College, Shahabad, Hardoi, India. For providing facilities, the authors extend their appreciation to the Deanship of Research and Graduate Studies at King Khalid University for funding this work through a Large Research project under grant number RGP2/8/46.

Conflict of Interest The authors declare no competing conflict of interest.

Declaration of Generative AI and AI-Assisted Technologies in the Writing Process While preparing this work, the authors used Grammarly to proofread the article. After using Grammarly, the authors reviewed and edited the content as needed and took full responsibility for the publication's content.

Funding The authors received no funding from an external source.Data AvailabilityNo data was generated during this study.

References

Aguilera JM (2019) The food matrix: implications in processing, nutrition and health. Crit Rev Food Sci Nutr 59(22):3612–3629
Amezaga V-D et al (2008) The Mediterranean food pattern: a good recipe for patients with the metabolic syndrome. Mediterr J Nutr Metab 1(1):3–1
Anastasiou CA et al (2017) Mediterranean diet and cognitive health: Initial results from the Hellenic Longitudinal Investigation of Ageing and Diet. PLoS One 12(8):e0182048
Araya Quintanilla F et al (2020) Effectiveness of omega-3 fatty acid supplementation in patients with Alzheimer disease: a systematic review and meta-analysis
Aridi YS et al (2017) The association between the Mediterranean dietary pattern and cognitive health: a systematic review. Nutrients 9(7):674
Armstrong RA (2013) What causes Alzheimer's disease? Folia Neuropathol 51(3):169–188
Armstrong RA (2019) Risk factors for Alzheimer's disease. Folia Neuropathol 57(2):87–105
Arriagada PV et al (1992) Distribution of Alzheimer-type pathologic changes in nondemented elderly individuals matches the pattern in Alzheimer's disease. Neurology 42(9):1681–1681
Arsene D et al (2011) Trace metals as factors involved in human neurodegeneration: real role or random association? An autopsy study. Trace Elements Electrolytes 28(4):221
Avallone R et al (2019) Omega-3 fatty acids and neurodegenerative diseases: new evidence in clinical trials. Int J Mol Sci 20(17):4256
Bach A et al (2006) The use of indexes evaluating the adherence to the Mediterranean diet in epidemiological studies: a review. Public Health Nutr 9(1a):132–146
Barrea L et al (2021) Nutrition and immune system: from the Mediterranean diet to dietary supplementary through the microbiota. Crit Rev Food Sci Nutr 61(18):3066–3090
Bartochowski Z et al (2020) Dietary interventions to prevent or delay Alzheimer's disease: what the evidence shows. Curr Nutr Rep 9:210–225
Bell MA, Ball MJ (1990) Neuritic plaques and vessels of visual cortex in aging and Alzheimer's dementia. Neurobiol Aging 11(4):359–370
Bhat AH et al (2015) Oxidative stress, mitochondrial dysfunction and neurodegenerative diseases; a mechanistic insight. Biomed Pharmacother 74:101–110
Bivona G et al (2019) Vitamin D and the nervous system. Neurol Res 41(9):827–835
Breteler MM (2000) Vascular risk factors for Alzheimer's disease:: an epidemiologic perspective. Neurobiol Aging 21(2):153–160
Charisis S et al (2021) Mediterranean diet and risk for dementia and cognitive decline in a Mediterranean population. J Am Geriatr Soc 69(6):1548–1559
Cheignon CM et al (2018) Oxidative stress and the amyloid beta peptide in Alzheimer's disease. Redox Biol 14:450–464
Christ A et al (2019) Western diet and the immune system: an inflammatory connection. Immunity 51(5):794–811
Christensen MG et al (2021) Use of ketogenic diets in the treatment of central nervous system diseases: a systematic review. Nord J Psychiatry 75(1):1–8
Cryan JF et al (2019) The microbiota-gut-brain axis. Physiol Rev 99:1877
Daulatzai MA (2014) Chronic functional bowel syndrome enhances gut-brain axis dysfunction, neuroinflammation, cognitive impairment, and vulnerability to dementia. Neurochem Res 39(4):624–644
Di Rosa G et al (2020) Healthspan enhancement by olive polyphenols in C. elegans wild type and Parkinson's models. Int J Mol Sci 21(11):3893

Doruk H et al (2010) The relationship between body mass index and incidental mild cognitive impairment, Alzheimer's disease, and Vascular Dementia in elderly. J Nutr Health Aging 14:834–838
El Gaamouch F et al (2021) Development of grape polyphenols as multi-targeting strategies for Alzheimer's disease. Neurochem Int 147:105046
Esposito K et al (2017) Mediterranean diet for type 2 diabetes: cardiometabolic benefits. Endocrine 56:27–32
Estrada Del Campo Y et al (2019) Feasibility and acceptability of a Mediterranean-style diet intervention to reduce cardiovascular risk for low income Hispanic American women. Ethn Health 24(4):415–431
Estruch R, Ros E (2020) The role of the Mediterranean diet on weight loss and obesity-related diseases. Rev Endocr Metab Disorders 21(3):315–327
Findeis MA (2007) The role of amyloid β peptide 42 in Alzheimer's disease. Pharmacol Ther 116(2):266–286
Finicelli M et al (2022) The Mediterranean diet: an update of the clinical trials. Nutrients 14(14):2956
Flint HJ et al (2012) The role of the gut microbiota in nutrition and health. Nat Rev Gastroenterol Hepatol 9(10):577–589
Garcia-Mantrana I et al (2018) Shifts on gut microbiota associated to mediterranean diet adherence and specific dietary intakes on general adult population. Front Microbiol 9:890
García-Montero C et al (2021) Nutritional components in Western diet versus Mediterranean diet at the gut microbiota–immune system interplay. Implications for health and disease. Nutrients 13(2):699
Gentleman SM et al (1993) β-Amyloid precursor protein (βAPP) as a marker for axonal injury after head injury. Neurosci Lett 160(2):139–144
George ES et al (2018) A Mediterranean diet model in Australia: strategies for translating the traditional Mediterranean diet into a multicultural setting. Nutrients 10(4):465
Georgoulis M et al (2022) Longitudinal trends, determinants, and Cardiometabolic impact of adherence to the Mediterranean diet among Greek adults. Foods 11(16):2389
Gilgun-Sherki Y et al (2001) Oxidative stress induced-neurodegenerative diseases: the need for antioxidants that penetrate the blood brain barrier. Neuropharmacology 40(8):959–975
Gillette-Guyonnet S et al (2013) Nutrition and neurodegeneration: epidemiological evidence and challenges for future research. Br J Clin Pharmacol 75(3):738–755
Gkotzamanis V, Panagiotakos D (2020) Dietary interventions and cognition: a systematic review of clinical trials. Psychiatriki 31(3):248–256
Godos J et al (2024) Underrated aspects of a true Mediterranean diet: Understanding traditional features for worldwide application of a "Planeterranean" diet. J Transl Med 22(1):294
Gombart AF et al (2020) A review of micronutrients and the immune system–working in harmony to reduce the risk of infection. Nutrients 12(1):236
Gomez-Pinilla F, Nguyen TT (2012) Natural mood foods: the actions of polyphenols against psychiatric and cognitive disorders. Nutr Neurosci 15(3):127–133
Grammatikopoulou MG et al (2020) To keto or not to keto? A systematic review of randomized controlled trials assessing the effects of ketogenic therapy on Alzheimer disease. Adv Nutr 11(6):1583–1602
Grubić Kezele T, Ćurko-Cofek B (2022) Neuroprotective panel of olive polyphenols: Mechanisms of action, anti-demyelination, and anti-stroke properties. Nutrients 14(21):4533
Guasch-Ferré M, Willett WC (2021) The Mediterranean diet and health: a comprehensive overview. J Intern Med 290(3):549–566
Gutierrez L et al (2021) Effects of nutrition on cognitive function in adults with or without cognitive impairment: a systematic review of randomized controlled clinical trials. Nutrients 13(11):3728
Heck AL, Handa RJ (2019) Sex differences in the hypothalamic–pituitary–adrenal axis' response to stress: an important role for gonadal hormones. Neuropsychopharmacology 44(1):45–58

Heyman A et al (1983) Alzheimer's disease: genetic aspects and associated clinical disorders. Ann Neurol 14(5):507–515

Hidalgo-Mora JJ, García-Vigara A, Sánchez-Sánchez ML, García-Pérez MÁ, Tarín J,Cano A (2020) The mediterranean diet: a historical perspective on food for health. Maturitas 132:65–69

Hoffman R, Gerber M (2013) Evaluating and adapting the Mediterranean diet for non-Mediterranean populations: a critical appraisal. Nutr Rev 71(9):573–584

Hoffman R, Gerber M (2014) Can rapeseed oil replace olive oil as part of a Mediterranean-style diet? Br J Nutr 112(11):1882–1895

Hong M et al (2022) Tea polyphenols as prospective natural attenuators of brain aging. Nutrients 14(15):3012

Jensen NJ et al (2020) Effects of ketone bodies on brain metabolism and function in neurodegenerative diseases. Int J Mol Sci 21(22):8767

Kaoutari AE et al (2013) The abundance and variety of carbohydrate-active enzymes in the human gut microbiota. Nat Rev Microbiol 11(7):497–504

Kaur G et al (2021) Dysregulation of the gut-brain axis, dysbiosis and influence of numerous factors on gut microbiota associated Parkinson's disease. Curr Neuropharmacol 19(2):233–247

Keys A et al (1986) The diet and 15-year death rate in the seven countries study. Am J Epidemiol 124(6):903–915

Khan H et al (2019) Neuroprotective effects of quercetin in Alzheimer's disease. Biomolecules 10(1):59

Kheirouri S, Alizadeh M (2022) MIND diet and cognitive performance in older adults: a systematic review. Crit Rev Food Sci Nutr 62(29):8059–8077

Kopp W (2019) How western diet and lifestyle drive the pandemic of obesity and civilization diseases. In: Diabetes, metabolic syndrome and obesity: targets and therapy, vol Volume 12, pp 2221–2236

Kouris-Blazos A et al (1999) Are the advantages of the Mediterranean diet transferable to other populations? A cohort study in Melbourne, Australia. Br J Nutr 82(1):57–61

Kowalski K, Mulak A (2019) Brain-gut-microbiota axis in Alzheimer's disease. J Neurogastroenterol Motility 25(1):48

Kozlowska A, Szostak-Wegierek D (2014) Flavonoids-food sources and health benefits. Rocz Panstw Zakl Hig 65(2):79

Krznaric Ž et al (2019) The Mediterranean diet and its association with selected gut bacteria. Curr Opin Clin Nutr Metab Care 22(5):401–406

Lahiri D, Maloney B (2010) The "LEARn" model integrates environmental risk factors and the developmental basis of Alzheimer's disease, and proposes remedial steps. Exp Gerontol 45:291

Larsson T (1963) Senile dementa: a clinical, sociomedical, and genetic study. Acta Psychiatr Scand 39(167):1–259

Lassale C, Batty GD, Baghdadli A, Jacka F, Sánchez-Villegas A, Kivimäki M, Akbaraly T (2019) Healthy dietary indices and risk of depressive outcomes: a systematic review and meta-analysis of observational studies. Mol Psychiatry 24(7):965–986

Limongi F et al (2020) The effect of adherence to the Mediterranean diet on late-life cognitive disorders: a systematic review. J Am Med Dir Assoc 21(10):1402–1409

Mafra D, Fouque D (2015) Gut microbiota and inflammation in chronic kidney disease patients. Oxford University Press 8:332–334

Maggini S et al (2009) Feeding the immune system: the role of micronutrients in restoring resistance to infections. CABI Reviews 2008:1–21

Martín S et al (2011) Neuroprotective properties of Spanish red wine and its isolated polyphenols on astrocytes. Food Chem 128(1):40–48

Martínez-González MÁ et al (2017) Transferability of the Mediterranean diet to non-Mediterranean countries. What is and what is not the Mediterranean diet. Nutrients 9(11):1226

Martínez-Lapiscina EH et al (2013) Mediterranean diet improves cognition: the PREDIMED-NAVARRA randomised trial. J Neurol Neurosurg Psychiatry 84(12):1318–1325

Marx W et al (2017) Nutritional psychiatry: the present state of the evidence. Proc Nutr Soc 76(4):427–436

Mattioli R et al (2020) Anthocyanins: a comprehensive review of their chemical properties and health effects on cardiovascular and neurodegenerative diseases. Molecules 25(17):3809
McEvoy CT et al (2022) Trial to Encourage Adoption and Maintenance of a MEditerranean Diet (TEAM-MED): a randomised pilot trial of a peer support intervention for dietary behaviour change in adults from a Northern European population at high CVD risk. Br J Nutr 128(7):1322–1334
Medina FX (2009) Mediterranean diet, culture and heritage: challenges for a new conception. Public Health Nutr 12(9A):1618–1620
Mojgani N et al (2025) Gut microbiota and postbiotic metabolites: biotic intervention for enhancing vaccine responses and personalized medicine for disease prevention. Probiot Antimicrob Proteins:1–23
Moreira SC et al (2020) Dietary interventions and cognition of Alzheimer's disease patients: a systematic review of randomized controlled trial. Dement Neuropsychol 14(03):258–282
Muralidharan J et al (2021) Effect on gut microbiota of a 1-y lifestyle intervention with Mediterranean diet compared with energy-reduced Mediterranean diet and physical activity promotion: PREDIMED-Plus Study. Am J Clin Nutr 114(3):1148–1158
Nishi SK et al (2021) Mediterranean, DASH, and MIND dietary patterns and cognitive function: the 2-year longitudinal changes in an older Spanish cohort. Front Aging Neurosci 13:782067
Osler M, Schroll M (1997) Diet and mortality in a cohort of elderly people in a north European community. Int J Epidemiol 26(1):155–159
Papadimitriou A et al (2023) Adherence to the Mediterranean Diet and Successful aging in Greeks living in Greece and abroad: The epidemiological Mediterranean Islands Study (MEDIS). Nutr Health 29(2):287–295
Pasinetti GM, Eberstein JA (2008) Metabolic syndrome and the role of dietary lifestyles in Alzheimer's disease. J Neurochem 106(4):1503–1514
Pecora F et al (2020) The role of micronutrients in support of the immune response against viral infections. Nutrients 12(10):3198
Picone P et al (2024) Enriching the Mediterranean diet could nourish the brain more effectively. Front Nutr 11:1489489
Praticò D (2002) Alzheimer's disease and oxygen radicals: new insights. Biochem Pharmacol 63(4):563–567
Ramesh BN et al (2010) Neuronutrition and Alzheimer's disease. J Alzheimer's Dis 19(4):1123–1139
Ranjan S et al (2021) Resveratrol as anti-cancer and cardio protective agent: a review. Onkologia i Radioterapia 15(12)
Rhee SH et al (2009) Principles and clinical implications of the brain–gut–enteric microbiota axis. Nat Rev Gastroenterol Hepatol 6(5):306–314
Romaguera D et al (2009) Adherence to the Mediterranean diet is associated with lower abdominal adiposity in European men and women. J Nutr 139(9):1728–1737
Ros E (2010) Health benefits of nut consumption. Nutrients 2(7):652–682
Sakurai K et al (2021) Consumption of oleic acid on the preservation of cognitive functions in Japanese elderly individuals. Nutrients 13(2):284
Salis C et al (2020) Olive oil polyphenols in neurodegenerative pathologies. In: GeNeDis 2018: genetics and neurodegeneration, pp 77–91
Sánchez-Moreno C et al (2006) Mediterranean vegetable soup consumption increases plasma vitamin C and decreases F2-isoprostanes, prostaglandin E2 and monocyte chemotactic protein-1 in healthy humans. J Nutr Biochem 17(3):183–189
Sánchez-Sánchez ML et al (2020) Mediterranean diet and health: a systematic review of epidemiological studies and intervention trials. Maturitas 136:25–37
Scheltens P (2009) Moving forward with nutrition in Alzheimer's disease. Eur J Neurol 16:19–22
Schepici G et al (2020) Efficacy of sulforaphane in neurodegenerative diseases. Int J Mol Sci 21(22):8637
Schwartz K, Boles BR (2013) Microbial amyloids–functions and interactions within the host. Curr Opin Microbiol 16(1):93–99

Serra-Majem L et al (2019) Benefits of the Mediterranean diet: Epidemiological and molecular aspects. Mol Asp Med 67:1–55

Shively CA et al (2019) Mediterranean versus Western diet effects on caloric intake, obesity, metabolism, and hepatosteatosis in nonhuman primates. Obesity 27(5):777–784

Silva Rui FM, Pogačnik L (2017) Food, polyphenols and neuroprotection. Neural Regen Res 12(4):582. https://doi.org/10.4103/1673-5374.205096

Singh RB et al (2002) Effect of an Indo-Mediterranean diet on progression of coronary artery disease in high risk patients (Indo-Mediterranean Diet Heart Study): a randomised single-blind trial. Lancet 360(9344):1455–1461

Sousa MJ, Guimarães J (2015) Prevenção da doença de Alzheimer: o papel da dieta Mediterrânica. Rev Nutr 28:691–703

Spadafranca A et al (2018) Adherence to the Mediterranean diet and serum adiponectin levels in pregnancy: Results from a cohort study in normal weight Caucasian women. Nutrients 10(7):928

Statovci D et al (2017) The impact of western diet and nutrients on the microbiota and immune response at mucosal interfaces. Front Immunol 8:838

Stein TD et al (2014) Chronic traumatic encephalopathy: a spectrum of neuropathological changes following repetitive brain trauma in athletes and military personnel. Alzheimer's Res Ther 6:1–11

Stilling RM et al (2014) Microbial genes, brain & behaviour–epigenetic regulation of the gut–brain axis. Genes Brain Behav 13(1):69–86

Stolfi C et al (2022) Implication of intestinal barrier dysfunction in gut dysbiosis and diseases. Biomedicine 10(2):289

Tomasik J et al (2015) Immunomodulatory effects of probiotic supplementation in schizophrenia patients: a randomized, placebo-controlled trial. Biomark Insights 10:BMI-S22007

Tosti V et al (2018) Health benefits of the Mediterranean diet: metabolic and molecular mechanisms. J Gerontol Ser A 73(3):318–326

Trichopoulou A et al (1995) Diet and overall survival in elderly people. BMJ 311:1457

Trichopoulou A et al (2003) Adherence to a Mediterranean diet and survival in a Greek population. N Engl J Med 348(26):2599–2608

Tsofliou F et al (2021) Barriers and facilitators to adoption of and adherence to a Mediterranean style diet in adults: a systematic review of observational and qualitative studies. Proc Nutr Soc 80(OCE5):E223

Valero-Vello M et al (2021) Searching for the antioxidant, anti-inflammatory, and neuroprotective potential of natural food and nutritional supplements for ocular health in the mediterranean population. Foods 10(6):1231

Vassallo N, Scerri C (2013) Mediterranean diet and dementia of the Alzheimer type. Curr Aging Sci 6(2):150–162

Weih M et al (2007) Non-pharmacologic prevention of Alzheimer's disease: nutritional and life-style risk factors. J Neural Transm 114:1187–1197

Westfall S, Pasinetti GM (2019) The gut microbiota links dietary polyphenols with management of psychiatric mood disorders. Front Neurosci 13:1196

Wojtunik-Kulesza KA et al (2016) The influence of common free radicals and antioxidants on development of Alzheimer's Disease. Biomed Pharmacother 78:39–49

Woodside J et al (2022) Culturally adapting the Mediterranean Diet pattern–a way of promoting more 'sustainable'dietary change? Br J Nutr 128(4):693–703

Wu D et al (2019) Nutritional modulation of immune function: analysis of evidence, mechanisms, and clinical relevance. Front Immunol 9:3160

Xu H et al (2014) Interactions of metals and Apolipoprotein E in Alzheimer's disease. Front Aging Neurosci 6:121

Yasmin S et al (2025) The role of ACE inhibitors and ARBs in preserving cognitive function via hypertension Management: A critical Update. Brain Res 1850:149400

Zhang J et al (2008) The role of lipocalin 2 in the regulation of inflammation in adipocytes and macrophages. Mol Endocrinol 22(6):1416–1426

Zhang C et al (2010) Interactions between gut microbiota, host genetics and diet relevant to development of metabolic syndromes in mice. ISME J 4(2):232–241
Zhang H et al (2020) Meat consumption, cognitive function and disorders: a systematic review with narrative synthesis and meta-analysis. Nutrients 12(5):1528
Zhou D-D et al (2021) Effects and mechanisms of resveratrol on aging and age-related diseases. Oxidative Med Cell Longev 2021(1):9932218

Prebiotics and Probiotics in Alzheimer's Disease

5

Soumya Mitra, Rakesh Kumar Dolai, Amit Kumar Halder, Nilanjan Ghosh, and Subhash Chandra Mandal

Abstract

Alzheimer's disease (AD) is a gradually advancing neurodegenerative condition that results in cognitive decline, memory impairment, and neuronal loss. The condition is characterized by hallmark pathological features, including extracellular amyloid-beta (Aβ) plaque accumulation, intracellular τ protein tangles, synaptic malfunction, and chronic neuroinflammation. While the precise aetiology of AD remains blurred, rising indications suggest a significant interplay between the central nervous system and the gut microbiome, collectively known as the gut-brain axis. Dysbiosis, or an imbalance in gut-microbiota, has been linked to increased neuroinflammation, oxidative stress, and blood-brain barrier disruption, all of which contribute to AD pathogenesis.

Recent investigations have emphasized on the potential influence of gut microbiota modulation in neuroprotection. Prebiotics and probiotics, which influence gut microbiota composition and function, have emerged as promising interventions for delaying or mitigating AD progression. Prebiotics, non-digestible dietary fibres, selectively promote the growth of beneficial gut bacteria, leading to the production of neuroactive metabolites such as short-chain fatty acids (SCFAs). SCFAs help regulate neuroinflammatory pathways, improve gut barrier integrity, and reduce oxidative stress, thereby mitigating neurodegenerative processes. Probiotics, live microorganisms that confer health benefits, play a

S. Mitra
Dr. B. C. Roy College of Pharmacy and AHS, Durgapur, West Bengal, India

Department of Pharmaceutical Technology, Jadavpur University, Kolkata, West Bengal, India

R. K. Dolai · A. K. Halder
Dr. B. C. Roy College of Pharmacy and AHS, Durgapur, West Bengal, India

N. Ghosh · S. C. Mandal (✉)
Department of Pharmaceutical Technology, Jadavpur University, Kolkata, West Bengal, India

J. Malik et al. (eds.), *Diet and Alzheimer's Disease: Let Food be Our Medicine*, Nutritional Neurosciences, https://doi.org/10.1007/978-981-95-2736-6_5

crucial role in restoring microbial balance. They enhance neurotransmitter synthesis, regulate immune responses, and modulate metabolic pathways, potentially reducing Aβ aggregation and inflammation associated with AD. Preclinical and early clinical studies suggest that probiotic supplementation may improve cognitive function, strengthen neuronal plasticity, and decrease neurodegenerative markers. These findings underscore the therapeutic potential of microbiota-targeted approaches in AD management.

Keywords

Alzheimer's disease · Gut-brain microbiome · Prebiotics · Probiotics · Dietary fibres

Abbreviations

AD	Alzheimer's disease
Aβ	Amyloid-beta
SCFA	Short-chain Fatty Acid
ACh	Acetylcholine
APP	Amyloid Precursor Protein
FAD	Familial Alzheimer's disease
PSEN	Presenilin
GMB	Gut microbiome
CNS	Central nervous system
BBB	Blood Brain Barrier
IL	Interleukin
TNF	Tumour Necrosis Factor
ROS	Reactive Oxygen Species
CAM	CNS-associated Macrophages
ISAPP	International Scientific Association of Probiotics and Prebiotics
FOS	Fructo-oligosaccharides
GOS	Galacto-oligosaccharides
RS	Resistant Starch
DP	Degree of Polymerization
GABA	Gamma amino butyric acid
IBD	Inflammatory bowel disease
GLP	glucagon-like peptide
HPA	Hypothalamic-pituitary-adrenal
AChE	Acetylcholinesterase

NFT	Neurofibrillary tangles
LTP	Long-term potentiation
BDNF	Brain-derived neurotrophic factor
GLUT	Glucose transporters
AMPK	AMP-activated protein kinase
SIRT	Sirtuin
NAD	Nicotinamide adenine dinucleotide
PARP	Polymerase
RAR	Retinoid activated receptor
ADAM	A Disintegrin and Metalloproteinase
FGF	Fibroblast growth factor
GIP	gastric inhibitory polypeptide
SOD	Superoxide dismutase
EPSP	Excitatory postsynaptic potentials
PPF	Paired-pulse facilitation
NO	Nitric oxide
MDA	Malondialdehyde
VLDL	Very low-density lipid
LDL	Low density lipid

5.1 Introduction

Neurodegenerative disorders, as the name suggests, can be indicated by their steady, irreversible progression, heading to the deterioration of neural tissues and a subsequent deterioration in cognitive skills. Among these, Alzheimer's disease (AD) stands out as the most dominant category of dementia. The origin of AD remains indistinct, but investigations indicate that the disease process begins 10–20 years before clinical symptoms become apparent (Sochocka et al. 2019). AD is marked by deteriorating impairments in memory, spatial awareness, and cognitive functions, alongside significant alterations in behaviour, such as increased aggression and shifts in personality (Knopman et al. 2021).

Aspects that may lead to the progress of the disorder, include routine habits such as dietary consumption, physical action, and educational background. Cognitive health and the natural aging process also play significant roles. Additionally, immunosenescence, the gradual decline of immune function with age, may increase susceptibility towards neurodegenerative diseases. Chronic infections, persistent inflammation, and latent infections are thought to further intensify the risk. Sleep disturbances and other related factors are also considered potential contributors to the disease's onset (Salminen 2025; Ooms et al. 2014). Epidemiological analysis has demonstrated that adherence to the Mediterranean dietary pattern, characterized by high consumption of plant-based foods is significantly correlated to a decreased incidence of AD and other neurodegenerative dementias (Lourida et al. 2013). Furthermore, Ozawa et al. (2014) conducted a 17-year study involving over 1000 Alzheimer's patients and found that higher intake of milk and dairy products significantly reduce incidence of the disease. These outcomes underscore the critical role of lifestyle factors in modulating AD's risk, suggesting that suboptimal dietary and behavioural habits might not only exacerbate neurodegenerative progression but also contribute to systemic health issues, including gut microbiota dysbiosis. Over the history of human race, the composition of symbiotic microorganisms has undergone significant changes due to agricultural, industrial, and global developments. These shifts have resulted in reduced microbial diversity and increased virulence, potentially facilitating the pathogenesis of AD. The historical evolution of microbial composition suggests a growing risk for Alzheimer's. Emerging evidence on the pathogenesis of AD suggests that lifelong dysbiosis of the microbiota, including nasal, oral and gastrointestinal communities, can provoke chronic systemic inflammation and modulate microglial immune activity of the CNS. A growing body experimental and clinical research underscores the pivotal contribution of gut microbiota imbalances and host microbiome interactions in the initiation and progression of neurodegenerative disorders (Kowalski and Mulak 2019).

Neurodegeneration typically originates in brain regions essential for memory consolidation, notably the entorhinal cortex and hippocampus, and progressively advance to involve broader cortical structures. This progression unsettles ACh-producing nerve cells, steering towards a steady deterioration in cognitive abilities over time. Recent investigations have also networked alterations in gut microbiota to a number of neurodegenerative illnesses, and there are suggestions indicating a contributory role of intestinal dysbiosis in the pathophysiology of neurodegenerative diseases (Ferreira-Vieira et al. 2016; Quigley 2017; Jiang et al. 2017).

5.1.1 Alzheimer's Disease (AD)

AD is a chronic, progressive neurodegenerative disorder characterized by the extracellular deposition of β-amyloid (Aβ) plaques and the intracellular aggregation of hyperphosphorylated tau (τ) protein in the form of neurofibrillary tangles. While the disease is most frequently associated with memory loss, particularly short-term memory impairment, it can also present with non-memory related cognitive deficits. These may include difficulties with expressive language, visuospatial skills, and executive functions such as mental flexibility and planning. While majority of the AD cases are not linked to dominant genetic inheritance, there is a multifaceted correlation among genetic elements and progression of the disorder in many individuals (Breijyeh and Karaman 2020).

5.1.2 Epidemiology

AD mostly affects older adults and has seen a dramatic increase in global prevalence over the years. In 1990, the worldwide prevalence of dementia was estimated at 20.3 million, but by 2016, this number had surged to 43.8 million, reflecting a striking 116% rise. Between 1990 and 2019, the occurrence of AD and different forms of dementia grew by 147.95% and 160.84%, respectively. Current predictions suggest that the figure could reach 150 million by the year 2050, marking almost fourfold growth compared to previous decades (Jessen et al. 2014; Li et al. 2022).

After 65, the incidence of AD increases each 5 years. Before 65, age-specific occurrence rates were not more than 1% annually; beyond the age of 85, they increased to 6% annually. After 65, the prevalence rates rise to 10%, and after age 85, they reach 40%. Women exhibit a somewhat greater occurrence of AD, particularly after the age of 85 (Qiu et al. 2009).

5.1.3 Pathophysiology

The defining neuropathological characteristics of Alzheimer's include the accumulation of abnormal neurofibrillary twists and neuritic plaques within the brain parenchyma. The lesions are accompanied by pronounced neuronal loss, with a particular vulnerability observed in the cholinergic neurons of the neocortex and basal forebrain. A schematic overview of the pathophysiological mechanisms underlying AD is illustrated in Fig. 5.1.

The prominent pathophysiological hypotheses behind the development of AD have been explained in the subsequent sections which include (i) the cholinergic hypothesis and (ii) the amyloid hypothesis.

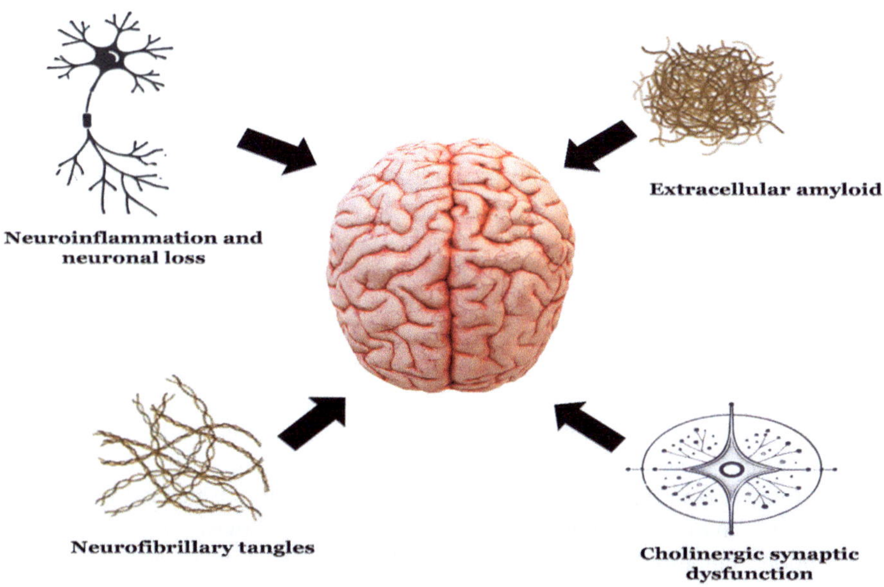

Fig. 5.1 Pathophysiological mechanisms behind Alzheimer's disease

5.1.4 The Cholinergic Hypothesis

One of the earliest pathological findings identified in the investigation of AD is a marked decline in the cerebral levels of ACh, a crucial neurotransmitter for learning and memory processes. This hypothesis posits that this decline, primarily resulting from the degeneration of neurons in the Nucleus Basalis of Meynert, plays a significant role in the pathogenesis of AD. This theory is supported by early and pronounced loss of cholinergic neurons observed in affected individuals, highlighting the essential role of ACh in mediating cognitive processes. Furthermore, β-amyloid has been implicated in the disruption of cholinergic neurotransmission, contributing to synaptic loss and impaired ACh release. The detrimental effects of anticholinergic drugs on memory performance in older adults further reinforce the critical involvement of cholinergic system in maintaining cognitive function (Hampel et al. 2018).

5.1.5 The Amyloid Hypothesis

One of the most accepted underlying mechanisms associated with AD is derived from the role of amyloids. Particularly relevant in inherited AD, this hypothesis proposes the contribution of amyloid beta (Aβ) peptide. The Aβ peptide is generated via sequential proteolytic cleavage of amyloid precursor protein (APP) by enzymes β-secretase (BACE1) and γ-secretase. Under physiological conditions, APP can also be processed by α-secretase, resulting in the formation of soluble, non-amyloidogenic fragments that do not contribute to plaque formation. Sequential fragmentation of APP through β-secretase and γ-secretase gives rise to Aβ42, a 42-amino acids long proteolytic peptide. Increased levels of Aβ42 promote the aggregation of amyloid, which is toxic to neurons. Aβ42 tends to favour the formation of fibrillary amyloid aggregates rather than the normal degradation of APP, contributing to the pathological processes observed in AD (Paroni et al. 2019; Morelli et al. 2012).

Genetics is also found to be responsible for AD. Early onset Familial Alzheimer's disease (FAD) often presents itself in children and young adults. This is caused by an additional copy of the APP gene, raising APP levels and causing amyloid-beta (Aβ) peptidase to be produced. This is further characterized during the occurrence of Down syndrome where a mutation of chromosome 21 can be seen. This results in existence of more than one pair of genes in many cells. This can subsequently result in occurrence of additional APP gene. Consequently, individuals with Down syndrome exhibit a marked risk of developing AD, largely due to the triplication of chromosome 21, which harbours the APP gene (Salehi et al. 2016). With almost total penetrance, AD can be inherited as an autosomal dominant condition. Mutations in three primary genes: APP on chromosome 21, Presenilin 1 (PSEN1) chromosome 14 and Presenilin 2 (PSEN2) in chromosome 1 are implicated in these hereditary variants. While APP mutations promote increased synthesis and deposition of Aβ peptides, mutations in PSEN1 and PSEN2 enhance γ-secretase activity, thereby accelerating Aβ generation and aggregation within the brain parenchyma (Hoogmartens et al. 2021).

5.1.6 Complex Relation Between Gut–Microbiome and AD

The gut microbiome (GMB), a highly diverse and dynamic consortium comprising trillions of bacteria, archaea, fungi, protozoa and viruses, is essential for preserving systemic physiological equilibrium. Increasingly, research has implicated the GMB in modulating neuroinflammatory processes and immune regulation across a range of neurological disorders, including AD, Parkinson's disease and multiple sclerosis. These effects are mediated through complex interactions along the GMB, involving microbial metabolites such as short-chain fatty acids (SCFAs), as well as immune and neuroendocrine signalling pathways. Dysbiosis within the GMB can perturb the production of these bioactive compounds and alter peripheral immune function, thereby influencing CNS immune responses and contributing to the pathophysiology and neurodegenerative conditions (Fukasawa et al. 2025).

Numerous studies have identified a distinct gut-microbiota profile in individuals with AD, marked by alterations in both microbial diversity and functional composition compared to cognitively healthy individuals. A prominent observation is the depletion of the beneficial butyrate producing bacteria; such as *Faecalibacterium* and *Roseburia*, which are essential for preserving intestinal epithelial integrity and mediating anti-inflammatory effect *via* the production of SCFAs. Concurrently, there is an enrichment of pro-inflammatory taxa, including specific members of *Proteobacteria* and *Bacteroidetes*, which are known to facilitate systemic inflammation (Sochocka et al. 2019; Schaible et al. 2025; Collins et al. 2012). This dysbiosis promotes the release if gut-derived inflammatory mediators capable of disrupting the blood-brain barrier (BBB), thereby permitting peripheral cytokines and microbial metabolites to access the CNS. In response, microglia, the resident immune cells of the CNS, become chronically activated and shift toward a pro-inflammatory phenotype. The activation is characterized by elevated secretion of interleukin-1β (IL-1β), tumour necrosis factor α (TNFα) and reactive oxygen species (ROS) (Wekerle 2016; Calsolaro and Edison 2016; Bonaz et al. 2018), all of which contribute to sustained neuroinflammation, synaptic impairment, and neuronal degeneration; hallmarks of AD progression. Beyond microglia, CNS-associated macrophages (CAM), including perivascular, leptomeningeal, and choroid plexus macrophages play a pivotal role in regulating neuroimmune homeostasis. These macrophages are actively involved in the clearance of Aβ across the BB and serve as key mediators of the bidirectional communication between the immune system and the brain. However, with disease progression, their efficiency in Aβ clearance diminishes, contributing to increased Aβ burden and amplifying neuroinflammatory cascades. Elucidating the complex interplay between the GBM, microglia and CAMs offers significant insight into the immunopathology of AD and identifies novel targets for therapeutic modulation. These finding underscore the importance of maintaining gut microbial balance for the CNS health and position the gut-brain axis as a promising focal point for the development of future AD interventions (Harach et al. 2017; Saji et al. 2019). A schematic representation of this intricate gut-brain relationship in provided in Fig. 5.2.

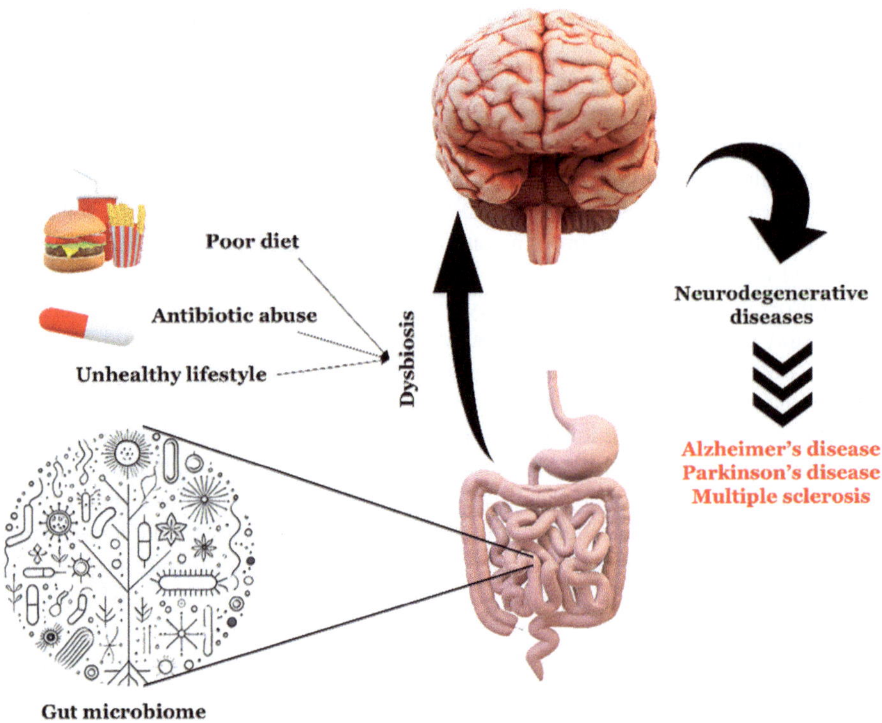

Fig. 5.2 Relationship between gut microbiota and CNS health

Following the findings about the role of gut-microbiota with AD, several researchers have opted for therapeutic advances involving dietary modifications. Dietary intake of prebiotics and probiotics have been established as major therapeutic tactic towards management of AD.

In the subsequent section of this chapter, we explore selected prebiotic and probiotic interventions that have demonstrated potential in modulating gut microbiota and contributing to the prevention or management of AD.

5.2 Prebiotics in Alzheimer's Disease

Building from emerging insights upon the function of the gut-microbiota in maintaining brain health, Glenn Gibson and Marcel Roberfroid introduced the term Prebiotics in 1995. Prebiotics are defined as non-digestible dietary components that confer health benefits to the host by selectively promoting the growth and/or metabolic activity of beneficial microbial populations, primarily within the colon, thereby contributing to the host's well-being (Davani-Davari et al. 2019). The initial definition of prebiotics primarily encompassed short and long-chain fructo-oligosaccharides (fructans) and lactulose, thereby limiting the scope of recognized prebiotic compounds. However, in 2008, the International Scientific Association for Probiotics and Prebiotics (ISAPP) expanded this definition, characterizing dietary prebiotics as selectively fermented substrates that induce specific alterations in the composition and/or activity of gastrointestinal microbiota, ultimately yielding health benefits to the host. To be classified as a prebiotic, a compound must fulfil the following criteria: (a) it must resist degradation by gastric acidity and mammalian digestive enzymes, and remain unabsorbed in the upper GIT, (b) it must be fermentable by resident intestinal microbiota, and (c) it must selectively stimulate the growth and/or metabolic activity of beneficial microbial taxa, thereby contributing to the host health (Gibson et al. 2017).

As discussed earlier, recent investigations have pointed out the beneficial contribution of healthy dietary components in alleviating AD progression. Dietary and prebiotic interventions that support the proliferation of beneficial gut microbial taxa, such as *Bifidobacteriun* spp., which are frequently diminished in individuals with AD; while concurrently suppressing the growth of potentially pathogenic genera such as *Bilophila*, may play a preventive role in AD pathogenesis. A key component in many prebiotic strategies is dietary fibre, which is broadly categorized based on its solubility in water. Insoluble fibre does not dissolve in water but instead retains water, increasing stool bulk and promoting gastrointestinal transit. In contrast, soluble fibre dissolves in water and is readily fermented by colonic microbiota, contributing to the production of SCFAs, enhancing satiety, and modulating metabolic and immune functions within the gut-brain axis. Recently, there has been a growing interest in classifying fibres based on functionality, which depends on their structural properties and fermentability. This shift in perspective is particularly relevant to Alzheimer's disease, as dietary fibres can influence gut microbiota composition, diversity, and richness. Certain fibres may promote beneficial microbial changes, resulting in formation of metabolites that enhance cognitive function and decrease neurotoxicity. A list of such dietary fibres which act as prebiotics and is beneficial in mitigating disease progression of AD has been provided in Table 5.1. In the next section, we discuss few dietary fibres, which can be used in the management of AD because of their established beneficial influence on the gut microbiota (Prosky et al. 1988).

5.2.1 Cellulose and Hemicellulose

Cellulose is a linear homopolysaccharide consisting of β-(1 → 4)-linked D-glucose monomers. As the most abundant organic polymer on Earth, it serves as a fundamental structural component of plant cell walls. Its extensive intra and intermolecular H-bonding network imparts high tensile strength and renders it as largely insoluble in water, contributing to the mechanical rigidity of plant tissues. In contrast, hemicellulose represents a structurally diverse class of branched heteropolysaccharides composed of various sugar residues such as xylose, mannose, arabinose and galactose. It has a more amorphous structure and is more easily degraded than cellulose (Klemm et al. 2005).

Table 5.1 A list of prebiotics linked with gut-microbiota

Prebiotic fibres	Primary function	Natural source	Associated bacteria
Cellulose	Provides shape and strength to cellular wall	As component of cell wall in leaves, barks, fruits among other plant organs	*Ruminococcus* spp. *Bacteroides* spp.
Hemicellulose	Imparts shape and mechanical rigidity of cell wall by interacting with cellulose and lignin in some cases	As component of cell wall in plants	*Butyrivibrio* spp. *Clostridium* spp. *Bacteroides* spp.
Lignin	Provides rigidity and hydrophobic protection to the cell walls. Also helps in transportation of water and minerals to the vascular bundles	Cell wall	*Bifidobacterium* spp.
Resistant starch	Acts as primary storage of food in plants. In human, it is fermented by microorganism in the colon	Seeds containing starch and cereal grains	*Bifidobacterium* spp. *Ruminococcus* spp.
Fructo-oligosaccharides (FOS)	Serves as important storage of carbohydrates	Bananas, onions, garlics, chicory, Jerusalem artichokes, asparagus	*Bifidobacterium* spp.
Inulin	Serves as important storage of carbohydrates	Agave, chicory roots, leeks, dandelions	*Bifidobacterium* spp.
Galacto-oligosaccharides (GOS)	Helps in sugar translocation and acts as carbohydrate reserve in vegetative organs	Legumes like beans, chickpeas and lentins	*Bifidobacterium* spp. *Lactobacillus* spp.
Beta (β) glucan	Development of cell wall, protection against pathogens	Cereals like oats and barley	*Bacteroides* spp. *Prevotella* spp. *Bifidobacterium* spp.
Pectin	Facilitates cellular adhesion, influences properties like porosity of the cell wall	Peels of citrus fruits, apples	*Bifidobacterium* spp. *Lactobacillus* spp. *Enterococcus* spp.

Certain commensal gut microorganisms, including *Ruminococcus* spp. and *Bacteroides* spp., are capable of producing cellulases; the enzymes that hydrolyze cellulose into smaller oligosaccharides and glucose. The carbohydrate products are subsequently fermented into SCFAs such as acetate, propionate and butyrate; which serve as key energy sources for colonocytes and contributes to intestinal health. In contrast to cellulose, hemicellulose is more readily degradable due to its amorphous and branched structure. It is metabolized by a wider range of gut bacteria, including

Butyrivibrio spp., *Clostridium* spp. and *Bacteroides* spp., which utilize enzymes such as xylanases and hemicellulases to break down hemicellulose into fermentable monosaccharides and oligosaccharides. The SCFAs not only support the gut barrier integrity and microbial homeostasis but also exhibit anti-inflammatory and neuroprotective properties, suggesting a mechanistic link between dietary fibre intake and brain health. The inclusion of cellulose and hemicellulose in the diet promotes expansion of SCFA-producing bacterial populations and enhances overall gut microbial diversity; factors that are closely associated with reduced systemic inflammation and improved metabolic and cognitive outcomes. Consequently, high-fibre diets may contribute to the attenuation of AD progression by modulating gut microbiota composition and reducing neuroinflammatory signalling through the gut-brain axis (Chassard et al. 2010, 2012; Robert et al. 2007; Wedekind et al. 1988).

5.2.2 Lignin

Lignin is a highly complex, water-insoluble biopolymer found in plant cell walls, where it functions as a structural support alongside cellulose and hemicellulose. Unlike these polysaccharides, lignin is not a carbohydrate but rather a cross-linked network of aromatic (phenylpropanoid) compounds, primarily composed of: ***a.*** Guaiacyl units (derived from coniferyl alcohol), ***b.*** Syringyl units (from sinapyl alcohol) and **c.** p-Hydroxyphenyl units (from p-coumaryl alcohol). This structural diversity makes lignin highly resistant to enzymatic degradation, contributing to plant rigidity, pathogen resistance, and reduced digestibility in many herbivores and omnivores. Lignin is considered non-fermentable due to its complex structure, which resists hydrolysis by most gut bacterial enzymes. Unlike cellulose and hemicellulose, which are broken down into SCFAs by gut microbes, lignin largely passes across the digestive system intact. Due to its heterogeneous structure and chemical stability, lignin's interaction with gut microbiota remains poorly understood. However, one study suggests that lignin may support the prolonged survival of *Bifidobacterium* spp. under in vitro conditions, indicating a potential role in maintaining gut microbial balance (Niemi et al. 2013; Aura et al. 2013).

5.2.3 Resistant Starch (RS)

RS is a form of water-insoluble starch polysaccharide that resists hydrolysis by the host digestive enzymes, particularly α-amylase, and consequently reaches the colon in an undigested form. Within the colon, RS serves as a fermentable substrate for specific gut microbiota, leading to the generation of SCFAs; predominantly butyrate. Butyrate plays a crucial role in maintaining gut epithelial integrity, exerting anti-inflammatory effects, and supporting colonocyte function. Notably, RS consumption has been associated with an increased Firmicutes-to-Bacteroides ratio, a microbial shift commonly linked to gut homeostasis and improved metabolic outcomes. In the context of neurodegeneration, butyrate has been shown to suppress the production of pro-inflammatory cytokines like IL-1β and TNF-α; thereby attenuating neuroinflammation, a central feature of AD pathology. Additionally, butyrate enhances the structural and functional integrity of BBB, reducing the translocation of peripheral inflammatory mediators into the CNS. At microbial level, RS selectively promotes the proliferation of beneficial genera, including *Bifidobacterium* and *Ruminococcus* spp., highlighting its potential role in modulating gut microbiota composition in a manner conducive to brain health (Ze et al. 2012; Crost et al. 2018).

5.2.4 Fructans

Fructans are carbohydrates composed of fructose chains, naturally occurring in various plants, as well as certain bacteria and fungi, where they serve as storage of carbohydrates and enhance resistance to cold and drought conditions. It is a group of water-soluble dietary fibres composed of fructose molecules linked together, often with a terminal glucose unit. They are non-starch polysaccharides that escape in the small intestine undigested and undergo fermentation by gut microbiota in the colon. Fructans are categorized based on their degree of polymerization (DP): (i) Short-chain fructans (Fructo-oligosaccharides, FOS) consisting of 2–10 fructose units and (ii) Long-chain fructans (Inulin) with more than 10 fructose units. These fibres naturally occur in food items such as onions, garlic, leeks, asparagus, chicory, wheat, and bananas (Roberfroid and Delzenne 1998; Franco-Robles and Lopez 2015; Liu et al. 2017).

Fructans have shown potential in modulating AD pathophysiology by influencing gut microbiota composition, attenuating neuroinflammation, and supporting CNS function *via* the gut-brain axis. These non-digestible oligosaccharides selectively stimulate the proliferation of beneficial bacterial genera like *Bifidobacterium* and *Lactobacillus*, which in turn facilitate the production of SCFAs, including butyrate. Butyrate, through its anti-inflammatory and neuroprotective effects maintains the BBB integrity, thereby limits the infiltration of peripheral toxins and inflammatory mediators into the brain parenchyma. In addition to supporting BBB function, fructan-derived SCFAs modulate microglial activation, shifting microglia from a pro-inflammatory to a homeostatic phenotype; mitigating chronic neuroinflammation. Furthermore, emerging evidence suggests that gut microbial fermentation of fructans may reduce cerebral accumulation of Aβ plaques and hyperphosphorylated tau (τ), the principal neuropathological markers of AD. Fructans have also been associated with improved insulin sensitivity and metabolic regulation, both of which are closely linked to cognitive decline and dementia risk. By promoting gut microbial balance, reducing systemic inflammation, and enhancing brain resilience, fructans offer a promising dietary approach for AD prevention and management (Salazar et al. 2015; Tarini and Wolever 2010).

Similar to FOS, galacto-oligosaccharides (GOS) are also prebiotic fibres composed of galactose molecules linked together, with a terminal glucose unit. They are non-digestible carbohydrates that escape enzymatic degradation in small intestine and undergo fermentation by colonic microbiota. While naturally present in trace amounts in human breast milk and certain dairy products, GOS shapes gut microbial ecology by selectively promoting the growth of *Bifidobacterium* and *Lactobacillus*. Upon fermentation by gut microbes, GOS are metabolized into SCFAs, such as butyrate, acetate and propionate. These generation of SCFAs promote neuroprotective effects similar to that explained in previous section (Wang et al. 2016; Cardelle-Cobas et al. 2011).

Furthermore, GOS support microglial function, ensuring that these brain-resident immune cells efficiently clear amyloid-β plaques while avoiding excessive inflammatory responses. They also influence neurotransmitter regulation, helping maintain optimal levels of serotonin, GABA, and dopamine, critical for mood stability, cognitive function, and memory formation. Moreover, GOS contribute to metabolic health by enhancing insulin sensitivity, reducing the risk of insulin resistance, known to be closely related to AD progression.

5.2.5 Beta (β) Glucan

β-glucan is a polysaccharide made up of β-D-glucose monomers linked *via* glycosidic bonds, with its physicochemical properties varying; based on source and structural conformation. The most physiologically relevant form, commonly derived from the bran of cereal grain such as oats and barley, is a linear, water-soluble β-glucan characterized by a mixed linkage of β-(1–3) and β-(1–4) glycosidic bonds. This unique arrangement imparts high viscosity and contributes to its ability to increase the bulk and moisture content of the faecal matter, thereby supporting healthy intestinal transit. This makes β-glucan an important dietary fibre for gut health.

Owing to its fermentable properties, β-glucan functions as a metabolic substrate for saccharolytic gut microbiota, thereby facilitating the proliferation of beneficial bacterial taxa within the colon. Experimental and clinical studies have demonstrated that the intake of high molecular weight β-glucan is associated with an increased relative abundance of *Bacteroides* and *Prevotella*; two genera known for their roles in fibre degradation and SCFA biosynthesis. Furthermore, dietary supplementation with whole grain oats or oats bran, both of which are rich in β-glucan, has been linked to elevated colonic production of SCFAs including butyrate, acetate and propionate, metabolites that are integral to gut homeostasis and systemic health. These SCFAs provide anti-inflammatory and neuroprotective benefits, reinforcing the importance of β-glucan in gut and overall metabolic health. Furthermore, oat-derived β-glucan exhibits a bifidogenic effect, stimulates the growth of *Bifidobacterium* spp., a key beneficial microbe known for its role in immune modulation and gut barrier maintenance. By modulating gut microbiota composition, enhancing SCFA production, and reducing systemic inflammation, β-glucan could be a valuable dietary component for preventing or managing neurodegenerative conditions such as Alzheimer's disease (Wang et al. 2016; Kristek et al. 2019).

5.2.6 Pectin

Pectin is a soluble dietary fibre, naturally present in apple skins, citrus fruits, and various vegetables, where it contributes to plant tissue structure and integrity. As a significant factor of the primary cell wall and middle lamella, pectin facilitates cell adhesion and provides mechanical strength to plant tissues. Structurally, pectin is a complex polysaccharide rich in galacturonic acid, with homogalacturonan being its most abundant form. This polymer consists of galacturonic acid units linked at β-1,4 positions, while different side-chain sugars contribute to structural variations in pectin, influencing its gel-forming ability, solubility, and physiological effects. Due to these properties, pectin is widely used as a gelling agent and stabilizer, while its health benefits make it an essential component of a fibre-rich diet (Yudishter and Dash 2025; Mohnen 2008).

Pectin, a fermentable prebiotic fibre, promotes the growth of advantageous gut microbes such as *Bifidobacterium*, *Lactobacillus* and *Enterococcus*, thereby supporting microbial equilibrium within the GIT. Its microbial degradation results in the production of SCFAs like butyrate, acetate and propionate; which are essential for reinforcing the intestinal barrier, modulating immune responses, and mitigating systemic inflammation. By influencing the gut-brain communication axis, pectin may play a neuroprotective role through the reduction of inflammatory and oxidative processes implicated in the progression of AD and other neurodegenerative disorders. Consistent consumption of pectin-rich foods may therefore aid in preserving cognitive function and enhancing overall health. Additionally, studies have shown that pectin can increase gut microbial diversity, particularly in individuals with ulcerative colitis, highlighting its potential therapeutic application in inflammatory bowel disease (IBD) management (Shinohara et al. 2010).

5.3 Probiotics in Alzheimer's Disease

The ISAPP describes probiotics as live microorganisms that provide health benefits to the host when consumed in sufficient quantities (Prosky et al. 1988; Ajayi et al. 2020). Probiotics exhibit neuroprotective properties in the context of AD; primarily through modulation of the gut-brain axis, attenuation of neuroinflammatory processes, and restoration of intestinal microbial balance. AD has been strongly associated with gut dysbiosis, which comprises gut-brain communication, increases BBB permeability, and exacerbates CNS inflammation. Probiotic strains, particularly those belonging to the genera *Lactobacillus* and *Bifidobacterium*, counteract these pathological effects through multiple mechanisms:

- Immune regulation: Probiotics downregulate pro-inflammatory cytokines such as IL-1β, interleukin-6 (IL-6) and TNF-α; while promoting the anti-inflammatory pathways, thereby reducing neuroinflammatory burden.
- SCFA production: The microbial synthesis of SCFAs; notably butyrate, acetate and propionate, enhances intestinal barrier function, supports BBB integrity, and

influences neurotransmitter biosynthesis, including serotonin and γ-amino butyric acid (GABA).
- Neuroprotective modulation: Probiotics have been shown to increase the release of glucagon-like peptide-1 (GLP-1), a neuroprotective hormone that inhibits neuronal apoptosis and improves cognitive performance through neurotransmitter modulation.
- Reduction of oxidative stress: By scavenging ROS and mitigating oxidative damage, probiotics contribute to the preservation of neuronal integrity and function.
- HPA axis regulation: Probiotic supplementation can attenuate stress-induced activation of hypothalamic-pituitary-adrenal (HPA) axis, thereby reducing glucocorticoid-mediated neurotoxicity, which is implicated in cognitive impairment (Verbeke et al. 2015; Correa-Oliveira et al. 2016; Vijay and Morris 2014; Chen et al. 2018; Yano et al. 2015).

Preclinical studies in animal models have been conducted to evaluate the potential of probiotics in modulating key pathological and cognitive features of AD. One such study assessed the impact of *Lactobacillus plantarum* MTCC 1325 in a D-galactose-induced AD mouse model, revealing significant reductions in neurofibrillary tangles (NFT) formation and Aβ deposition, along with elevated ACh concentration in both the hippocampus and cerebral cortex. These neurochemical enhancements were positively correlated with marked improvements in spatial learning and memory performance in probiotic-treated group. The observed neuroprotective effects of L. plantarum MTCC 1325 are attributed to its capacity to synthesize ACh, capable of showing antioxidant activity. This increase in antioxidant capacity appears to Na^+/K^+-ATPase activity by inhibiting acetylcholinesterase (*AChE*), thereby sustaining cholinergic signalling and supporting cognitive function. Moreover, the elevated ACh levels contribute in stabilizing the soluble α-helical conformation of Aβ (25–35), effectively inhibiting its pathological aggregation into oligomeric and fibrillar forms, which are central to Aβ plaque formation, a hallmark of AD neuropathology (Nimgampalle and Kuna 2017). Additional evidence supporting the neuroprotective effects of probiotics comes from a study involving Aβ-injected rat models, where supplementation with *Lactobacillus* and *Bifidobacterium* strains over a period of 56 days led to significant neurological improvements. Notably, probiotic treatment restored hippocampal synaptic plasticity, an effect associated with the normalization of long-term potentiation (LTP), a critical process in enhancing neuronal signal transduction and memory encoding. Moreover, the intervention resulted in a reduction of microglial activation, indicating a dampened neuroinflammatory response. Simultaneously, levels of brain-derived neurotrophic factor (BDNF) and synapsin were significantly upregulated, both of which are key modulators of neuronal health, synaptic integrity, and cognitive function. These findings were strongly associated with improvements in spatial learning and memory performance, further emphasizing the therapeutic potential of probiotic supplementation of AD-related cognitive decline (Rezaei Asl et al. 2019).

Further reinforcing the therapeutic potential of probiotics in AD, studies by Bonfili et al. revealed that supplementation with the SLAB51; a probiotic formulation, significantly reduced Aβ aggregation in transgenic AD (Tg-AD) mice, thereby delaying disease onset and progression. Treatment with SLAB51 led to upregulation of glucose transporters GLUT1 and GLUT3, as well as insulin-like growth factor receptor (IGF-IR), while concurrently suppressing the phosphorylation of AMP-activated protein kinase (AMPK) and Akt. Although AMPK activation typically promotes autophagy, its reduced phosphorylation in this context appears to reflect modulated autophagic flux, aiding in the clearance of Aβ aggregates and attenuation of τ pathology. Additionally, probiotic intake enhanced Sirtuin-1 (SIRT1) levels; an NAD^+ dependent deacetylase known for its neuroprotective and antioxidant functions. SIRT1 plays a pivotal role in reducing ROS and promoting neuronal survival by counteracting p53-mediated apoptosis. This protective effect is mediated through SIRT1-p53 binding, which inhibits apoptotic pathways, and through the deacetylation of PARP-1, further reducing cell death. Moreover, SLAB51 probiotics increased expression of retinoic acid receptor β (RARβ), which is instrumental in shifting the APP processing towards the non-amyloidogenic pathway. This shift is facilitated through SIRT1-dependent deacetylation, enhancing ADAM10 (disintegrin and metalloproteinase 10) transcription, a key α-secretase responsible for cleaving APP in a non-pathogenic manner. Consequently, this pathway leads to reduced Aβ production and deposition, highlighting a multifaceted mechanism through which probiotics exert protective effects in AD (Bonfili et al. 2018).

Another study by Bonfili et al. comprised of the investigations demonstrated the neuroprotective effects of probiotics in AD mouse models. The findings showed amplified levels of cathepsin L, beclin-1, LC3-II, fibroblast growth factor 9 (FGF9), and cortical thickness, along with elevated gut levels of ghrelin, leptin, glucagon-like peptide-1 (GLP-1), and gastric inhibitory polypeptide (GIP) after 4 months of probiotic supplement. Moreover, there was a noteworthy drop in chymotrypsin-like, trypsin-like, peptidyl-glutamyl peptide-hydrolyzing activities, and p62 levels in brain homogenates. FGF9 plays an essential part in promoting cell proliferation and reducing neuronal death under oxidative stress, while also enhancing neurotrophic factors and downregulating apoptotic markers such as caspase-3. Probiotic-induced cortical thickening in the hippocampus aligns with findings that *Lactococcus lactis* consumption supports cortical neural progenitor cell proliferation and increases cortical neuron density. Beyond their neuroprotective role, probiotics also regulate gut microbiota, which is intimately connected to neurodegenerative processes. Modifying the gut microbiome through probiotics has been associated with increased leptin levels in serum and higher plasma ghrelin levels. Ghrelin, prominent for its involvement in glucose and lipid metabolism, also impacts mitochondrial respiration and has neuroprotective properties. Along with leptin, ghrelin stimulates neurotrophic factor secretion, which helps prevent amyloid-β (Aβ) oligomer toxicity and reduces phosphorylated τ accumulation in AD. Additionally, probiotics restore cathepsin B, a cysteine protease linked to amyloid plaques, by enhancing β-secretase activity, leading to a reduction in Aβ levels. The regulation of autophagy-related markers further indicates improved cellular waste clearance. Increased beclin-1 and LC3-II levels, along with decreased p62 levels, suggest enhanced autophagic flux, promoting the removal of toxic protein aggregates and mitigating AD pathology (Bonfili et al. 2020; Teglas et al. 2020).

A 2018 study by Azm et al. further validated the neuroprotective benefits of probiotics in AD. The researchers assessed the impact of a probiotic formulation containing *Lactobacillus acidophilus* 1688FL431-16LA02, *Lactobacillus fermentum* ME3, *Bifidobacterium lactis* 1195SL609-16BS01, and *Bifidobacterium longum* 1152SL593-16BL03 on $A\beta_{1-42}$-injected rats. The administration of 1×10^{10} CFU/g of this probiotic mixture produced outcomes similar to previous in vivo studies, including a decrease in Aβ plaque deposition followed by improvements in cognitive abilities. One notable finding was the significant increase in superoxide dismutase (SOD) levels. As a primary antioxidant enzyme, SOD takes important part in neutralizing oxidative stress, which is a major factor in neurotoxicity, impaired neuronal signalling, neuroinflammation, and Aβ aggregation. Oxidative stress contributes to cognitive dysfunction and accelerates AD pathology. By enhancing the body's antioxidant defence, probiotics help mitigate neuroinflammation, facilitate the clearance of Aβ plaques, and ultimately reduce both their number and size. These findings emphasize the ability of probiotics as a beneficial intervention for AD by targeting oxidative damage and plaque accumulation (Athari Nik Azm et al. 2018).

A separate study conducted by Rezaeiasl et al. (2019) underscored the cognitive-enhancing potential of probiotics in AD. The researchers evaluated the effects of a probiotic formulation comprising *Lactobacillus acidophilus*, *Bifidobacterium bifidum* and *Bifidobacterium longum* in a mouse model of AD induced by $A\beta_{1-42}$ administration. The treatment resulted in a notable increase in excitatory postsynaptic potentials (EPSPs) within the hippocampus, indicating enhanced synaptic responsiveness. Moreover, a reduction in paired-pulse facilitation (PPF) and serum nitric oxide (NO) levels were also observed. Given that PPF is a physiological marker of presynaptic neurotransmitter release and is closely tied to LTP, its decrease suggests improved presynaptic function and heightened neurotransmitter release. The concomitant increase in LTP further supports the hypothesis that probiotic supplementation strengthens synaptic plasticity, thereby facilitating learning and memory processes. Collectively, these findings provide additional evidence for the therapeutic promise of probiotics in mitigating cognitive decline associated with AD, through modulation of synaptic efficiency and neurochemical signalling (Rezaeiasl et al. 2019).

Additional research by Patel et al. demonstrated that a combined intervention of curcumin and *Lactobacillus rhamnosus* holds promise as an adjunctive therapeutic strategy for dementia. In a scopolamine-induced mouse model of cognitive impairment, this combination was shown to enhance memory retention, improve learning performance and elevate the activity of endogenous antioxidant enzymes, suggesting synergistic neuroprotective effects (Patel et al. 2020). Cogliati et al. explored the neuroprotective role of *Bacillus subtilis* in a *Caenorhabditis elegans* model of AD. Using the CL2120 strain, which expresses the neurotoxic peptide $A\beta_{1-42}$ pan-neuronally, the study demonstrated that colonization by *B. subtilis* conferred resistance to Aβ-induced behavioural deficits. Importantly, the findings highlighted that the formation of a gut-associated biofilm was critical for mediating the anti-Alzheimer's effects of *B. subtilis*, suggesting that probiotic-host interactions at the mucosal surface play a key role in modulating disease-related neurotoxicity (Cogliati et al. 2020). Kobayashi et al. reported that supplementation with *Bifidobacterium breve* A1 strain and *Bifidobacterium infantis* resulted in reduced Aβ deposition and decreased levels of pro-inflammatory cytokines in the hippocampus of AD mice. Concurrently, there was a significant elevation in the SOD activity, indicating increased antioxidant defence. These probiotic interventions also improved cognitive performance and modulated immune signalling by elevating plasma acetate levels, which is directly correlated with the downregulation of immune-reactive gene expression in the hippocampus. These outcomes underscore the anti-inflammatory and neuroprotective potential of *Bifidobacterium* strains in mitigating Aβ-associated neurotoxicity (Kobayashi et al. 2017). Supporting these findings, Mehrabadi et al. conducted a complementary study involving a five-group rat model, comprising a control group, a sham group, an AD model (injected intrahippocampally with $A\beta_{1-40}$), a probiotic-treated AD group and an AD + rivastigmine group (treated with 0.6 mg/kg rivastigmine). The probiotic group received a daily oral dose of 2 g (10^{10} CFU) containing *Lactobacillus reuteri*, *Lactobacillus rhamnosus* and *Bifidobacterium infantis* for 10 weeks. Behavioural assessment with Morris Water Maze revealed significant enhancements in spatial learning and memory in the group that received probiotic supplementation. Biochemically, the probiotic intervention led to a reduction in Aβ plaque burden, lowered malondialdehyde (MDA) levels, and an increase in SOD activity, reflecting both anti-amyloidogenic and antioxidative effects. Additionally, a reduced level of pro-inflammatory IL-1 and TNF-α concentrations further confirmed anti-inflammatory action of the probiotic mixture (Mehrabadi and Sadr 2020).

The potential role of probiotics in managing AD has also been explored in clinical setup. In a study conducted by Agahi et al., participants were divided into a control group and a probiotic intervention group to evaluate the effects of probiotic supplementation on cognitive function and systemic inflammation. The probiotic formulation administered consisted of *Bifidobacterium bifidum*, *Lactobacillus acidophilus*, *Lactobacillus plantarum* and *Lactobacillus fermentum*, with the treatment group receiving one pill daily. Cognitive abilities were assessed using standardized memory tests, and blood samples were analysed both before and after the intervention to measure levels of inflammatory cytokines and oxidative stress indicators, including MDA, glutathione and 8-hydroxy-2′-deoxyguanosine (8-OHdG). Although some biomarker fluctuations were observed, the intervention did not yield statistically significant improvements in cognitive outcomes among individuals with advanced AD (Agahi et al. 2019). In contrast, another randomized controlled trial involving 79 AD patients assessed the synergistic effects of probiotics and selenium co-supplementation over 12 weeks long period. Participants were allocated to receive either a placebo or a combined daily regimen of 200 μg of selenium with 2 x 10^9 CFU of probiotics containing *Bifidobacterium bifidum*, *Bifidobacterium longum* and *Lactobacillus acidophilus*. The combination therapy group demonstrated notable improvements in cognitive function, as reflected by enhanced mini-mental state examination (MMSE) scores. Furthermore, this group exhibited significant increases in total glutathione concentration and overall antioxidant capacity, alongside marked reductions in inflammatory markers (e.g. high-sensitivity C-reactive protein) and lipid profile parameters including triglycerides, LDL, VLDL, and total cholesterol. These findings suggest that probiotics, particularly when combined with antioxidants like selenium, may offer cognitive and metabolic benefits in individuals with mild to moderate AD (Tamtaji et al. 2019).

In a separate clinical investigation, Leblhuber et al. examined the impact of probiotic supplementation on gut-related immune responses in a group of 20 outpatient individuals with AD. The study assessed gut microbiota composition, inflammatory markers, and immune activation indicators both prior to and following the intervention. The results revealed a notable elevation in circulating kynurenine levels, which the authors attributed to macrophage activation; a mechanism potentially involved in the clearance of Aβ aggregates. Additionally, a strong positive correlation between the kynurenine-tryptophan (*Kyn/Trp*) ratio and neopterin levels was observed, suggesting that innate immune cells, particularly macrophages and dendritic cells, were actively engaged in modulating the neuroimmune response during probiotic treatment (Leblhuber et al. 2018).

5.4 Conclusion and Future Scope

AD is a multifactorial neurodegenerative disorder for which a definitive cure remains elusive. However, recent advances in research underscore the pivotal influence of the gut microbiota in modulating disease onset and progression. Both probiotics and prebiotics are emerging as promising adjunct therapeutic strategies aimed at restoring gut-brain axis homeostasis, alleviating neuroinflammation, enhancing antioxidant capacity, and supporting cognitive performance. Evidence from preclinical and clinical studies suggests that specific probiotic strains, particularly those belonging to the *Lactobacillus* and *Bifidobacterium* genera, can attenuate Aβ deposition, regulate immune signalling, and promote the expression of neurotrophic factors, collectively contributing to the deceleration of neurodegenerative processes. Likewise, prebiotics, through the selective stimulation of commensal gut bacteria, enhance intestinal barrier integrity and mitigate systemic inflammatory responses, further supporting brain health in the context of AD.

Despite these encouraging findings, further research is needed to fully understand the precise mechanisms by which probiotics and prebiotics influence AD pathology. Future studies should focus on large-scale clinical trials to confirm their effectiveness, refine strain selection, and determine optimal dosages for therapeutic use. Additionally, investigating the combined effects of probiotics with dietary interventions, pharmaceutical treatments, and lifestyle modifications could enhance their benefits. Advancements in microbiome research, including personalized probiotic therapies tailored to individual gut profiles, hold great potential for developing targeted interventions against AD. Moreover, assessing the long-term safety and efficacy of probiotic supplementation is essential before incorporating them into standard AD treatment protocols. In summary, probiotics and prebiotics offer a promising, non-invasive approach to managing Alzheimer's disease by modulating gut microbiota and reducing neuroinflammation. Continued interdisciplinary research in microbiome science and neurodegenerative disorders may open new avenues for delaying or preventing AD.

References

Agahi A, Hamidi GA, Daneshvar R, Hamdieh M, Soheili M, Alinaghipour A et al (2019) Corrigendum: does severity of Alzheimer's disease contribute to its responsiveness to modifying gut microbiota? A double blind clinical trial. Front Neurol 10:31. https://doi.org/10.3389/fneur.2019.00031

Ajayi A, Jolaiya T, Smith S (2020) Evolving technologies in gastrointestinal microbiome era and their potential clinical applications. J Clin Med 9. https://doi.org/10.3390/jcm9082565

Athari Nik Azm S, Djazayeri A, Safa M, Azami K, Ahmadvand B, Sabbaghziarani F et al (2018) Lactobacilli and bifidobacteria ameliorate memory and learning deficits and oxidative stress in beta-amyloid (1-42) injected rats. Appl Physiol Nutr Metab 43:718–726. https://doi.org/10.1139/apnm-2017-0648

Aura AM, Niemi P, Mattila I, Niemela K, Smeds A, Tamminen T et al (2013) Release of small phenolic compounds from brewer's spent grain and its lignin fractions by human intestinal microbiota in vitro. J Agric Food Chem 61:9744–9753. https://doi.org/10.1021/jf4024195

Bonaz B, Bazin T, Pellissier S (2018) The Vagus nerve at the Interface of the microbiota-gut-brain Axis. Front Neurosci 12:49. https://doi.org/10.3389/fnins.2018.00049

Bonfili L, Cecarini V, Cuccioloni M, Angeletti M, Berardi S, Scarpona S et al (2018) SLAB51 probiotic formulation activates SIRT1 pathway promoting antioxidant and neuroprotective effects in an AD mouse model. Mol Neurobiol 55:7987–8000. https://doi.org/10.1007/s12035-018-0973-4

Bonfili L, Cecarini V, Gogoi O, Berardi S, Scarpona S, Angeletti M et al (2020) Gut microbiota manipulation through probiotics oral administration restores glucose homeostasis in a mouse model of Alzheimer's disease. Neurobiol Aging 87:35–43. https://doi.org/10.1016/j.neurobiolaging.2019.11.004

Breijyeh Z, Karaman R (2020) Comprehensive review on Alzheimer's disease: causes and treatment. Molecules 25. https://doi.org/10.3390/molecules25245789

Calsolaro V, Edison P (2016) Neuroinflammation in Alzheimer's disease: current evidence and future directions. Alzheimers Dement 12:719–732. https://doi.org/10.1016/j.jalz.2016.02.010

Cardelle-Cobas A, Corzo N, Olano A, Pelaez C, Requena T, Avila M (2011) Galactooligosaccharides derived from lactose and lactulose: influence of structure on lactobacillus, streptococcus and Bifidobacterium growth. Int J Food Microbiol 149:81–87. https://doi.org/10.1016/j.ijfoodmicro.2011.05.026

Chassard C, Delmas E, Robert C, Bernalier-Donadille A (2010) The cellulose-degrading microbial community of the human gut varies according to the presence or absence of methanogens. FEMS Microbiol Ecol 74:205–213. https://doi.org/10.1111/j.1574-6941.2010.00941.x

Chassard C, Delmas E, Robert C, Lawson PA, Bernalier-Donadille A (2012) Ruminococcus champanellensis sp. nov., a cellulose-degrading bacterium from human gut microbiota. Int J Syst Evol Microbiol 62:138–143. https://doi.org/10.1099/ijs.0.027375-0

Chen XY, Du YF, Chen L (2018) Neuropeptides exert neuroprotective effects in Alzheimer's disease. Front Mol Neurosci 11:493. https://doi.org/10.3389/fnmol.2018.00493

Cogliati S, Clementi V, Francisco M, Crespo C, Arganaraz F, Grau R (2020) Bacillus subtilis delays neurodegeneration and behavioral impairment in the Alzheimer's disease model Caenorhabditis Elegans. J Alzheimer's Dis 73:1035–1052. https://doi.org/10.3233/JAD-190837

Collins SM, Surette M, Bercik P (2012) The interplay between the intestinal microbiota and the brain. Nat Rev Microbiol 10:735–742. https://doi.org/10.1038/nrmicro2876

Correa-Oliveira R, Fachi JL, Vieira A, Sato FT, Vinolo MA (2016) Regulation of immune cell function by short-chain fatty acids. Clin Transl Immunol 5:e73. https://doi.org/10.1038/cti.2016.17

Crost EH, Le Gall G, Laverde-Gomez JA, Mukhopadhya I, Flint HJ, Juge N (2018) Mechanistic insights into the cross-feeding of Ruminococcus gnavus and Ruminococcus bromii on host and dietary carbohydrates. Front Microbiol 9:2558. https://doi.org/10.3389/fmicb.2018.02558

Davani-Davari D, Negahdaripour M, Karimzadeh I, Seifan M, Mohkam M, Masoumi SJ et al (2019) Prebiotics: definition, types, sources, mechanisms, and clinical applications. Foods 8. https://doi.org/10.3390/foods8030092

Ferreira-Vieira TH, Guimaraes IM, Silva FR, Ribeiro FM (2016) Alzheimer's disease: targeting the cholinergic system. Curr Neuropharmacol 14:101–115. https://doi.org/10.2174/1570159x13666150716165726

Franco-Robles E, Lopez MG (2015) Implication of fructans in health: immunomodulatory and antioxidant mechanisms. ScientificWorldJournal 2015:289267. https://doi.org/10.1155/2015/289267

Fukasawa N, Tsunoda J, Sunaga S, Kiyohara H, Nakamoto N, Teratani T et al (2025) The gut-organ axis: clinical aspects and immune mechanisms. Allergol Int. https://doi.org/10.1016/j.alit.2025.01.004

Gibson GR, Hutkins R, Sanders ME, Prescott SL, Reimer RA, Salminen SJ et al (2017) Expert consensus document: the international scientific Association for Probiotics and Prebiotics (ISAPP) consensus statement on the definition and scope of prebiotics. Nat Rev Gastroenterol Hepatol 14:491–502. https://doi.org/10.1038/nrgastro.2017.75

Hampel H, Mesulam MM, Cuello AC, Farlow MR, Giacobini E, Grossberg GT et al (2018) The cholinergic system in the pathophysiology and treatment of Alzheimer's disease. Brain 141:1917–1933. https://doi.org/10.1093/brain/awy132

Harach T, Marungruang N, Duthilleul N, Cheatham V, Mc Coy KD, Frisoni G et al (2017) Reduction of Abeta amyloid pathology in APPPS1 transgenic mice in the absence of gut microbiota. Sci Rep 7:41802. https://doi.org/10.1038/srep41802

Hoogmartens J, Cacace R, Van Broeckhoven C (2021) Insight into the genetic etiology of Alzheimer's disease: a comprehensive review of the role of rare variants. Alzheimers Dement (Amst) 13:e12155. https://doi.org/10.1002/dad2.12155

Jessen F, Amariglio RE, van Boxtel M, Breteler M, Ceccaldi M, Chetelat G et al (2014) A conceptual framework for research on subjective cognitive decline in preclinical Alzheimer's disease. Alzheimers Dement 10:844–852. https://doi.org/10.1016/j.jalz.2014.01.001

Jiang C, Li G, Huang P, Liu Z, Zhao B (2017) The gut microbiota and Alzheimer's disease. J Alzheimer's Dis 58:1–15. https://doi.org/10.3233/JAD-161141

Klemm D, Heublein B, Fink HP, Bohn A (2005) Cellulose: fascinating biopolymer and sustainable raw material. Angew Chem Int Ed Engl 44:3358–3393. https://doi.org/10.1002/anie.200460587

Knopman DS, Amieva H, Petersen RC, Chetelat G, Holtzman DM, Hyman BT et al (2021) Alzheimer disease. Nat Rev Dis Primers 7:33. https://doi.org/10.1038/s41572-021-00269-y

Kobayashi Y, Sugahara H, Shimada K, Mitsuyama E, Kuhara T, Yasuoka A et al (2017) Therapeutic potential of Bifidobacterium breve strain A1 for preventing cognitive impairment in Alzheimer's disease. Sci Rep 7:13510. https://doi.org/10.1038/s41598-017-13368-2

Kowalski K, Mulak A (2019) Brain-gut-microbiota Axis in Alzheimer's disease. J Neurogastroenterol Motil 25:48–60. https://doi.org/10.5056/jnm18087

Kristek A, Wiese M, Heuer P, Kosik O, Schar MY, Soycan G et al (2019) Oat bran, but not its isolated bioactive beta-glucans or polyphenols, have a bifidogenic effect in an in vitro fermentation model of the gut microbiota. Br J Nutr 121:549–559. https://doi.org/10.1017/S0007114518003501

Leblhuber F, Steiner K, Schuetz B, Fuchs D, Gostner JM (2018) Probiotic supplementation in patients with Alzheimer's dementia – an explorative intervention study. Curr Alzheimer Res 15:1106–1113. https://doi.org/10.2174/1389200219666180813144834

Li X, Feng X, Sun X, Hou N, Han F, Liu Y (2022) Global, regional, and national burden of Alzheimer's disease and other dementias, 1990-2019. Front Aging Neurosci 14:937486. https://doi.org/10.3389/fnagi.2022.937486

Liu F, Li P, Chen M, Luo Y, Prabhakar M, Zheng H et al (2017) Fructooligosaccharide (FOS) and Galactooligosaccharide (GOS) increase Bifidobacterium but reduce butyrate producing bacteria with adverse glycemic metabolism in healthy young population. Sci Rep 7:11789. https://doi.org/10.1038/s41598-017-10722-2

Lourida I, Soni M, Thompson-Coon J, Purandare N, Lang IA, Ukoumunne OC et al (2013) Mediterranean diet, cognitive function, and dementia: a systematic review. Epidemiology 24:479–489. https://doi.org/10.1097/EDE.0b013e3182944410

Mehrabadi S, Sadr SS (2020) Assessment of probiotics mixture on memory function, inflammation markers, and oxidative stress in an Alzheimer's disease model of rats. Iran Biomed J 24:220–228. https://doi.org/10.29252/ibj.24.4.220

Mohnen D (2008) Pectin structure and biosynthesis. Curr Opin Plant Biol 11:266–277. https://doi.org/10.1016/j.pbi.2008.03.006

Morelli L, Perry G, Tagliavini F (2012) The contribution of the amyloid hypothesis to the understanding of Alzheimer's disease: a critical overview. Int J Alzheimers Dis 2012:709613. https://doi.org/10.1155/2012/709613

Niemi P, Aura AM, Maukonen J, Smeds AI, Mattila I, Niemela K et al (2013) Interactions of a lignin-rich fraction from brewer's spent grain with gut microbiota in vitro. J Agric Food Chem 61:6754–6762. https://doi.org/10.1021/jf401738x

Nimgampalle M, Kuna Y (2017) Anti-Alzheimer properties of probiotic, lactobacillus plantarum MTCC 1325 in Alzheimer's disease induced albino rats. J Clin Diagn Res 11:KC01-KC5. https://doi.org/10.7860/JCDR/2017/26106.10428

Ooms S, Overeem S, Besse K, Rikkert MO, Verbeek M, Claassen JA (2014) Effect of 1 night of total sleep deprivation on cerebrospinal fluid beta-amyloid 42 in healthy middle-aged men: a randomized clinical trial. JAMA Neurol 71:971–977. https://doi.org/10.1001/jamaneurol.2014.1173

Ozawa M, Ohara T, Ninomiya T, Hata J, Yoshida D, Mukai N et al (2014) Milk and dairy consumption and risk of dementia in an elderly Japanese population: the Hisayama study. J Am Geriatr Soc 62:1224–1230. https://doi.org/10.1111/jgs.12887

Paroni G, Bisceglia P, Seripa D (2019) Understanding the amyloid hypothesis in Alzheimer's disease. J Alzheimer's Dis 68:493–510. https://doi.org/10.3233/JAD-180802

Patel C, Pande S, Acharya S (2020) Potentiation of anti-Alzheimer activity of curcumin by probiotic lactobacillus rhamnosus UBLR-58 against scopolamine-induced memory impairment in mice. Naunyn Schmiedeberg's Arch Pharmacol 393:1955–1962. https://doi.org/10.1007/s00210-020-01904-3

Prosky L, Asp NG, Schweizer TF, DeVries JW, Furda I (1988) Determination of insoluble, soluble, and total dietary fiber in foods and food products: interlaboratory study. J Assoc Off Anal Chem 71:1017–1023

Qiu C, Kivipelto M, von Strauss E (2009) Epidemiology of Alzheimer's disease: occurrence, determinants, and strategies toward intervention. Dialogues Clin Neurosci 11:111–128. https://doi.org/10.31887/DCNS.2009.11.2/cqiu

Quigley EMM (2017) Microbiota-brain-gut Axis and neurodegenerative diseases. Curr Neurol Neurosci Rep 17:94. https://doi.org/10.1007/s11910-017-0802-6

Rezaei Asl Z, Sepehri G, Salami M (2019) Probiotic treatment improves the impaired spatial cognitive performance and restores synaptic plasticity in an animal model of Alzheimer's disease. Behav Brain Res 376:112183. https://doi.org/10.1016/j.bbr.2019.112183

Rezaeiasl Z, Salami M, Sepehri G (2019) The effects of probiotic lactobacillus and Bifidobacterium strains on memory and learning behavior, long-term potentiation (LTP), and some biochemical parameters in beta-amyloid-induced rat's model of Alzheimer's disease. Prev Nutr Food Sci 24:265–273. https://doi.org/10.3746/pnf.2019.24.3.265

Roberfroid MB, Delzenne NM (1998) Dietary fructans. Annu Rev Nutr 18:117–143. https://doi.org/10.1146/annurev.nutr.18.1.117

Robert C, Chassard C, Lawson PA, Bernalier-Donadille A (2007) Bacteroides cellulosilyticus sp. nov., a cellulolytic bacterium from the human gut microbial community. Int J Syst Evol Microbiol 57:1516–1520. https://doi.org/10.1099/ijs.0.64998-0

Saji N, Niida S, Murotani K, Hisada T, Tsuduki T, Sugimoto T et al (2019) Analysis of the relationship between the gut microbiome and dementia: a cross-sectional study conducted in Japan. Sci Rep 9:1008. https://doi.org/10.1038/s41598-018-38218-7

Salazar N, Dewulf EM, Neyrinck AM, Bindels LB, Cani PD, Mahillon J et al (2015) Inulin-type fructans modulate intestinal Bifidobacterium species populations and decrease fecal short-chain fatty acids in obese women. Clin Nutr 34:501–507. https://doi.org/10.1016/j.clnu.2014.06.001

Salehi A, Ashford JW, Mufson EJ (2016) The link between Alzheimer's disease and down syndrome. A historical perspective. Curr Alzheimer Res 13:2–6. https://doi.org/10.2174/1567205012999151021102914

Salminen A (2025) The role of inhibitory immune checkpoint receptors in the pathogenesis of Alzheimer's disease. J Mol Med (Berl) 103:1–19. https://doi.org/10.1007/s00109-024-02504-x

Schaible P, Henschel J, Erny D (2025) How the gut microbiota impacts neurodegenerative diseases by modulating CNS immune cells. J Neuroinflammation 22:60. https://doi.org/10.1186/s12974-025-03371-0

Shinohara K, Ohashi Y, Kawasumi K, Terada A, Fujisawa T (2010) Effect of apple intake on fecal microbiota and metabolites in humans. Anaerobe 16:510–515. https://doi.org/10.1016/j.anaerobe.2010.03.005

Sochocka M, Donskow-Lysoniewska K, Diniz BS, Kurpas D, Brzozowska E, Leszek J (2019) The gut microbiome alterations and inflammation-driven pathogenesis of Alzheimer's disease-a critical review. Mol Neurobiol 56:1841–1851. https://doi.org/10.1007/s12035-018-1188-4

Tamtaji OR, Heidari-Soureshjani R, Mirhosseini N, Kouchaki E, Bahmani F, Aghadavod E et al (2019) Probiotic and selenium co-supplementation, and the effects on clinical, metabolic and genetic status in Alzheimer's disease: a randomized, double-blind, controlled trial. Clin Nutr 38:2569–2575. https://doi.org/10.1016/j.clnu.2018.11.034

Tarini J, Wolever TM (2010) The fermentable fibre inulin increases postprandial serum short-chain fatty acids and reduces free-fatty acids and ghrelin in healthy subjects. Appl Physiol Nutr Metab 35:9–16. https://doi.org/10.1139/H09-119

Teglas T, Abraham D, Jokai M, Kondo S, Mohammadi R, Feher J et al (2020) Exercise combined with a probiotics treatment alters the microbiome, but moderately affects signalling pathways in the liver of male APP/PS1 transgenic mice. Biogerontology 21:807–815. https://doi.org/10.1007/s10522-020-09895-7

Verbeke KA, Boobis AR, Chiodini A, Edwards CA, Franck A, Kleerebezem M et al (2015) Towards microbial fermentation metabolites as markers for health benefits of prebiotics. Nutr Res Rev 28:42–66. https://doi.org/10.1017/S0954422415000037

Vijay N, Morris ME (2014) Role of monocarboxylate transporters in drug delivery to the brain. Curr Pharm Des 20:1487–1498. https://doi.org/10.2174/13816128113199990462

Wang Y, Ames NP, Tun HM, Tosh SM, Jones PJ, Khafipour E (2016) High molecular weight barley beta-Glucan alters gut microbiota toward reduced cardiovascular disease risk. Front Microbiol 7:129. https://doi.org/10.3389/fmicb.2016.00129

Wedekind KJ, Mansfield HR, Montgomery L (1988) Enumeration and isolation of cellulolytic and hemicellulolytic bacteria from human feces. Appl Environ Microbiol 54:1530–1535. https://doi.org/10.1128/aem.54.6.1530-1535.1988

Wekerle H (2016) The gut-brain connection: triggering of brain autoimmune disease by commensal gut bacteria. Rheumatology (Oxford) 55:ii68–ii75. https://doi.org/10.1093/rheumatology/kew353

Yano JM, Yu K, Donaldson GP, Shastri GG, Ann P, Ma L et al (2015) Indigenous bacteria from the gut microbiota regulate host serotonin biosynthesis. Cell 161:264–276. https://doi.org/10.1016/j.cell.2015.02.047

Yudishter SR, Dash KK (2025) Polysaccharide nanoparticles as building blocks for food processing applications: a comprehensive review. Food Sci Biotechnol 34:527–546. https://doi.org/10.1007/s10068-024-01695-w

Ze X, Duncan SH, Louis P, Flint HJ (2012) Ruminococcus bromii is a keystone species for the degradation of resistant starch in the human colon. ISME J 6:1535–1543. https://doi.org/10.1038/ismej.2012.4

Emerging Trends in Therapeutic Potential of Curcumin in Alzheimer's Disease

6

Saima Bashir, Areena Suhail Khan, Anwar-ul-Hassan Gilani, and Touqeer Ahmed

Abstract

Alzheimer's disease (AD) is a neurodegenerative disorder leading to dementia. It is characterized by a progressive cognitive deficit and formation of pathological Aβ plaques and neurofibrillary tangles. The available potential therapeutic options for AD provide symptomatic relief and struggle to slow down its progression. The available therapies are associated with limitations and demonstrate low potential for curing disease pathogenesis. Therefore, an alternative approach, such as natural products and their active compounds, are being explored as a neuroprotective agent. Curcumin alone and as a mixture of its derivatives and as hybrids with several AD drugs are being investigated as a potential agent to manage AD. Several in vivo and in vitro experiments demonstrated that the mechanism of curcumin action is mainly by inhibiting Aβ and NFTs, modulating several inflammatory pathways and microglial activity, mitochondrial dysfunction, decreasing oxidative stress and lowering acetylcholinesterase activity. Several studies depict the therapeutic potential of curcumin and its collective contribution to cognitive improvement in animal models of AD. Although curcumin possesses potential in effectively combating AD pathology, further research is required to improve its pharmacokinetic properties in pre-clinical and clinical trials. There is a need to study and overcome limitations associated with curcumin in terms of bioavailability, stability, and therapeutic efficacy for its development as a promising therapeutic agent in managing neurological and other disorders.

S. Bashir · A. S. Khan · T. Ahmed (✉)
Neurobiology Laboratory, Department of Biomedicine, Atta-ur-Rahman School of Applied Biosciences (ASAB), National University of Sciences and Technology (NUST), Islamabad, Pakistan
e-mail: touqeer.ahmed@asab.nust.edu.pk

Anwar-ul-Hassan Gilani
Higher Education Commission, H-9, Islamabad, Pakistan

J. Malik et al. (eds.), *Diet and Alzheimer's Disease: Let Food be Our Medicine*, Nutritional Neurosciences, https://doi.org/10.1007/978-981-95-2736-6_6

Keywords

Alzheimer's disease · Neurodegenerative disorder · Aβ plaques · NFTs · Curcumin

Abbreviations

AD	Alzheimer's disease
Aβ	Amyloid beta
APP	Amyloid Precursor Protein
BBB	Blood Brain Barrier
Cdk5	Cyclin dependent kinase 5
ETC	Electron Transfer Chain
FDA	Food and Drug Administration
GSK-3β	Glycogen synthase kinase-3 beta
LMTX	Leucomethylhioninium
LPS	Lipopolysaccharide
NFT	Neuro Fibrillary Tangles
NSAIDS	Non-steroidal anti-inflammatory drugs
PKA	Protein Kinase A
ROS	Reactive Oxygen Species

6.1 Alzheimer's Disease

Alzheimer's disease (AD), a known neurodegenerative disorder that accounts for about 60–80% of dementia cases in the elderly. Globally, it is estimated to affect about 100 million people by 2050. AD is characterized by culminating cognitive deficits and pathological formation of neurofibrillary tangles (NFTs) and Aβ plaques (Shao et al. 2023; Wang et al. 2022). The multifactorial disease pathogenesis mainly incorporates: (a) an Aβ cascade via formation of varied lengths of Aβ fragments. The β and γ secretase produces the major toxic peptide Aβ 42 (responsible for insoluble plaque formation) via enzymatic clearance of amyloid precursor protein (APP); (b) a hyperphosphorylated tau which disrupts the normal functioning of microtubule assembly, synaptic and axonal regulation, resulting in decreased microtubule binding affinity, leading to aggregation and ultimately NFTs formation; (c) the decline in cholinergic system functioning via reduced choline acetyltransferase activity in the basal nucleus of Meynert (Cakmak et al. 2022). To date, despite the advancements in the treatment of AD, the available therapies are ineffective in targeting the complete disease profile and only aim to impede the disease progression.

6.2 Available Potential Therapeutic Options for Alzheimer's Disease

Available treatment strategies are struggling to overcome the AD progression. To name a few drugs approved by the Food and Drug Administration (FDA) are rivastigmine, aducanumab, donepezil, memantine, and galantamine, all of which solely aim to manage the AD symptomology. Due to unclear AD pathogenesis, a substantial number of pharmacological interventions remain speculative and perhaps far off from targeting the main nosogenesis of the disease (Pardo-Moreno et al. 2022).

A number of potential therapeutic approaches such as RAGE inhibitors (Azeliragon; TTP488), σ-1 receptor agonist (blarcamesine; ANAVEX2-73), NMDA (memantine) and 5-HT receptor antagonist, gut microbiota (sodium oligomannate; GV-971) and lipid metabolism regulators, leucomethylthioninium (LMTX), active (AN1792, UB-311) and passive immunotherapies (monoclonal antibodies) are undergoing investigation and are being currently evaluated for their adequacy, long term safety and efficacy. They are still in phase III and IV clinical trials and are expected to be completed by 2030, clarifying whether there is hope to cure AD. Therapies such as liposomal and stem cells are being employed by targeting lysosomal and Notch or Wnt/b-catenin pathways, respectively, in providing effective treatment. Futuristic approaches like biomolecule-loaded nanoparticles, CRISPR/cas9, specialized therapy, as well as evolving therapeutic approaches like anti-sense therapy all have proven effectiveness and are known to target AD-related genes (PSEN1, PSEN2, APP, APOE). Additionally, the use of off-label drugs such as non-steroidal anti-inflammatory drugs (NSAIDS), e.g., ibuprofen, anti-diabetic medication (pioglitazone), and mood stabilizers (lithium), and some other pharmacological interventions are being explored to help manage the outcome in AD (Ahmad et al. 2024).

6.3 Challenges in Alzheimer's Disease Therapies

Several potential remedial interventions for Alzheimer's disease therapy offer positive findings in addition to certain limitations (Ahmad et al. 2024). The ability to cross the blood-brain barrier remains a major concern, as 98% of molecules, despite their smaller size, fail to pass through this barrier (Pardridge 2012). Target selection associated side effects, mechanism of drug delivery, and an early diagnosis need to be resolved, which will pave the way for the development of various immunotherapeutic approaches. Several above-mentioned therapies are currently undergoing clinical trials, like aducanumab, lecanemab, and various vaccines, like amyloid beta epitope-based and DNA vaccines. Furthermore, limitations such as the animal model's inability to accurately replicate the conditions in the human brain, particularly the age-related clinical symptoms appearance in AD, are not the same in animal models, even after decades of studies. Therefore, findings extrapolated from animal studies for AD treatment require careful consideration (Ahmad et al. 2024). The poor prognosis in synthetic drug research led to the development of natural products as anti-AD therapies, opening new avenues for further investigation (Long et al. 2021).

6.4 Natural Products as an Alternate Approach for Alzheimer's Treatment

Since currently available FDA-approved medications for AD provide symptomatic relief, requiring life-long use resulting in multiple side-effects and high cost, alternative treatments employ natural products and bioactive compounds from natural sources for the treatment of AD (Bhat et al. 2022). Natural products have gained attention due to the advantages they offer, including a wide variety, their source, and an action spectrum (Long et al. 2021) and are being explored for their various medicinal bioactivities as they all come with diverse pharmacological benefits, including anti-tumor, antioxidant, antibacterial, anti-inflammatory, antidiabetic, and neuroprotective (Long et al. 2021; Shao et al. 2023).

Natural products and their active compounds have been observed to display significant neuroprotective effects in AD management, with some adverse effects (Chen et al. 2021). The irreversible neurological damage to brain cells and the molecular mechanisms related to Alzheimer have highlighted the importance for development of novel therapies with better efficacy (Bhat et al. 2022). Natural products are structurally categorized into phenylpropanoids, flavonoids, glycosides, alkaloids, and others. They are generally found in herbs, vegetables, and fruits. They are considered safe and are becoming an alternative source for AD treatment. Such as curcumin, dihydromyricetin and salidroside, icariin, schizandrol A have been established to inhibit abnormal tau phosphorylation and Aβ deposition, inhibit mTOR signaling, and activate P13K/AKT signaling pathways (Long et al. 2021).

Mohamed Yusof and Mohd Fauzi (2024) reviewed natural product-based compounds tested for clinical trials for AD treatment and reported to have observed 13 compounds exhibiting cognitive improvement; however, they have not yet been approved by the FDA. These products are categorized as supplements rather than drugs, producing synergistic effects and offering advantages, with physiochemical properties in desirable ranges and a pharmacological profile relatively better than synthetic compounds (Mohamed Yusof and Mohd Fauzi 2024).

6.5 Curcumin Alone and as a Mixture of Its Derivatives

Curcumin [1, 7-bis (4-hydroxy-3-methoxyphenyl)-1, 6-hepatadiene-3,5-dione] also known as diferuloylmethane, is a naturally extracted polyphenol acquired from the roots of "*Curcuma longa*" (turmeric) plant. It is extensively used in Asian countries for managing various ailments due to its anti-oxidation, antimicrobial, anti-inflammatory, antimutagenic, and neuro-protective effects (Lou et al. 2024; Shao et al. 2023; Wang et al. 2022).

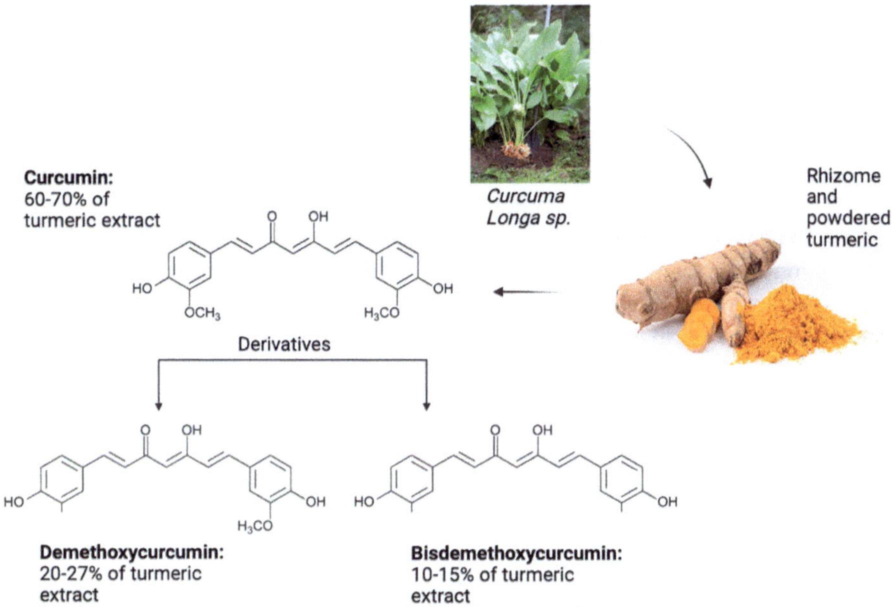

Fig. 6.1 Chemical structure of curcumin and its derivatives

Curcumin is a diketone- brightly colored orange-yellow crystalline powder. The biological function of curcumin is related to its keto or enol form, which is related to its solubility in different solvents. It forms a predominant keto-enol tautomeric compound in acidic and neutral media and a predominant enol form in alkaline media. The potential action of the parent plant (*Curcuma longa)* against diseases such as arthritis, diabetes, Alzheimer's, different allergies, and several other chronic diseases is primarily due to the active constituent, curcumin (curcumin I). However, there are other components; the curcuminoids group, including demethoxycurcumin (curcumin II) and bis-demethoxycurcumin (curcumin III) present as well (Ahmed and Gilani 2014; Pulido-Moran et al. 2016). While curcumin constitutes 77% of the total weight (the most abundant component of *Curcuma longa*), curcuminoids comprise only about 5%. Their chemical forms are shown in Fig. 6.1, and the methoxy groups present on the phenyl rings of curcumin are vital for its beneficial pharmacological profile in various disorders (Long et al. 2021; Xu et al. 2018).

6.6 Curcumin as a Hybrid Molecule

AD and other complex disorders are known to affect multiple targets, including organs and tissues. Hence, to improve drug efficacy, researchers are focusing on multi-target therapies for optimal treatment outcomes. As a result, hybridization techniques have been employed to combine the active compound found in natural products like curcumin and other AD-related drugs, resulting in the development of hybrids that possess the ability to exhibit enhanced therapeutic efficacy (Bansal and Silakari 2014).

Several compounds exhibited neuroprotective capacity, mainly by inhibiting Aβ and tau aggregation and directly inhibiting cholinesterase and butyrylcholinesterase activity, and further demonstrating antioxidant and metal chelation effects by binding to Cu^{+2} and Fe^{+2}. Several compounds presented themselves as promising candidates in vitro with their ability to cross the blood-brain barrier (BBB) and enhanced absorption capacity as shown in Table 6.1. However, a need exists to explore the in vivo efficacy, pharmacokinetics, and toxicity profiles. Drug-design using computational tools can help to further optimize the interactions of hybrid molecules with multiple targets, since these mixtures hold considerable potential as novel drugs for AD treatment (Shao et al. 2023; Wang et al. 2022).

6.7 Curcumin Targeting Multiple Therapeutic Pathways of Alzheimer's Disease

Curcumin exhibits a wide range of significant therapeutic properties for Alzheimer's pathology due to its anti-amyloidogenic, anti-tau, anti-inflammatory, and antioxidant properties as shown in Fig. 6.2 (Ahmed and Gilani 2014; del Velasco-Rodríguez et al. 2024). Existing literature showed that curcumin's mechanisms of action in AD are multifaceted. It mainly targets two hallmarks of AD: extracellular Aβ plaques and intracellular NFTs formed by tau hyperphosphorylation in the brain, along with the modulation of several inflammatory pathways and microglial activity, reducing oxidative stress, lowering acetylcholinesterase activities, and regulating the insulin signaling pathway (Tang et al. 2017), which collectively leads to the pathogenesis of AD. Research indicates that curcuminoid mixture and their constituents showed cognitive improvements in AD. A study led by Ahmed et al. (2010) demonstrated that individual constituents of curcuminoids enhanced the spatial memory in Aβ-peptide-infused rat model of AD (Ahmed et al. 2010).

6.7.1 Curcumin and Aβ Inhibition

Curcumin inhibits the formation and aggregation of Aβ plaques, a distinguishing feature of AD pathology, and promotes the clearance of already formed Aβ fibrils and plaques (Doytchinova et al. 2020) and interacts with the Aβ protein through its keto-enol functional groups, disrupting its β-sheet structure, thereby preventing the development and aggregation of Aβ plaques and promoting their subsequent degradation (den Haan et al. 2018). Previous in vitro and in vivo research showed that curcumin treatment prevented the aggregation of Aβ plaques (McClure et al. 2017; Mei et al. 2020; Zou et al. 2021). Furthermore, mutation in APP, PSEN1, and PSEN2 causes Aβ accumulation, leading to neurodegeneration in AD (Paroni et al. 2019). The findings of Jaroonwitchawan et al. (2017) indicated that curcumin treatment decreased the expression of APP in SH-SY5Y cells exposed to paraquat,

Table 6.1 Summary of pharmacological activities of curcumin and AD drug hybrids for multiple target therapies

Curcumin hybrids with different molecules	Compound no.	Structure	Pharmacological activity	References
Melatonin	5	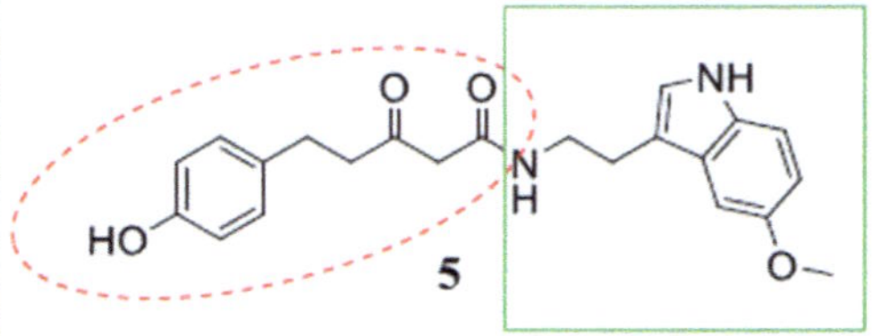 **5**	Reduces Aβ-plaque accumulation in hippocampus and cortex, and suppresses inflammation and oxidation	Chojnacki et al. (2014)
Tacrine	8	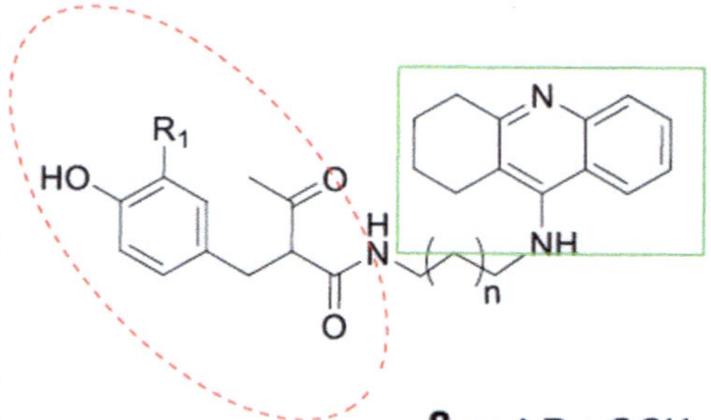 **8** n=4 R_1=OCH_3	AChE and BuChE inhibition, ROS elimination, and metal chelation	Liu et al. (2017)
Donepezil	10	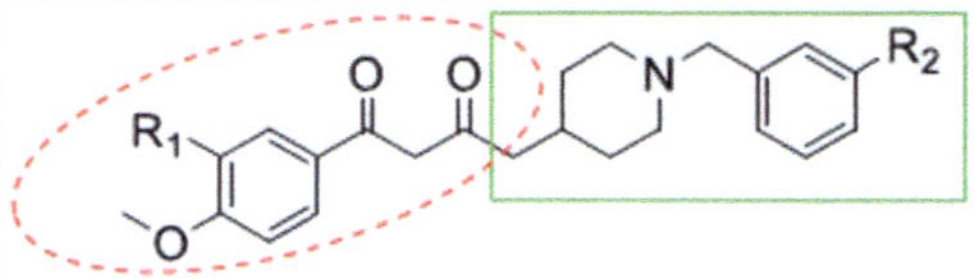 **10**: R_1=OH R_2=H	Inhibition of Aβ 42, and AChE, antioxidation, and metal chelation properties	Yan et al. (2017)

(continued)

Table 6.1 (continued)

Curcumin hybrids with different molecules	Compound no.	Structure	Pharmacological activity	References
Rivastigmine	20	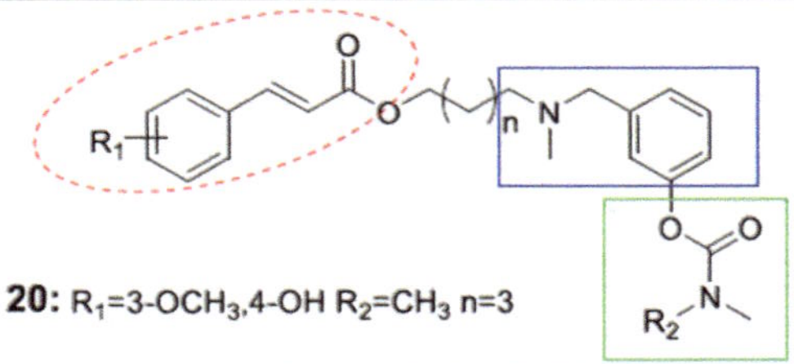 **20:** R_1=3-OCH_3,4-OH R_2=CH_3 n=3	Inhibition of Aβ 42 aggregation and AChE, antioxidation and neuroprotection activity	Lan et al. (2020)
Galantamine	22	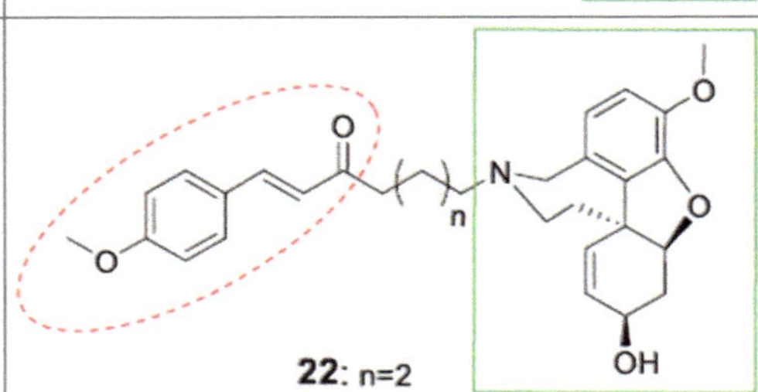 **22**: n=2	AChE inhibition and antioxidant activity	Stavrakov et al. (2020)
Coumarin	24	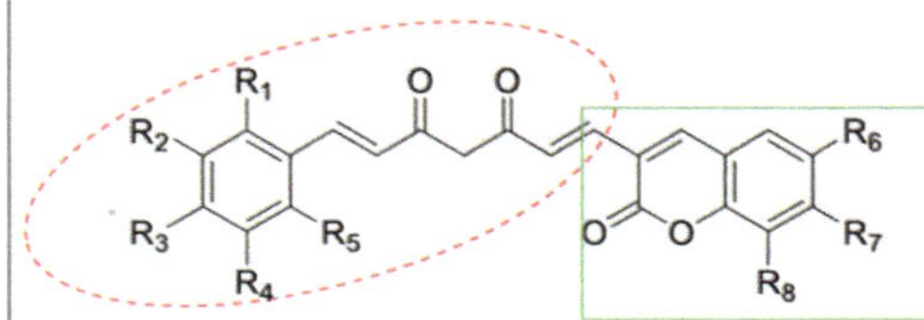 **24**: R_2=R_3=R_5=R_7=R_8=OH;R_1=R_4=R_6=H	Inhibition of monoamine and acetylcholine (ACh) degradation BuChE inhibition	Quezada et al. (2021)

Piperlongumine	33	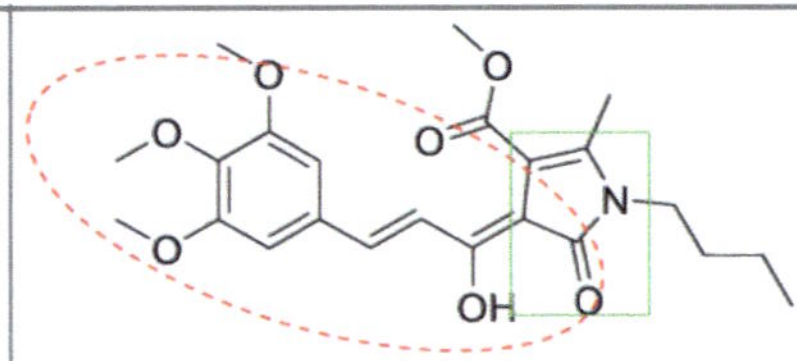 **33**	Neuroprotection and tau aggregation inhibition	Cores et al. (2022)
Resveratrol	36	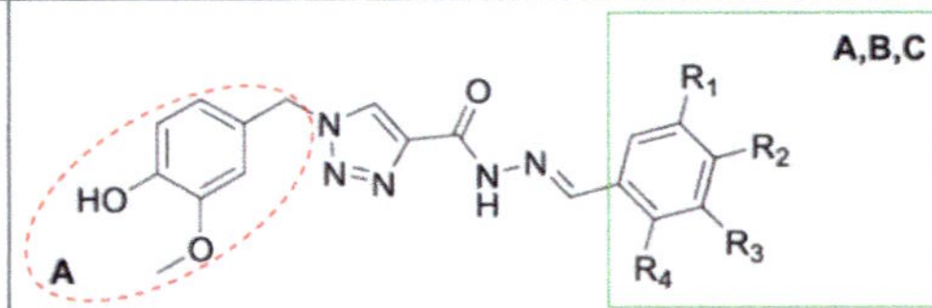 **36**: R_1=OCH_3,R_2=OH,R_3=R_4=H	AChE inhibition, antioxidant activity	De Freitas Silva et al. (2020)
Silibinin	39b	**39a** 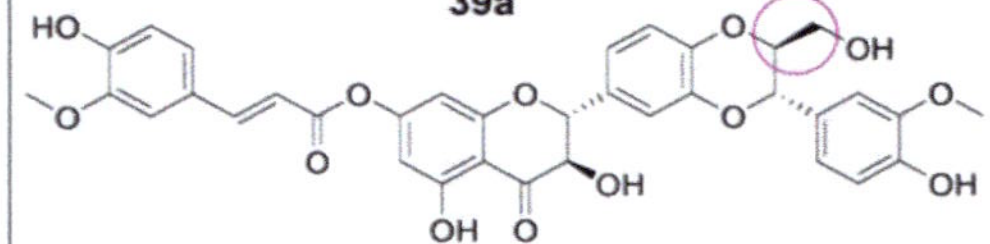 **39b**	Tau and Aβ aggregation inhibition and neuroprotection	Schramm et al. (2018)

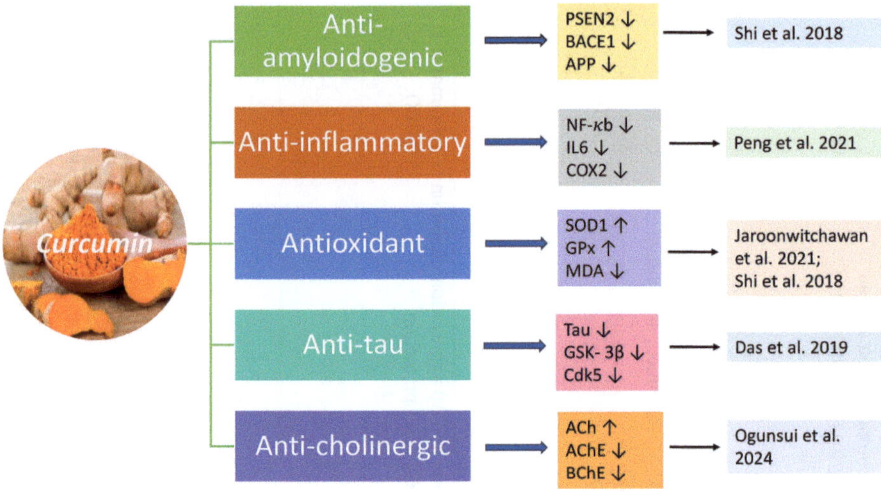

Fig. 6.2 Flow chart illustrating the effects of curcumin on various biological systems

which upregulates the expression of APP (Jaroonwitchawan et al. 2017). Furthermore, Shi et al. (2018) showed that curcumin effectively rescued the acrolein-induced altered expression of key proteins (BACE-1, APP, RAGE, and ADAM-10), involved in AD pathology in HT22 mouse hippocampal neuronal cells (Shi et al. 2018).

6.7.2 Curcumin and Tau-Inhibition

Tau protein self-assembles into insoluble fibrillar structures and accumulates in the brain to form neurofibrillary tangles, which cause neuronal death and cognitive deficits contributing to the development of AD (Oakley et al. 2020). Tau has a high number of proline-mediated serine/threonine residues and is regarded as a good target for several kinases, mainly glycogen synthase kinase 3 beta (GSK-3β), protein kinase A (PKA), and cyclin-dependent kinase 5 (Cdk5), which cause abnormal tau hyperphosphorylation (Liu et al. 2016). Curcumin has the potential to inhibit tau hyperphosphorylation. A study showed that curcumin treatment decreased the tau phosphorylation in a scopolamine-induced AD rat model by reducing the levels of kinases such as GSK-3β, Cdk5, and activators of Cdk5 (P35, P25), which regulate abnormal tau phosphorylation (Das et al. 2019).

6.7.3 Role of Curcumin in Neuroinflammation and Astrogliosis

Neuroinflammation, production of glial cells, and inflammatory markers, including chemokines and proinflammatory cytokines, are key drivers in AD pathogenesis (Zhang et al. 2024). Curcumin regulates several inflammatory pathways by altering the expression of inflammatory markers. A review demonstrated that curcumin binds with several proteins, such as JAK/STAT, TLRs, AP-1, MAPK, and other signaling pathways, thereby regulating the levels of proinflammatory cytokines and enzymes, such as IL6, TNF-α, IL1-β, COX-2, and NO, by modulating the NF-κB pathway in different inflammatory diseases (Peng et al. 2021). Research indicates that oral curcumin administration attenuated memory deficits in AD mice by inhibiting the HMGB1-RAGE/TLR4-NF-κB inflammatory pathway (Han et al. 2021). Furthermore, administration of curcumin alone and a mixture of its constituents at 30 mg/kg significantly decreased the expression of IL1-β and GFAP in the cortex and hippocampus of Aβ plus ibotenic acid-infused rat model mimicking AD (Ahmed and Gilani 2011). Another study reported that curcumin inhibits the production of inflammatory markers (TNF-α, NO, and PGE2) in PAM3CSK4-stimulated microglial cells by increasing HO-1 expression through the Nrf2 signaling pathway and reduces neuroinflammation by inhibiting NF-κB and MAPK signaling pathways (Jin et al. 2018).

6.7.4 Curcumin as an Antioxidant

Curcumin also exhibits strong antioxidant properties, which help to mitigate oxidative stress, a key feature of AD pathogenesis and progression. Different experimental studies reported that treatment with curcumin mitigates oxidative stress by reducing the production of reactive oxygen species (ROS) and significantly increasing the expression of SOD, GSH, GPx, and decreasing the expression of MDA (Jaroonwitchawan et al. 2017; Shi et al. 2018). Buccarello et al. (2020) explored the impact of oxidative stress induced by H_2O_2 on AD-associated proteins (tau, SUMO-1, and JNK) in SH-SY5Y cell lines. Oxidative stress induced by H_2O_2 led to decreased cell viability, increased cell death, and altered protein degradation marker levels. Curcumin treatment reduced oxidative stress by restoring JNK, tau, and SUMO-1 protein functions, preventing their accumulation in nuclear bodies, and at a concentration of 5 μM, curcumin improved cell viability, reduced cell death, and normalized levels of protein degradation markers (Buccarello et al. 2020).

6.7.5 Neuro-Modulatory Effects of Curcumin in Mitochondrial Dysfunction

Mitochondrial dysfunction is reported as one of the contributing factors in the pathogenesis of AD (Wang et al. 2020). Oxidative stress and increased levels of ROS are associated with mitochondrial dysfunction. Curcumin can maintain the antioxidant system to prevent mitochondrial function by regulating ROS levels since it leads to damage of mDNA, mito proteins, less effective Electron Transfer Chain (ETC), and impaired quality control in mitophagy (van der Rijt et al. 2020). Several studies reported that curcumin upregulates CISD2 expression levels, therefore preventing mitochondrial dysfunction in lipopolysaccharide (LPS) challenged neural cells (Lin et al. 2019). Another in vivo study reported curcumin treatment with 30 mg/kg for 30 days, to elevate mood and memory functions in male Sprague-Dawley rats model of Gulf War illness. In addition, the rats also showed enhanced hippocampal neurogenesis, enhanced antioxidant gene expression, and normalized the expression of genes associated with mitochondrial respiratory chain (Kodali et al. 2018). Furthermore, Rastogi et al. (2014) demonstrated neuroprotective effects of curcuminoid (100 mg/kg) to prevent age-related mitochondrial dysfunction in the brain of female Wistar rats by increasing the activity of mitochondrial enzymes, specifically NADH dehydrogenase, isocitrate dehydrogenase, and cytochrome c oxidase (Rastogi et al. 2014).

6.7.6 Role of Curcumin in Modulating Cholinergic Pathways

Cholinergic pathways are associated with cognitive functions, and their dysfunction in Alzheimer's patients highlights the vital role of cholinergic neurons in cognitive integrity. Ogunsuyi et al. (2024) reported that the coadministration of curcumin and donepezil exhibits a better rescuing effect on the cholinesterase (AChE and BChE) function in the rat brain cortex and hippocampus, when compared to the donepezil alone-treated group (Ogunsuyi et al. 2024). Another study reported synthesized curcumin analogs (1d and 1c) to demonstrate higher cholinesterase inhibitory function, promising enhancement of cognitive functions in vivo and ex vivo experiments, which was further strengthened by molecular docking studies in mice models, highlighting their potential as valuable therapeutic candidates for neurological disorders (Hussain et al. 2022).

The cholinergic system plays a pivotal role in cognitive functions by regulating the levels of acetylcholine. The decreased ACh levels can be managed by their inhibitors, such as donepezil, galantamine, rivastigmine, and tacrine (Hussain et al. 2022), many of which are combined with neuroprotective curcumin, and their effects on critical brain regions are investigated (Ogunsuyi et al. 2024). An investigation led by Ahmed and Gilani (2009) reported that curcuminoid treatment enhanced memory and AChE inhibitory effects in both in vitro assays and the Scopolamine-induced amnesia rat model of AD (Ahmed and Gilani 2009). Another study reported that the co-administration of three phytonutrients, known as curcumin, carnosic acid, and trans-resveratrol as a bioavailable complex (CGM+), enhanced memory function and regulated neurotransmitter levels to support neurogenesis. Further biochemical and molecular analysis reveals a bioavailable complex to modulate mitochondrial bioenergetics, promoting recovery in traumatic brain injury mice (Mohan et al. 2025).

6.8 Therapeutics of Curcumin in Animal Model of Alzheimer's Disease

Several in vivo studies have shown the therapeutic potential of curcumin in alleviating AD-like cognitive impairments by mitigating Aβ plaques, tau hyperphosphorylation, and inflammatory pathways, which collectively contribute to the cognitive impairment in AD (Cakmak et al. 2022; Liu et al. 2021). A systematic review of 33 studies showed that curcumin significantly improves cognitive impairments in animal models of AD (Fan and Zhang 2024). Furthermore, Zhang et al. (2015) reported that chronic treatment of curcumin (50, 100, and 200 mg/kg) improved cognitive deficits induced by Aβ1-42 mimicking the AD rodent model through activation of the BDNF/ERK signaling pathway (Zhang et al. 2015). Another study reported that oral administration of high dose (150 mg/kg/day) of curcumin in AD mice improved cognitive function, reduced apoptosis, neuronal loss, Aβ1-42 levels, and inflammatory markers including IL-6, TNF-α, IL1-β, and increased superoxide dismutase activity through activating the AMPK pathway (Shao et al. 2023). On the other hand, low-dose intake of dietary curcumin may also delay the onset of AD. A study reported that long-term, low-dose oral administration of curcumin reduced amyloid-beta deposition, tau hyperphosphorylation, brain inflammation, and improved short-term memory in an AD mouse model (Maruyama et al. 2024). Furthermore, a recent study demonstrated that the oral administration of derivative 27 (50 mg/kg/day) to APP/PS1 transgenic mice for 28 days resulted in significant improvement in short-term memory, reduction in hippocampal and plasma levels of IL1-β, APP, and Aβ (Rodrigues et al. 2025).

6.9 Pharmacokinetic Properties of Curcumin

Curcumin has poor water solubility (11 ng/ml), poor absorption, high metabolism in the intestine and liver, rapid elimination from the body, resulting in poor bioavailability (<1%), which is the major limitation for its therapeutic potential (Hegde et al. 2023). Upon first-pass metabolism in liver, curcumin produces a series of metabolites, including inactive metabolites such as sulfates and glucuronides, and is cleared metabolically through glucuronidation, sulfation, and reduction (Zhou et al. 2022). Currently, various strategies, such as nanoparticle encapsulation, such as niosomes, gold nanoparticles, liposomes, polymeric micelles, nanogel formulations, and coadministration with adjuvants like piperine, are being widely used to enhance the bioavailability and pharmacokinetic properties of curcumin (Askarizadeh et al. 2020; Genchi et al. 2024; Wu and Yang 2023). Recent studies reported that curcumin nanoparticles effectively reduced the behavioral, neurochemical, and histopathological changes caused by intracerebroventricular injection of streptozotocin in the rat model of AD and thus serve as a promising therapeutic approach for AD management (Noor et al. 2022). Thus, further research is required to improve the curcumin delivery system for more promising therapeutic results in pre-clinical and clinical trials.

6.10 Limitations Associated with Curcumin

Curcumin is generally safe, non-toxic, and non-mutagenic to humans, even at high doses (≥6 g/day orally) (Soleimani et al. 2018; Zang et al. 2024). However, it may have some limitations, such as gastrointestinal disturbance including nausea and diarrhea, contact urticaria or dermatitis, and hepatotoxicity at high doses (Chaudhari et al. 2015; Zahra et al. 2024). Another limitation associated with curcumin is poor bioavailability and limited BBB permeation (Hegde et al. 2023). Curcumin therapy showed promising results in the in vitro and in vivo experimental studies, but positive results of curcumin have not been fully replicated in clinical trials (Mohseni et al. 2021), due to poor curcumin delivery into the brain. Placebo-controlled and double-blind clinical trials of oral administration of curcumin in AD patients showed no difference in the improvement of cognitive functions and levels of Aβ plaques and tau protein in both the treatment and placebo groups (Rainey-Smith et al. 2016; Ringman et al. 2012). Clinical studies exploring the effects of curcumin in AD did not employ any drug delivery system to improve bioavailability. Therefore, further research is required to improve the bioavailability, stability, and therapeutic efficacy of curcumin for developing curcumin as a promising therapeutic agent in treating neurological and other diseases.

6.11 Conclusion

Curcumin has been speculative in preventing and treating Alzheimer's, one of the most common disorders of the brain. However, research demonstrating its specific mechanism of action has been elucidated in recent decades. Curcumin primarily targets AD's two key histopathological hallmarks: intracellular NFTs and amyloid-β plaques. Moreover, it is established to influence different risk factors, physiological processes, and biomarkers associated with AD pathology. Hence, the diverse biological activities and disease-modifying characteristics of curcumin may have a vital impact on AD treatment in the future. The current mainstream drug therapies only target a single disease pathology, do not delay the onset of the disease or alter its progression, and exhibit multiple adverse effects. Furthermore, their effectiveness declines as neurons begin to degenerate. The different categories of available drugs to manage AD clinical symptoms come with adverse effects. Promising approaches like CRISPR/Cas9 and stem cell therapy are being explored; however, logistical and patient-centric factors such as cost of treatment, patient consent, and probability of failure need to be addressed before it gets approved for global use. Even though humanized animal models have significantly improved our understanding of neurological disorders, it is still not enough. In contrast, the effects of curcumin, being a natural compound, are multi-targeted, and it has the potential to alter the disease progression and is considered relatively safe. It can be stated that the curcumin nano preparations, hybrids, derivatives, and analogs have proven to be more beneficial in terms of the therapeutic potential and have shown an improved pharmacological and pharmacokinetic profile as compared to traditional monotherapy. Further research is required to enhance the bioavailability, therapeutic efficacy, and translational applicability of curcumin from preclinical to clinical trials for the treatment of AD.

References

Ahmad F, Karan A, Sharma R, Sharma NS, Sundar V, Jayaraj R, Mukherjee S, DeCoster MA (2024) Evolving therapeutic interventions for the management and treatment of Alzheimer's disease. Ageing Res Rev 95(September 2023):102229. https://doi.org/10.1016/j.arr.2024.102229

Ahmed T, Gilani AH (2009) Inhibitory effect of curcuminoids on acetylcholinesterase activity and attenuation of scopolamine-induced amnesia may explain medicinal use of turmeric in Alzheimer's disease. Pharmacol Biochem Behav 91(4):554–559. https://doi.org/10.1016/j.pbb.2008.09.010

Ahmed T, Gilani AH (2011) A comparative study of curcuminoids to measure their effect on inflammatory and apoptotic gene expression in an Aβ plus ibotenic acid-infused rat model of Alzheimer's disease. Brain Res 1400:1–18. https://doi.org/10.1016/j.brainres.2011.05.022

Ahmed T, Gilani AH (2014) Therapeutic potential of turmeric in Alzheimer's disease: curcumin or curcuminoids? In: Phytotherapy research, vol 28. Wiley, pp 517–525. https://doi.org/10.1002/ptr.5030

Ahmed T, Enam SA, Gilani AH (2010) Curcuminoids enhance memory in an amyloid-infused rat model of Alzheimer's disease. Neuroscience 169(3):1296–1306. https://doi.org/10.1016/j.neuroscience.2010.05.078

Askarizadeh A, Barreto GE, Henney NC, Majeed M, Sahebkar A (2020) Neuroprotection by curcumin: a review on brain delivery strategies. Int J Pharm 585. https://doi.org/10.1016/j.ijpharm.2020.119476

Bansal Y, Silakari O (2014) Multifunctional compounds: smart molecules for multifactorial diseases. Eur J Med Chem 76:31–42. https://doi.org/10.1016/j.ejmech.2014.01.060

Bhat BA, Almilaibary A, Mir RA, Aljarallah BM, Mir WR, Ahmad F, Mir MA (2022) Natural therapeutics in aid of treating Alzheimer's disease: a green gateway toward ending quest for treating neurological disorders. Front Neurosci 16. https://doi.org/10.3389/fnins.2022.884345

Buccarello L, Dragotto J, Iorio F, Hassanzadeh K, Corbo M, Feligioni M (2020) The pivotal role of SUMO-1-JNK-Tau axis in an in vitro model of oxidative stress counteracted by the protective effect of curcumin. Biochem Pharmacol 178. https://doi.org/10.1016/j.bcp.2020.114066

Cakmak G, Kaplan DS, Yildirim C, Ulusal H, Tarakcioglu M, Ozturk ZA (2022) Improvement of cognitive deficit of curcumin on scopolamine-induced Alzheimer's disease models. Caspian J Intern Med 13(1):16–22. https://doi.org/10.22088/cjim.13.1.16

Chaudhari SP, Tam AY, Barr JA (2015) Curcumin: a contact allergen. J Clin Aesthet Dermatol 8(11):43–48

Chen X, Drew J, Berney W, Lei W (2021) Neuroprotective natural products for alzheimer's disease. Cells 10(6). https://doi.org/10.3390/cells10061309

Chojnacki JE, Liu K, Yan X, Toldo S, Selden T, Estrada M, Rodríguez-Franco MI, Halquist MS, Ye D, Zhang S (2014) Discovery of 5-(4-hydroxyphenyl)-3-oxo-pentanoic acid [2-(5-methoxy-1H-indol-3-yl)-ethyl]-amide as a neuroprotectant for Alzheimer's disease by hybridization of curcumin and melatonin. ACS Chem Neurosci 5(8):690–699

Cores Á, Carmona-Zafra N, Martín-Cámara O, Sánchez JD, Duarte P, Villacampa M, Bermejo-Bescós P, Martín-Aragón S, León R, Menéndez JC (2022) Curcumin-piperlongumine hybrids with a multitarget profile elicit neuroprotection in in vitro models of oxidative stress and hyperphosphorylation. Antioxidants 11(1). https://doi.org/10.3390/antiox11010028

Das TK, Jana P, Chakrabarti SK, Abdul Hamid MRW (2019) Curcumin downregulates GSK3 and Cdk5 in scopolamine-induced Alzheimer's disease rats abrogating Aβ40/42 and tau hyperphosphorylation. J Alzheimer's Dis Rep 3(1):257–267. https://doi.org/10.3233/ADR-190135

De Freitas Silva M, Lima ET, Pruccoli L, Castro NG, Guimarães MJR, Da Silva FMR, Nadur NF, De Azevedo LL, Kümmerle AE, Guedes IA, Dardenne LE, Gontijo VS, Tarozzi A, Viegas C (2020) Design, synthesis and biological evaluation of novel triazole N-acylhydrazone hybrids for alzheimer's disease. Molecules 25(14):1–18. https://doi.org/10.3390/molecules25143165

del Velasco-Rodríguez LC, García HS, Rascón-Díaz MP (2024) Curcumin and omega-3 polyunsaturated fatty acids as bioactive food components with synergistic effects on Alzheimer's disease. Psychogeriatrics 24(3):701–718. https://doi.org/10.1111/psyg.13115

den Haan J, Morrema THJ, Rozemuller AJ, Bouwman FH, Hoozemans JJM (2018) Different curcumin forms selectively bind fibrillar amyloid beta in post mortem Alzheimer's disease brains: implications for in-vivo diagnostics. Acta Neuropathol Commun 6(1):75. https://doi.org/10.1186/s40478-018-0577-2

Doytchinova I, Atanasova M, Salamanova E, Ivanov S, Dimitrov I (2020) Curcumin inhibits the primary nucleation of amyloid-beta peptide: a molecular dynamics study. Biomolecules 10(9):1–14. https://doi.org/10.3390/biom10091323

Fan L, Zhang Z (2024) Therapeutic potential of curcumin on the cognitive decline in animal models of Alzheimer's disease: a systematic review and meta-analysis. Naunyn Schmiedeberg's Arch Pharmacol 397(7):4499–4509. https://doi.org/10.1007/s00210-024-02946-7

Genchi G, Lauria G, Catalano A, Carocci A, Sinicropi MS (2024) Neuroprotective effects of curcumin in neurodegenerative diseases. Foods 13(11). https://doi.org/10.3390/foods13111774

Han Y, Chen R, Lin Q, Liu Y, Ge W, Cao H, Li J (2021) Curcumin improves memory deficits by inhibiting HMGB1-RAGE/TLR4-NF-κB signalling pathway in APPswe/PS1dE9 transgenic mice hippocampus. J Cell Mol Med 25(18):8947–8956. https://doi.org/10.1111/jcmm.16855

Hegde M, Girisa S, BharathwajChetty B, Vishwa R, Kunnumakkara AB (2023) Curcumin formulations for better bioavailability: what we learned from clinical trials thus far? ACS Omega 8(12):10713–10746. https://doi.org/10.1021/acsomega.2c07326

Hussain H, Ahmad S, Shah SWA, Ullah A, Ali N, Almehmadi M, Ahmad M, Khalil AAK, Jamal SB, Ahmad H, Halawi M (2022) Attenuation of scopolamine-induced amnesia via cholinergic modulation in mice by synthetic curcumin analogs. Molecules 27(8). https://doi.org/10.3390/molecules27082468

Jaroonwitchawan T, Chaicharoenaudomrung N, Namkaew J, Noisa P (2017) Curcumin attenuates paraquat-induced cell death in human neuroblastoma cells through modulating oxidative stress and autophagy. Neurosci Lett 636:40–47. https://doi.org/10.1016/j.neulet.2016.10.050

Jin M, Park SY, Shen Q, Lai Y, Ou X, Mao Z, Lin D, Yu Y, Zhang W (2018) Anti-neuroinflammatory effect of curcumin on Pam3CSK4-stimulated microglial cells. Int J Mol Med 41(1):521–530. https://doi.org/10.3892/ijmm.2017.3217

Kodali M, Hattiangady B, Shetty GA, Bates A, Shuai B, Shetty AK (2018) Curcumin treatment leads to better cognitive and mood function in a model of gulf war illness with enhanced neurogenesis, and alleviation of inflammation and mitochondrial dysfunction in the hippocampus. Brain Behav Immun 69:499–514. https://doi.org/10.1016/j.bbi.2018.01.009

Lan JS, Zeng RF, Jiang XY, Hou J (2020) Design, synthesis and evaluation of novel ferulic acid derivatives as multi-target-directed ligands for the treatment of Alzheimer's disease. Bioorg Chem 94(October 2019):103413. https://doi.org/10.1016/j.bioorg.2019.103413

Lin CC, Chiang TH, Sun YY, Lin MS (2019) Protective effects of cisd2 and influence of curcumin on cisd2 expression in aged animals and inflammatory cell model. Nutrients 11(3). https://doi.org/10.3390/nu11030700

Liu SL, Wang C, Jiang T, Tan L, Xing A, Yu JT (2016) The role of Cdk5 in Alzheimer's disease. Mol Neurobiol 53(7):4328–4342. https://doi.org/10.1007/s12035-015-9369-x

Liu Z, Fang L, Zhang H, Gou S, Chen L (2017) Design, synthesis and biological evaluation of multifunctional tacrine-curcumin hybrids as new cholinesterase inhibitors with metal ions-chelating and neuroprotective property. Bioorganic Med Chem 25(8):2387–2398. https://doi.org/10.1016/j.bmc.2017.02.049

Liu M, Zhang X, Wang Y (2021) Curcumin alleviates Aβ42-induced neuronal metabolic dysfunction via the Thrb/SIRT3 Axis and improves cognition in APPTG mice. Neurochem Res 46(12):3166–3178. https://doi.org/10.1007/s11064-021-03414-x

Long HZ, Cheng Y, Zhou ZW, Luo HY, Wen DD, Gao LC (2021) PI3K/AKT signal pathway: a target of natural products in the prevention and treatment of Alzheimer's disease and Parkinson's disease. Front Pharmacol 12. https://doi.org/10.3389/fphar.2021.648636

Lou S, Gong D, Yang M, Qiu Q, Luo J, Chen T (2024) Curcumin improves neurogenesis in Alzheimer's disease mice via the upregulation of Wnt/β-catenin and BDNF. Int J Mol Sci 25(10). https://doi.org/10.3390/ijms25105123

Maruyama H, Ooizumi T, Kawakami F, Lwin TT, Akita H, Kunii T, Shirai R, Takeda T (2024) Long-term oral administration of curcumin is effective in preventing short-term memory deterioration and prolonging lifespan in a mouse model of Alzheimer's disease. Adv Tradit Med 24(1):373–385. https://doi.org/10.1007/s13596-023-00706-1

McClure R, Ong H, Janve V, Barton S, Zhu M, Li B, Dawes M, Jerome WG, Anderson A, Massion P, Gore JC, Pham W (2017) Aerosol delivery of curcumin reduced amyloid-β deposition and improved cognitive performance in a transgenic model of Alzheimer's disease. J Alzheimer's Dis 55(2):797–811. https://doi.org/10.3233/JAD-160289

Mei X, Zhu L, Zhou Q, Li X, Chen Z (2020) Interplay of curcumin and its liver metabolism on the level of Aβ in the brain of APPswe/PS1dE9 mice before AD onset. Pharmacol Rep 72(6):1604–1613. https://doi.org/10.1007/s43440-020-00116-z

Mohamed Yusof NIS, Mohd Fauzi F (2024) Nature's toolbox for Alzheimer's disease: a review on the potential of natural products as Alzheimer's disease drugs. Neurochem Int 176. https://doi.org/10.1016/j.neuint.2024.105738

Mohan MC, Anjana AS, Hilmi Jaufer TA, Deepti A, Krishnakumar IM, Baby Chakrapani PS (2025) Co-delivery of curcumin-resveratrol-carnosic acid complex promotes neurogenesis and cognitive recovery in a rodent model of repeated mild traumatic brain injury. Biomed Pharmacother 183:117818. https://doi.org/10.1016/j.biopha.2025.117818

Mohseni M, Sahebkar A, Askari G, Johnston TP, Alikiaii B, Bagherniya M (2021) The clinical use of curcumin on neurological disorders: an updated systematic review of clinical trials. Phytother Res 35(12):6862–6882. https://doi.org/10.1002/ptr.7273

Noor NA, Hosny EN, Khadrawy YA, Mourad IM, Othman AI, Aboul Ezz HS, Mohammed HS (2022) Effect of curcumin nanoparticles on streptozotocin-induced male Wistar rat model of Alzheimer's disease. Metab Brain Dis 37(2):343–357. https://doi.org/10.1007/s11011-021-00897-z

Oakley SS, Maina MB, Marshall KE, Al-Hilaly YK, Harrington CR, Wischik CM, Serpell LC (2020) Tau filament self-assembly and structure: tau as a therapeutic target. Front Neurol 11. https://doi.org/10.3389/fneur.2020.590754

Ogunsuyi OB, Ogunruku OO, Umar HI, Oboh G (2024) Effect of curcumin–donepezil combination on spatial memory, astrocyte activation, and cholinesterase expressions in brain of scopolamine-treated rats. Mol Biol Rep 51(1). https://doi.org/10.1007/s11033-024-09712-1

Pardo-Moreno T, González-Acedo A, Rivas-Domínguez A, García-Morales V, García-Cozar FJ, Ramos-Rodríguez JJ, Melguizo-Rodríguez L (2022) Therapeutic approach to Alzheimer's disease: current treatments and new perspectives. Pharmaceutics 14(6):1–20. https://doi.org/10.3390/pharmaceutics14061117

Pardridge WM (2012) Drug transport across the blood-brain barrier. J Cereb Blood Flow Metab 32(11):1959–1972. https://doi.org/10.1038/jcbfm.2012.126

Paroni G, Bisceglia P, Seripa D (2019) Understanding the amyloid hypothesis in Alzheimer's disease. J Alzheimer's Dis 68(2):493–510. https://doi.org/10.3233/JAD-180802

Peng Y, Ao M, Dong B, Jiang Y, Yu L, Chen Z, Hu C, Xu R (2021) Anti-inflammatory effects of curcumin in the inflammatory diseases: status, limitations and countermeasures. Drug Des Devel Ther 15:4503–4525. https://doi.org/10.2147/DDDT.S327378

Pulido-Moran M, Moreno-Fernandez J, Ramirez-Tortosa C, Ramirez-Tortosa MC (2016) Curcumin and health. Molecules 21(3). https://doi.org/10.3390/molecules21030264

Quezada E, Rodríguez-Enríquez F, Laguna R, Cutrín E, Otero F, Uriarte E, Viña D (2021) Curcumin–coumarin hybrid analogues as multitarget agents in neurodegenerative disorders. Molecules 26(15). https://doi.org/10.3390/molecules26154550

Rainey-Smith SR, Brown BM, Sohrabi HR, Shah T, Goozee KG, Gupta VB, Martins RN (2016) Curcumin and cognition: a randomised, placebo-controlled, double-blind study of community-dwelling older adults. Br J Nutr 115(12):2106–2113. https://doi.org/10.1017/S0007114516001203

Rastogi M, Ojha RP, Sagar C, Agrawal A, Dubey GP (2014) Protective effect of curcuminoids on age-related mitochondrial impairment in female Wistar rat brain. Biogerontology 15(1):21–31. https://doi.org/10.1007/s10522-013-9466-z

Ringman JM, Frautschy SA, Teng E, Begum AN, Bardens J, Beigi M, Gylys KH, Badmaev V, Heath DD, Apostolova LG, Porter V, Vanek Z, Marshall GA, Hellemann G, Sugar C, Masterman DL, Montine TJ, Cummings JL, Cole GM (2012) Oral curcumin for Alzheimer's disease: tolerability and efficacy in a 24-week randomized, double blind, placebo-controlled study. Alzheimer's Res Ther 4(5). https://doi.org/10.1186/alzrt146

Rodrigues B, Ventura E, Moreira P, Resende R, Bicker J, Santos AE, Pereira CF, Cruz MT, Robalo MP, Silva A, Silva S (2025) New low-dose curcumin derivative with therapeutic potential in Alzheimer's disease: results from an in vitro and in vivo study in mice. Neurobiol Aging 147:105–123. https://doi.org/10.1016/j.neurobiolaging.2024.12.005

Schramm S, Huang G, Gunesch S, Lang F, Roa J, Högger P, Sabaté R, Maher P, Decker M (2018) Regioselective synthesis of 7-O-esters of the flavonolignan silibinin and SARs lead to compounds with overadditive neuroprotective effects. Eur J Med Chem 146:93–107. https://doi.org/10.1016/j.ejmech.2018.01.036

Shao S, Ye X, Su W, Wang Y (2023) Curcumin alleviates Alzheimer's disease by inhibiting inflammatory response, oxidative stress and activating the AMPK pathway. J Chem Neuroanat 134. https://doi.org/10.1016/j.jchemneu.2023.102363

Shi LY, Zhang L, Li H, Liu TL, Lai JC, Wu ZB, Qin J (2018) Protective effects of curcumin on acrolein-induced neurotoxicity in HT22 mouse hippocampal cells. Pharmacol Rep 70(5):1040–1046. https://doi.org/10.1016/j.pharep.2018.05.006

Soleimani V, Sahebkar A, Hosseinzadeh H (2018) Turmeric (Curcuma longa) and its major constituent (curcumin) as nontoxic and safe substances: review. Phytother Res 32(6):985–995. https://doi.org/10.1002/ptr.6054

Stavrakov G, Philipova I, Lukarski A, Atanasova M, Zheleva D, Zhivkova ZD, Ivanov S, Atanasova T, Konstantinov S, Doytchinova I (2020) Galantamine-curcumin hybrids as dual-site binding acetylcholinesterase inhibitors. Molecules 25(15). https://doi.org/10.3390/molecules25153341

Tang M, Taghibiglou C, Liu J (2017) The mechanisms of action of curcumin in Alzheimer's disease. J Alzheimer's Dis 58(4):1003–1016. https://doi.org/10.3233/JAD-170188

van der Rijt S, Molenaars M, McIntyre RL, Janssens GE, Houtkooper RH (2020) Integrating the hallmarks of aging throughout the tree of life: a focus on mitochondrial dysfunction. Front Cell Dev Biol 8. https://doi.org/10.3389/fcell.2020.594416

Wang W, Zhao F, Ma X, Perry G, Zhu X (2020) Mitochondria dysfunction in the pathogenesis of Alzheimer's disease: recent advances. Mol Neurodegener 15(1). https://doi.org/10.1186/s13024-020-00376-6

Wang Y, Zhang H, Hua L, Wang Z, Geng S, Zhang H, Zeng Z, Zhao J, Wang X, Wang Y (2022) Curcumin prevents Alzheimer's disease progression by upregulating JMJD3. Am J Transl Res 14(8):5243

Wu X, Yang Y (2023) Research progress on drug delivery systems for curcumin in the treatment of gastrointestinal tumors. World J Gastrointest Oncol 15(8):1342–1348. https://doi.org/10.4251/wjgo.v15.i8.1342

Xu XY, Meng X, Li S, Gan RY, Li Y et al (2018) Bioactivity, health benefits, and related molecular mechanisms of curcumin: current progress, challenges, and perspectives. Nutrients 10(10). https://doi.org/10.3390/nu10101553

Yan J, Hu J, Liu A, He L, Li X, Wei H (2017) Design, synthesis, and evaluation of multitarget-directed ligands against Alzheimer's disease based on the fusion of donepezil and curcumin. Bioorganic Med Chem 25(12):2946–2955. https://doi.org/10.1016/j.bmc.2017.02.048

Zahra M, Hadi F, Maqbool T, Sultana H, Abid F, Aslam MA, Ahmad M, Muhammad S, Qideer M, ul Hassan. (2024) Curcumin (turmeric): a carcinogenic, miscarriage and cirrhosis causing agent. J Health Rehabil Res 4(2):1738–1743. https://doi.org/10.61919/jhrr.v4i2.1159

Zang WB, Wei HL, Zhang WW, Ma W, Li J, Yao Y (2024) Curcumin hybrid molecules for the treatment of Alzheimer's disease: structure and pharmacological activities. In. Eur J Med Chem 265. https://doi.org/10.1016/j.ejmech.2023.116070

Zhang L, Fang Y, Xu Y, Lian Y, Xie N, Wu T, Zhang H, Sun L, Zhang R, Wang Z (2015) Curcumin improves amyloid β-peptide (1-42) induced spatial memory deficits through BDNF-ERK signaling pathway. PLoS One 10(6). https://doi.org/10.1371/journal.pone.0131525

Zhang Q, Yang G, Luo Y, Jiang L, Chi H, Tian G (2024) Neuroinflammation in Alzheimer's disease: insights from peripheral immune cells. Immun Ageing 21(1). https://doi.org/10.1186/s12979-024-00445-0

Zhou X, Venigalla M, Raju R, Münch G (2022) Pharmacological considerations for treating neuroinflammation with curcumin in Alzheimer's disease. J Neural Transm 129(5–6):755–771. https://doi.org/10.1007/s00702-022-02480-x

Zou X, Himbert S, Dujardin A, Juhasz J, Ros S, Stöver HDH, Rheinstädter MC (2021) Curcumin and Homotaurine suppress amyloid-β25-35Aggregation in synthetic brain membranes. ACS Chem Neurosci 12(8):1395–1405. https://doi.org/10.1021/acschemneuro.1c00057

Tea Catechins: A Useful Chemical Weapon Against Alzheimer's Disease

7

Hafize Yuca, Aleyna Kopuz, and Songül Karakaya

Abstract

Tea (*Camellia sinensis*) is among the most widely consumed beverages worldwide, valued for its rich history, cultural significance, and numerous health benefits. One of its key bioactive components, catechins, has attracted considerable scientific interest due to its strong antioxidant and neuroprotective properties. Among these, epigallocatechin-3-gallate (EGCG) has been extensively studied for its potential role in preventing and managing neurodegenerative diseases, particularly Alzheimer's disease (AD).

Alzheimer's disease is a progressive neurodegenerative disorder characterized by cognitive decline, amyloid-beta plaque accumulation, and oxidative stress. Several in vitro, in vivo, and clinical studies have highlighted the neuroprotective effects of catechins. EGCG, in particular, has demonstrated the ability to inhibit amyloid-beta aggregation, modulate key signaling pathways such as Nrf2/ARE, PI3K/Akt, and MAPK, and reduce oxidative stress and neuroinflammation. Furthermore, EGCG has been shown to enhance synaptic plasticity, improve mitochondrial function, and regulate cholinergic neurotransmission, which are crucial for maintaining cognitive function.

Preclinical studies have demonstrated that catechins reduce neuronal apoptosis, protect against amyloid toxicity, and enhance memory performance in AD models. Clinical evidence further suggests that regular tea consumption may

H. Yuca (✉)
Department of Pharmacognosy, Faculty of Pharmacy, Atatürk University, Erzurum, Türkiye
e-mail: hafize.yuca@atauni.edu.tr

A. Kopuz
Department of Graphic Arts, Faculty of Fine Arts, Atatürk University, Erzurum, Türkiye

S. Karakaya
Department of Pharmaceutical Botany, Faculty of Pharmacy, Atatürk University, Erzurum, Türkiye

J. Malik et al. (eds.), *Diet and Alzheimer's Disease: Let Food be Our Medicine*, Nutritional Neurosciences, https://doi.org/10.1007/978-981-95-2736-6_7

lower the risk of cognitive impairment and slow age-related cognitive decline. However, challenges such as low bioavailability and metabolic instability limit their therapeutic potential. Novel delivery systems, including nanoformulations and bioenhancers, are being explored to optimize their efficacy.

Overall, tea catechins, particularly EGCG, hold promise as a natural therapeutic strategy for AD prevention and management. Further clinical studies are needed to validate their long-term effects and establish optimal dosages for neuroprotection. This chapter provides a comprehensive overview of tea catechins' biochemical properties, mechanisms of action, and their potential in combating Alzheimer's disease.

Keywords

Camellia sinensis · Tea catechins · Alzheimer's disease · Neuroprotection · Oxidative stress

7.1 Tea's Journey Through Time: Origins, Trade, and Culture

The plant, known by its Latin name *Camellia sinensis* (L.) Kuntze (formerly *Thea sinensis*), is a type of camellia and belongs to the Theaceae family. As indicated, the tea plant is classified under the *Camellia* genus. The distinguished botanist Georg Joseph Kamel, often credited as the father of the *Camellia* genus, was born on April 21, 1661, in Brno, Czech Republic (Aykaç et al. 2014).

Tea originates from the Chinese word "ça," which has influenced the terminology for tea in many languages. While the Chinese use "çai" to refer to tea, the term is expressed differently in various languages. For instance, the Turkish language uses "çay," while Russians and Iranians call it "chay." In Arabic, it is also referred to as "çay," and in Greek, it is "şay." The Japanese use "tsa-i," while Portuguese and Romanian speakers say "cha." In other languages, such as Czech, Serbian, and Croatian, tea is called "ceai," and in English and Hungarian, it is known as "cai." The French term is "tea," while in Denmark and Sweden, it is called "thee," and Jewish speakers use "te" (Altaş 2018). Tea—known as 荼/茶 in Chinese, чай in Russian, *thé* in French, *Tee* in German, お茶/茶の木 in Japanese, شاي in Arabic, چائے in Urdu, *chai* in Hindi, and *çay* in Turkish—is a beverage of great cultural, economic, and health significance worldwide. Originating in China, tea is derived from the processed leaves of the perennial plant *C. sinensis*. Various processing techniques influence the characteristics of different tea varieties, shaping its diverse flavors and uses across cultures (Pan et al. 2022).

Shen Nung, a legendary figure who is believed to have lived around 5000 years ago, holds a significant place in Chinese history, mythology, and cultural heritage. Known as the "Red Emperor," he is considered one of the three great rulers of classical Chinese medicine. He is credited with compiling *Pen T-Sao* (2800 BC), an ancient compendium that documented the medicinal properties of 365 plants, which he is said to have personally tested. According to legend, the discovery of tea dates back to

2737 BC, when leaves from a tea plant accidentally fell into a pot of boiling water prepared by Shen Nung. The resulting infusion had a pleasing aroma and taste, leading to the gradual adoption of tea as a favored drink. Over time, tea spread from China to other parts of the world, and today, *C. sinensis*—the tea plant—is the most widely consumed beverage after water. The leading tea-producing nations include China, India, Sri Lanka, Japan, and Taiwan. There are multiple theories regarding the origins of tea. According to Indian mythology, tea is believed to have first emerged in northern India before being introduced to China's Sichuan Province in the 1600s. Beyond being a simple beverage, tea plays an essential role in social customs and traditions across various cultures. Its preparation differs by region; in India, it is commonly brewed with milk and sugar, while in North Africa, it is infused with fresh mint. In ancient China, tea was regarded as one of the seven daily essentials and was initially valued for its medicinal benefits before becoming a widely enjoyed drink. Early tea preparations involved steaming and grinding the leaves, which were then mixed with ingredients such as rice, ginger, salt, orange peel, spices, milk, and even onion. In Chinese tradition, premium tea leaves are described poetically as being "as dark as the boots of Tatar horsemen, as curved as the horns of a strong ox, and as bright as a lake touched by a gentle breeze." Tea is traditionally consumed in small, wide-mouthed cups, which enhance both the drinking experience and the visual appreciation of the infusion. The practice of steeping the same tea leaves multiple times is highly regarded in China and is often considered an art. When tea was introduced to Japan, it was refined into a structured and highly symbolic ritual, forming the foundation of the Japanese tea ceremony. Rooted in Taoist, Buddhist, and Zen philosophies, these ceremonies emphasize mindfulness, hospitality, and grace. Some even argue that the cultural and spiritual significance of the Japanese tea ritual surpasses that of wine in Catholic ceremonies, where it symbolizes the blood of Christ. Tea has been consumed as a beverage in China for approximately 4000 years. By the classical era, it was recognized for its refreshing and stimulating effects, making it a valuable aid for Buddhist monks during meditation. Its deep association with Buddhism helped shape elite tastes in China and played a crucial role in influencing broader Asian culture. The pursuit of an ideal vessel that would not alter the delicate flavors of fine teas led to the development of Chinese porcelain, which became a key component of global trade and industry. During the Tang (618–907) and Song (960–1279) dynasties, tea became an integral part of Chinese society, inspiring poetry, art, and the establishment of teahouses. Books dedicated to tea cultivation and preparation began to appear, with the earliest known text being *The Classic of Tea*, written by Lu Yu between 760 and 780. In 805, the Buddhist monk Saichō is believed to have introduced the first tea seedlings to Japan, laying the groundwork for a tea culture that remains one of Japan's most defining traditions. Modern research and historical records indicate that all tea plants cultivated globally trace their origins to southwest China. Cluster analysis has revealed that due to the tea plant's ability to hybridize easily, numerous intermediate hybrids and spontaneous polyploid varieties exist. This suggests that *Camellia sinensis* likely originated from a single region, believed to be the border area of Yunnan and Sichuan provinces in China, extending into parts of India and northern Burma (Awasom 2011; Aykaç et al. 2014; Gilbert 2008; Han 2007; Üstün and Demirci 2013; Vishnoi et al. 2018; Yan et al. 2020; Pan et al. 2022; Weisburger 1997).

Tea arrived in Korea during the Three Kingdoms period (55–668 AD) and was cultivated during the Silla Unified Period (618–935 AD). A Chinese Taoist in the eighth century praised its quality. In China, tea's cultural significance grew under Buddhism, Confucianism, and Daoism, becoming an art during the Tang dynasty (618–907 AD). By the sixteenth century, it was deeply rooted in East Asia. The Portuguese introduced tea to Europe around 1580, but they were mainly traders. The Dutch dominated the early tea trade until the British took over in the eighteenth century. By 1820, tea was a British staple. Despite early access via Macao, Portugal saw tea as a luxury, mostly for elites. The Azores had records of tea use by the eighteenth century, likely introduced earlier by Jesuits. By the nineteenth century, tea became widespread in Portugal, shifting from luxury to necessity, aided by smuggling in the Azores (Moura 2015).

7.2 Morphological and Anatomical Features of *C. sinensis*

Tea plants, classified under *Camellia* sect. *Thea* (Theaceae), are globally significant for beverage production. The sect. *Longissima* is now considered a synonym of the sect. *Thea*. A taxonomic review identifies 11 natural species and seven infraspecific taxa, including *C. costata*, *C. crispula*, *C. crispula* var. *multiplex*, *C. fangchengensis*, *C. gymnogyna*, *C. kwangsiensis*, *C. kwangsiensis* var. *kwangnanica*, *C. leptophylla*, *C. longissima*, *C. ptilophylla*, *C. sinensis*, *C. sinensis f. formosensis*, *C. sinensis* var. *assamica*, *C. sinensis* var. *dehungensis*, *C. sinensis* var. *pubilimba*, *C. tachangensis*, *C. tachangensis* var. *remotiserrata*, and *C. taliensis* (Zhao 2024). Tea (*Camellia sinensis*), a member of the Theaceae family, is cultivated across Asia, Africa, Latin America, and Oceania. Believed to have originated in northeastern India, northern Myanmar, and southwestern China, tea plants are classified into three types: shrubs, arbors, and minor arbors. Wild tea trees can grow up to 10–15 meters, while cultivated ones are pruned to 0.6–1.5 meters. Tea leaves are alternate, leathery, and lanceolate, measuring 3.8–30 cm in length and about 4 cm in width. Mature leaves are dark green and glossy, while young ones are pubescent. The plant produces fragrant white flowers, 2.5–4 cm in diameter, either solitary or in clusters, with five sepals, five to nine petals, and numerous stamens with yellow anthers. The fruit is a small, smooth, three-chambered capsule containing one to four seeds, each about the size of a small nut. Young leaves and buds are covered in trichomes, which decrease as they mature. During harvest, the shoot typically includes the bud and the two youngest leaves. Fresh leaves are typically harvested from late March to early April through July each year. Green tea accounts for roughly 20% of the dried tea produced annually and is predominantly consumed in Asian countries like Japan, largely due to its affordability and relative safety compared to other beverages (Mahmood et al. 2010; Wang et al. 2022).

The study investigated the anatomy of *C. sinensis*, highlighting key tissue structures within the Theaceae family. The stem features a protective cuticle, a cortex composed of collenchymatous and parenchymatous cells, and a continuous vascular cylinder with collateral bundles. The leaf anatomy includes stomata on the lower surface, with the mesophyll divided into palisade and spongy parenchyma, and crescent-shaped vascular bundles. Trichomes, more prevalent in young leaves, assist in protection and water retention. The petiole shares a similar structural composition, with embedded sclereids and a well-defined vascular system. Notably, anatomical differences were observed in the thickness of cell layers and the number and size of vascular bundles across leaves, stems, and roots. Sclereids were also present in the ground tissues of petioles and midribs. While the vascular bundles in the stems were of the collateral type, those in the roots were radial and polyarch (Aboulwafa et al. 2019; Das et al. 2023; Naing et al. 2019).

7.3 The Cultural and Health Significance of Tea: A Timeless Global Tradition

Archaeological evidence suggests that the practice of steeping tea leaves in hot water may date back hundreds of thousands of years. Botanical studies indicate that tea cultivation first began in regions such as India and China. Today, tea is consumed by millions of people worldwide, with green tea (*Camellia sinensis*) gaining increasing recognition for its potential health benefits. Green tea has a long-standing history of use in countries like India, China, Japan, and Thailand. In traditional Chinese and Indian medicine, it has been valued for its stimulant, diuretic, and astringent properties, as well as its role in promoting cardiovascular health. Additionally, green tea has been traditionally used to aid digestion, regulate body temperature and blood sugar levels, relieve bloating, and enhance cognitive function (Chopade et al. 2008).

Tea has long been used as a traditional remedy, with its physiological effects extensively studied in Asian countries, particularly China and Japan. According to data from the UK National Diet and Nutrition Survey, which includes over 7000 adults, 77% of individuals drink tea, with an average daily intake of 2.3 mugs (540 mL). Tea is thought to offer potential health benefits in preventing and treating various diseases and is often linked to increased longevity (Wang et al. 2012).

Tea is a global social staple, with each culture adding its own twist. In India, it is brewed with milk and sugar, while in North Africa adds fresh mint. In ancient China, tea started as medicine and became a daily ritual, often brewed with spices. Japanese tea ceremonies are rooted in Taoism, Buddhism, and Zen, focusing on hospitality and enlightenment. The Mongols drink tea with oil, salt, flour, and sometimes dried lamb for strength in harsh conditions. In France, tea is served in fine cups with light treats. Russians brew tea in a samovar with sugar and lemon, while Americans associate it with the Revolution. The British made afternoon tea a tradition. Tea is enjoyed worldwide for both its cultural significance and health benefits (Üstün and Demirci 2013).

T'Len Yi Heng described tea as a way to "forget the noise of the world," while Reimertz compared it to a gentle alarm clock, offering both pleasure and meditation. Drinking tea has become a calming habit, promoting relaxation both physically and mentally. Beyond meeting a need, tea fosters social interaction and helps individuals adapt to their surroundings (Erşahin 2021).

Tea, made from the leaf buds, leaves, and internodes of various *Camellia sinensis* cultivars, is processed in different ways and prepared by steeping the cured leaves in hot water. It is the second most consumed beverage globally, after water, and originated in Southeast China before spreading to India, Pakistan, and other countries. Tea offers numerous health benefits, such as preventing dental decay, reducing cholesterol buildup, aiding in heart disease prevention, and combating intestinal bacterial infections. In Pakistan, tea has become a key part of social life, with the country being the third-largest importer of tea. Pakistan consumes around 170 million kg of black tea annually, primarily from Kenya, India, and Sri Lanka (Naveed and Hameed 2014).

The illustration (Fig. 7.1) in this book chapter was created by Aleyna Kopuz to provide visual support for the content. The illustration was created using an Apple iPad Pro 11-inch (fourth generation) tablet and an Apple Pencil (second generation) stylus. The Procreate application was chosen for its extensive digital drawing and painting capabilities.

During the illustration process, an initial sketch phase was conducted to establish the fundamental composition. Subsequently, a layer-based working method was adopted for detailing and coloring. Light, shadow, and texture applications were utilized to add depth to the illustration. The pressure-sensitive stylus technology of the graphic tablet was employed to achieve organic and fluid lines throughout the drawing process.

The illustration was designed in accordance with the predetermined concept and content to enhance the expressive power of the work in both artistic and academic contexts. Its digital format allows for adjustments and adaptation to different formats as needed.

The story behind the illustration is provided in Fig. 7.1.

7.4 Exploring Tea Varieties: Characteristics, Health Benefits, and Cultural Significance

7.4.1 Tea Comes in Different Varieties, Each with Its Unique Characteristics

Green tea is made from unfermented leaves, unlike oolong and black teas, which undergo partial and full fermentation, respectively. It is rich in antioxidants, particularly polyphenols, and has less caffeine than black tea. Its taste is often a sweet bitterness. White tea is made from young tea leaves and buds that are minimally processed. It contains the highest level of antioxidants and the least caffeine compared to other teas. Black tea is the most widely produced type, making up 72% of global tea consumption. Though it has fewer antioxidants due to fermentation, it still contains flavonoids, which help detoxify the body. It has a stronger flavor, often consumed with sugar or milk. Oolong tea is partially fermented, blending characteristics of both green and black teas. It is rich in antioxidants, promoting healthy skin and slowing aging. Pu'erh tea comes from a large-leaf tea plant and is aged for 50

Fig. 7.1 The global journey of tea: cultural and scientific perspectives
The origins of tea trace back to ancient China, where it was initially used for medicinal purposes. According to legend, Emperor Shen Nong (2737 BCE) discovered tea when leaves accidentally fell into boiling water. Over time, tea became an integral part of Chinese culture, leading to the development of distinct processing techniques, such as green, black, oolong, and white tea. Tea spread beyond China during the Tang Dynasty (618–907 CE), reaching Japan through Buddhist monks. This led to the development of Japan's renowned tea ceremony, *Chanoyu*. By the ninth century, tea had also reached Korea. In the sixteenth century, European explorers and traders, particularly the Portuguese and Dutch, introduced tea to the West. By the seventeeth century, the British East India Company played a pivotal role in popularizing tea in England, leading to the

to 100 years, which gives it a unique flavor and profile. It is processed similarly to black tea. Rooibos tea is made from a shrub native to South Africa. It is caffeine-free and high in antioxidants, making it a good choice for those who avoid caffeine. In general, green and oolong teas are popular in Asia, while black tea is favored in Western countries (Mahmood et al. 2010; Patel 2005; Sharangi 2009).

For centuries, green tea has been an integral part of daily life in India, China, Japan, and Thailand. In traditional Indian and Chinese medicine, it has been recognized for its stimulating, diuretic, and astringent properties, as well as its potential benefits for cardiovascular health and wound healing. Furthermore, green tea has been traditionally used to alleviate bloating, regulate body temperature and blood sugar levels, support digestion, and improve cognitive function (Chopade et al. 2008).

Fig. 7.1 (continued) stablishment of large-scale plantations in India and Sri Lanka (Ceylon). This marked the beginning of industrialized tea production. Tea was introduced to England in the seventeenth century by Portuguese princess Catherine of Braganza. Initially a luxury for the aristocracy, tea became a staple for all social classes by the nineteenth century, largely due to British colonial trade. Key traditions include (Afternoon Tea: Introduced by Anna, Duchess of Bedford, in the early nineteenth century, featuring tea served with sandwiches, scones, and pastries. English Breakfast Tea: A strong black tea blend, often consumed with milk and sugar.) Turkey is a Nation of Tea Lovers. Although historically overshadowed by coffee in the Ottoman Empire, tea gained popularity in Turkey during the twentieth century due to economic shifts and coffee shortages. Today, Turkey ranks among the world's top tea consumers. The Black Sea region, particularly Rize, is the country's primary tea-growing area. Brewed in a two-tiered teapot (*çaydanlık*), Turkish tea is served in small tulip-shaped glasses without milk, often accompanied by sugar cubes. It is a central element of hospitality and daily social interactions. India is both a leading producer and consumer of tea. British colonial rule in the early nineteenth century saw the establishment of tea plantations in Assam, Darjeeling, and Nilgiri to reduce reliance on Chinese imports. (Assam Tea: A strong, malty black tea grown in the humid Assam region. Darjeeling Tea: A light, floral tea, often called the "Champagne of Teas. Masala Chai: A popular spiced tea made with black tea, milk, and spices such as cardamom, ginger, and cinnamon. Green and Herbal Teas: Gaining popularity for their health benefits. India's tea industry supports millions of workers and plays a vital role in the national economy, with major exports worldwide. Tea arrived in Russia in the seventeenth century via trade with China along the Silk Road. Initially a luxury, it became a staple beverage by the nineteenth century. Tea is brewed in a *samovar*, a large metal urn that keeps water hot for extended periods. *Zavarka*, a strong tea concentrate, is brewed separately and diluted with hot water. Black tea is the most commonly consumed variety, often accompanied by lemon, sugar, jam (*varenye*), or honey. Tea is an essential part of Russian hospitality, symbolizing warmth and community. Recent technological advancements have led to the development of functional tea-based products. Innovative extraction techniques such as supercritical CO_2 extraction and nanoencapsulation have enabled the creation of nutraceuticals, anti-aging formulations, and bioactive skincare applications. The journey of tea from an ancient medicinal drink to a staple beverage worldwide underscores its cultural, economic, and scientific significance. While the United Kingdom emphasizes ritual and refinement, Turkey integrates tea into daily life, India thrives as a major producer with diverse flavors, and Russia maintains deep-rooted traditions with its *samovar*. Ongoing scientific studies continue to explore tea's bioactive compounds and health benefits, reinforcing its role in modern wellness trends. As consumption patterns evolve and sustainability practices improve, tea remains not just a beverage but a historical and scientific phenomenon, connecting cultures and generations across the globe. The tale of tea: a comprehensive history of tea from prehistoric times to the present day (Ashby 2014; Beard et al. 2013; Chakrabarti 2022; Fleming 2010; Ger and Kravets 2020; Güneş 2012; Honda 2012; Lutgendorf 2012; Barry 1986; Morgan 2003; Peterson 1994; Sadler 2011; Serinsu 2022; Takako 2018; Turay 2001; Wohl 2017; Young 2001)

7.5 Tea Catechins

Catechins are a class of polyphenolic flavonoids predominantly found in tea, cocoa, and certain fruits, known for their strong antioxidant and anti-inflammatory properties. Structurally, catechins exist in various forms, including (+)-catechin, (−)-epicatechin, (−)-epigallocatechin (EGC), and (−)-epigallocatechin gallate (EGCG), with EGCG being the most bioactive and abundant in green tea (Fig. 7.2). These compounds play a significant role in human health by neutralizing free radicals, thereby reducing oxidative stress and preventing cellular damage. Catechins have been extensively studied for their cardioprotective effects, as they help lower blood pressure, reduce LDL oxidation, and enhance endothelial function, mitigating the risk of atherosclerosis and cardiovascular diseases. Additionally, their anti-inflammatory properties stem from their ability to modulate key signaling pathways, such as NF-κB inhibition and cytokine regulation, contributing to their potential in managing chronic inflammatory conditions. Catechins also exhibit anti-cancer properties by inducing apoptosis, inhibiting tumor cell proliferation, and preventing metastasis through the modulation of multiple molecular targets. Moreover, they are beneficial in metabolic health, improving insulin sensitivity, regulating lipid metabolism, and supporting weight management, making them promising candidates for the prevention and management of obesity and diabetes. Despite these benefits, catechin bioavailability remains a critical factor, as their absorption and metabolism are influenced by dietary components and individual physiological factors. Understanding these mechanisms is essential for optimizing their therapeutic applications in disease prevention and overall health promotion (Isemura 2019).

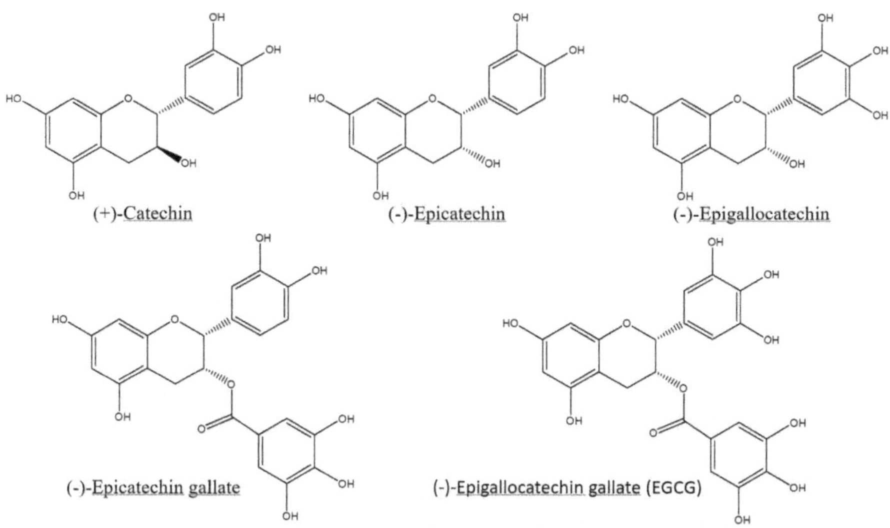

Fig. 7.2 Chemical structures of several catechins

7.6 Alzheimer's Disease

In 1907, Alois Alzheimer described a 51-year-old woman with rapid memory loss and psychiatric symptoms, marking the first identification of Alzheimer's disease (AD). AD was initially divided by age of onset: presenile dementia for those under 65 and senile dementia for older individuals. Although these classifications are still used, AD is now recognized as a single condition with a sharp increase in prevalence with age (Castellani et al. 2010).

AD, the leading cause of dementia, is rapidly becoming one of the most expensive and burdensome diseases of the twenty-first century. Recent advancements have improved our understanding of its pathology, including the identification of causative genes, new biomarkers, and early signs of potential treatment benefits. Clinical presentations vary, with some cases showing early genetic markers, others presenting with language difficulties instead of memory loss, and the typical amnestic variant is more common in older individuals. As the disease progresses, individuals often become increasingly dependent on others for care (Scheltens et al. 2021).

Dementia, primarily AD, is a cognitive decline that impacts daily life, especially in those over 65. AD leads to memory, language, and reasoning impairments and raises the risk of complications. While there's no cure, treatments focus on symptom management. The disease progresses through stages, with memory loss followed by difficulties in judgment and motor skills. Key risk factors include age, genetics, cardiovascular disease, and obesity, while lifestyle changes like exercise may reduce risk. Dementia cases are rising globally, with projections reaching 150 million by 2050. AD is marked by amyloid plaques and neurofibrillary tangles (Kumar et al. 2024).

The FDA has approved two main types of drugs for AD: cholinesterase inhibitors (ChEIs) for early stages and NMDA receptor blockers for moderate-to-severe stages, often used together. These medications have shown some cognitive benefits, but their effectiveness could improve with advanced brain-targeted delivery systems. Recently approved antibodies, aducanumab and lecanemab, target amyloid plaques, but their role in disease modification is debated. Challenges like medication adherence and side effects remain, especially in elderly patients. Long-acting injectables are being explored to improve adherence and reduce relapse. This review discusses current AD treatments, their innovative formulations, and the potential for enhanced drug delivery systems to improve efficacy and reduce side effects. Currently, only five treatments are approved for AD in the USA, with memantine being the most recent. Despite this, most treatments developed in the last 15 years have failed, and AD remains under-treated. Given the complexity of AD, combining therapies targeting different disease mechanisms might be more effective than single treatments, as seen in other diseases like cancer. Combination trials, distinct from add-on trials, test two drugs together and individually, helping distinguish their individual and combined effects. These combinations can target symptoms or modify the disease biology. Sequential combinations and multifunctional molecules are also being explored. Current clinical studies focus on adding new disease-modifying therapies to existing treatments to assess their combined impact. Although AD has a significant impact, current treatments mainly address symptoms. Two approved drug classes for AD are N-methyl-D-aspartate (NMDA) antagonists and cholinesterase (ChE) inhibitors (Cummings et al. 2019; Parikh et al. 2024; Singh et al. 2024).

7.7 Tea Catechins against Alzheimer's Disease

Catechins, particularly epigallocatechin-3-gallate (EGCG), activate the Nrf2/ARE pathway, promoting the expression of antioxidant and detoxifying enzymes such as heme oxygenase-1 (HO-1). This activation strengthens cellular defense mechanisms against oxidative stress and inflammation, playing a protective role in neurodegenerative diseases. Additionally, EGCG influences signaling pathways like MAPK, PI3K/Akt, and AMPK, contributing to cell survival, reducing apoptosis, and enhancing neuronal function (Mandel et al. 2004).

7.8 In Vitro Studies

A study evaluated the cholinesterase inhibitory potential of five catechin compounds—(−)-epicatechin (EC), catechin, (−)-epicatechin-3-gallate (ECG), (−)-epigallocatechin (EGC), and (−)-epigallocatechin-3-gallate (EGCG)—using the Ellman assay. Among them, only EGCG demonstrated significant competitive inhibition of both AChE and BuChE, with IC_{50} values of 0.0148 μmol/mL and 0.0251 μmol/mL, respectively (Okello and Mather 2020).

A study explored the effects of green and black tea extracts, along with flavan-3-ols, on β-amyloid (Aβ)-induced toxicity in a primary rat hippocampal cell model of Alzheimer's disease (AD). Given the detrimental role of Aβ in AD pathology, it was proposed that tea consumption might lower the risk of age-related neurodegenerative diseases. Both green and black tea extracts (5–25 μg/mL) demonstrated neuroprotective properties against Aβ toxicity. Among the tested compounds, gallic acid (1–20 μM), epicatechin gallate (ECG; 1–20 μM), and epigallocatechin gallate (EGCG; 1–10 μM) exhibited significant protective effects, with gallic acid showing the highest potency. In contrast, epicatechin and epigallocatechin did not exhibit protective effects at the same concentrations. Notably, ECG, EGCG, and gallic acid inhibited Aβ25–35-induced apoptotic processes, while EGCG and gallic acid also prevented Aβ aggregation and the formation of toxic oligomeric species. These findings suggest that catechin gallates, particularly ECG and EGCG, play a key role in the neuroprotective effects of green and black tea. Moreover, EGCG's ability to inhibit Aβ fibril and oligomer formation supports the notion that both tea types may help mitigate the risk of neurodegenerative diseases such as AD (Bastianetto et al. 2006).

A study examined the impact of small-molecule drugs, either approved for human use or in clinical trials (phase I – III), on amyloid fibril formation by human apolipoprotein C-II (apoC-II). The findings revealed that several of these compounds influenced the rate of amyloid formation. Notably, epigallocatechin gallate (EGCG), a green tea catechin, effectively inhibited apoC-II fibril formation, while the antipsychotic drug fluphenazine·HCl acted as a potent activator. Both compounds exhibited concentration-dependent effects, bound to apoC-II fibrils with high affinity, and competitively reduced thioflavin T binding. EGCG significantly altered fibril size distribution by promoting lateral association and the formation of larger aggregates. In contrast, fluphenazine·HCl had minimal impact on fibril size distribution but appeared to induce a distinct population of small, rod-like fibrils (Zlatic et al. 2015).

A study emphasized the increasing interest in the protective effects of phenolic compounds, particularly catechins, in mitigating age-related diseases, including neurodegenerative disorders. Although the precise mechanisms are not fully understood, polyphenols have been found to activate defense genes via the xenobiotic response. Recent findings from the authors' research demonstrated that epigallocatechin-3-gallate (EGCG), the predominant catechin in green tea, upregulated HO-1 expression through the Nrf2/ARE pathway in hippocampal neurons. This activation played a crucial role in safeguarding neurons against oxidative damage (Scapagnini et al. 2011).

Another study explored the cytoprotective effects of (−)-epigallocatechin-3-gallate (EGCG) against oxidative stress in rat neuronal cells. The findings revealed that EGCG (25–100 μM) upregulated HO-1 expression at both the mRNA and protein levels, likely via Nrf2 pathway activation. Pre-treatment with EGCG (50 μM) enhanced neuronal resistance to oxidative damage induced by glucose oxidase; however, this protective effect was significantly diminished when heme oxygenase activity was inhibited by zinc protoporphyrin IX. These results suggest that EGCG confers neuroprotection by activating the Nrf2/HO-1 pathway, highlighting its potential as a therapeutic agent for oxidative stress-related neuronal damage (Romeo et al. 2009).

A study examined the impact of epigallocatechin gallate (EGCG) on amyloid-β (Aβ) aggregates linked to Alzheimer's disease. The results demonstrated that EGCG is membrane-active, spontaneously integrating into synthetic brain membranes. Moreover, EGCG effectively disrupted peptide aggregates, leading to a reduction in cross-β and β-sheet structures, which are characteristic of toxic Aβ assemblies (Gastaldo et al. 2020).

A study explored the neuroprotective effects of various polyphenolic compounds, including epigallocatechin gallate (EGCG), against quinolinic acid (QUIN)-induced excitotoxicity in human neurons. QUIN toxicity is driven by increased Ca^{2+} influx, nitric oxide (NO)-induced oxidative stress, DNA damage, and cell death. Among the tested compounds, EGCG, catechin hydrate, and curcumin exhibited the strongest protective effects, significantly reducing Ca^{2+} influx and neuronal nitric oxide synthase (nNOS) activity. Notably, while catechin hydrate did not influence Ca^{2+} influx, it effectively inhibited nNOS activity, suggesting a distinct protective mechanism. All polyphenols mitigated oxidative stress, preventing NAD+ depletion and subsequent cell death (Braidy et al. 2010).

A study examined the neuroprotective role of epigallocatechin gallate (EGCG) in Alzheimer's disease (AD) by evaluating its interaction with the PPARγ pathway. The accumulation of β-amyloid (Aβ), a hallmark of AD, is regulated by BACE1, the enzyme responsible for Aβ production. PPARγ, a nuclear receptor, is known to suppress BACE1 expression, thereby reducing Aβ levels. Using N2a/APP695 cells, researchers found that EGCG downregulated BACE1 transcription and translation, mirroring the effects of the PPARγ agonist pioglitazone. This effect was reversed by the PPARγ inhibitor GW9662, suggesting a PPARγ-dependent mechanism. Additionally, EGCG upregulated PPARγ mRNA and protein levels, decreased pro-apoptotic markers (Bax, caspase-3), oxidative stress indicators (ROS, MDA), and inflammation (NF-κB activity), while enhancing the antioxidant enzyme MnSOD. The attenuation of these protective effects upon GW9662 co-treatment further supports the involvement of PPARγ activation in EGCG's neuroprotective action against AD (Zhang et al. 2017).

A study explored the role of catechins in mitigating the toxicity of intrinsically disordered protein (IDP) oligomers, which contribute to Alzheimer's disease by interacting with biological membranes. Researchers utilized a catechin library with diverse natural chemical modifications and employed a coupled NMR-statistical approach to map catechin interactions with amyloid-beta (Aβ) oligomers and model membranes at atomic resolution. The results identified multiple affinity determinants for catechin binding and demonstrated that certain covalent modifications, while individually reducing affinity, could enhance binding through positive cooperativity. Interestingly, this cooperative effect was more pronounced for Aβ oligomers than for membrane interactions, with distinct structural determinants influencing catechin recognition for each target. Notably, the gallate moiety played a key regulatory role by recruiting previously unengaged substituents into the binding interface, resulting in a more compact and stabilized receptor-bound conformation (Ahmed et al. 2022).

A study focused on the rational design of potential therapeutic compounds for Alzheimer's disease (AD), a neurodegenerative disorder characterized by neuroinflammation and tau protein aggregation. Researchers examined the effects of epicatechin and catechin derivatives (EPIC-PYR, CAT-PYR, CAT-PhG) alongside N-acetylcysteine (NAC) on tau aggregation and neuritogenesis. In vitro assays assessed tau aggregation, while confocal microscopy was used to analyze neuritogenesis in mouse neuroblastoma (N2a) cells. The findings revealed that EPIC-PYR, CAT-PYR, and CAT-PhG effectively inhibited tau aggregation and promoted neuritogenesis in a dose-dependent manner. Among them, the catechin derivative modified with a phloroglucinol group (CAT-PhG) exhibited the strongest neuroprotective effect, suggesting an enhancement in activity due to this functional modification. Additionally, NAC also promoted neuritogenesis and inhibited tau self-aggregation, potentially via a distinct mechanism (Andrade et al. 2020).

7.9 In Vivo Studies

This study underscored the critical neuroprotective role of HO-1 in mitigating oxidative stress and neuronal damage in neurodegenerative diseases. By activating the Nrf2/ARE pathway, HO-1 induction enhances cellular defenses, reducing inflammation and apoptosis—key factors in neurodegenerative disease progression. Investigating the effects of epicatechin (EC), a flavanol from cocoa and tea, researchers found that EC conferred significant neuroprotection against ischemic brain injury in mice. Oral pretreatment with EC (5, 15, or 30 mg/kg) before middle cerebral artery occlusion (MCAO) led to reduced brain infarcts and improved neurological deficit scores. Similarly, posttreatment with 30 mg/kg EC at 3.5 hours post-MCAO improved outcomes. In NMDA-induced excitotoxicity models, EC also reduced lesion volumes. In vitro, EC protected neuronal cultures from oxidative stress, but its neuroprotective effects were abolished in mice lacking HO-1 or Nrf2, confirming the pathway's essential role in EC-mediated neuroprotection (Shah et al. 2010).

This study highlighted the multifaceted neuroprotective effects of EGCG in a preclinical Alzheimer's disease (AD) and type 2 diabetes (T2DM) model. Using transgenic APP/PS1 mice fed a high-fat diet (HFD), EGCG treatment improved insulin sensitivity and enhanced liver insulin signaling. Additionally, it alleviated memory deficits by increasing synaptic markers and CREB phosphorylation. EGCG also reduced amyloid-β (Aβ) production and plaque burden by upregulating α-secretase (ADAM10), while its anti-inflammatory effects were evidenced by reduced astrocyte reactivity and lower toll-like receptor 4 (TLR4) expression. Furthermore, EGCG modulated the unfolded protein response (UPR), decreasing ATF4 and PTP1B levels, which may contribute to its overall neuroprotective properties (Ettcheto et al. 2020).

This study underscored the role of endoplasmic reticulum (ER) stress in EGCG's neuroprotective effects against Alzheimer's disease-related neurotoxicity. In SH-SY5Y cells treated with Aβ1–42, EGCG improved cell viability and reduced apoptosis in a dose-dependent manner by suppressing ER-stress-associated proteins, including GRP78, CHOP, cleaved-caspase-12, and cleaved-caspase-3. Furthermore, EGCG counteracted ER stress-induced toxicity caused by tunicamycin (TM) and thapsigargin (TG), two known ER stress activators. In APP/PS1 transgenic mice, EGCG inhibited cortical neuronal apoptosis, alleviated ER structural abnormalities, and downregulated ER-stress-related proteins, reinforcing its neuroprotective potential in vivo (Du et al. 2018).

This study highlighted the potential of green tea catechin (GTC) in mitigating oxidative stress-induced abnormalities in Alzheimer's disease (AD). Transgenic mice, aged 7 months, were treated with low (1 mg) or high (10 mg) doses of GTC for 6 months. The treatment significantly improved AD-related phenotypes, including reduced behavioral impairment and decreased Aβ-42 production. It also lowered APP-C99/89 levels, γ-secretase components, Wnt proteins, γ-secretase activity, and MAPK activation. Additionally, GTC increased APP-C83 levels, α-secretase activity, neprilysin, and Pin1 expression. Notably, GTC treatment improved lipid metabolism by reducing total and LDL cholesterol levels while increasing HDL cholesterol, suggesting its multifaceted neuroprotective effects (Lim et al. 2013).

This study introduced a novel approach for discovering and utilizing new tea catechins by constructing tea component transformation networks (TCTN), detecting novel molecules via LC-ESI-HRMS/MS, and confirming their structures through synthesis. A previously unidentified precursor ion (546.1749 [M-H]-) was detected in Yingde green tea, indicating the presence of a cinnamoylated flavoalkaloid derived from theanine, p-hydroxycinnamic acid, and epicatechin via a nucleophilic reaction. Three isomers (1–3) with the same molecular weight (547) were synthesized and identified in 18 different green tea cultivars. Compounds 1 and 2 (50 μM) exhibited anti-Alzheimer's disease (anti-AD) potential in the transgenic *Caenorhabditis elegans* strain CL4176, and compound 2 (200 μM) significantly extended the lifespan of wild-type *C. elegans* by 66%, highlighting its promising neuroprotective and longevity-enhancing properties (Qi et al. 2024).

This study investigated the neuroprotective effects of catechin-loaded chitosan-alginate nanoparticles (NPs) in an aluminum chloride ($AlCl_3$)-induced rat model of Alzheimer's disease (AD). The nanoparticles were synthesized using the ionotropic gelation (IG) method and characterized through Zetasizer Nano system, scanning electron microscopy, and Fourier transform infrared spectroscopy. Male Wistar rats were divided into six groups (control, $AlCl_3$-treated, catechin-only, nanocarrier-only, $AlCl_3$ + catechin, and $AlCl_3$ + nanocarrier) and treated for 21 days. Cognitive function was assessed using the Morris water maze (MWM) test, while oxidative stress markers and acetylcholinesterase (AChE) activity were analyzed. $AlCl_3$ exposure led to cognitive impairment, increased AChE activity, and decreased catalase (CAT) and total antioxidant capacity (TAC) levels. Treatment with both catechin and catechin-loaded NPs significantly improved AChE activity, oxidative stress markers, spatial memory, and learning performance, highlighting the potential of nanoparticle-based catechin delivery for AD therapy (Mohammadbaghban et al. 2024).

This study explored the neuroprotective effects of catechin hydrate (CH), a flavonoid with antioxidant and anti-inflammatory properties, in an intracerebroventricular streptozotocin (ICV-STZ)-induced rat model of Alzheimer's disease (AD). Male Wistar rats were pretreated with CH (10 and 20 mg/kg, orally) for 21 days before receiving bilateral ICV-STZ injections (3 mg/kg), while a sham group received a vehicle solution. Two weeks post-infusion, cognitive function was assessed using the Morris water maze (MWM) test, followed by biochemical and histopathological evaluations. CH treatment significantly improved antioxidant status, prevented memory impairment, and restored choline acetyltransferase (ChAT) expression, which was reduced in the ICV-STZ group. Moreover, CH markedly attenuated inflammatory markers, including TNF-α, IL-1β, and iNOS expression, suggesting its potential as a neuroprotective agent in AD therapy (Ahmed et al. 2013).

7.10 Clinical Studies

This study examined the relationship between green tea consumption and cognitive function in older adults. While experimental and animal studies suggest green tea has neuroprotective properties, human data were previously lacking. Researchers analyzed cross-sectional data from a 2002 community-based Comprehensive Geriatric Assessment (CGA) of 1003 Japanese individuals aged 70 and older. Cognitive function was assessed using the Mini-Mental State Examination (MMSE), with cognitive impairment defined by various cutoff scores (<28, <26, and < 24). The findings revealed that higher green tea consumption was associated with a lower prevalence of cognitive impairment. After adjusting for confounders, individuals consuming ≥2 cups of green tea per day had significantly lower odds of cognitive impairment (OR = 0.46, 95% CI: 0.30, 0.72, P for trend = 0.0006) compared to those drinking ≤3 cups per week. Notably, no significant associations were found for black or oolong tea or coffee consumption, suggesting that green tea may have unique cognitive benefits in aging populations (Kuriyama et al. 2006).

This study explored the link between tea consumption and cognitive impairment or decline in older Chinese adults. While laboratory research suggests tea has neuroprotective effects, human evidence remains inconclusive. Researchers analyzed data from the Singapore Longitudinal Ageing Studies cohort, which included 2501 participants aged 55 and older. Cognitive function was assessed using the Mini-Mental State Examination (MMSE) at baseline and after 1–2 years. Cognitive impairment was defined as an MMSE score of $\leq$23, while cognitive decline was classified as a decrease of $\geq$1 point. The findings showed that higher tea consumption was significantly associated with a lower prevalence of cognitive impairment, independent of other risk factors. Compared to those with rare or no tea intake, participants with low, medium, and high tea consumption had lower odds of cognitive impairment, with ORs of 0.56, 0.45, and 0.37, respectively (P for trend <0.001). Similarly, increased tea intake was linked to a reduced risk of cognitive decline (P for trend = 0.042). These protective effects were most evident for black and oolong teas, which were the predominant types consumed in this population. Notably, no significant association was observed between coffee consumption and cognitive status (Ng et al. 2008).

This cross-sectional study, part of the Singapore Longitudinal Aging Studies (SLAS), investigated the relationship between tea consumption and cognitive function in older adults. The study included 716 community-living Chinese adults aged 55 and older, who self-reported their tea consumption habits, including frequency and type of tea consumed. Cognitive function was assessed using a battery of neuropsychological tests, with composite scores calculated for attention, memory, executive function, and information processing speed. Global cognitive function was measured using the Mini-Mental State Examination (MMSE). After adjusting for potential confounders, higher total tea consumption was independently associated with better cognitive performance across multiple domains, including global cognition ($p = 0.03$), memory ($p = 0.01$), executive function ($p = 0.009$), and information processing speed ($p = 0.001$). Both black/oolong and green tea consumption contributed to these positive associations, whereas coffee consumption showed no significant relationship with cognitive function. These findings suggest that regular tea consumption is linked to better cognitive performance in older adults, with protective effects observed across different types of tea (Feng et al. 2010).

This study examined the association between tea consumption and cognitive impairment (CoI) in older adults in Suzhou, China, using data from the Weitang Geriatric Diseases Study. A total of 4579 individuals aged 60 and above were assessed for tea consumption habits and cognitive function through questionnaires and the Abbreviated Mental Test (AMT). Participants were categorized into a normal cognitive function group (AMT score $\geq$ 8) and a CoI group (AMT score $\leq$ 7). Logistic regression analysis revealed an inverse association between tea consumption and the prevalence of CoI (odds ratio = 0.74, $p = 0.032$). Individuals who seldom consumed tea had lower AMT scores compared to habitual tea drinkers.

Notably, the protective effect of tea was more pronounced among never smokers (odds ratio = 0.63), whereas no significant benefit was observed in current or former smokers (odds ratio = 1.10). In never smokers, a higher frequency of tea consumption was significantly linked to a lower risk of CoI (p for trend = 0.010). These findings suggest that regular tea consumption may be associated with a reduced risk of cognitive impairment in older adults, particularly among those who have never smoked (Gu et al. 2018).

7.11 Toxicity

7.11.1 Acute Toxicity

Acute toxicity tests, typically conducted on mice and rats, help estimate lethal dose (LD50), identify target organs, and assess species sensitivity. The LD50 classification ranges from dangerously toxic (<1 mg/kg) to practically nontoxic (>15,000 mg/kg). In one study, a single high dose of EGCG (1868 mg/kg) in rats led to lethargy, labored breathing, and deaths within 72 hours, with histopathological analysis revealing organ damage in deceased animals. However, a lower dose (186.8 mg/kg) caused no toxicity. Other studies highlight EGCG's potential for liver toxicity compared to other tea phenolics. In mice, 100 mg/kg EGCG induced liver damage, while 150–200 mg/kg doses resulted in death within 24 hours. Conversely, tea flower extract exhibited an oral LD50 of >12.0 g/kg in rats, suggesting significantly lower toxicity (Bedrood et al. 2018).

7.11.2 Eye and Skin Toxicity

EGCG exhibited significant eye irritation in rabbits. When applied topically to mice at concentrations of 1%, 3%, and 10% (w/w/day) for 30 days, it was generally well tolerated. However, its combination with depilatory products led to increased skin toxicity. High doses in female BALB/c mice resulted in erythema and lesions, whereas no such effects were observed in hairless SKH1 mice. Ingredients derived from *C. sinensis* leaves are considered safe for cosmetic use at current concentrations when formulated to minimize sensitization risks. However, there is insufficient data to evaluate the safety of non-leaf-derived components (Becker et al. 2019; Bedrood et al. 2018).

7.11.3 Subacute Toxicity

The potential toxic effects of high-dose green tea from concentrated extracts remain incompletely understood. Subacute toxicity, which involves repeated dosing over 14–28 days, is typically assessed in both male and female rats and dogs to establish safe dosage levels for extended studies. In mice, the no observed adverse effect level (NOAEL) for green tea extract over a 28-day period is 2500 mg/kg/day. In rats, heat-sterilized green tea catechins (GTCs) induced gastric effects at 1000 mg/kg/day, whereas the non-sterilized form was considered safe up to 2000 mg/kg/day (Bedrood et al. 2018; Hsu et al. 2011).

7.11.4 Subchronic Toxicity

Subchronic toxicity studies (30–90 days) evaluate the effects of repeated exposure, typically in rodents and dogs, to establish NOAEL and LOAEL for long-term studies. Green tea extract (up to 1000 mg/kg for 14 weeks) induced liver and nasal toxicity in rats and mice, with a NOAEL of 500 mg/kg for liver effects. Pu-erh green tea extract influenced body weight, organ weights, and blood parameters in rats, with a NOAEL of 2500 mg/kg/day. Lipid-soluble green tea extract (up to 2.3 g/kg/day for 30 days) and tea flower aqueous extract (up to 4 g/kg/day for 13 weeks) showed no toxicity. Green tea catechins (up to 5% in diet) impacted lipid metabolism, organ weights, and liver enzymes, with NOAELs of 764 mg/kg/day in males and 820 mg/kg/day in females. EGCG (500 mg/kg/day for 13 weeks) was non-toxic in rats, but in dogs, fasting increased its toxicity, which was mitigated when administered with food. Polyphenon E® studies confirmed lower toxicity when consumed with meals (Bedrood et al. 2018). In a subchronic study, Sprague Dawley rats received oral doses of Pu-erh green tea extract (PGTE) at 0, 1250, 2500, and 5000 mg/kg/day for 91 days. Clinical evaluations, including survival, hematology, biochemistry, urinalysis, and histopathology, were conducted to assess adverse effects. Results showed that high-dose PGTE led to reduced body weight gain and liver dysfunction, with a NOAEL of 2500 mg/kg/day for both sexes (Wang et al. 2012).

7.11.5 Chronic Toxicity

Chronic toxicity studies evaluate the long-term effects of repeated exposure over 3 months or more. The NOAEL for heat-sterilized green tea catechins (GTC) after 6 months was 1200 mg/kg/day in males and 400 mg/kg/day in females. In a 12-month rat study, no toxicity was observed at doses of 1922.9 mg/kg/day for males and 2525.7 mg/kg/day for females. However, toxicity increased under fasting conditions. In dogs, administration of Polyphenon E® at doses of 200–1000 mg/kg/day for 6.5 months on an empty stomach resulted in severe adverse effects (Bedrood et al. 2018).

7.11.6 Mutagenicity, Genotoxicity, Carcinogenicity, and Developmental Toxicity

Mutagenicity, genotoxicity, carcinogenicity, and developmental toxicity are assessed using methods such as the Ames test, comet assay, and MTT assay. Green tea catechins (GTCs) generally exhibit no mutagenic or genotoxic effects in vivo, though certain in vitro assays suggest DNA damage at high concentrations. EGCG demonstrated dose-dependent DNA damage in comet and MTT assays but showed no clastogenic effects in vivo. Polyphenon E® was non-genotoxic in bacterial and mammalian tests, with some toxicity observed in the mouse lymphoma assay. While oxidative stress-related DNA damage was noted in vitro, findings were inconsistent in vivo. Developmental toxicity studies in rats reported no significant adverse effects of EGCG on fetal development, though high doses led to reduced offspring weight. EGCG has been associated with both improvements in sperm quality and reduced reproductive fitness in fruit flies. High green tea consumption during pregnancy was linked to lower serum folate levels but showed no definitive association with birth defects. Some studies suggest potential protective effects of green tea during pregnancy, including reduced risks of transplacental carcinogenesis and metabolic disorders (Bedrood et al. 2018).

Green tea is a popular beverage, especially in China and Japan, with no significant toxicity reported from regular consumption. Research conducted by Chengelis et al. (2008) on green tea catechins (GTC) indicated no adverse effects at doses reaching 2000 mg/kg/day for both heat-sterilized (GTC-H) and non-heat-sterilized (GTC-UH) forms. Additional studies confirmed that GTC-H did not negatively impact embryonic development in rats, establishing the same NOAEL of 2000 mg/kg/day. Similarly, Hsu et al. observed no toxic effects in ICR mice given 2500 mg/kg for 28 days.

Despite its general safety, green tea extracts (GTE) are often promoted for weight loss, and excessive intake has been linked to liver toxicity in some cases. A review by the United States Pharmacopoeia (USP) identified liver injury as a frequent adverse event associated with high-dose GTE consumption. In 2010, Lambert et al. reported significant liver damage in mice administered high doses of EGCG (1500 mg/kg for 7 days), suggesting that elevated EGCG levels might induce oxidative stress in the liver. However, these doses were substantially higher than those typically consumed through tea. Based on toxicological evaluations, Hu et al. (2018) recommended a daily EGCG intake limit of 338 mg for adults. Further investigations by Furukawa et al. revealed that EGCG could trigger oxidative DNA damage in vitro, potentially contributing to mutations and an increased cancer risk.

A separate study demonstrated that EGCG caused DNA damage in human lymphocytes and Nalm6 cells in a dose-dependent manner, with higher concentrations leading to reduced cell viability. Excessive consumption of both black and green tea may result in adverse effects due to three primary factors: caffeine content, aluminum accumulation, and the influence of tea polyphenols on iron absorption. Although green tea contains less caffeine than coffee and is known to enhance cognitive function, it can still cause side effects such as nervousness, headaches, and irregular heart rhythms, particularly in caffeine-sensitive individuals. Moderation is advised, especially for pregnant women and those with heart conditions. Black tea, in particular, has a higher tendency to accumulate aluminum, which may pose risks for individuals with kidney impairments and could contribute to neurological disorders. Additionally, tea consumption has been linked to reduced iron absorption, potentially exacerbating iron deficiency anemia, particularly when consumed alongside meals. To minimize this effect, drinking tea between meals or adding lemon to tea is recommended (Cabrera et al. 2006; Zhao et al. 2022).

References

Aboulwafa MM, Youssef FS, Gad HA, Altyar AE, Al-Azizi MM, Ashour ML (2019) A comprehensive insight on the health benefits and phytoconstituents of *Camellia sinensis* and recent approaches for its quality control. Antioxidants 8(10):455

Ahmed ME, Khan MM, Javed H, Vaibhav K, Khan A, Tabassum R, Islam F (2013) Amelioration of cognitive impairment and neurodegeneration by catechin hydrate in rat model of streptozotocin-induced experimental dementia of Alzheimer's type. Neurochem Int 62(4):492–501

Ahmed R, Huang J, Lifshitz R, Pomier KM, Melacini G (2022) Structural determinants of the interactions of catechins with Aβ oligomers and lipid membranes. J Biol Chem 298(2):101502

Altaş A (2018) Tüm Yönleriyle Çay ve Çay Turizmi: Gastronomi ve Mutfak Sanatları (1. baskı)

Andrade V, Cortés N, Pastor G, Gonzalez A, Ramos-Escobar N, Pastene E, Maccioni RB (2020) N-acetyl cysteine and catechin-derived polyphenols: a path toward multi-target compounds against Alzheimer's disease. J Alzheimer's Dis 75(4):1219–1227

Ashby BC (2014) Tea culture and the British Empire, 1600–1900. Master's thesis, The University of North Carolina at Charlotte

Awasom I (2011) Tea. J Agric Food Inf 12(1):12–22

Aykaç G, Uzun MB, Özçelikay G (2014) Tea in social aspect *camellia sinensis*. Mersin Üniversitesi Tıp Fakültesi Lokman Hekim Tıp Tarihi ve Folklorik Tıp Dergisi 4(1):1–5

Barry MJ (1986) The samovar: history and use. MJP Barry, Anchorage, Alaska

Bastianetto S, Yao ZX, Papadopoulos V, Quirion R (2006) Neuroprotective effects of green and black teas and their catechin gallate esters against β-amyloid-induced toxicity. Eur J Neurosci 23(1):55–64

Beard JJ, Ochoa R, Childers CC, Bauchan GR, Shepard M (2013) Travelling with tea: a Tuckerella's tale. Exp Appl Acarol 59:177–202

Becker LC, Bergfeld WF, Belsito DV, Hill RA, Klaassen CD, Liebler DC et al (2019) Safety assessment of *Camellia sinensis*–derived ingredients as used in cosmetics. Int J Toxicol 38(3_suppl):48S–70S

Bedrood Z, Rameshrad M, Hosseinzadeh H (2018) Toxicological effects of *Camellia sinensis* (green tea): a review. Phytother Res 32(7):1163–1180

Braidy N, Grant R, Adams S, Guillemin GJ (2010) Neuroprotective effects of naturally occurring polyphenols on quinolinic acid-induced excitotoxicity in human neurons. FEBS J 277(2):368–382

Cabrera C, Artacho R, Giménez R (2006) Beneficial effects of green tea – a review. J Am Coll Nutr 25(2):79–99

Castellani RJ, Rolston RK, Smith MA (2010) Alzheimer disease. Dis Mon 56(9):484

Chakrabarti M (2022) Indian tea culture with special reference to Kolkata. *Available at SSRN 4314358*

Chengelis CP, Kirkpatrick JB, Regan KS, Radovsky AE, Beck MJ, Morita O, Suzuki H (2008) 28-Day oral (gavage) toxicity studies of green tea catechins prepared for beverages in rats. Food Chem Toxicol 46(3):978–989

Chopade V, Phatak A, Upaganlawar A, Tankar A (2008) Green tea (*Camellia sinensis*): chemistry, traditional, medicinal uses and its pharmacological activities – a review. Pharmacogn Rev 2(3):157

Cummings JL, Tong G, Ballard C (2019) Treatment combinations for Alzheimer's disease: current and future pharmacotherapy options. J Alzheimer's Dis 67(3):779–794

Das P, Chettri V, Ghosh S, Ghosh C (2023) Micromorphological studies of the leaf and stem of *Camellia sinensis* (L.) Kuntze with reference to their taxonomic significance. Microsc Res Tech 86(4):465–472
Du K, Liu M, Zhong X, Yao W, Xiao Q, Wen Q, Wei M (2018) Epigallocatechin gallate reduces amyloid β-induced neurotoxicity via inhibiting endoplasmic reticulum stress-mediated apoptosis. Mol Nutr Food Res 62(8):1700890
Erşahin R (2021) Shall we have a tea? Tea in a socio-cultural perspective. Tourism Recreation 3(1):55–65
Ettcheto M, Cano A, Manzine PR, Busquets O, Verdaguer E, Castro-Torres RD, Camins A (2020) Epigallocatechin-3-Gallate (EGCG) improves cognitive deficits aggravated by an obesogenic diet through modulation of unfolded protein response in APPswe/PS1dE9 mice. Mol Neurobiol 57:1814–1827
Feng L, Gwee XY, Kua EH, Ng TP (2010) Cognitive function and tea consumption in community-dwelling older Chinese in Singapore. J Nutr Health Aging 14:433–438
Fleming S (2010) Time for Tea: The Cultural Significance of Tea in the British Atlantic World, 1730–1750
Gastaldo IP, Himbert S, Ram U, Rheinstädter MC (2020) The effects of resveratrol, caffeine, β-carotene, and epigallocatechin gallate (EGCG) on amyloid-β 25–35 aggregation in synthetic brain membranes. Mol Nutr Food Res 64(22):2000632
Ger G, Kravets O (2020) Special and ordinary times: tea in motion. In: Time, consumption and everyday life. Routledge, pp 189–202
Gilbert M (2008) Chinese tea in world history. Educ About Asia 13(2):11
Gu YJ, He CH, Li S, Zhang SY, Duan SY, Sun HP, Pan CW (2018) Tea consumption is associated with cognitive impairment in older Chinese adults. Aging Ment Health 22(9):1237–1243
Güneş S (2012) Turkish tea culture and its products. Millî Folklor 93
Han KPL (2007) Tracing the history of tea culture. In: Tea and tourism: tourists, traditions and transformations, pp 23–37
Honda Y (2012) History of Turkish tea drinking habits. In: Proceedings of Turkish tea culture lecture meeting
Hsu YW, Tsai CF, Chen WK, Huang CF, Yen CC (2011) A subacute toxicity evaluation of green tea (*Camellia sinensis*) extract in mice. Food Chem Toxicol 49(10):2624–2630
Hu J, Webster D, Cao J, Shao A (2018) The safety of green tea and green tea extract consumption in adults–results of a systematic review. Regul Toxicol Pharmacol 95:412–433
Isemura M (2019) Catechin in human health and disease. Molecules 24(3):528
Kumar A, Sidhu J, Lui F, Tsao JW (2024) Alzheimer disease. In *StatPearls*. StatPearls Publishing
Kuriyama S, Hozawa A, Ohmori K, Shimazu T, Matsui T, Ebihara S, Tsuji I (2006) Green tea consumption and cognitive function: a cross-sectional study from the Tsurugaya project. Am J Clin Nutr 83(2):355–361
Lim HJ, Shim SB, Jee SW, Lee SH, Lim CJ, Hong JT, Hwang DY (2013) Green tea catechin leads to global improvement among Alzheimer's disease-related phenotypes in NSE/hAPP-C105 Tg mice. J Nutr Biochem 24(7):1302–1313
Lutgendorf P (2012) Making tea in India: chai, capitalism, culture. Thesis Eleven 113(1):11–31
Mahmood T, Akhtar N, Khan BA (2010) The morphology, characteristics, and medicinal properties of *Camellia sinensis* tea. J Med Plants Res 4(19):2028–2033
Mandel S, Weinreb O, Amit T, Youdim MB (2004) Cell signaling pathways in the neuroprotective actions of the green tea polyphenol (−)-epigallocatechin-3-gallate: implications for neurodegenerative diseases. J Neurochem 88(6):1555–1569
Mohammadbaghban E, Taravati A, Najafzadehvarzi H, Khaleghzadeh-Ahangar H, Tohidi F (2024) Oral administration of encapsulated catechin in chitosan-alginate nanoparticles improves cognitive function and neurodegeneration in an aluminum chloride-induced rat model of Alzheimer's disease. Physiol Rep 12(13):e16095
Morgan P (ed) (2003) Japanese tea culture: art, history, and practice. Routledge
Moura M (2015) Tea: a journey from the east to mid-Atlantic. Eur Sci J 11(29)

Naing YY, Khaing WW, Kywe T (2019) Morphological and anatomical characteristics of Camellia sinensis (L.) Kuntze from two different localities. Doctoral dissertation, MERAL Portal

Naveed S, Hameed A (2014) Consumption of tea in professionals and non-professionals. SOJ Pharm Pharmaceut Sci 1(3):1–4

Ng TP, Feng L, Niti M, Kua EH, Yap KB (2008) Tea consumption and cognitive impairment and decline in older Chinese adults. Am J Clin Nutr 88(1):224–231

Okello EJ, Mather J (2020) Comparative kinetics of acetyl- and butyryl-cholinesterase inhibition by green tea catechins: relevance to the symptomatic treatment of Alzheimer's disease. Nutrients 12(4):1090

Pan SY, Nie Q, Tai HC, Song XL, Tong YF, Zhang LJF, Liang C (2022) Tea and tea drinking: China's outstanding contributions to the mankind. Chin Med 17(1):27

Parikh NH, Parikh PK, Rao HJ, Shah K, Dave BP, Prajapati BG (2024) Current trends and updates in the treatment of Alzheimer's disease. In: Alzheimer's disease and advanced drug delivery strategies. Academic Press, pp 373–390

Patel SH (2005) *Camellia sinensis*: historical perspectives and future prospects. J Agromedicine 10(2):57–64

Peterson DE (1994) "Samovar life": Russian nurture and Russian nature in the rural prose of Valentin Rasputin. Russ Rev 53(1):81–96

Qi Y, Yang Y, Ke JP, Chen CH, Liu Z, Yang Z, Bao GH (2024) Identification, anti-Alzheimer's disease and antiaging activities of cinnamoylated flavoalkaloids from green tea. Food Biosci 58:103765

Romeo L, Intrieri M, D'Agata V, Mangano NG, Oriani G, Ontario ML, Scapagnini G (2009) The major green tea polyphenol (−)-epigallocatechin-3-gallate induces heme oxygenase in rat neurons and acts as an effective neuroprotective agent against oxidative stress. J Am Coll Nutr 28(sup4):492S–499S

Sadler AL (2011) Cha-no-yu: the Japanese tea ceremony. Tuttle Publishing

Scapagnini G, Sonya V, Nader AG, Calogero C, Zella D, Fabio G (2011) Modulation of Nrf2/ARE pathway by food polyphenols: a nutritional neuroprotective strategy for cognitive and neurodegenerative disorders. Mol Neurobiol 44:192–201

Scheltens P, De Strooper B, Kivipelto M, Holstege H, Chételat G, Teunissen CE, van der Flier WM (2021) Alzheimer's disease. Lancet 397(10284):1577–1590

Serinsu BA (2022) Hindistan'da çay kültürü ve terakota bhar fincanları. Medeniyet Sanat Dergisi 8(1):67–82

Shah ZA, Li RC, Ahmad AS, Kensler TW, Yamamoto M, Biswal S, Doré S (2010) The flavanol (−)-epicatechin prevents stroke damage through the Nrf2/HO1 pathway. J Cereb Blood Flow Metab 30(12):1951–1961

Sharangi AB (2009) Medicinal and therapeutic potentialities of tea (*Camellia sinensis* L.) – a review. Food Res Int 42(5–6):529–535

Singh B, Day CM, Abdella S, Garg S (2024) Alzheimer's disease current therapies, novel drug delivery systems and future directions for better disease management. J Control Release 367:402–424

Takako M (2018) Tea drinking culture in Russia. J Int Econ Stud 32:57–74

Turay M (2001) The social significance of the Japanese tea ceremony. Jpn Stud Assoc J 3:57–72

Üstün Ç, Demirci N (2013) The plant of tea (*Camellia sinensis* L.) historical development and medical evaluation. Lokman Hekim J 3(3):5–12

Vishnoi H, Bodla RB, Kant R, Bodla RB (2018) Green tea (*Camellia sinensis*) and its antioxidant property: a review. Int J Pharm Sci Res 9(5):1723–1736

Wang D, Meng J, Xu K, Xiao R, Xu M, Liu Y, Liu L (2012) Evaluation of oral subchronic toxicity of Pu-erh green tea (*Camellia sinensis* var. *assamica*) extract in Sprague Dawley rats. J Ethnopharmacol 142(3):836–844

Wang C, Han J, Pu Y, Wang X (2022) Tea (*Camellia sinensis*): a review of nutritional composition, potential applications, and omics research. Appl Sci 12(12):5874

Weisburger JH (1997) Tea and health: a historical perspective. Cancer Lett 114(1–2):315–317

Wohl S (2017) The Turkish tea garden: exploring a "third space" with cultural resonances. Space Cult 20(1):56–67

Yan W, Ge Z, Xiong L (2020) Research on the influence of Chinese tea technology on the world tea industry. Am J Ind Bus Manag 10(1):135

Young H (2001) Tea culture in Britain, 1660–1800. *ICOS*

Zhang ZX, Li YB, Zhao RP (2017) Epigallocatechin gallate attenuates β-amyloid generation and oxidative stress involvement of PPARγ in N2a/APP695 cells. Neurochem Res 42:468–480

Zhao DW (2024) Botany and taxonomy of tea (*Camellia sinensis*, Theaceae) and its relatives. In: The tea plant genome. Springer Nature Singapore, Singapore, pp 13–37

Zhao T, Li C, Wang S, Song X (2022) Green tea (*Camellia sinensis*): a review of its phytochemistry, pharmacology, and toxicology. Molecules 27(12):3909

Zlatic CO, Mao Y, Ryan TM, Mok YF, Roberts BR, Howlett GJ, Griffin MD (2015) Fluphenazine·HCl and epigallocatechin gallate modulate the rate of formation and structural properties of apolipoprotein C-II amyloid fibrils. Biochemistry 54(24):3831–3838

Mindful Eating: Ayurvedic Insights into Diet and Alzheimer's Disease Management

8

Jyoti Dahiya, Anupam K. Mangal, Ankita Ghosh, Sudipta Roy, and G. Babu

Abstract

Alzheimer's disease (AD) is a progressive neurodegenerative disorder that primarily affects the elderly population. The traditional Indian medical system, known as Ayurveda, provides a distinctive viewpoint on diagnosing and treating AD. Ayurveda states that a *Vata* dosha imbalance, along with *Majja Dhatu* (bone marrow or nervous tissue) in the cerebral cortex, causes the cortex and hippocampus to shrink and amyloid β plaques to develop, which progressively leads to AD. For optimum health, Ayurveda stresses the significance of eating a balanced diet (*Ahara*) and having a healthy digestive system (*Agni*). To guarantee balanced nutrition, the Ayurvedic diet adheres to particular dietary standards (*Aharvidhivisheshayatan*) and integrates six tastes (*Shadrasa*). Additionally, according to their main dosha (*Vata*, *Pitta*, or *Kapha*). Ayurveda also categorizes individuals into three subgroups based on their dominant dosha (*Vata*, *Pitta*, or *Kapha*), each requiring specific dietary recommendations. The best diet for AD is said to be *Sattvic*, which is high in fresh fruits, vegetables, complete grains, dairy, and nuts. Because of their neuroprotective and cognitive-boosting qualities, whole foods like grains, legumes, fruits, nuts, spices, ghee, milk, and herbal supplements are advised. A balanced diet, spiritual activities, lifestyle changes, and Ayurvedic principles can all be included to provide a comprehensive approach to controlling AD and fostering mental health. By finding balance between three states (*Vata*, *Pitta*, or *Kapha*), people can achieve spiritual development, mental and emotional health, and a holistic approach to mental well-being.

J. Dahiya (✉) · A. K. Mangal · A. Ghosh · S. Roy · G. Babu
Central Ayurveda Research Institute (CARI), Central Council for Research in Ayurvedic Sciences (CCRAS), Kolkata, West Bengal, India

J. Malik et al. (eds.), *Diet and Alzheimer's Disease: Let Food be Our Medicine*, Nutritional Neurosciences, https://doi.org/10.1007/978-981-95-2736-6_8

Keywords

Alzheimer's disease · Ayurvedic diet · Sattvic diet · Wholesome food · Viruddha ahaara

Abbreviations

AD	Alzheimer's disease
Ach	Acetylcholine
AChE	Acetylcholinesterase
BChE	Butyrylcholinesterase and NMDA: N-methyl-D-aspartate

8.1 Introduction

Alzheimer's disease (AD) is a progressive and irreversible neurodegenerative disorder that affects mostly the elderly. It is distinguished by memory and executive function declines, as well as personality changes (Breijyeh and Karaman 2020; Jiang et al. 2013). Alzheimer's disease affects roughly two-thirds of the population. Prof. Alois Alzheimer, a German neuropsychiatrist, first described Alzheimer's disease in 1906 as a complex, multifaceted, and progressive disorder associated with a diminished level of acetylcholine in the brain (Jiang et al. 2013). According to a recent study, dementia caused by Alzheimer's affects over 4.7 million people aged 65 and up in the United States, and approximately 496,000 people in the United Kingdom. India has 1.37 billion people, and currently, somewhere around 8.8 million Indians over the age of 60 suffer from dementia (Lee et al. 2023).

In Alzheimer's disease brain condition worsens over time and it is characterized by neurological changes that cause the accumulation of specific proteins. Alzheimer's disease (AD) shrinks the brain and eventually kills the cells (Kumar et al. 2014). In order to enter the synapse and attach to a receptor, acetylcholine is typically stored in nerve terminals called vesicles and distributed from them on depolarization. The brain enzymes acetylcholinesterase (AChE) and butyrylcholinesterase (BChE), which are present in neurofibrillary tangles and neuritic plaques, are elevated in Alzheimer's disease patients, and the acetylcholine that is released has a relatively short half-life. These enzymes cause acetylcholine (Ach) molecules to lose their stimulatory activity by hydrolyzing their ester bond to choline and acetate and cause the brain's hippocampus region to produce senile plaques, which are clusters of amyloid-beta protein that damage and kill neurons, impairing cell-to-cell communication (Cras et al. 1991; Perl 2010; Hampel et al. 2021). The decreased availability of acetylcholine, a neurotransmitter involved in memory and learning, contributes significantly to the development of Alzheimer's disease. The enzyme acetylcholinesterase degrades ACh before it can perform its function, resulting in neuronal death and cognitive deficits (Chen et al. 2022). The most prevalent kind of dementia, Alzheimer's disease, is characterized by a progressive deterioration in thinking, behaviour, memory, and social skills. These changes make it harder for a person to perform daily duties (Breijyeh and Karaman 2020; Jiang et al. 2013). Neurofibrillary tangles (NFTs) are aberrant tau protein deposits that impair normal brain function. Tau protein must function properly in order to carry important nutrients and minerals throughout the brain. However, in this disorder, the tau protein's thread twists into abnormal tangles, leading the transport mechanism to malfunction. This causes cortical shrinkage, which results in the death of brain cells (Brion et al. 2001; Yang et al. 2024). Alzheimer's disease that develops early, usually after the age of 60, is linked to genetic abnormalities (Williamson et al. 2009). Risk factors include gene mutations in APOE, APP, PSEN1, and PSEN2 (Hoogmartens et al. 2021). People with diabetes, high blood pressure, high cholesterol, and who smoke are more likely to develop Alzheimer's. Mood swings, learning difficulties, and short-term memory loss are early symptoms (Bature et al. 2017; Li et al. 2014).

One of the most important clinical advancements in the treatment of Alzheimer's disease has been the use of AChE and BChE inhibitors to increase ACh levels in the brain, despite cholinergic agents having nicotinic and muscarinic agonist properties. BChE cleaves ACh, affecting glial cells and neuronal pathways. Studies show slower cognitive decline in dementia patients with BChE polymorphisms. Allopathic therapy uses pharmaceutical and non-pharmacologic treatments, including NMDA antagonists (like memantine), cholinergic activators (like tacrine, rivastigmine, donepezil, and galantamine), and various other cerebroactive drugs (like piracetam, pyritinol, dihydro-ergotoxine, and piribedil) are among the valuable anti-Alzheimer's drugs. Other treatments include insulin, secretase inhibitors, anti-inflammatory drugs, and antihypertensive drugs. Drugs like etanercept and brain-derived neurotropic factor can also be used (Singh et al. 2024). Regrettably, the advantages of currently available medications are sometimes outweighed by their exorbitant price, transient effectiveness, serious side effects, and patient non-compliance. This has accelerated the search for complementary or alternative therapies that are more patient-friendly and less hazardous new treatment strategies based on medicinal plants are currently being investigated (Leucht et al. 2015).

Naturopathic medicine is a very individualized line of work. Treatment plans cover a wide range of topics, from fundamental lifestyle changes to more complex therapies. A crucial part of any procedure is guaranteeing the determinants of health, which include a clean environment, food, and air (Myers and Vigar 2019). Detoxification and lifestyle changes may be necessary to remove barriers to recovery. Because they address the full person, physical medicine, homeopathy, and hydrotherapy are valuable resources to think about utilizing (Timko et al. 2016).

Using Ayurvedic principles to treat and prevent Alzheimer's disease is the other and most advantageous strategy.

8.2 Ayurvedic Approach

The traditional Indian medical system, Ayurveda, holds that everything is made up of the five components known as "*Mahabhutas*"—air (*Vayu*), fire (*Tejas*), water (*Jala*), earth *(Prithivi*), and space (*Akasha*)—in a unique ratio that results in different doshas. Additionally, Ayurveda emphasizes the need to determine the root causes of ailments in the first place (Jaiswal and Williams 2016).

वायुः पित्तं कफश्चेति त्रयो दोषाः समासतः। ६।
विकृताविकृता देहं घ्नन्ति ते वर्तयन्ति च। ७।

Astanga hrudaya, Sutrasthana, Chap. 1; The three *doshas* of the body are *vata*, *pitta*, and *kapha*. Health results from a perfect balance of the three *doshas*; diseases result from an imbalance in the *tridosha*

Ayurveda identifies three bio-energies (*Dosha*), namely *Vata* (all types of movement), *Pitta* (digestive and metabolic activities), and *Kapha* (growth, union, stability, and composition). *Doshas* are the psychophysiological principles that regulate the functions of the body. Also, seven distinct body tissues (*Dhatu*) that include *rasa* (plasma tissue, lymph), *rakta* (blood, red blood cells), *mamsa* (muscle), *meda* (adipose tissue, fat), *asthi* (bone and cartilage), *majja* (nerve, marrow, connective tissue), and *shukra/artava* (male and female reproductive system), and three wastes (*Mala*) which are crucial for sustaining a healthy life of human beings (Debnath et al. 2015). Together, the elements of air and space make up *Vata* dosha, which stands for the idea of mobility. And regulates functions associated with movement and communication in the body, like nerve transmission and breathing. The *Pitta* dosha, which is made up of water and fire, is in charge of transformation and metabolism. The elements of earth and water make up the *Kapha* dosha, which is responsible for solidity and structure (Kachhawaha et al. 2021). Of these *Doshas*, *Vata* is considered as "lead" *Dosha* on account of the quality of movement and leads *Pitta* and *Kapha*, which cannot move on their own. Each individual has a unique ratio of *Vatta*, *Pitta*, and *Kapha*; this ratio defines the *Prakriti* or psychophysiology of that individual. Any disturbance in the *Prakriti* causes the state of disease. And the imbalance state is known as *Vikriti* (Sharma and Chandola 2011).

Of all the *Dhatus*, movement, intelligence, and communication are governed by the *Majja Dhatu*, which is either bone marrow or nervous tissue. When *Dosha* and *Majja Dhatu* are out of balance, neurological disorders develop. When combined with *Majja Dhatu* in the cerebral cortex, the dry and cold *Vata* predominance causes shrinkage in the cortex and hippocampus as well as the development of amyloid β plaques, which over time lead to a number of neurological conditions, including Alzheimer's disease (Jaiswal and Williams 2016; Majhi et al. 2020). However, there is not a disorder in Ayurveda that is clinically comparable to Alzheimer's disease. But among the *Vatavyadhi* (neurodegenerative disorders) category, there is a condition called *Prana Vrtta Saman Vatavyadhi* that includes a number of clinical characteristics and symptoms that appear to be related to Alzheimer's disease (Bredesen and Rao 2017).

According to a recent interpretation of the metabolic profile of Alzheimer's disease patients, Ayurveda separates individuals into three subgroups depending on their doshas (*Vata, Pitta*, and *Kapha*):

1. **The *Vata* type AD** is characterized by trophic support withdrawal and hormonal drop, similar to type 2 AD. It can cause memory loss, anxiety, and confusion. It is distinguished by concentration issues, memory restlessness, and cognitive decline. Ayurvedic therapies for *Vata*-type AD focus on promoting stability,

calming the mind, and fortifying the neurological system. *Rasayana* and *Vajikarana* treatments boost resistance and immunity (Bredesen 2015; Sharma and Chandola 2011).

2. ***Pitta*** **Subtype**: *Pitta*-type and type 1 AD (inflammatory AD) have similar metabolic profiles, but can occur in *Pitta*-typed individuals due to a Pitta-vitiating lifestyle that degenerates cortical cells and the hippocampus. This imbalance can cause cognitive decline and inflammation in the brain. Ayurvedic treatments, such as *Rasayana* and *Vajikarana*, focus on cooling and calming the body and mind to reduce inflammation and improve cognitive function (Mishra et al. 2001; Akiyama et al. 2000).
3. ***Kapha*** **subtype:** The *Krimi* subtype of Alzheimer's disease (AD) is linked to pathogens and toxic chemicals, causing neurodegenerative processes. Treatment focuses on detoxification, digestive system cleansing, and balance restoration to address *Krimi*-related cognitive deterioration (Bredesen 2015; Bredesen 2016).

Other concepts addressed in the Charaka Samhita and Sushrutha Samhita state that the *Manas* (mind), which is closely related to the *Sharira* (body) and the *Aatma* (soul), as the key to mental wellness.

धीधृतिस्मृतिविभ्रंशः सम्प्राप्तिः कालकर्मणाम्।
असात्म्यार्थागमश्चेति ज्ञातव्या दुःखहेतवः।।९८।।

Charaka Samhita, Sharira sthana, Chap. 1; *Mana*, characterized by *sattva*, is pure and influenced by *rajas* and *tamas* in the mind. Misery can result from actions, unhealthful contact, aging, *dhriti*, memory, and intellect impairment, contributing to misery

A person's psychological state is determined by their balance of mental, emotional, and behavioural tendencies, represented by the three *Manas Gunas*: *Rajas* (hyperactivity, restlessness, desire), *Sattva* (clarity, intelligence, and emotional stability), and *Tamas* (inertia, lethargy, ignorance, and emotional numbness). Inadequate interaction between these *Gunas* can lead to mental illness, such as Alzheimer's disease, which is considered an advanced form of dementia. *Rajas* and *Tamas* are two states of mind that deplete dhatus and degrade *Sattva Guna*, a mental quality representing perception and purity. *Sattva* is attentive, active, and intellectual, promoting mental clarity, emotional stability, and compassion. Ayurveda suggests *Sattvic* minds promote tranquillity, happiness, and mental clarity, while *Rajas* is an attachment-dominated state characterized by anger, passion, and delusions (Akter and Nahar 2024; Sajjan 2022).

Ayurveda emphasizes the importance of *Sattva* in mental purity (*Manas Shuddhi*) and emotional resilience (*Dhiarya*). Imbalances in *Rajas* and *Tamas* can lead to psychiatric diseases. Ayurvedic treatment aims to restore balance by regulating these two elements and enhancing *Sattva*. Mental health is a dynamic equilibrium between the *Gunas*, affecting behaviour, emotions, lifestyle, and spiritual well-being. Ayurvedic therapies aim to decrease excessive *Rajas* and *Tamas* and boost *Sattva* through spiritual activities, herbal medicines, diet, and lifestyle (Seshan et al. 2024; Mangalasseri and Chandran 2017; Patwardhan and Datta 2021). Therefore,

Ayurvedic treatment entails panchakarma, the use of medicinal plants, internal medications, and adhering to the Ayurvedic diet. Of these, following Ayurvedic principles for healthy dietary conditions and the Ayurvedic diet is the most crucial, and the current review is compiling all the information needed for this.

8.3 Ayurvedic Diet

In Indian culture and Ayurveda lot of value is placed on *Ahara*, the Diet. The Ayurvedic diet is a set of rules that takes into account the gut environment, bio-individuality, circadian cycles, individual preferences, and humans' ability to absorb nutrients from food. The Ayurvedic diet is straightforward yet effective. Even within a few weeks, an Ayurvedic diet improves sleep, overall health, and digestive symptoms (Dhanya et al. 2019). According to Ayurvedic principles, food is more than just a source of nutrients for our bodies. However, Ayurvedic wisdom states that even before the diet provides nourishment, it is critical to preserve and improve the gut environment (*Agni*). A food that is thought to be abundant in nutrients but causes gut dysfunction will eventually damage digestion, which will reduce the benefits of all other foods. Toxin accumulation is caused by ingesting unhealthy and incompatible combinations of foods (such as fish and milk), as well as eating too much sour, spicy, and processed food without first digesting it. Indulging in excessive non-vegetarian foods also causes blockages in the body's small channels. This causes an increase in *Pitta Dosha*, which affects cells in an inflammatory manner. Uncontrollable growth will result from the impaired cell division caused by the increased *Vata Dosha*. This comprises anomalies in blood (*Rakta Dhatu*) and bone marrow (*Majja Dhatu*), in addition to deterioration in adipose tissue (*Meda Dhatu*), circulatory fluids (*Rasa Dhatu*), and bones (*Asthi Dhatu*). Additionally, it could result in the emergence of illnesses like Alzheimer's. As a result, according to Ayurveda, maintaining a healthy diet is essential. Consuming a plant-based, whole-food diet rich in antioxidants and healthy fats is advised by the Ayurvedic diet. Supplements that contain beta-carotene, tocopherol, and B vitamins are also advised because they may offer extra neuroprotective and antioxidant advantages (Sharma and Chandola 2015; Dhanya et al. 2019).

8.4 The Concept of *Agni* for Proper Digestion

For optimum health, proper food digestion is also essential. The many digestive and metabolic processes are involved in the Ayurvedic idea of *Agni,* or fire. And transformational concepts that operate at every level of physiology (Fernández-Tomé 2024). Food digestion, the conversion of its different components into the proper *Dhatu* (tissue) elements, and the breakdown and utilization of certain constituents for the generation of energy needed for essential functions are all included in this. *Jatharagni*, the digestive fire that breaks down food, is the primary of the 13 varieties of Agni. Thermogenic mechanisms in *Jatharagni* are in charge of the body and provide support for every other Agni. One of the earliest classical scriptures of Ayurveda, the Charaka Samhita, states that maintaining *Jatharagni* properly contributes to a long and healthy life (Samhita 1949).

8.5 *Shadrasa:* The Concept of Six Tastes in Ayurveda Diet

Ayurveda describes six types of *Rasa* (taste): *Madhura* (sweet), *Amla* (sour/acidic), *Lavana* (salty), *Katu* (pungent), *Tikta* (bitter), and *Kashaya* (astringent). All six tastes should be included in meals, according to the Ayurvedic diet. Ayurveda does not define balance as the battle to maintain an appropriate carbohydrate, fat, and protein intake ratio. Rather, Ayurveda makes meals more flavourful and balanced by incorporating all six tastes. A meal that includes all six tastes facilitates elimination, nourishment, absorption, and digestion. Additionally, by using taste to predict how a particular food will affect your body, you can become your own health expert (Sharma and Chandola 2015; Wagh and Bhalerao 2020; Katariya and Mishra 2021).

1. ***Madhura*/Sweet Taste:** It refers to naturally sweet and moist foods like rice, ghee, milk, and so on, as opposed to processed sweets. Even though these foods are typically hefty, they are vital since they supply the building blocks for the body's new cell formation. Root vegetables, juicy sweet veggies, healthful fats, and moist grains are examples of foods.
2. ***Amla*/Sour Taste:** Sour meals release digestive enzymes and increase hunger. Additionally, sour meals naturally warm the body and aid in the breakdown of other foods. Foods like tamarind, lemon/lime, pickled ginger, and amla juice are examples.
3. ***Lavana*/Salty Taste:** According to Ayurvedic principles, it scrapes and prevents food from adhering to the stomach lining. However, too much salt can throw off the body's water equilibrium. Himalayan salt is suggested by Ayurveda since it is believed to have less adverse effects. Olives, spinach, black salt, and Himalayan pink salt are a few examples of meals.
4. **Pungent Taste:** Pungent foods should be consumed in small amounts because they are warm in nature, which aids in the breakdown of other foods, particularly those with a sweet taste. The most often recommended pungent foods are spices. The mucosal lining may get damaged if overly strong foods are consumed without other balancing flavours. This explains why people have never eaten turmeric on its own. Spices, ginger, cloves, mustard seeds, peppers, and parsley are a few examples of foods.
5. ***Tikta*/Bitter Taste:** Bitter food causes dry mouth and makes digestion violent and slimy. By cooling the intestines and removing extra mucus, bitter meals help to maintain equilibrium. Examples of foods include chard, bitter melon, fenugreek leaves, arugula, dandelion greens, spinach, curry leaves, kale, and broccoli.
6. ***Kashaya*/Astringent Taste:** In addition to dehydrating digestive enzymes, some astringent meals also serve as a warning to stop eating. To prevent the mucosal lining from drying out, astringent foods should always be consumed with other flavours. Examples of foods include cruciferous vegetables, beans, chickpeas, lentils, buttermilk, turmeric, and green plants.

Therefore, Sweet flavours are beneficial to the sensory organs, promote happiness, vitality, and contentment, and aid in the development of all body tissues. Acidic or sour foods boost energy, soothe the senses, enlighten the mind, spark the digestive fire, and regulate the peristaltic flow of *Vata*. Foods with strong flavours arouse the stomach fire and increase the sensitivity of the sense organs. Numerous *Vata* disorders might result from consuming an excessive amount of meals with strong, bitter, or astringent flavours (Sharma and Chandola 2015; Wagh and Bhalerao 2020; Katariya and Mishra 2021).

The following are some fundamental Ayurvedic dietary guidelines that nourish the body and support *Agni* (the digestive fire). The emphasis on how to take the diet is just as important as the concept of *Ahar* itself. Therefore, in Ayurveda, the instructions for eating are just as significant and stressed as the digestive process (Fernández-Tomé 2024). We refer to this as "*Aharvidhivisheshayatan*" here, *Ahar* means food; *Vidhi* means method, *Vishesh* is peculiar, and *Ayatanani* is abode. In Ayurveda, it is also known as dietetics or dietary rules for eating. The practice of eating with discipline is equally important in this *Ahar* concept. When eaten properly, food acts as a medicine to balance our *Dosha*, *Dhatu*, and *Malas*. Otherwise, food becomes dangerous and can lead to a lot of illnesses (Agnivesha 2005: Chap. 1/21).

Charaka Samhita described eight *Ahara vidhi* principles, known as *Aharvidhivisheshayatan*, to highlight the significance of various aspects to take into account when consuming *Ahar*. The eight guiding concepts are:

1. ***Prakriti* of food:** the quality of food (*Ahar*) is denoted as *Guru and Laghu*, it denotes food of natural quality. The term "*Laghu Ahar*" refers to foods that are easy to digest, such as green grams. The food that takes the longest time for digestion is *Guru Ahar.* It is essential to take food material *Prakriti into* account while eating, and other elements rely on food's *Prakriti*. Food type determines the amount, and timing of consumption. Thus, in *Aharvidhivisheshayatan*, it comes first. The consumer's *Prakriti* (nature/constitution) should also be taken into account. The *Ahar* of a particular individual is chosen based on the individual's constitution (Agnivesha 2005: Chap. 1/21).
2. ***Karan* (*Sanskar*):** *Sanskara/Karana,* refers to the processing of materials that modifies the materials' intrinsic qualities. Food undergoes changes in quality after processing, such as roasting, frying, and boiling. This change results from the implementation of clarification, emulsification, heat (vaporization, distillation, and sublimation), storage, maturation, flavouring, impregnation, preservation, and the material of the container. In the same way that roasted grain flour turns heavy when processed into a cooked bolus (Agnivesha 2005: Chap. 27/339).
3. ***Samayoga* (combination of two or more):** *Samayoga,* the combination of two or more things, none of the components by themselves may have the special qualities that this combination does. Therefore, even though it is very beneficial but can become harmful sometimes, for example, fish with milk, banana with milk is an unwholesome diet as per Ayurveda. As fish is *ushna viryatmak* and

milk is *sheet viryatmak*, eating both together is an unwholesome diet. When combined, the incompatible diet (*Viruddhahar*) causes *Ama* to form and vitiate blood and obstruct channels. *Ama* is thought to play a significant role in the emergence of certain diseases. A major factor in the etiopathogenesis of numerous immune-mediated illnesses is the immunological reaction that Ama may cause. Fish and milk are both excellent sources of protein, and when they are combined, they may produce novel protein molecules that have the ability to imitate molecules and cause autoimmunity (Agnivesha 2005: Chap. 26/84; Agnivesha 2005: Chap. 27/9).

4. ***Rashi* (Quantity):** Quantity/*Matra* is the amount of overall food consumed; it measures the mass of each component as well as the total mass. *Sarvagraha* and *Parigraha* are the terminology used for quantitative analysis. *Sarvagraha* refers to the total amount of food consumed. *Parigraha* is the amount of each of its components. The amount of food depends on the capacity of the digestive system (*Jatharagni*) of the individual. Each person has a unique digestive capacity. The right amount of food is that which is digested at the right time and does not interfere with the person's daily routine (Agnivesha 2005: Chap. 5/4). *Vagbhat* states that in order to allow *Vata* to flow freely, half the stomach should be filled with solid food, one quarter with liquids, and the remaining quarter left empty (Agnivesha 2005: Chap. 1/21–3).
5. ***Desha* (Habitat):** Two types of *Desha* are seen *Deha* (body) and *Bhumi*. When ingesting food, one should take into account both *Bhumi Desha* and *Deha Desha*. *Bhumi Desha* is divided into three groups based on the local climate, which influences *Dosha* dominance of the individuals who live there: *Jangla*, *Anupa*, and *Sadharana Desha*. According to Charaka, the *Vata* and *Pitta doshas* predominate in the *Jangla Desha*, whereas the *Vata* and *Kapha doshas* predominate in the *Anupa Desha*. As stated by Geographical gradients, food quality varies according to the place of origin. It shows that different soil and ambient conditions result in different amounts of chemicals. One must choose a diet taking into account the climate of *Bhumidesh* and the *doshik* predominance of *Deha Desha*. For example, the *Desha* determines a consumer's healthy diet; in South India, including curd in a daily diet is beneficial, but in East India, it causes illnesses. In Punjab and Rajasthan, a person can readily digest more *Ghrita* and other *Guru Dravyas*, while in South India, the same individual finds it difficult to digest the same foods (Agnivesha 2005: Chap. 2/7).
6. ***Kala* (Time):** There are two meanings for this phrase. *Awasthik Kala* and *Nityag Kala*. *Nityag Kala* refers to seasonal *awastha*, which takes into account both *Ritu* and day/night. Therefore, a person's *Doshawastha* varied according to their biological clock. *Ritu Ahara* and the day's *Doshawastha* should be taken in accordance with *Dincharya* and *Ritucharya*, which support the body's ability to adapt to its surroundings and maintain its health. *Vyadhi Avastha* is thought to be *Avasthika Kala*. For instance, *Langhana* should be performed for 7 days in *Naveen Jwara*, whereas *Ghrit* is recommended if *Jwara* is *Jeerna* (Agnivesha 2005: Chap. 1/21–6).

7. ***Upyoga Samstha*** **(Dietetic Rules):** It simply refers to the rules of dietetics. In the *Vimansthan* first *Adhyay*, *Acharya Charak* listed ten dietary guidelines that *Upayokta* (consumers) should abide by. It includes all of the information about diet, including when, how, and what to eat (Agnivesha 2005: Chap. 1/21–7). These are described later under dietetic rules.
8. ***Upyokta*** **(The Consumer):** *Upayokta* determines all food, whether it is nutritious or not. Since every food serves as medicine, the *Upyokta's Prakriti* determines whether a food is wholesome or unwholesome for them (Agnivesha 2005: Chap. 1/24–1).

Similarly, Sushruta outlines 12 considerations for food consumption, which he calls *Dvadasha Asana Vidhi*. These are specifically mentioned for those who are ill. One should choose his food according to his condition (Maharshi Sushruth 2004: Chap. 46/3).

1. *Sheet Ahara* for weakness, burning, thirst, alcoholism, and *Raktapitta.*
2. *Ushna Ahara for* patients with *Kapha and Pitta* imbalance.
3. *Snigdha Ahara for* an individual with *Vata Prakrati.*
4. *Ruksha Ahara is* helpful for diabetics, obese people with *Kapha Prakrati.*
5. *Drava Ahara* used on weak and dehydrated patients.
6. *Shushka Ahara* for people with diabetes mellitus and dermatological conditions.
7. *Ek-kalika Ahara,* it is recommended for people with impaired *Agni* (digestive power) once daily.
8. *Dvi- Kalika Ahara*, this is indicated in patients with appropriate *Agni* twice a day.
9. *Aushadha Yukta Ahara*, patients who are unable to take unpalatable medications should be treated with a diet supplemented with necessary medications.
10. *Matraheen Ahara,* for both the sick person and the person with impaired *Agni* should recommend consuming the diet in smaller amounts.
11. *Prashamaka Ahara,* which balances all the *Doshas*, is recommended based on seasonal variations.
12. *Vrattiprayojaka Ahara*is is recommended for healthy people in order to support psychosomatic health.
13. *Viruddhara*, Ayurveda claims that some dietary combinations have antagonistic effects, interfere with the metabolic process, and prevent tissues from properly forming with regard to the tissues. These pairings are referred to as "incompatible foods," or "*Viruddha Ahara*."

यत् किञ्चिद्दोषमास्राव्यन निर्हरति कायतः।
आहारजातं तत् सर्वमहितायोपपद्यते।।८५।।

Charaka Samhita. Sutrasthana, Chap. 26; According to *Charakacharya,* the foods (diet) and drugs (medicines) which dislodge the *Doshas* from its proper place, but do not expel them from the body are to be regarded as *Viruddha* or incompatible or unwholesome diet

Viruddha Ahara is made up of a variety of incompatibilities, such as the method used to process or prepare the food, the quantity of particular foods consumed together, the combination of particular foods, and so forth. Numerous systemic illnesses are thought to be primarily caused by *Viruddha Ahara*, which also has an impact on neurological health. Some examples include: It is not recommended to consume *dahi* (yogurt) at night or with sour foods like bananas, fish, tea, or coffee. When using honey, it should not be heated. Bananas, sour fruits, and fermented foods should not be consumed with milk, nor should they be consumed with main meals (Sharma and Chandola 2015).

8.6 Dietetic Rules

1. ***Ushnam Asniyat*/Eat Warm Foods:** Ayurveda emphasizes the importance of healthy saliva and gastric juice secretions for proper digestion. Warm food stimulates the digestive environment, while cold foods can dwindle *Agni*, which is vital for life. To maintain *Agni* activity, all foods should be consumed warm, within 48 minutes of cooking, and not refrigerated. Ayurveda also advises consuming only cooked foods, as raw foods require a more significant breakdown process, potentially leading to long-term gut dysfunction. Fruits, however, are precooked by the sun and can be consumed directly (Agnivesha 2005: Chap. 1/24–1).
2. ***Snigdham Ashniyat*:** The meal should be *Snigdha* (unctuous); *Agnideepana* happens when *Snigdha Ahara* creates the right taste sensation. It facilitates early digestion, *Vata Anulomna*, and facilitates the smooth passage of digestion. It increases the brightness and the vitality of the sense organs (Agnivesha 2005: Chap. 1/24–2).
3. ***Matravat Ashniyat*:** *Matravat* indicates sufficient food intake, as indicated by *Lakshna*, which includes no heart pain, stomach heaviness, sensory satisfaction, alleviation of thirst/hunger, and comfort. It is recommended to consume the necessary amount. Food is similar to an *Agni*, and adding the right amount of *Samidha* to *Agni* allows proper digestion. Over-adding *Samidha* disrupts *Agni* function and causes disease formation (Agnivesha 2005: Chap. 1/24–3).
4. ***Jirne Ashniyat*:** Consuming food before the previous meal is digested can cause undigested *Ahara Rasa* from the previous meal to mix with the *Ahara Rasa* from the subsequent meal, causing *Dushti* in every *Dosha* (Agnivesha 2005: Chap. 1/24–4).
5. ***Virya Aviruddham Ashniyata:*** Consume foods that are not harmful to your *Virya* (potency). The body suffers when foods with antagonistic potencies are consumed. It causes a lot of illnesses. For instance, eating milk and fish simultaneously causes *Raktadhatu Dushti* because they are antagonistic in strength (Agnivesha 2005: Chap. 1/24–5).
6. ***Ishta Deshe Ashniyata:*** Eating in a friendly setting prevents mental stress. As a result, eating should take place in a welcoming environment. The space needs to be clean and well-lit. The *Sushruta Samhita* states that stated that there is a connection between *Bhojan* and *Bhojya Desha*. It should be a well-built, levelled, and comfortable space (Agnivesha 2005: Chap. 26/65).
7. ***Ishta Sarva Upakarnam Ashniyata:*** Similarly, utensils used for eating and cooking are equally important. Never prepare or consume *Amla Rasatmak Ahar Dravyas* using

copper or any other metallic utensils because there is a risk of a chemical reaction between the acids in *Amla Rasa*, and metallic utensils (Agnivesha 2005: Chap. 1/24–6).

8. ***Natidrutam Ashniyata Nativilambitam Ashniyata:*** Not to eat quickly. Eating quickly can impede the passage of food and cause nausea. Additionally, it is not recommended to eat very slowly. Eating slowly can take a long time, and the thali's food will cool down over time, which interferes with the digestive process (Agnivesha 2005: Chap. 1/24–7).
9. ***Ajalpam, Ahasan, Tanmana Bhunjeet***: Excessive talking or laughter during eating is not recommended, as it diverts attention from the meal. Ayurveda emphasizes engaging the mind during consumption, as the nervous system triggers the release of neurotransmitters and enzymes (Agnivesha 2005: Chap. 1/24–8/9).
10. ***Atmanamabhisamikshya Samyak:*** One should eat appropriately while keeping in mind the aforementioned considerations. Consider the consumer's *Prakriti* when selecting a diet. Consider the meal to be *Purnabrahma* at all times. Everything we eat will aid in our self-replenishment. Also, aligning food with circadian rhythm is crucial for our *Agni*, a diurnal mammal that responds to the sun's cues. Light breakfasts are best during the morning, lunches should be heaviest during midday, and dinners should be light and simple (Agnivesha 2005: Chap. 1/24–10).

8.7 Specific Diets According to *Prakriti* Type

1. ***Vata* type dosha (*Vata*-AD subtype):** It is advised to follow a diet that minimizes the dry, chilly, and light characteristics of *Vata*. Foods must therefore be warm, freshly prepared, greasy (moist), and mildly seasoned. *Vata* imbalances can be lessened by eating foods that have sweet, sour, and salty flavours. It is ideal to have a diet rich in whole grains, fresh veggies, almonds, ghee, coconut, or sesame oil. The intake of vitamins, minerals, and protective antioxidants will be maximized if you eat fruits and vegetables of all colours. In addition to being high in fibre, fruits and vegetables also aid in the relief of constipation and other digestive problems. The brain-protective routine should include daily servings of fruits, vegetables, and greens. Regular meals should be consumed at the same time every day. Additionally, Ayurveda advises fasting for at least 12 hours between the previous day's last meal and the first meal of the next day. According to recent research, daily fasting periods cause the brain to become hypoglycaemics, which promotes autophagy, lipid catabolism, and the release of ketone bodies that protect the brain. These processes all contribute to increased brain plasticity and improved cognition (Bredesen and Rao 2017; Ninivaggi 2010; Murphy et al. 2014; McCarty et al. 2015).
2. ***Pitta* type dosha (*Pitta*-AD subtype):** It is advised to follow a diet that minimizes *Pitta*'s drying and hot characteristics. *Pitta* dominance needs sweet and inherently cooling foods to remain in balance. Foods must therefore be warm, freshly prepared, greasy (wet), and mildly seasoned with cooling spices. Since they mainly bring in moist, cool, and weighty properties, foods that combine sweet, bitter, and astringent flavours aid in calming the *Pitta*. Diets that are pungent, sour, or salty should be avoided because they can exacerbate the underlying

inflammation and heat. Foods that are extensively fermented or pickled should be avoided. Whole grains, bitter greens, dairy products, cooling oils like ghee, coconut, and olive, and raw salads are all recommended components of a healthy diet. Eating fruits that are completely ripe will help you get as much of the anti-inflammatory and antioxidants. It is also advised to fast briefly in between meals (Bredesen and Rao 2017; Ninivaggi 2010).

3. ***Kapha* type dosha (*Kapha*-AD subtype):** It is advised to follow a diet that emphasizes foods that are cool, light, and moist, and that have detoxifying and chelating qualities. *Kapha* dominance needs warm, stimulating, and light foods to remain in balance. Pectin, which is abundant in fruits including citrus fruits, bananas, green apples, and grapes, is a known chelator of heavy metals and other harmful substances from the bloodstream. Parsley, coriander seeds, and cilantro leaves are great meals to help the body rid itself of heavy metals, including lead, aluminium, and mercury. Antioxidants found in cruciferous vegetables like broccoli, kale, and cabbage stimulate the body's detoxification enzymes. Foods high in sulphur, such as broccoli, cabbage, cauliflower, eggs, Brussels sprouts, onions, and garlic, help the body get rid of heavy metals like lead (Bredesen and Rao 2017; Ninivaggi 2010).

8.8 Types of Diet According to *Gunas*

One of the most crucial elements in preserving dosha balance and a sattvic mind is dietary intake. For AD, Ayurveda suggests a Sattvic diet since it includes the foods that are highest in nutrients(Sharma and Chandola 2015; Akter and Nahar 2024).

1. ***Sattvic* cuisine is** easy, vegetarian, and made with ingredients that are in season and fresh. Consuming organic and genetically modified-species-free *Sattvic* foods is essential to avoiding exposure to environmental contaminants and manipulated genetic components that could exacerbate illness. Common examples include ripe fruits, cereal grains, butter, ghee, almonds, squash, green vegetables, and milk. Fresh fruits, vegetables, whole grains, dairy, and nuts make up a *Sattvic* diet, which is thought to support emotional stability and mental clarity. Given the global obesity crisis, it may be more acceptable to emphasize whole grains, fruits, and vegetables rather than using fattier *Sattvic* foods. Cardamom, fennel, turmeric, and cinnamon are among the mild seasonings used in *sattvic* cuisine.
2. ***Rajasic* foods:** are foods that are harsh, pungent, bitter, sour, salty, oily, spicy, or highly caffeinated. They create suffering, sorrow, and illness.
3. ***Tamasic* foods:** *Tamasic* foods like processed, stale, or heavy foods are avoided, as they increase lethargy and mental dullness.

8.9 Wholesome Diet

Ayurveda states that maintaining optimum health requires a balanced diet, or *Pathya Ahara*. The food consumed determines the body's structural and functional integrity. A healthy diet fosters growth and strength that lasts throughout one's entire life. It stimulates the intellect. When used correctly, it supports vitality, clarity of the sense organs, and the appropriate distribution of bodily materials. According to Ayurveda, eating a healthy diet is crucial to reaching and preserving optimum health. Food is, therefore, frequently thought of as medication. In general, foods that are thick, greasy, and sweet are good for neurological health and the brain. The foods that are advised are described in the sections that follow:

1. **Grains and legumes:** *Godhuma* (wheat) and *Yava* (barley) possess the qualities of *Brihmana* (restorative), *Jivana* (vitalizing), and *Sthairyakara* (creating steadiness). Black gram, or *masa*, is a kind of lentil and *Kapikacchu* (*Mucuna pruriens*, a velvet bean) are great for reducing *Vata*. They are greasy and enervating. Studies have shown that *kapikacchu* can effectively treat neurological diseases (Sharma and Chandola 2015; Katzenschlager et al. 2004; Esquivel 2022).
2. **Fruits and nuts:** Many researchers found that fruits and nuts also play a significant role in maintaining mental health. Fruits like *Kharjura* (dates) and *Phalgu* (common fig) are soothing and delightful. Coconut is restorative, oily, soothing, strengthening, and sweet (Sharma and Chandola 2015). *Mardvika* (grapes) are restorative. Grape seed extract is rich in proanthocyanidins, resveratrol, and other beneficial polyphenols, and has been shown to significantly reduce brain amyloid formation in Alzheimer's disease. These compounds can help reduce cognitive decline and prevent Aβ plaque formation by inhibiting protofibril formation, pre-protofibril oligomerization, and protein secondary structure transitions (Wang et al. 2009; Wang et al. 2008; Pasinetti et al. 2015). *Vatama* (almond), *Akshota* (walnut), and *Mukulaka* (pistachio) are restorative (Sharma and Chandola 2015).
3. **Spices:** Use of spices can stimulate digestion (*Agni*). Each meal should ideally include both a spice (the warmth) and healthy fat (the fuel) to maintain the *Agni*. Spices can be sweeter, like fennel, cinnamon, and cardamom, or more pungent-like ginger and black pepper (Mirmosayyeb et al. 2017).
4. **Curcumin from curcuma longa:** Curcumin, a phenolic molecule found in turmeric, has been linked to a lower risk of Alzheimer's disease on account of reduced oxidized proteins, plaque load, insoluble Aβ, soluble Aβ, and interleukin 1β (a proinflammatory cytokine) levels. Consuming curcumin, a phenolic molecule similar to resveratrol, may explain why there are fewer Alzheimer's patients in the Indian population. Curcumin's cell reinforcement and anticholinesterase activity make it effective in preventing the progression of Alzheimer's disease (Chen et al. 2018).
5. **Ghee:** Cow's ghee (*Ghrita* in Sanskrit), or clarified butter, enhances *Ojas*, *Agni* (digestive fire), memory, and intelligence. It promotes *Kapha* and fat balance, alleviates *Vata* and *Pitta* imbalances, and treats conditions like toxins, madness, wasting diseases, and fever. It is delicious and refreshing. Ghee that has aged is

beneficial and essential in the treatment of neurological conditions and fainting. Ghee is advantageous for the healthy operation of the sensory organs. It is also *Vayasthapana* (rejuvenating) and enhances *Ayu* (longevity) and *Bala* (strength). Also, an excellent *Anupana* (therapeutic vehicle) for transportation of herbs into the deeper tissues on account of the lipophilic nature of ghee (Sharma and Chandola 2015; Kataria and Singh 2024).

6. **Oil:** Sesame oil is sweet, spicy, invasive, and easily absorbed. Many antioxidants are present in it. It is great for reducing *Vata* and raising *Pitta*. The skin, brain, and stomach fire are all strengthened and improved by sesame oil (Fukuda et al. 1986).
7. **Milk:** *Kshira*, or cow's milk, is regarded as healthy for all living things. It is cooling, lactation-promoting, sweet, greasy, revitalizing, fortifying, and nervine tonic (*Medhya*). It quenches thirst, aids in proper digestion, is calming, and cleansing. The ten qualities of cow's milk are as follows: *Shrikshna* (smooth), *Pichchila* (sticky), *Guru* (heavy), *Sheeta* (cooling), *Mridu* (soft), *Snigdha* (oily), *Bahala* (viscous), *Madhura* (sweet), and *Prasanna* (pure). *Ojas*, which is referred to in Ayurveda as a vital component that sustains the body's power and immunity, possesses these same attributes. As a result, cow's milk is considered the best vitalizer and rejuvenator and raises *Ojas* (Sharma and Chandola 2015).

8.10 Herbal Supplements

The emergence of herbal supplements and medications in the medical market reflects the growing trend of using plant-based products to treat Alzheimer's disease. These products are frequently marketed as having the ability to improve brain health, slow cognitive decline, and increase memory. *Medharasayanas* include a collection of therapeutic plants with multiple benefits, particularly for memory and cognition through *Prabhava* (specific action), namely *Mandukaparni, Yastimadhu, Guduchi, Ashwagandha, Brahmi, Vacha, Jatamansi Shankhapushpi*. These are well-documented in Ayurveda and other classical texts (Sharma and Chandola 2015; Kulkarni et al. 2016).

In addition to this, Ayurveda also offers spiritual practices like meditation, *Pranayama*, and *Yogasana* help calm the *Rajasic* mind and purify the *Tamasic* mind, while yogic practices promote physical and mental discipline. Ayurvedic principles can be integrated into modern therapeutic frameworks, offering a more holistic approach to mental health recovery.

8.11 Conclusion

The current study came to the conclusion that the three bio-energies (Doshas) that govern bodily functions—*Pitta*, *Kapha*, and *Vata*—were out of balance in Alzheimer's disease (AD), a progressive neurodegenerative illness. The significance of diet (*Ahar*) in preserving health is emphasized by this study. Six tastes

(*Shadrasa*) and particular directions (*Aharvidhivisheshayatan*) to enable good digestion (*Agni*) are included in the detailed analysis of the Ayurvedic diet. Additionally, diets depending on a person's main *Dosha*, *Gunas*, and *Prakriti* (constitution) are examined and documented. Fresh, vegetarian, and high in nutrients, sattvic foods are thoroughly researched and suggested for AD. Incorporating nutritious meals such as grains, legumes, fruits, nuts, spices, ghee, milk, and herbal supplements is essential for maintaining brain health. For a comprehensive approach to mental health rehabilitation, contemporary therapy frameworks might incorporate Ayurvedic principles, lifestyle considerations, and spiritual activities.

References

Agnivesha CS (2005) Ed. of Vidyotini hindi Commentary by Pt. Kashinatha Shastri and Dr. Gorakha Nath Chaturvedi, Chaukhambha Bharati Academy, Varanasi, Vimansthan: Reprint

Akiyama H, Barger S, Barnum S, Bradt B, Bauer J, Cole GM, Cooper NR, Eikelenboom P, Emmerling M, Fiebich BL, Finch CE, Frautschy S, Griffin WS, Hampel H, Hull M, Landreth G, Lue L, Mrak R, Mackenzie IR, McGeer PL, O'Banion MK, Pachter J, Pasinetti G, Plata-Salaman C, Rogers J, Rydel R, Shen Y, Streit W, Strohmeyer R, Tooyoma I, Van Muiswinkel FL, Veerhuis R, Walker D, Webster S, Wegrzyniak B, Wenk G, Wyss-Coray T (2000) Inflammation and Alzheimer's disease. Neurobiol Aging 21(3):383–421

Akter I, Nahar F (2024) Sattva, rajas, and Tamas as tools for understanding psychological disorders. J Ayurvedic Maulik Siddhant 1(1):06–08

Bature F, Guinn BA, Pang D, Pappas Y (2017) Signs and symptoms preceding the diagnosis of Alzheimer's disease: a systematic scoping review of literature from 1937 to 2016. BMJ Open 7(8):e015746

Bredesen DE (2015) Metabolic profiling distinguishes three subtypes of Alzheimer's disease. Aging 7(8):595–600

Bredesen DE (2016) Inhalational Alzheimer's disease: an unrecognized-and treatable epidemic. Aging 8(2):304–313

Bredesen DE, Rao RV (2017) Ayurvedic profiling of Alzheimer's disease. Altern Ther Health Med 23(3):46–50

Breijyeh Z, Karaman R (2020) Comprehensive review on Alzheimer's disease: causes and treatment. Molecules 25(24):5789

Brion JP, Anderton BH, Authelet M, Dayanandan R, Leroy K, Lovestone S, Octave JN, Pradier L, Touchet N, Tremp G (2001) Neurofibrillary tangles and tau phosphorylation. Biochem Soc Symp 67:81–88

Chen M, Du ZY, Zheng X, Li DL, Zhou RP, Zhang K (2018) Use of curcumin in diagnosis, prevention, and treatment of Alzheimer's disease. Neural Regen Res 13(4):742–752

Chen ZR, Huang JB, Yang SL, Hong FF (2022, 1816) Role of cholinergic signaling in Alzheimer's disease. Molecules 27(6)

Cras P, Kawai M, Lowery D, Gonzalez-DeWhitt P, Greenberg B, Perry G (1991) Senile plaque neurites in Alzheimer disease accumulate amyloid precursor protein. Proc Natl Acad Sci 88(17):7552–7556

Debnath PK, Banerjee S, Debnath P, Mitra A, Mukherjee PK (2015) Ayurveda-opportunities for developing safe and effective treatment choices for the future. In: Evidence-based validation of herbal medicine. Elsevier, pp 427–454

Dhanya S, Ramesh NV, Mishra A (2019) Traditional methods of food habits and dietary preparations in Ayurveda – the Indian system of medicine. J Ethn foods 6(14):1–9

Esquivel MK (2022) Nutrition benefits and considerations for whole foods plant-based eating patterns. Am J Lifestyle Med 16(3):284–290

Fernández-Tomé S (2024) Role of food digestion and digestive system in the nutritional, functional and health properties of food bioactives. Nutrients 16(5):712

Fukuda Y, Nagata M, Osawa T, Namiki M (1986) Contribution of lignan analogues to antioxidative activity of refined unroasted sesame seed oil. J Americ Oil Chemists' Soc 63(8):1027–1031

Hampel H, Hardy J, Blennow K, Chen C, Perry G, Kim SH, Villemagne VL, Aisen P, Vendruscolo M, Iwatsubo T, Masters CL, Cho M, Lannfelt L, Cummings JL, Vergallo A (2021) The amyloid-β pathway in Alzheimer's disease. Mol Psychiatry 26(10):5481–5503

Hoogmartens J, Cacace R, Van Broeckhoven C (2021) Insight into the genetic etiology of Alzheimer's disease: a comprehensive review of the role of rare variants. Alzheimer's Dement (Amst) 13(1):e12155

Jaiswal YS, Williams LL (2016) A glimpse of Ayurveda–The forgotten history and principles of Indian traditional medicine. J Tradit Complement Med 7(1):50–53

Jiang T, Yu JT, Tian Y, Tan L (2013) Epidemiology and etiology of Alzheimer's disease: from genetic to non-genetic factors. Curr Alzheimer Res 10(8):852–867

Kachhawaha S, Sharma RK, Sharma DC (2021) Importance of Vata Dosha and its Aatma Roopas (cardinal symptoms) in maintaining health and balance in life. Intern Res J Ayurveda and Yoga 4(7):75–80

Kataria D, Singh G (2024) Health benefits of ghee: review of Ayurveda and modern science perspectives. J Ayurveda Integr Med 15(1):100819

Katariya KS, Mishra BR (2021) Shad rasa and their effect on the body. J Ayurveda Integr Med Sci 6(6):124–130

Katzenschlager R, Evans A, Manson A, Patsalos PN, Ratnaraj N, Watt H, Timmermann L, Van der Giessen R, Lees AJ (2004) Mucuna pruriens in Parkinson's disease: a double blind clinical and pharmacological study. J Neurol Neurosurg Psychiatry 75(12):1672–1677

Kulkarni R, Shetty SK, Rajarajeshwari NM, Rao PN, Nayan J (2016) Rasayana herbs of Ayurveda to treat age related cognitive decline: An update. Pharm J 8(5):411–423

Kumar DR, Shankar MS, Reddy PP, Kumar BR, Sumalatha N (2014) A review on Alzheimer's disease. Res J Pharmacol Pharmacodyn 6(1):59–63

Lee J, Meijer E, Langa KM, Ganguli M, Varghese M, Banerjee J, Khobragade P, Angrisani M, Kurup R, Chakrabarti SS, Gambhir IS, Koul PA, Goswami D, Talukdar A, Mohanty RR, Yadati RS, Padmaja M, Sankhe L, Rajguru C, Gupta M, Kumar G, Dhar M, Chatterjee P, Singhal S, Bansal R, Bajpai S, Desai G, Rao AR, Sivakumar PT, Muliyala KP, Bhatankar S, Chattopadhyay A, Govil D, Pedgaonkar S, Sekher TV, Bloom DE, Crimmins EM, Dey AB (2023) Prevalence of dementia in India: national and state estimates from a nationwide study. Alzheimers Dement 19(7):2898–2912

Leucht S, Helfer B, Gartlehner G, Davis JM (2015) How effective are common medications: a perspective based on meta-analyses of major drugs. BMC Med 13:253

Li XL, Hu N, Tan MS, Yu JT, Tan L (2014) Behavioral and psychological symptoms in Alzheimer's disease. BioMed Res Intern 2014(1):1–9

Maharshi Sushruth (2004) Sushruth Samhita Ed. of Ayurveda-Tattva-Sandipika Hindi Commentary by Kaviraja Ambikadutta Shastri, Chaukhambha Sanskrit Sansthan, Varanasi, 14th edition, Sutra Sthana, 46th Chapter, Shloka 3

Majhi V, Choudhury B, Paul S (2020) Prevention and treatment of Alzheimer's disease in the light of Ayurveda. In: Phytomedicine and Alzheimer's disease. CRC Press, pp 85–96

Mangalasseri P, Chandran S (2017) A brief insight into the pathogenesis and management of Alzheimer's disease in Ayurvedic parlance. Intern J Green Pharm 11(01):S81–S85

McCarty MF, DiNicolantonio JJ, O'Keefe JH (2015) Ketosis may promote brain macroautophagy by activating Sirt1 and hypoxia-inducible factor-1. Med Hypotheses 85(5):631–639

Mirmosayyeb O, Tanhaei A, Sohrabi HR, Martins RN, Tanhaei M, Najafi MA, Safaei A, Meamar R (2017) Possible role of common spices as a preventive and therapeutic agent for Alzheimer's disease. Intern J Prev Med 8:5

Mishra L, Singh BB, Dagenais S (2001) Healthcare and disease management in Ayurveda. Altern Ther Health Med 7(2):44–50

Murphy T, Dias GP, Thuret S (2014) Effects of diet on brain plasticity in animal and human studies: mind the gap. Neural Plast 2014(1):563160
Myers SP, Vigar V (2019) The state of the evidence for whole-system, multi-modality naturopathic medicine: a systematic scoping review. J Altern Complement Med 25(2):141–168
Ninivaggi FJ (2010) A comprehensive guide to traditional Indian medicine for the west. Rowman & Littlefield Publishers, pp 153–184
Pasinetti GM, Wang J, Ho L, Zhao W, Dubner L (2015) Roles of resveratrol and other grape-derived polyphenols in Alzheimer's disease prevention and treatment. Biochim Biophys Acta 1852(6):1202–1208
Patwardhan B, Datta HS (2021) Ayurveda and brain health. In: Nutraceuticals in brain health and beyond. Academic Press, pp 441–453
Perl DP (2010) Neuropathology of Alzheimer's disease. Mt Sinai J Med 77(1):32–42
Sajjan S (2022) An exploratory study on critical analysis and understanding of Medhya Rasayana (intelligence enhancer) in dementia. RGUHS J Ayush Sci 9(1):1–7
Samhita C (1949) Charaka Samhita, vol I-VI. Shree Gulab Kunverba Ayurvedic Society, Jamnagar
Seshan S, Sharma MM, Soni N, Meena S (2024) A scientific evaluation of mental health through Ayurveda. J Ayurveda Integr Med Sci 9(2):141–148
Sharma H, Chandola HM (2011) Prameha in Ayurveda: correlation with obesity, metabolic syndrome, and diabetes mellitus: part 1-etiology, classification, and pathogenesis. J Altern Complement Med 17(6):491–496
Sharma H, Chandola HM (2015) Ayurvedic approach to food and dietary supplements for the brain and neurologic health. In: Bioactive nutraceuticals and dietary supplements in neurological and brain disease: prevention and therapy. Academic Press, pp 173–177
Singh B, Day CM, Abdella S, Garg S (2024) Alzheimer's disease current therapies, novel drug delivery systems and future directions for better disease management. J Control Release 367:402–424
Timko C, Schultz NR, Britt J, Cucciare MA (2016) Transitioning from detoxification to substance use disorder treatment: facilitators and barriers. J Subst Abus Treat 70:64–72
Wagh K, Bhalerao S (2020) Traditional foods, Ayurveda, and diet. In: Nutritional and health aspects of food in south Asian countries. Academic Press, pp 99–111
Wang J, Ho L, Zhao W, Ono K, Rosensweig C, Chen L, Humala N, Teplow DB, Pasinetti GM (2008) Grape-derived polyphenolics prevent Abeta oligomerization and attenuate cognitive deterioration in a mouse model of Alzheimer's disease. J Neurosci 28(25):6388–6392
Wang YJ, Thomas P, Zhong JH, Bi FF, Kosaraju S, Pollard A, Fenech M, Zhou XF (2009) Consumption of grape seed extract prevents amyloid-β deposition and attenuates inflammation in brain of an Alzheimer's disease mouse. Neurotox Res 15(1):3–14
Williamson J, Goldman J, Marder KS (2009) Genetic aspects of Alzheimer disease. Neurologist 15(2):80–86
Yang J, Zhi W, Wang L (2024) Role of tau protein in neurodegenerative diseases and development of its targeted drugs: a literature review. Molecules 29(12):2812

The Role of Nuts in Alzheimer's Prevention and Management

9

Gamze GÖGER and Nagehan SALTAN

Abstract

A common cause of dementia is Alzheimer's disease (AD), which is a serious public health concern because of its rising prevalence around the globe. This degenerative neurological disease causes a steady decline in cognitive performance, behavior, and function. While the precise process of AD remains unclear, a number of variables, such as amyloid-beta (Aβ) deposition, tau protein alterations, neuroinflammation, oxidative stress, and transition metal imbalance, are believed to play a role in the disease's development. The goals of AD treatment are to reduce symptoms, slow the development of the illness, and enhance patients' quality of life. Treatment approaches include pharmacological therapies, various nanoparticle formulations, stem cell therapies, and lifestyle modifications. The study investigates the potential of nuts in managing and preventing AD. Nuts are foods high in nutrients that have antioxidant and anti-inflammatory properties. They are rich in bioactive ingredients, including phenolic compounds, fiber, vitamins, minerals, phytosterols, and unsaturated fatty acids. These components are known to have positive effects on brain function through their neurotrophic, antioxidant, and anti-inflammatory properties. Nuts are thought to exert neuroprotective effects through various mechanisms such as reducing oxidative stress, reducing inflammation, increasing synaptic plasticity, and reducing amyloid beta deposition.

G. GÖGER (✉)
Department of Pharmacognosy, Faculty of Pharmacy, Afyonkarahisar Health Sciences University, Afyonkarahisar, Türkiye
e-mail: gamze.goger@afsu.edu.tr

N. SALTAN
Department of Pharmaceutical Botany, Faculty of Pharmacy, Anadolu University, Eskişehir, Türkiye
e-mail: ndagdeviren@anadolu.edu.tr

J. Malik et al. (eds.), *Diet and Alzheimer's Disease: Let Food be Our Medicine*, Nutritional Neurosciences, https://doi.org/10.1007/978-981-95-2736-6_9

This study examines in detail the nutritional contents, phytochemical constituents, and potential effects on AD of various nuts such as almonds (*Prunus dulcis* Mill. DA Webb), Brazil nuts (*Bertholletia excelsa* Humb. & Bonpl), chestnuts (*Castanea sativa* Mill.), pine nuts, (*Pinus pinea* L.), hazelnuts (*Corylus avellana* L.), walnuts (*Juglans regia* Linn.), and Pistachios (*Pistacia vera* L.). It is emphasized that each of these nuts may target different aspects of AD through their specific constituents and play a valuable role in the prevention and management of the disease.

Keywords

Alzheimer's disease · Nuts · Prevention and management

9.1 Introduction

Alzheimer's disease (AD) is becoming a serious public health issue because of its rising prevalence globally (Pribis and Shukitt-Hale 2014). A persistent decline in cognitive performance, behavior, and function is a hallmark of AD. Short-term memory, executive, and visuospatial functioning, as well as praxis (movement abilities), are frequently impaired in people with AD (Hu et al. 2013; McGirr et al. 2020). An insidious course of cognitive impairment, frequently focused on memory loss, is the most prevalent sign of AD. Early on, cognitive abilities unrelated to memory, like word recognition, visual/spatial issues, and poor judgment or thinking, also deteriorate. Neuropsychiatric symptoms are present in nearly all AD patients at some point during their illness, with the most common early symptoms being apathy and depression (McGirr et al. 2020). Cognitive impairments become more noticeable and pervasive as the illness worsens, eventually impacting everyday living tasks (Tarawneh and Holtzman 2012).

9.1.1 Alzheimer Disease Mechanisms

Although the precise cause of Alzheimer's disease is unknown, a number of factors are believed to play a role in its onset. Some of the factors are discussed in Fig. 9.1.

9.1.2 AD Treatments

Approaches to treating Alzheimer's disease (AD) attempt to alleviate symptoms, delay disease progression, and improve patients' quality of life. Treatment of AD involves an integrated strategy, and medicine alone is insufficient. Various

1. Amyloid cascade theory :According to this theory, a major contributing cause to the illness is an increase of amyloid-beta (Aβ) peptide in the brain. Amyloid precursor protein (APP) is processed to produce the peptide Aβ. The Aβ peptide builds up as plaques in the brain when it folds incorrectly. It is believed that these plaques affect neuronal transmission and ultimately result in cell death.

2. The tau hypothesis: In neurons, the protein tau aids in microtubule stabilization. Tau protein becomes increased phosphorylation in AD, which causes it to detach from microtubules and form insoluble fibers called neurofibrillary tangles (NFTs). These tangles cause transit problems in neurons, which ultimately result in cell death.

3. **Neuroinflammation:** Microglia are immune cells in the brain that assist eliminate injured neurons and other debris. In Alzheimer's disease, microglia are continuously active and emit inflammatory chemicals, which can cause neuronal damage and cell death.

4. Oxidative stress: One theory about the origin of Alzheimer's disease is oxidative stress.When the body builds up harmful substances called reactive oxygen species (ROS), oxidative stress results. Studies show that increased oxidative stress can result from tau hyperphosphorylation, neuroinflammation, and Aβ buildup.

5. Dysregulated transition metals: Zinc (Zn), copper (Cu), and iron (Fe) are examples of transition metals that catalyze several enzymes and are necessary for many biological processes in the human body, including brain function. When these metals are deregulated, dangerous free radicals may emerge.

Fig. 9.1 Mechanism of Alzheimer's disease. (Zhao and Zhao 2013; Maccioni et al. 2018; Srivastava et al. 2021)

Table 9.1 The therapeutic strategies for Alzheimer's disease

1. Pharmacologic therapies (Srivastava et al. 2021)	Antipsychotics; AChE inhibitors; N-methyl-D-aspartate receptor antagonists; pimavanserin; other drugs; escitalopram, lithium, and vitamin D
2. Various nanoparticle formulations (Srivastava et al. 2021)	Rapamycin-loaded solid lipid nanoparticles; berberine-loaded multi-walled carbon nanotubes; mitochondria-targeted serum nanoparticles
3. Stem cell therapies (Fouad 2019)	Neural stem cells, mesenchymal stem cells, and stem cell-based therapy
4. Lifestyle changes and other approaches	Social interaction, exercise, nutrition, and cognitive activities

therapeutic interventions have been used for the treatment of AD (Table 9.1). Treatment decisions should be personalized to the the patient's unique illness, symptoms, and treatment goals. New treatment options and drugs are continually being explored and developed. These treatments seek to address the complexities of AD while also mitigating the disease's impacts.

9.1.3 Bioactive Compounds of Nuts and Effect Mechanisms

Nuts are foods that contain a lot of nutrients, including anti-inflammatory and antioxidant components. The physiologically active substances and nutrients found in

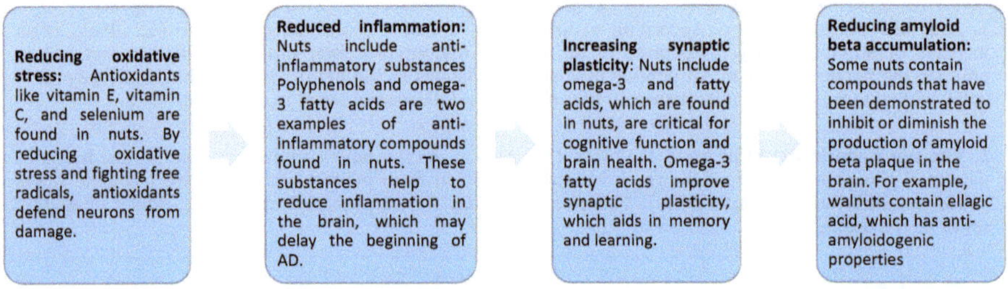

Fig. 9.2 Nuts' neuroprotective effects mediated by several ways. (Bahaeddin et al. 2017; Muthaiyah et al. 2011; Muthaiyah et al. 2014)

nuts, including fiber, unsaturated fatty acids, a variety of vitamins and essential minerals, phytosterols, phenolic compounds, and high-quality protein, may have positive benefits on brain function (Theodore et al. 2021). Because of their neurotrophic, anti-inflammatory, and antioxidant qualities (Fig. 9.2), nuts are a great source of bioactive chemicals that have been demonstrated to have protective benefits against AD (Gorji et al. 2018).

9.2 Almond (Rosaceae): *Prunus dulcis* Mill. DA Webb

9.2.1 Nutrients and Phytochemical Components of Almond

The Rosaceae family includes almonds (*Prunus dulcis* Miller DA Webb, known as sweet almonds) (Fig. 9.3), which have long been used as a natural dietary supplement. Fatty acids, lipids, proteins, carbohydrates, amino acids, vitamins, minerals, and secondary metabolites are among the many nutritious components found in almonds (Roncero et al. 2020). Fibre, minerals, vitamins (Niacin and E), and fatty acids (linoleic acid, oleic acid, palmitic, and stearic acids) make up the majority of the lipid profile (Alasalvar and Shahidi 2008). Polyphenols and phytosterols, as well as flavonoids, phenolic acids, hydrolyzable tannins, lignans, and stilbenes, are among the substances that have been shown to have health-promoting qualities (Colic et al. 2017; Barreca et al. 2020).

9.2.2 Alzheimer's Disease Prevention and Management of Almonds

Neuroprotective fatty acids that have been shown to lower Aβ levels include oleic acid, linoleic acid, and α-linolenic acid (Gorji et al. 2018). By reducing cholesterol and altering secretases, phytosterols, including β-sitosterol, campesterol, and stigmasterol, have been demonstrated to contribute to AD pathogenesis (Moreau et al.

Fig. 9.3 Almond (https://pixabay.com/photos/almond-nuts-healthy-eating-food-1474961/)

2002). Vitamins like folic acid, vitamin E, niacin (B3), and thiamine are found in almonds. These vitamins may help prevent AD by increasing neurogenesis, decreasing inflammation, and lowering oxidative stress (Boccardi et al. 2016; Xia and Mo 2016). In one study, rats that received (150–300 and 600 mg/kg) of almond supplements for seven and fourteen days in a row showed a significant recovery from scopolamine-induced amnesia. In rats' brains, almonds were demonstrated to decrease choline esterase activity, cut triglycerides and cholesterol, and significantly increase glucose levels-all of which may be helpful in the treatment of AD (Kulkarni et al. 2010). Rats given 400 mg/kg/day of walnut and almond supplements for 28 days showed decreased cadmium-induced memory impairment, most likely as a result of cholinergic and antioxidant effects (Batool et al. 2019). The almonds protect against endothelial dysfunction by increasing NO production and decreasing endothelial nitric oxide synthase inhibition (Jamshed and Gilani 2014). Almonds inhibit the oxidation of LDL cholesterol. Polyphenol antioxidants found in almond skin reduce cholesterol oxidation in test tubes and animals (Amarowicz and Pegg 2017; Chen et al. 2019).

A human study with 27 individuals found that eating almonds for 1 month reduced oxidized LDL cholesterol levels by 14% (Jenkins et al. 2002). Consuming almonds enhances vascular function and decreases cholesterol, according to a clinical study. Almond consumption was found to raise urine NO and decrease vascular cell adhesion molecule-1 in patients with coronary artery disease in one clinical investigation (Chen et al. 2015)]. To determine how an almond intervention affected the cognitive abilities of healthy middle-aged/older adults (50–75 years old), a 6-month, single-blind, randomized-controlled study was carried out. Almonds did not considerably improve cognitive performance over a 6-month period in middle-aged and older individuals who were cognitively intact. However, after 6 months, cognitive tests showed a significant improvement when three ounces of almonds were consumed daily. Longer-term studies in participants at risk for age-related cognitive deterioration are required (Mustra Rakic et al. 2022).

9.3 Brazil Nut (Lecythidaceae): *Bertholletia excelsa* Humb. & Bonpl

9.3.1 Nutrients and Phytochemical Components of Brazil Nut

A plant belonging to the Lecythidaceae family that is indigenous to the Amazon rainforest produces edible seeds known as Brazil nuts (*Bertholletia excelsa* HBK) (Fig. 9.4). The leading producers of BNs are Bolivia, Peru, and Brazil. Brazil nuts are regarded as a high-nutrient food that has drawn interest recently due to its advantageous health effects (Vasquez-Rojas et al. 2021). These benefits are caused by the biological traits, nutrients, and phytochemical components of this nut, including its high content of dietary fiber, phenolic compounds, minerals, unsaturated fatty acids, tocopherols, phytosterols, and selenoamino acids. It is different from other nuts because of its selenium concentration (Cardoso et al. 2017). The most prevalent fatty acids in Brazil nuts are linoleic acid, oleic acid, and α-linolenic acid (Ryan et al. 2006). Brazilian nuts have roughly 14.3 grams of protein per 100 grams. The abundance of sulfur-containing amino acids in these proteins promotes the absorption of selenium and other minerals (Yang 2009). Important micronutrients like calcium, iron, potassium, zinc, copper, thiamine, niacin, magnesium, phosphorus, vitamin E, vitamin B6, and arginine are also found in Brazil nuts.

9.3.2 Alzheimer's Disease Prevention and Management of Brazil Nut

The best source of selenium (Se), which is essential for immune system regulation, antioxidant defense system formation, and the prevention of age-related illnesses, is found in Brazilian nuts. (Alehagen et al. 2021). The control group and the Brazil nut group were assigned at random to thirty-one elderly individuals with mild cognitive impairment (MCI). For 6 months, the treatment group ate one Brazil nut daily (calculated to be 288.75 μg). Measurements were made of malondialdehyde, erythrocyte glutathione peroxidase (GPx) activity, blood selenium levels, and oxygen radical absorbance capacity. The selenium levels were measured following the intervention, while the control group showed no change. Among the antioxidant system metrics, only the change in erythrocyte GPx activity differed significantly

Fig. 9.4 Brazil nut (https://pixabay.com/photos/brazil-nuts-brazil-nut-kernels-nuts-6582873/)

across the groups. After 6 months, the supplemented group experienced significantly higher gains in constructional praxis ($p = 0.031$) and verbal fluency ($p = 0.007$) than the control group (Rita Cardoso et al. 2016). Da Silva et al. (2022) demonstrated in a recent systematic review that 15 of the 24 included studies were non-randomized. Brazil nuts were consumed as part of a regular, free-living diet in all of the studies. Higher levels of selenium and/or glutathione peroxidase activity in plasma were observed to increase antioxidant status in all studies. Subjects with coronary artery disease reported inconsistent results on the change in a lipid peroxidation marker, whereas older adults with mild cognitive impairment demonstrated improvements in verbal fluency and constructional praxis (da Silva et al. 2022). According to a recent analysis, when consistently ingested and combined with a well-balanced diet, Brazil nuts and Brazil nut oil can enhance lipid profiles, reduce oxidative stress, and lower inflammatory responses by increasing antioxidant enzyme activity and gene expression. Nevertheless, more investigation is required to elucidate the processes of action (de Araújo et al. 2024).

9.4 Chestnut (Fagaceae): *Castanea sativa* Mill.

9.4.1 Nutrients and Phytochemical Components of Chestnut

The fruit of the sweet chestnut (Fig. 9.5), *Castanea sativa* Mill., is a plant that is primarily grown in Mediterranean Europe and member of the genus Castanea and family Fagaceae. Chestnuts are a healthy food and a wonderful source of energy because of their high fiber content and very low fat level, which is primarily made up of unsaturated fatty acids. Chestnuts are rich in micronutrients (Mn, Fe, Zn, and Cu) and macronutrients (K, P, Mg, Ca, and Na) as well as vitamin C, fiber, and starch are all present in considerable levels in chestnut fruit (Gonçalves et al. 2010). Additionally, Ribeiro et al. (2007) have characterized the seven organic acids-quinic, cis-aconitic, citric, oxalic, ascorbic, malic, and fumaric acids-that are found in Portuguese cultivars. Vitamin E is an antioxidant that has also been found in chestnut fruits (Barreira et al. 2009). The fruits are also rich in polyphenols, with the most common hydrolyzable and condensed tannins being gallic and ellagic acids (De Vasconcelos et al. 2007; Choupina 2019). Chestnut fruits include zeaxanthin,

Fig. 9.5 Chesnut (https://pixabay.com/photos/chestnut-nuts-fruits-harvest-5621532/)

lutein, and beta-carotene (De Vasconcelos et al. 2010). The following flavonoids have been discovered from the methanolic extract of fruits from several chestnut cultivars including flavanols (gallocatechin, afzelechin, epiafzelechin, and taxifolin), flavanones, chalcones (phlorizin, phloretin), eriodictyol, quercitrin, isoquercetin, and hyperoside (Ciucure et al. 2022). Numerous health benefits for people are associated with the bioactive properties of these compounds, which include anticancer, anti-inflammatory, cardio-protective, anti-aging, antimicrobial, and antiviral properties.

9.4.2 Alzheimer's Disease Prevention and Management of Chestnut

In vitro research was done on the anti-amnesic qualities of the chestnut inner skin. These characteristics include neural defense against cell death caused by amyloid β (Aβ) and the ability of the primary phenolics (gallic acid, catechin, and epicatechin) to inhibit acetylcholinesterase (AChE). When it came to AChE, gallic acid exhibited the most dose-dependent and competitive inhibition. However, catechins showed neuroprotective benefits against neurotoxicity caused by Aβ (Kim et al. 2011). Using lactate dehydrogenase 3-[4,5-dimethythiazol-2-yl]-2,5-diphenyl tetrazolium bromide and 20,70 dichlorofluorescein diacetate, Jeong et al. investigated the neuronal protective properties of an ethyl acetate fraction from the inner skin of chestnuts (Jeong et al. 2012). An investigation into the neuroprotective properties of *Castanea sativa* wood extract was carried out using in vitro models of oxidative stress-mediated damage. Glutamate (50 mM, 24 h) or hydrogen peroxide (25 mM, 1 hour followed by 24 h with medium) was administered to human neuroblastoma SH-SY5Y cells. The findings demonstrated that whereas adding extract (1–50 mg/mL) to cell media after the injury was unsuccessful, performing this before neuronal damage resulted in neuroprotection in both experimental types. According to their findings, *Castanea sativa* wood extract may be a useful dietary supplement since it combines strong antioxidant activity with beneficial neuroprotective benefits (Brizi et al. 2016). Products made from chestnuts, particularly chestnut honey, seem to have potential benefits in preventing Alzheimer's disease. Among the many advantages of chestnut honey are its antiviral, antibacterial, immunomodulatory, anti-inflammatory, and antioxidant properties. In an early study, the ethyl acetate fraction of chestnut honey was examined for its potential to mitigate the effects of glutamate-exposed neurotoxicity in HT22 cells and scopolamine-induced cognitive deterioration in rats. Through the activation of Nrf-2 and TrkB/CREB signaling pathways, the ethyl acetate fraction of chestnut honey can improve cognitive impairment and protect neurons from oxidative damage, restoring the decrease in cell viability caused by glutamate exposure (Jeong et al. 2012). At a non-toxic dosage, the antioxidant activity of chestnut honey was evaluated using the DPPH and CUPRAC methods. Two dosages of chestnut honey, 150 mg/kg and 300 mg/kg, were used for the in vivo investigations. The neurodegeneration was brought on by a combination of 100 mg/kg of aluminum chloride and 120 mg/kg of D-galactose.

The brain was examined histopathologically after the neurological examinations. The honey's CUPRAC value was 0.57 ± 0.27 mM TEAC/g, and its DPPH scavenging activity was at IC_{50} = 10.98 ± 1.20 g/mL. According to neurological tests, learning ability significantly improved. Histopathological findings confirmed these findings, showing that honey significantly decreased hippocampus atrophy and neuronal death in a brain tissue segment (Yagoub et al. 2024).

9.5 Pine Nut (Pinaceae): *Pinus pinea* L

9.5.1 Nutrients and Phytochemical Components of Pine Nut

The edible seeds of pine trees (*Pinus* spp.) of the Pinaceae family are known as pine nuts (Fig. 9.6). The most important edible pine nut species worldwide are *Pinus pinea, Pinus koraiensis, Pinus gerardiana*, and *Pinus sibirica*, out of the 29 species listed by the Food and Agriculture Organization. *Pinus pinea* L. or "Stone pine," is a Mediterranean-native tree species that is the primary supplier of pine nuts with significant commercial value. Pine nuts are mostly produced economically in Portugal, Spain, Türkiye, Italy, and Tunisia (Evaristo et al. 2010; Loewe-Muñoz and Noel 2021). It is also grown in Southern Hemisphere regions such as Australia, Argentina, New Zealand, and Chile. Pine nuts are utilized in many different dishes and hold a significant position in Mediterranean food (Awan and Pettenella 2017). One of the most important components of *P. pinea* nuts is oil, which is present in proportions ranging from 42.2% to 46.9%. These fats usually contain healthy unsaturated fatty acids 8 (Nasri et al. 2005). Polyunsaturated fatty acids (PUFA), particularly linoleic acid, are abundant in it. Oleic acid and other monounsaturated fatty acids are also present (Evaristo et al. 2010; Lutz et al. 2017). Linoleic acid has an important role at the vascular level and can increase blood clotting (Awan and

Fig. 9.6 Pine nut (https://pixabay.com/photos/pine-nuts-healthy-food-natural-3091390/)

Pettenella 2017). Pine nuts also contain high amounts of protein, ranging from 32% to 35% and contain a certain amount of fiber. Pinolenic acid is present in *P. pinea* nuts, in contrast to those of other pine species. *P. pinea* has a lower abundance of this acid than other pine species (Wolff and Bayard 1995). *Pinus koraiensis* species has notably high levels of pinolenic acid, which can be used as a marker to differentiate between the two species (Evaristo et al. 2010). The α-linoleic acid, eicosadienoic acid, and gadoleic acid are among the fatty acids that are less abundant in *P. pinea* and *P. koraiensis* seeds (Nergiz and Dönmez 2004; Evaristo et al. 2010). *P. pinea* nuts are rich in phytosterols, which are present in plant cell membranes. β-sitosterol is the most abundant phytosterol (Marianc Lutz et al. 2017). Because phytosterols prevent the intestinal absorption of cholesterol, they may help reduce LDL and total plasma cholesterol levels (Rudkowska 2010). In addition, they have antioxidant, anti-cancer, anti-atherogenic and anti-inflammatory properties (Rudkowska 2010; Ryan et al. 2006]. It contains vitamin E, and there are also vitamin B1 and found had a 1.50 mg/100 g edible portion level (Nergiz and Dönmez 2004). Nuts also contain vitamin C, although not as much as other nutrients (Lutz et al. 2017). It contains minerals such as calcium, iron, manganese, sulfur, magnesium, potassium, phosphorus, and zinc. The mineral that is most prevalent is phosphorus (Nergiz and Dönmez 2004). The average phenolic content was 0.34 ± 0.05 mg gallic acid equivalents/g for nut (Mariane Lutz et al. 2017).

9.5.2 Alzheimer's Disease Prevention and Management of Pine Nut

Research on the potential biological activities of *P. pinea* on Alzheimer's disease suggests that some of its components may exert neuroprotective effects. The pathophysiology of Alzheimer's disease is known to be significantly influenced by oxidative stress. Pine nuts have polyunsaturated fatty acids, phenolic compounds, and vitamins, particularly vitamin E all support antioxidant action. Particularly, PUFAs contain elements that are critical to brain function. By preserving neuronal function and cell signaling, PUFAs can help the nervous system (Godos and Grosso 2019). Additionally, *P. pinea* nuts include phytosterols, which lower cholesterol and support heart health. Given the connection between cardiovascular health and cognitive function, this could indirectly promote brain health (Ros et al. 2021). In a study, Pine nuts was showed an average antioxidant capacity (ORAC value) at 8.54 ± 0.03 μmoL TE/g (Lutz et al. 2017). In mice with scopolamine-induced memory impairment, a recent study examined the effects of a pine nut albumin hydrolysate and its short peptide derivative [Trp-Tyr-Pro-Gly-Lys] on synaptic plasticity and memory function, as well as the possible underlying mechanism in PC12 cells. The findings demonstrate that fraction (<3 kDa) and peptid derivative enhance memory and learning capacity through SIRT3-induced synaptic plasticity both in vitro and (Lu et al. 2021). In a recent study, Wojdylo et al. (2022) demonstrated that polyphenols were absent from pine nuts. This information helps to explain the low polyphenol concentration because these nuts are skinless. However, the same study also discovered that pine nuts had the highest quantity of triterpene (690.3 mg/100 g), with oleanolic acid making up almost half of the total triterpene. One of the most effective nut extracts against AChE was found to be found in pine nuts (27%) (Wojdyło et al. 2022). It has been demonstrated that the pine nut peptide-zinc chelate from *P. koraiensi* improves memory and learning, inhibits brain aging and damage, and controls gut flora in rats. It may function through enhanced ChAT and decreased AChE in the central cholinergic system, endogenous thiol antioxidants (enzymes) in the cerebral cortex, and bacterial flora in the gut (Zhang et al. 2023).

9.6 Hazelnut (Betulaceae): *Corylus avellana* L.

9.6.1 Nutrients and Phytochemical Components of Hazelnut

Corylus avellana L. (Betulaceae), known as hazelnut (Fig. 9.7), reaches a height of 3–10 m and with oblong and serrated leaves, can be found in the temperate zones (Dobhal et al. 2018; Balta et al. 2023). Even though *C. avellana* grows in temperate regions like Türkiye, Spain, and Italy, its natural range is limited to the northern hemisphere (Bottone et al. 2019). A nut is a spherical, edible, oily seed with a hard shell (Dobhal et al. 2018). Hazelnuts, widely used in various products such as chocolates, baked goods, and processed foods, are rich in unsaturated fats and bioactive substances and contain substances with neuroprotective properties such as vegetable protein, minerals, tocopherols, phytosterols, and phenolic compounds (Ros 2010; Di Renzo et al. 2014; Silvestri et al. 2021). Unsaturated fatty acids make up around 92% of the total fatty acid makeup (Alasalvar and Shahidi 2008). Hazelnut kernels have been found to contain a variety of phenolic acids. Gallic acid, one of the several phenolic acids identified in hazelnut kernels, represents the main compound of *C. avellana* due to its presence in different parts (Prosperini et al. 2009; Pelvan et al. 2018; Yuan et al. 2018)]. Quercetin, myricetin, and hyperoside have been determined in hazelnut kernels and shells (Pelvan et al. 2018; Yuan et al. 2018). Gallotannins in hazelnuts are distinguished by molecules in which galloyl units are linked to catechin units (Bottone et al. 2019). A hazelnut flavonoid with

Fig. 9.7 Hazelnut (https://pixabay.com/photos/nut-nuts-hazelnuts-shell-brown-65292/)

neuroprotective qualities, epigallocatechin gallate (EGCG) can enhance neuronal cell function, reduce oxidative stress and neuroinflammation, and prevent neurodegenerative illnesses (Daglia et al. 2014; Singh et al. 2015; Sun et al. 2019).

9.6.2 Alzheimer's Disease Prevention and Management of Hazelnut

Persian medicine emphasized hazelnut, offering noteworthy documented therapeutic experiences in cognitive illnesses, including dementia and amnesia. Some Persian medical publications place particular emphasis on how hazelnuts preserve brain tissue, stop brain atrophy, and enhance memory (Gorji et al. 2018; Talebi et al. 2024). Based on their clinical observations and experience, certain Persian medical writings suggest hazelnuts as a great neuroprotective meal to enhance memory (Bahaeddin et al. 2017; Gorji et al. 2018). Recent research indicates that foods high in flavonoids, such as hazelnuts, may help prevent neurodegenerative illnesses by acting as antioxidants and anti-inflammatory agents (Joseph et al. 2009; Bahaeddin et al. 2017). Vitamins have been found to have therapeutic effects in certain neurodegenerative disorders due to their anti-inflammatory and antioxidant properties (Rai et al. 2021). Tocotrienols and tocopherols abundant in hazelnuts have been investigated for their possible preventive and therapeutic roles and possible molecular pathways in AD (Boccardi et al. 2016; Xia and Mo 2016). Researchers recommend hazelnuts as a good source of vitamin E, which may help reduce the incidence of AD (Luchsinger and Mayeux 2004). Taking into account the different potential causes of AD, vitamin E, an antioxidant, has been demonstrated to be beneficial due to its anti-inflammatory, lipid peroxidation-reducing, and Aβ-associated reactive oxygen species-preventing qualities. It has been demonstrated that tocopherol increases the activity of the enzyme protein phosphatase 2A (PP2A), which is linked to the pathophysiology of AD (Voronkov et al. 2011). Some studies have indicated that vitamin E concentrations are increased in the brains of AD patients and decreased in their plasma, suggesting a possible compensatory response to oxidative damage (Harandi et al. 2015). According to reports, vitamin E can reduce lipid peroxidation sensitivity in AD patients by 60% when compared to controls. Additionally, postmortem cerebrospinal fluid examination revealed a positive correlation between α-tocopherol levels and patients' AD pathology and perceptual speed. Also, vitamin E administration decreased oxidative stress markers in rat neuronal cultures and prevented the formation of ROS related to Aβ (Harandi et al. 2015; Li et al. 2017a, b; Carey et al. 2012). MUFA and PUFA make up the majority of the nutrient-dense, rich lipid profile found in hazelnuts (Fernandes et al. 2017). The most prevalent unsaturated fatty acids found in almonds and hazelnut are oleic and linoleic acids (Alasalvar and Shahidi 2008). The frontal cortex and hippocampus of AD brains have been shown to exhibit a substantial reduction in MUFAs, including oleic acid (Youn et al. 2014). Oleic acid has been demonstrated to dramatically inhibit prolyl endopeptidase activity, which is elevated in the AD brain (Park et al. 2006). Due to the multifactorial nature of Alzheimer's disease, it is important to find multitargeted ligands, produce new interventions with natural products and medical foods, and identify valuable sources of bioactive compounds for drug discovery. Research indicates that free radicals enhance the aggregation of tau and amyloid beta peptides ($A\beta$), and the pathogenesis of AD is primarily characterized by the oxidation of proteins, lipids, and DNA (Nunomura et al. 2006; Gella and Durany 2009). Hazelnuts also have the intriguing ability to regulate oxidative stress at the genomic level since eating them causes the genes that produce pro-oxidants to be down-regulated and the genes that produce antioxidants to be up-regulated. One important factor in the onset and development of oxidative stress in Alzheimer's disease is oxidized LDLs. A Mediterranean diet high in hazelnuts also dramatically lowers the quantity of oxidized LDL, which lowers the risk of Alzheimer's and other neurological diseases, according to some research (Talebi et al. 2024; Di Renzo et al. 2014). Hazelnut samples grown in the Piedmont region of Italy and the Black Sea region of Turkey were evaluated with in vitro enzyme inhibition properties and ex vivo experiments to investigate their potential neuroprotective effects as well as the potential benefit of hazelnuts in Alzheimer's neurodegenerative disease (Mollica et al. 2018). In particular, Turkish hazelnut methanol extracts showed greater enzyme inhibition potential than Italian hazelnut methanol extracts in vitro enzyme inhibition tests. Ex vivo tests have shown that both hazelnut types can reduce neurochemical changes in Alzheimer's neurodegeneration in rodents (Mollica et al. 2018; Bahaeddin et al. 2017). In recent in vivo studies, mice given a diet high in hazelnuts were tested for anxiety-related behaviors. It was discovered that supplementing mice with hazelnuts could enhance their memory by lowering acetylcholinesterase activity. Furthermore, it has been suggested that hazelnuts may be able to lower brain levels of caspase-3, which may preserve neural integrity and stop neuronal apoptosis (Olofinnade et al. 2021; Adriano et al. 2018). A study found that rats fed 800 mg/kg of hazelnut kernels daily had better memory and less anxiety, as well as a healing impact on amyloid-β ($A\beta$)-induced neuroinflammation and apoptosis (Bahaeddin et al. 2017). Moreover, a Mediterranean diet high in PUFA and MUFA may help lessen cognitive decline brought on by aging, while PUFA may reverse reduce the risk of AD due to its anti-inflammatory and neuroprotective role in the brain (Solfrizzi et al. 2006; Boudrault et al. 2009). Additionally, selenium, another component in nuts that exhibits neuroprotective properties, may reduce neuronal death in AD patients due to reduced Aβ production (Gwon et al. 2010). The neuroprotective qualities of the hazelnut shell extracts were examined in a 2019 study by assessing their inhibition of tyrosinase, butyrylcholinesterase, and acetylcholinesterase enzymes that contribute to the pathogenesis of Parkinson's disease, AD, and other neurological disarrays (Gezici 2019).

9.7 Walnut (Juglandaceae): *Juglans regia* Linn

9.7.1 Nutrients and Phytochemical Components of Walnut

Juglans regia Linn. (Juglandaceae), commonly referred to as English or Persian walnut (Fig. 9.8), is a medicinal plant that may be used to cure several illnesses in traditional medicine (Panth et al. 2016). Flavonoids (kaempferol, myricetin, quercetin, coumaric acid, quercetin 3-galactoside), phenolic acid (ellagic acid, gallic acid, protocatechuic acid, ferulic acid), sterols, steroids, terpenoids, vitamins, minerals, omega-3 fatty acids, melatonin, juglone, proanthocyanidins, and flavonoids are among the many antioxidants found in walnuts (Gorji et al. 2018; Adarmanabadi et al. 2023; Rahbardar and Hosseinzadeh 2024). The neuroprotective effects of walnuts are known to be due to the bioactive compounds they contain, including hydrolysates, peptides, polyphenols, and oligopeptides. Walnuts are referred to be "brain food" due to their high capacity of minerals, proteins, flavonoids, and unsaturated fatty acids (Rahbardar and Hosseinzadeh 2024; Adarmanabadi et al. 2023; Luan et al. 2021).

9.7.2 Alzheimer's Disease Prevention and Management of Walnut

The most abundant compounds in various parts of the walnut, whose fruit consists of four main parts (seed, shell, shell and green shell), and which have some pharmacological effects (immunomodulatory, anti-inflammatory, neuroprotective, hepatoprotective, antidiabetic, antifertility, antidiarrheal, astringent, antiviral and antimicrobial) are quinones, phenolics, triterpenoids and diarylheptanoids (Rahbardar and Hosseinzadeh 2024; Luan et al. 2021). Delaying the beginning and progression of cognitive decline, boosting mitochondrial activity, autophagy, and neurogenesis, and lowering oxidative stress, apoptosis, proteolysis, and β-amyloid

Fig. 9.8 Walnut (https://pixabay.com/photos/nuts-walnuts-meal-healthy-food-4637557/)

plaque deposition in Alzheimer's disease are all benefits of long-term walnut consumption. A study has also indicated that the extracts of *J. regia* leaves and fruits reduce the risk of AD due to their antioxidant activities (Orhan et al. 2011). In vitro experiments on the evaluation of neuroprotective effects of walnut leaf and fruit extracts have determined that the extracts protect against cell death and DNA damage by reducing the formation of free radicals (Muthaiyah et al. 2011). In vivo studies demonstrated that supplementing with walnuts reduced memory impairment caused by scopolamine, Alzheimer's disease, and aging by increasing acetylcholine levels and antioxidant enzyme activity (Rahbardar and Hosseinzadeh 2024). In a study evaluating the therapeutic potential of ethnomedicinal plants with preservative effects against Beta-Amyloid Peptide (A)1–42 against AD, it was reported that phytoextract prepared from *Juglans regia* (leaves) species had inhibitory activity (Althobaiti et al. 2022). The study conducted by Fotuhi and Do (2010) investigated the effects of walnuts in enhancing learning in a transgenic mouse model of AD. Control mice without walnuts in their diet exhibited anxiety-related behaviors and had reduced learning ability. However, mice on the walnut diet reported a notable improvement in memory and learning capacity (Fotuhi and Do 2010). Aβ contributes to producing free radicals that induce oxidative stress and is a major cause of amyloid plaques in AD patients. The beneficial impacts of walnuts in preventing AD are as follows: Walnuts may mitigate beta-amyloid-dependent cytotoxicity by decreasing free radical generation and minimizing damage to DNA and cell membranes. A walnut-rich diet reduces oxidative stress and Aβ fibrillation, which helps prevent AD-related dementia (Chauhan and Chauhan 2012). Due to the substances in walnuts, numerous studies have been conducted on their use as a natural source in the treatment of Alzheimer's. Joseph et al. (2009) exhibited that walnuts can affect aging and health span, and accordingly, walnut consumption can slow down the process of brain degeneration. Sala-Vila et al. (2020) showed that walnut consumption had positive impacts on brain phenotypes and regional GMV (gray matter volume) in trials conducted on 187 middle-aged persons at risk of AD, based on the idea that eating walnuts can improve brain function. In the study by Zou et al. (2016), mice fed a diet high in concentrated walnut peptides improved their memory and cognitive function and had positive effects on recovering levels of antioxidant enzymes and inflammatory mediators (Zou et al. 2016). Using walnut powder extracts at 800 and 1600 mg/kg/day for 6 weeks, Alzheimer's disease-afflicted mice showed improvements in memory impairment caused by AD-galactose (Lv et al. 2021). In a study that examined the impact of walnuts on improving cognitive impairments and decreasing Aβ aggregation, an active peptide termed PW5 (Pro-Pro-Lys-Asn-Trp) that is present in walnuts demonstrated anti-Aβ42 accumulation activity, indicating its impacts on cognitive improvement (Wang et al. 2019)]. According to Muthaiyah et al. (2014), AD-Tg mice given a diet enriched with 6% or 9% walnuts showed good improvements in memory stages and retained data in mind longer than controls (Muthaiyah et al. 2014). Recent studies have shown that walnuts prevent the formation of fibrillar Aβ-protein due to the components they contain (such as vitamin E, melatonin, ellagitannins, and urolithins) and prevent oxidative stress and protein damage caused by the enzyme that plays a role in Aβ

fibrils (Djedjibegovic et al. 2020; Pandareesh et al. 2018; Muthaiyah et al. 2011). The findings imply that it might be crucial in halting DNA damage and neuronal cell death brought on by Aβ Li et al. (2017a, b) evaluated the antioxidant effects of walnut protein hydrolysate by in vitro experiments, and also determined that the ameliorative effect in scopolamine-induced dementia mice increased in relation to the amount of acetylcholine receptors by in vivo tests. Considering the brain-nourishing feature of defatted walnut powder in Chinese medicine, the effects of its consumption on AD model mice were evaluated. Forty-three metabolic pathways modulated by defatted walnut powder, which can alleviate cognitive dysfunction, were identified, and its consumption was reported to improve AD progression (Xu et al. 2024a, b). According to a study, organic acids from walnut seeds have a major therapeutic impact on AD. The study assessed the binding relationship between the active ingredients and target proteins using brain tissue biochemistry testing and histopathological examination (Xu et al. 2024b).

9.8 Pistachio (Anacardiaceae): *Pistacia vera* L.

9.8.1 Nutrients and Phytochemical Components of Pistachio

Pistacia vera (Anacardiaceae) has substantial commercial value and is well-known worldwide for its tasty and healthy berries. The fruit's seeds (Fig. 9.9) contain oleic acid, which has been demonstrated to offer protection against AD and other neurological conditions (Moeini et al. 2019). It has been found that the Pistachios (*P. vera*) fresh, unripe fruits, which are high in oxygenated sesquiterpenes and monoterpenes, yield more essential oil than the leaves. The primary constituents of *P. vera* were identified as bornyl acetate, terpinolene, α-pinene, and β-pinene also oleanonic acid and Merulinic acid A were isolated (Tsokou et al. 2007; Rauf et al. 2017; Duru et al. 2003). Vitamin E, γ-tocopherol, phytomelatonin, polyphenols, fatty acids, MUFA, and PUFA are among the many potentially neuroprotective compounds found in pistachios that can effectively control brain activities and protect against neurodegeneration (Pribis and Shukitt-Hale 2014). Likewise, they are a great source of

Fig. 9.9 Pistachio (https://pixabay.com/photos/pistachio-seeds-shelled-oilseeds-7574181/)

γ-tocopherol and other antioxidants, and they have a lot of resveratrol in their shells (Mandalari et al. 2021). Bioactive substances such as terpenoids, polyphenolic chemicals, and volatile oils (EOs) are abundant in its leaves.

9.8.2 Alzheimer's Disease Prevention and Management of Pistachio

Pictaciha vera seeds, commonly known as pistachio, are one of the plant-based foods recommended by Traditional Persian Medicine (TPM) for treating dementia and memory impairment (Iranshahy and Javadi 2019). Traditional Chinese Medicine (TCM), Ayurveda (India), Iranian folk medicine, and other old pharmacopeias have all referenced the medicinal applications of plants in the *Pistacia* genus, which are thought to be among the natural resources with neuroprotective potential (Bibi et al. 2015). Antioxidant, anti-neuroinflammatory, anti-A aggregation, AChE inhibitory activity, and modulation of several other cellular pathways are thought to be associated with the species' neuroprotective properties (Moeini et al. 2019). Because of their analgesic, astringent, antitussive, diuretic, and expectorant qualities, *P. vera* leaves have long been utilized. Additionally, proanthocyanidins have been shown to have pharmacological and therapeutic qualities linked to antioxidant and radical scavenging, and pistachios are the only nuts known to possess anthocyanins, which are the pigments that give many fruits and vegetables their color (Tomaino et al. 2010; Bagchi et al. 2000; Cos et al. 2004). Pistachios, known to be rich in fat-soluble antioxidants such as tocopherols and carotenoids, have been investigated for their benefits against high-fat diet-induced neurodegeneration in a mouse model. It has been reported that peanut supplementation to a diet strong in fat prevents obesity-associated neurodegeneration by reducing lipid metabolism disorder and mitochondrial dysfunction (Nuzzo et al. 2020). Recent studies have demonstrated that the hydroalcoholic extract of *P. vera* seeds has an anxiolytic effect in female rats and its aqueous methanol extract can inhibit AChE activity to a certain extent (Golchin et al. 2015; Rostampour et al. 2016). It has also been recounted that *P. vera* gum extract can reduce malondialdehyde (MDA) levels in rats with oxidative damage and increase the antioxidant capacity of the brain (Mansouri et al. 2005, 2011). Hosseinkhani et al. (2017) reported that *P. vera* (fruit) is one of the plants used in the treatment of AD discussed under the term Nesyan in traditional Persian books. Several investigations have been conducted to correlate the widespread traditional use of the plant with its pharmacological effects. The first signs of AD include memory loss and other developing cognitive issues. It is also believed to be connected to inflammation, oxidative stress, and acetylcholine (ACh) depletion. After 5 weeks of application, *P. vera* kernel has been shown to correct spatial memory impairments from neurotoxic anticancer medications in rats and block cognitive and motor impairments created by vincristine or cisplatin. According to the report, this is attributed to the high flavonoid and phenolic content found in the seed extract (Golchin et al. 2015).

The seed oil enhanced memory and cognitive impairment, according to another in vivo investigation (Tavakoli et al. 2019). In light of this, it may be maintained that pistachio fruit has high flavonoid and phenolic content may also support the effects of cognitive enhancement (Golchin et al. 2015). A pilot study assessing the effects of pistachio supplementation on cognitive function and mood in overweight youth people indicated a favorable modulation of mood states, including anxiety, anger-hostility, and sadness-depression (Landaverde-Mejia et al. 2024). In a rat model of AD, it was found that pistachio hydroalcoholic extract dramatically enhanced memory and learning (Abdi Gorabi et al. 2020). In one experiment, rats given pistachio extract (10, 50, and 100 mg/kg/day for 14 days) had a markedly higher delay to enter the dark room than the control group. Additionally, pistachios reduced the amount of time spent in the dark room, suggesting that they have positive impacts on memory and learning (Salari et al. 2014). A variety of AChE inhibitors are used to treat AD symptoms. The plant's stem aqueous methanol extract was found to have the ability to inhibit AChE in vitro in a study assessing its effects (Gholamhoseinian et al. 2009). According to a further study, *P. vera* is a natural product that may be employed in treating glaucoma, AD, diabetes, and epilepsy owing to the polyphenolics it contains (Karagecili et al. 2023).

9.9 Conclusion

The role of nuts for the prevention and management of Alzheimer's disease is increasingly recognized. Nuts are nutrient-rich foods with anti-inflammatory and antioxidant properties and contain bioactive substances such as unsaturated fatty acids, fiber, minerals, vitamins, phytosterols, and phenolic compounds. Because of their anti-inflammatory, neurotrophic, and antioxidant properties, these substances are known to improve brain function. Because of the variations in their composition, different kinds of nuts can target different parts of Alzheimer's disease and help prevent and control the condition. Walnuts contain ellagic acid with anti-amyloidogenic properties. Selenium, which is abundant in Brazil nuts, is necessary for the development of antioxidant defense systems and the control of the immune system. Almonds contain vitamins such as folic acid, vitamin E, and niacin, which can help prevent Alzheimer's disease by increasing neurogenesis, reducing inflammation, and lowering oxidative stress. Hazelnuts are rich in flavonoids, which may help prevent neurodegenerative diseases by acting as antioxidants and anti-inflammatory agents. Pistachios contain neuroprotective substances such as vitamin E, folic acid, phytomelatonin, antioxidant polyphenols, and fatty acids that can regulate neuronal function and protect against neurodegeneration. Therefore, consuming a variety of nuts as part of a balanced diet can help prevent Alzheimer's disease.

References

Abdi Gorabi S, Mohammadzadeh H, Rostampour M (2020) The effects of ripe pistachio hulls Hydroalcoholic extract and aerobic training on learning and memory in Streptozotocin-induced diabetic male rats. Basic Clin Neurosci 11:525–534. https://doi.org/10.32598/bcn.9.10.400

Adarmanabadi SMHH, Gilavand HK, Taherkhani A, Rafiei SKS, Shahrokhi M, Faaliat S, Deravi N (2023) Pharmacotherapeutic potential of walnut (Juglans spp.) in age-related neurological disorders. IBRO Neurosci Rep 14:1–20

Adriano M, Gokhan Z, Azzurra S, Claudio F, Luigi M, Giustino O, Olakunle JO (2018) Nutraceutical potential of *Corylus avellana* daily supplements for obesity and related dysmetabolism. J Funct Foods 47:562–574. https://doi.org/10.1016/j.jff.2018.06.016

Alasalvar C, Shahidi F (2008) Tree nuts: composition, phytochemicals, and health effects: an overview. CRC Press, pp 15–24

Alehagen U, Opstad TB, Alexander J, Larsson A, Aaseth J (2021) Impact of selenium on biomarkers and clinical aspects related to ageing. A review. Biomolecules 11:1478. https://doi.org/10.3390/biom11101478

Althobaiti NA, Menaa F, Dalzell JJ, Albalawi AE, Ismail H et al (2022) Ethnomedicinal plants with protective effects against Beta-amyloid peptide (Aβ)1-42 indicate therapeutic potential in a new *in vivo* model of Alzheimer's disease. Antioxidants (Basel) 11(10):1865. https://doi.org/10.3390/antiox11101865

Amarowicz R, Pegg RB (2017) The potential protective effects of phenolic compounds against low-density lipoprotein oxidation. Curr Pharm Des 23(19):2754–2766. https://doi.org/10.2174/1381612823666170329142936

Awan HUM, Pettenella D (2017) Pine nuts: a review of recent sanitary conditions and market development. Forests 8(10):367. https://doi.org/10.3390/f8100367

Bagchi D, Bagchi M, Stohs SJ (2000) Free radicals and grape seed proanthocyanidin extract: importance in human health and disease prevention. Toxicology 148:187–197. https://doi.org/10.1016/s0300-483x(00)00210-9

Bahaeddin Z, Yans A, Khodagholi F, Hajimehdipoor H, Sahranavard S (2017) Hazelnut and neuroprotection: improved memory and hindered anxiety in response to intra-hippocampal Aβ injection. Nutr Neurosci 20(6):317–332. https://doi.org/10.1080/1028415X.2015.1126954

Balta C, Herman H, Ciceu A, Mladin B, Rosu M, Sasu A, Peteu VE, Voicu SN, Balas M, Gherghiceanu M et al (2023) Phytochemical profiling and anti-fibrotic activities of the Gemmotherapy bud extract of *Corylus avellana* in a model of liver fibrosis on DiabeticMice. Biomedicine 11:1771. https://doi.org/10.3390/biomedicines11061771

Barreca D, Nabavi SM, Sureda A, Rasekhian M, Raciti R, Silva AS, Mandalari G (2020) Almonds (*Prunus dulcis* Mill. DA Webb): a source of nutrients and health-promoting compounds. Nutrients 12(3):672. https://doi.org/10.3390/nu12030672

Barreira JCM, Alves RC, Casal S, Ferreira ICFR, Oliveira MBPP, Pereira JA (2009) Vitamin E profile as a reliable authenticity discrimination factor between chestnut (*Castanea sativa* Mill.) cultivars. J Agric Food Chem 57:5524–5528

Batool Z, Agha F, Tabassum S, Siddiqui T, Haider S (2019) Prevention of cadmium-induced neurotoxicity in rats by essential nutrients present in nuts. Acta Neurobiol Exp 79(2):169–183. https://doi.org/10.21307/ane-2019-015

Bibi Y, Zia M, Qayyum A (2015) Review-an overview of Pistacia integerrima medicinal plant species: ethnobotany, biological activities, and phytochemistry. Pak J Pharm Sci 28:1009–1013

Boccardi V, Baroni M, Mangialasche F, Mecocci P (2016) Vitamin E family: role in the pathogenesis and treatment of Alzheimer's disease. Alzheimer's Dement Transl Res Clin Interven 2(3):182–191. https://doi.org/10.1016/j.trci.2016.08.002

Boudrault C, Bazinet RP, Ma DWL (2009) Experimental models and mechanisms underlying the protective effects of n-3 polyunsaturated fatty acids in Alzheimer's disease. J Nutr Biochem 20(1):1–10. https://doi.org/10.1016/j.jnutbio.2008.05.016

Brizi C, Santulli C, Micucci M, Budriesi R, Chiarini A, Aldinucci C, Frosini M (2016) Neuroprotective effects of *Castanea sativa* Mill. Bark extract in human neuroblastoma cells subjected to oxidative stress. J Cell Biochem 117(2):510–520. https://doi.org/10.1002/jcb.25302

Bottone A, Cerulli A, D'Urso G, Masullo M, Montoro P, Napolitano A, Piacente S (2019) Plant specialized metabolites in hazelnut (Corylus avellana) kernel and byproducts: an update on chemistry, biological activity, and analytical aspects. Planta Med 85(11/12):840–855

Cardoso BR, Duarte GBS, Reis BZ, Cozzolino SMF (2017) Brazil nuts: nutritional composition, health benefits and safety aspects. Food Res Int 100:9–18. https://doi.org/10.1016/j.foodres.2017.08.036

Carey AN, Poulose SM, Shukitt-Hale B (2012) The beneficial effects of tree nuts on the aging brain. Nutrition and Aging 1(1):55–67

Chauhan A, Chauhan V (2012) Potential beneficial effects of a diet with walnuts in aging and Alzheimer's disease. In: Brain aging and therapeutic interventions. Springer, pp 239–252. https://doi.org/10.1007/978-94-007-5237-5_16

Chen CO, Holbrook M, Duess MA, Dohadwala MM, Hamburg NM, Asztalos BF, Vita JA (2015) Effect of almond consumption on vascular function in patients with coronary artery disease: a randomized, controlled, cross-over trial. Nutr J 14:1–11. https://doi.org/10.1186/s12937-015-0049-5

Chen CO, Milbury PE, Blumberg JB (2019) Polyphenols in almond skins after blanching modulate plasma biomarkers of oxidative stress in healthy humans. Antioxidants (Basel) 8(4):95. https://doi.org/10.3390/antiox8040095. PMID: 30974789; PMCID: PMC6523744

Choupina AB (2019) Nutritional and health potential of European chestnut. Revista de Ciências Agrárias 42(3):801–807. https://doi.org/10.19084/rca.17701

Ciucure CT, Geana EI, Sandru C, Tita O, Botu M (2022) Phytochemical and nutritional profile composition in fruits of different sweet chestnut (*Castanea sativa* mill.) cultivars grown in Romania. Separations 9(3):66. https://doi.org/10.3390/separations9030066

Colic SD, Aksic MMF, Lazarević KB, Zec GN, Gašić UM, Zagorac DČD, Natić MM (2017) Fatty acid and phenolic profiles of almond grown in Serbia. Food Chem 234:455–463

Cos P, De Bruyne T, Hermans N, Apers S, Berghe DV, Vlietinck AJ (2004) Proanthocyanidins in health care: current and new trends. Curr Med Chem 11(10):1345–1359. https://doi.org/10.2174/0929867043365288

da Silva A, Silveira BKS, de Freitas BVM, Hermsdorff HHM, Bressan J (2022) Effects of regular Brazil nut (*Bertholletia excelsa* HBK) consumption on health: a systematic review of clinical trials. Foods 11(18):2925. https://doi.org/10.3390/foods11182925

Daglia M, Di Lorenzo A, Nabavi F, Talas ZS, Nabavi ZM (2014) Polyphenols: well beyond the antioxidant capacity: gallic acid and related compounds as neuroprotective agents: you are what you eat! Curr Pharm Biotechnol 15(4):362–372

de Araújo BAN, de Campos LMTDS, Souza GT, Abreu GDM, de Almeida CDORP, Figueiredo MS (2024) Brazil nut (Bertholletia excelsa HBK) and Brazil nut oil: effects on oxidative stress and potential therapeutic health benefits from human, animal, and cell culture models. Nutr Rev nuae016. https://doi.org/10.1093/nutrit/nuae016

De Vasconcelos MDCBM, Bennett RN, Rosa EA, Cardoso JVF (2007) Primary and secondary metabolite composition of kernels from three cultivars of Portuguese chestnut (*Castanea sativa* Mill.) at different stages of industrial transformation. J Agric Food Chem 55(9):3508–3516

De Vasconcelos MDCB, Nunes F, Viguera CG, Bennett RN, Rosa EA, Ferreira-Cardoso JV (2010) Industrial processing effects on chestnut fruits (*Castanea sativa* Mill.). Minerals, free sugars, carotenoids and antioxidant vitamins. Int J Food Sci Technol 45(3):496–505. https://doi.org/10.1111/j.1365-2621.2009.02155.x

Di Renzo L, Alberto C, Botta R, Contessa C, Sartor C, Iacopino AM et al (2014) Nutrient analysis critical control point (NACCP): hazelnut as a prototype of nutrigenomic study. Food Nutr Sci 5(1):79–88

Djedjibegovic J, Marjanovic A, Panieri E, Saso L (2020) Ellagic acid-derived urolithins as modulators of oxidative stress. Oxidative Med Cell Longev 2020:5194508. https://doi.org/10.1155/2020/5194508

Dobhal K, Singh N, Semwal A, Negi A (2018) A brief review on: hazelnuts. Int J Recent Scient Res 9(1):23680–23684

Duru ME, Cakir A, Kordali S, Zengin H, Harmandar M, Izumi S, Hirata T (2003) Chemical composition and antifungal properties of essential oils of three *Pistacia* species. Fitoterapia 74(1–2):170–176. https://doi.org/10.1016/s0367-326x(02)00318-0

Evaristo I, Batista D, Correia I, Correia P, Costa R (2010) Chemical profiling of Portuguese *Pinus pinea* L. nuts. J Sci Food Agric 90(6):1041–1049. https://doi.org/10.1002/jsfa.3914

Fernandes GD, Gómez-Coca RB, Pérez-Camino M, Moreda W, Barrera-Arellano D (2017) Chemical characterization of major and minor compounds of nut oils: almond, hazelnut, and pecan nut. J Chem 2017:1–11. https://doi.org/10.1155/2017/2609549

Fotuhi M, Do D (2010) Factors determining the size of the hippocampus. Alzheimers Dement 4(6):S69. https://doi.org/10.1016/j.jalz.2010.05.204

Fouad GI (2019) Stem cells as a promising therapeutic approach for Alzheimer's disease: a review. Bull Natl Res Cent 43(1):52

Gella A, Durany N (2009) Oxidative stress in Alzheimer disease. Cell Adhes Migr 3(1):88–93. https://doi.org/10.4161/cam.3.1.7402

Gezici S (2019) Cancer preventive and neuroprotective potentials of red hulls, kernels and oleo-gum resins from pistachio. Int J Agric Environ Food Sci 3(3):137–143. https://doi.org/10.31015/jaefs.2019.3.4

Gholamhoseinian A, Moradi MN, Sharifi-Far F (2009) Screening the methanol extracts of some Iranian plants for acetylcholinesterase inhibitory activity. Res Pharm Sci 4:105–112

Golchin L, Shabani M, Harandi S, Razavinasab M (2015) Pistachio supplementation attenuates motor and cognition impairments induced by cisplatin or vincristine in rats. Adv Biomed Res 11(4):92. https://doi.org/10.4103/2277-9175.156656

Gonçalves B, Borges O, Costa HS, Bennett R, Santos M, Silva AP (2010) Metabolite composition of chestnut (Castanea sativa mill.) upon cooking: proximate analysis, fibre, organic acids and phenolics. Food Chem 122(1):154–160. https://doi.org/10.1016/j.foodchem.2010.02.032

Gorji N, Moeini R, Memariani Z (2018) Almond, hazelnut and walnut, three nuts for neuroprotection in Alzheimer's disease: a neuropharmacological review of their bioactive constituents. Pharmacol Res 129:115–127. https://doi.org/10.1016/j.phrs.2017.12.003

Gwon AR, Park JS, Park JH, Baik SH, Jeong HY, Hyun DH et al (2010) Selenium attenuates Aβ production and Aβ-induced neuronal death. Neurosci Lett 469(3):391–395. https://doi.org/10.1016/j.neulet.2009.12.035

Godos J, Castellano S, Galvano F, Grosso G (2019) Linking omega-3 fatty acids and depression. In Omega fatty acids in brain and neurological health Academic Press pp 199–212

Harandi S, Golchin L, Ansari M, Moradi A, Shabani M, Sheibani V (2015) Antiamnesic effects of walnuts consumption on scopolamine-induced memory impairments in rats. Basic Clin Neurosci 6(2):91–99

Hosseinkhani A, Sahragard A, Namdari A, Zarshenas MM (2017) Botanical sources for Alzheimer's: a review on reports from traditional Persian medicine. Am J Alzheimers Dis Other Dement 32(7):429–437. https://doi.org/10.1177/1533317517717013

Hu N, Yu JT, Tan L, Wang YL, Sun L, Tan L (2013) Nutrition and the risk of Alzheimer′ s disease. Biomed Res Int 2013(1):524820. https://doi.org/10.1155/2013/524820

Iranshahy M, Javadi B (2019) Diet therapy for the treatment of Alzheimer's disease in view of traditional Persian medicine: a review. Iran J Basic Med Sci 22:1102–1117. https://doi.org/10.22038/ijbms.2019.36505.8694

Jamshed H, Gilani AH (2014) Almonds inhibit dyslipidemia and vascular dysfunction in rats through multiple pathways. J Nutr 144(11):1768–1774. https://doi.org/10.3945/jn.114.198721

Jenkins DJ, Kendall CW, Marchie A, Parker TL, Connelly PW, Qian W, Spiller GA (2002) Dose response of almonds on coronary heart disease risk factors: blood lipids, oxidized low-density lipoproteins, lipoprotein (a), homocysteine, and pulmonary nitric oxide: a randomized, controlled, crossover trial. Circulation 106(11):1327–1332. https://doi.org/10.1161/01.CIR.0000028421.91733.20

Jeong HR, Jo YN, Jeong JH, Jin DE, Song BG, Choi SJ, Heo HJ (2012) Antiamnesic effects of ethyl acetate fraction from chestnut (Castanea crenata var. dulcis) inner skin on Aβ25–35-induced cognitive deficits in mice. J Med Food 15(12):1051–1056. https://doi.org/10.1089/jmf.2011.210

Joseph J, Cole G, Head E, Ingram D (2009) Nutrition, brain aging, and neurodegeneration. J Neurosci 29(41):12795–12801. https://doi.org/10.1523/JNEUROSCI.3520-09.2009

Karagecili H, Yilmaz MA, Alwasel SH, Arık M, Gulcin I (2023) Comprehensively revealing the profile of *Pistacia vera* L. cv. Siirt turpentine – antioxidant, antidiabetic, anti-Alzheimer, and Antiglaucoma effects. Records Nat Prod 17(5):918–937. https://doi.org/10.25135/rnp.410.2305.2787

Kim JH, Choi GN, Kwak JH, Jeong HR, Jeong CH, Heo HJ (2011) Neuronal cell protection and acetylcholinesterase inhibitory effect of the phenolics in chestnut inner skin. Food Sci Biotechnol 20:311–318. https://doi.org/10.1007/s10068-011-0044-3

Kulkarni KS, Kasture SB, Mengi SA (2010) Efficacy study of *Prunus amygdalus* (almond) nuts in scopolamine-induced amnesia in rats. Indian J Pharmacol 42(3):168–173. https://doi.org/10.4103/0253-7613.66841

Landaverde-Mejia K, Dufoo-Hurtado E, Camacho-Vega D, Maldonado-Celis ME, Mendoza-Diaz S, Campos-Vega R (2024) Pistachio (*Pistacia vera* L.) consumption improves cognitive performance and mood in overweight young adults: a pilot study. Food Chem 1(457):140211. https://doi.org/10.1016/j.foodchem.2024.140211

Li W, Zhao T, Zhang J, Xu J, Sun-Waterhouse D, Zhao M et al (2017a) Effect ofwalnut protein hydrolysate on scopolamine-induced learning and memory deficits in mice. J Food Sci Technol 54(10):3102–3110

Li W, Zhao T, Zhang J, Xu J, Sun-Waterhouse D, Zhao M et al (2017b) Effect of walnut protein hydrolysate on scopolamine-induced learning and memory deficits in mice. J Food Sci Technol 54(10):3102–3110. https://doi.org/10.1007/s13197-017-2746-x

Loewe-Muñoz V, Noel D (2021) Mediterranean Pinus pinea L. nuts from southern hemisphere provenances. Rendiconti Lincei Scienze Fisiche e Naturali 32:181–189. https://doi.org/10.1007/s12210-021-00980-8

Lu H, Fang L, Wang J, Zhao F, Liu C, Gao Y, Min W (2021) Pine nut antioxidant peptides ameliorate the memory impairment in a scopolamine-induced mouse model via SIRT3-induced synaptic plasticity. Food Funct 12(17):8026–8036. https://doi.org/10.1039/D1FO01817E

Luan F, Wang Z, Yang Y, Ji Y, Lv H, Han K et al (2021) *Juglans mandshurica* maxim.: a review of its traditional usages, phytochemical constituents, and pharmacological properties. Front Pharmacol 11:569800. https://doi.org/10.3389/fphar.2020.569800

Luchsinger JA, Mayeux R (2004) Dietary factors and Alzheimer's disease. Lancet Neurol 3(10):579–587. https://doi.org/10.1016/S1474-4422(04)00878-6

Lutz M, Álvarez K, Loewe V (2017) Chemical composition of pine nut (*Pinus pinea* L.) grown in three geographical macrozones in Chile. CyTA – Journal of Food 15(2):284–290. https://doi.org/10.1080/19476337.2016.1250109

Lv H, Chen D, Xu C et al (2021) Positive effects of walnut meal extracts on memory impairment of Alzheimer's disease mice induced by D-galactose. Research Square. https://doi.org/10.21203/rs.3.rs-209172/v1

Maccioni RB, González A, Andrade V, Cortes N, Tapia JP, Guzmán-Martínez L (2018) Alzheimer s disease in the perspective of Neuroimmunology. Open Neurol J 12:50. https://doi.org/10.2174/1874205X01812010050

Mansouri SMT, Naghizadeh B, Hosseinzadeh H (2005) The effect of *Pistacia vera* L. gum extract on oxidative damage during experimental cerebral ischemia-reperfusion in rats. Iran Biomed J 9(4):181–185

Mansouri S, Naghizadeh B, Hosseinzadeh H (2011) Evaluation the neuroprotective effect of *Pistacia vera* L. gum extract against cerebral ischemia-reper fusion model of rats. Toxicol Lett 205:S297. https://doi.org/10.1016/j.toxlet.2011.05.1005

Mandalari G, Barreca D, Gervasi T, Roussell MA, Klein B, Feeney MJ, Carughi A (2021) Pistachio nuts(Pistacia vera L.): production, nutrients, bioactives and novel health effects. Plants 11(1):18

McGirr S, Venegas C, Swaminathan A (2020) Alzheimers disease: a brief review. *Journal of.* Exp Neurol 1(3):89–98. https://doi.org/10.33696/Neurol.1.015

Moeini R, Memariani Z, Asadi F, Bozorgi M, Gorji N (2019) *Pistacia* genus as a potential source of neuroprotective natural products. Planta Med 85(17):1326–1350. https://doi.org/10.1055/a-1014-1075

Mollica A, Zengin G, Stefanucci A, Ferrante C, Menghini L, Orlando G, Brunetti L et al (2018) Nutraceutical potential of *Corylus avellana* daily supplements for obesity and related dysmetabolism. J Funct Foods 47:562–574. https://doi.org/10.1016/j.jff.2018.06.016

Moreau RA, Whitaker BD, Hicks KB (2002) Phytosterols, phytostanols, and their conjugates in foods: structural diversity, quantitative analysis, and health-promoting uses. Prog Lipid Res 41(6):457–500. https://doi.org/10.1016/S0163-7827(02)00006-1

Mustra Rakic J, Tanprasertsuk J, Scott TM, Rasmussen HM, Mohn ES, Chen CYO, Johnson EJ (2022) Effects of daily almond consumption for six months on cognitive measures in healthy middle-aged to older adults: a randomized control trial. Nutr Neurosci 25(7):1466–1476. https://doi.org/10.1080/1028415X.2020.1868805

Muthaiyah B, Essa MM, Chauhan V, Chauhan A (2011) Protective effects of walnut extract against amyloid beta peptide-induced cell death and oxidative stress in PC12 cells. Neurochem Res 36(11):2096–2103. https://doi.org/10.1007/s11064-011-0533-z

Muthaiyah B, Essa MM, Lee M, Chauhan V, Kaur K, Chauhan A (2014) Dietary supplementation of walnuts improves memory deficits and learning skills in transgenic mouse model of Alzheimer's disease. J Alzheimer's Dis 42(4):1397–1405

Nasri N, Khaldi A, Fady B, Triki S (2005) Fatty acids from seeds of *Pinus pinea* L.: composition and population profiling. Phytochemistry 66(14):1729–1735. https://doi.org/10.1016/j.phytochem.2005.05.023

Nergiz C, Dönmez I (2004) Chemical composition and nutritive value of *Pinus pinea* L. seeds. Food Chem 86(3):365–368. https://doi.org/10.1016/j.foodchem.2003.09.009

Nunomura A, Castellani RJ, Zhu X, Moreira PI, Perry G, Smith MA (2006) Involvement of oxidative stress in Alzheimer disease. J Neuropathol Exp Neurol 65(7):631–641. https://doi.org/10.1097/01.jnen.0000228136.58062.bf

Nuzzo D, Galizzi G, Amato A, Terzo S, Picone P et al (2020) Regular intake of pistachio mitigates the deleterious effects of a high fat-diet in the brain of obese mice. Antioxidants 9(4):317. https://doi.org/10.3390/antiox9040317

Olofinnade AT, Onaolapo AY, Onaolapo OJ, Olowe OA, Mollica A, Zengin G, Stefanucci A (2021) *Corylus avellana* L. modulates neurobehaviour and brain chemistry following high-fat diet. Front Biosci (Landmark Ed) 26(3):537–551

Orhan IE, Suntar IP, Akkol EK (2011) *In vitro* neuroprotective effects of the leaf and fruit extracts of *Juglans regia* L.(walnut) through enzymes linked to Alzheimer's disease and antioxidant activity. Int J Food Sci Nutr 62(8):781–786. https://doi.org/10.3109/09637486.2011.585964

Pandareesh MD, Chauhan V, Chauhan A (2018) Walnut supplementation in the diet reduces oxidative damage and improves antioxidant status in transgenic mouse model of Alzheimer's disease. J Alzheimer's Dis 64(4):1295–1305. https://doi.org/10.3233/JAD-180361

Panth N, Paudel KR, Karki R (2016) Phytochemical profile and biological activity of *Juglans regia*. J Integr Med 14(5):359–373. https://doi.org/10.1016/S2095-4964(16)60274-1

Park YS, Jang HJ, Lee KH, Hahn TR, Paik YS (2006) Prolyl endopeptidase inhibitory activity of unsaturated fatty acids. J Agric Food Chem 54(4):1238–1242. https://doi.org/10.1021/jf052521h

Pelvan E, Olgun EO, Karadag A, Alasalvar C (2018) Phenolic profiles and antioxidant activity of Turkish Tombul hazelnut samples (natural, roasted, and roasted hazelnut skin). Food Chem 244:102–108. https://doi.org/10.1016/j.foodchem.2017.10.011

Pribis P, Shukitt-Hale B (2014) Cognition: the new frontier for nuts and berries. Am J Clin Nutr 100:347S–352S. https://doi.org/10.3945/ajcn.113.071506

Prosperini S, Ghirardello D, Scursatone B, Gerbi V, Zeppa G (2009) Identification of soluble phenolic acids hazelnut (*Corylus avellana* L.) kernel. Acta Hortic 845:677–680. https://doi.org/10.17660/ActaHortic.2009.845.106

Rai SN, Singh P, Steinbusch HW, Vamanu E, Ashraf G, Singh MP (2021) The role of vitamins in neurodegenerative disease: an update. Biomedicine 9(10):1284. https://doi.org/10.3390/biomedicines9101284

Rahbardar MG, Hosseinzadeh H (2024) Neuroprotective effects of walnut (Juglans regia L.) in nervous system disorders: A comprehensive review. Iran J Basic Med Sci 27(12):1492. https://doi.org/10.22038/ijbms.2024.79854.17302

Rauf A, Patel S, Uddin G, Siddiqui BS, Ahmad B, Muhammad N et al (2017) Phytochemical, ethnomedicinal uses and pharmacological profile of genus *Pistacia*. Biomed Pharmacother 86:393–404. https://doi.org/10.1016/j.biopha.2016.12.017

Ribeiro B, Rangel J, Valentao P, Andrade PB, Pereira JA, Bölke H, Seabra RM (2007) Organic acids in two Portuguese chestnut (*Castanea sativa* Miller) varieties. Food Chem 100(2):504–508. https://doi.org/10.1016/j.foodchem.2005.09.073

Rita Cardoso B, Apolinário D, da Silva Bandeira V, Busse AL, Magaldi RM, Jacob-Filho W, Cozzolino SMF (2016) Effects of Brazil nut consumption on selenium status and cognitive performance in older adults with mild cognitive impairment: a randomized controlled pilot trial. Eur J Nutr 55:107–116. https://doi.org/10.1007/s00394-014-0829-2

Roncero JM, Álvarez-Ortí M, Pardo-Giménez A, Rabadán A, Pardo JE (2020) Review about non-lipid components and minor fat-soluble bioactive compounds of almond kernel. Foods 9(11):1646. https://doi.org/10.3390/foods9111646

Ros E (2010) Health benefits of nut consumption. Nutrients 2(7):652–682. https://doi.org/10.3390/nu2070652

Ros E, Singh A, O'Keefe JH (2021) Nuts: natural pleiotropic nutraceuticals. Nutrients 13(9):3269

Rostampour M, Hadipour E, Oryan S, Soltani B, Saadat F (2016) Anxiolytic-like effect of hydroalcoholic extract of ripe pistachio hulls in adult female Wistar rats and its possible mechanisms. Res Pharm Sci 11:454–460. https://doi.org/10.4103/1735-5362.194870

Rudkowska I (2010) Plant sterols and stanols for healthy ageing. Maturitas 66:158–162. https://doi.org/10.1016/j.maturitas.2009.12.015

Ryan E, Galvin K, O'connor TP, Maguire AR, O'brien NM (2006) Fatty acid profile, tocopherol, squalene and phytosterol content of Brazil, pecan, pine, pistachio and cashew nuts. Int J Food Sci Nutr 57(3–4):219–228. https://doi.org/10.1080/09637480600768077

Sala-Vila A, Crous-Bou M et al (2020) Eating a weekly serving of walnuts relates to beneficial brain imaging phenotypes in a cohort at increased risk of Alzheimer's disease. Curr Dev Nutr 4(Supplement_2):1234. https://doi.org/10.1093/cdn/nzaa057_050

Salari E, Baloochi M, Shamsizedeh A, Ayoobi F, Allahtavakoli M, Taghavi Y, Ravari A (2014) Effect of the hydroalcoholic extract of pistachio on avoidance learning in male Wistar rats.

Silvestri C, Bacchetta L, Bellincontro A, Cristofori V (2021) Advances in cultivar choice, hazelnut orchard management, and nut storage to enhance product quality and safety: an overview. J Sci Food Agric 101(1):27–43. https://doi.org/10.1002/jsfa.10557

Singh NA, Mandal AKA, Khan ZA (2015) Potential neuroprotective properties of epigallocatechin-3-gallate (EGCG). Nutr J 15:1–17. https://doi.org/10.1186/s12937-016-0179-4

Solfrizzi V, Colacicco AM, D'Introno A, Capurso C, Torres F, Rizzo C et al (2006) Dietary intake of unsaturated fatty acids and age-related cognitive decline: a 8.5- year follow-up of the Italian longitudinal study on aging. Neurobiol Aging 27(11):1694–1704

Srivastava S, Ahmad R, Khare SK (2021) Alzheimer's disease and its treatment by different approaches: a review. Eur J Med Chem 216:113320. https://doi.org/10.1016/j.ejmech.2021.113320

Sun W, Shahrajabian MH, Cheng Q (2019) Epigallocatechin gallate, a unique natural compound with tremendous health benefits. Res Crop Ecophysiol 14(1):52–64

Talebi S, Khodagholi F, Bahaeddin Z, Dezfouli MA, Zeinaddini-Meymand A, Kankam SB, Foolad F, Alijaniha F, Piranghar FF (2024) Does hazelnut consumption affect brain health and function against neurodegenerative diseases? Nutr Neurosci 27(9):1008–1024. https://doi.org/10.1080/1028415X.2023.2296164

Tarawneh R, Holtzman DM (2012) The clinical problem of symptomatic Alzheimer disease and mild cognitive impairment. Cold Spring Harb Perspect Med 2(5):a006148. https://doi.org/10.1101/cshperspect.a006148

Tavakoli J, Hajpour Soq K, Yousefi A, Estakhr P, Dalvi M, Mousavi Khaneghah A (2019) Antioxidant activity of Pistacia atlantica var mutica kernel oil and it's unsaponifiable matters. J Food Sci Technol 56:5336–5345. https://doi.org/10.1007/s13197-019-04004-0

Theodore LE, Kellow NJ, McNeil EA, Close EO, Coad EG, Cardoso BR (2021) Nut consumption for cognitive performance: a systematic review. Adv Nutr 12(3):777–792. https://doi.org/10.1093/advances/nmaa153

Tomaino A, Martorana M, Arcoraci T, Monteleone D, Giovinazzo C, Saija A (2010) Antioxidant activity and phenolic profile of pistachio (*Pistacia vera* L., variety Bronte) seeds and skins. Biochimie 92:1115–1122. https://doi.org/10.1016/j.biochi.2010.03.027

Tsokou A, Georgopoulou K, Melliou E, Magiatis P, Tsitsa E (2007) Composition and enantiomeric analysis of the essential oil of the fruits and the leaves of *Pistacia vera* from Greece. Molecules 12:1233–1239

Vasquez-Rojas WV, Martín D, Miralles B, Recio I, Fornari T, Cano M (2021) P composition of Brazil nut (*Bertholletia excels* HBK), its beverage and by-products: a healthy food and potential source of ingredients. Foods 10:–3007

Voronkov M, Braithwaite SP, Stock JB (2011) Phosphoprotein phosphatase 2A: a novel druggable target for Alzheimer's disease. Future Med Chem 3:821–833. https://doi.org/10.4155/fmc-2023-0039

Wang M, Amakye WK, Guo L, Gong C, Zhao Y, Yao M et al (2019) Walnut-derived peptide PW5 ameliorates cognitive impairments and alters gut microbiota in APP/PS1 transgenic mice. Mol Nutr Food Res 63(18):190032. https://doi.org/10.1002/mnfr.201900326

Wojdyło A, Turkiewicz IP, Tkacz K, Nowicka P, Bobak Ł (2022) Nuts as functional foods: variation of nutritional and phytochemical profiles and their *in vitro* bioactive properties. Food Chemistry: X 15:100418. https://doi.org/10.1016/j.fochx.2022.100418

Wolff RL, Bayard CC (1995) Fatty acid composition of some pine seed oils. J Am Oil Chem Soc 72(9):1043–1046

Xia W, Mo H (2016) Potential of tocotrienols in the prevention and therapy of Alzheimer's disease. J Nutr Biochem 31:1–9. https://doi.org/10.1016/j.jnutbio.2015.10.011

Xu X, Ding Y, Liu M, Zhang X, Wang D, Pan Y, Ren S, Liu X (2024a) Neuroprotective mechanisms of defatted walnut powder against scopolamine-induced Alzheimer's disease in mice revealed through metabolomics and proteomics analyses. J Ethnopharmacol 319(2024):117107. https://doi.org/10.1016/j.jep.2023.117107

Xu X, Song Y, Jiang M, Liu M, Zhang X, Wang D, Pan Y, Ren S, Liu X (2024b) Screening of the active substances for the assessment of walnut kernel in the treatment of scopolamine-induced AD animals. Mol Nutr Food Res 68(3) https://doi.org/10.1002/mnfr.202200816

Yagoub A, Oskay GS, Nazli A (2024) Chestnut 'castanea sativa mill.'honey potential in preventing memory decline on alzheimer's disease model mice: *in vitro* enzyme activities, behavioural, memorial and histopathological studies. https://doi.org/10.21203/rs.3.rs-3948865/v1

Yang J (2009) Brazil nuts and associated health benefits: a review. LWT-Food Sci Technol 42(10):1573–1580. https://doi.org/10.1016/j.lwt.2009.05.019

Youn K, Yun EY, Lee J, Kim JY, Hwang JS, Jeong WS et al (2014) Oleic acid andlinoleic acid from Tenebrio molitor larvae inhibit BACE1 activity *in vitro*:molecular docking studies. J Med Food 17(2):284–289. https://doi.org/10.1089/jmf.2013.2968

Yuan B, Lu M, Eskridge KM, Isom LD, Hanna MA (2018) Extraction, identification, and quantification of antioxidant phenolics from hazelnut (*Corylus avellana* L.) shells. Food Chem 244:7–15

Zhang Z, Sun J, Li Y, Yang K, Wei G, Zhang S (2023) Ameliorative effects of pine nut peptide-zinc chelate (Korean pine) on a mouse model of Alzheimer'sdisease. Exp Gerontol 183:112308. https://doi.org/10.1016/j.exger.2023.112308

Zhao Y, Zhao B (2013) Oxidative stress and the pathogenesis of Alzheimer′ s disease. Oxidative Med Cell Longev 2013(1):316523. https://doi.org/10.1155/2013/316523

Zou J, Cai PS, Xiong CM, Ruan JL (2016) Neuroprotective effect of peptides extracted from walnut (*Juglans sigilata* Dode) proteins on Aβ25-35-induced memory impairment in mice. J Huazhong Univ Sci Technolog Med Sci 36(1):21–30

Role of Omega-3 and Omega-6 Fatty Acids Against Alzheimer's Disease

10

Sayani Banerjee, Nitish Nirala, Pradeep Kumar, and Sugato Banerjee

Abstract

Alzheimer's disease (AD) is a neurodegenerative age-related illness with a complex etiology that includes overlapping environmental and genetic risk factors. There are no available treatments to change the condition, whereas only symptomatic treatments are available in current situations. Omega-3 and omega-6 fatty acids play crucial roles in neuronal maintenance. Neuronal tissues in the brain have a very high lipid composition. The human brain holds about 60% lipids, with a 10–14% level in the grey matter and a proportionally lesser component in the white matter being DHA. Through other pathways, DHA also protects against a number of cognitive risk factors associated with AD. It suppresses a number of signal transduction pathways triggered by Aβ, including two major kinases that phosphorylate the microtubule-associated protein tau and encourage neurofibrillary tangle pathology. It also restricts the production and accumulation of the amyloid β peptide, a major cause of the disease. Arachidonic acid (AA), an omega-6 fatty acid, is vital for proper brain development and function. It supports neuronal growth, protection, and repair. Recent studies focus on the part of optimal omega-3 vs omega-6 ratio in the brain. According to current research, these essential fatty acids could be important in AD pathology neuroprotection. In this chapter, we explored the effect of omega-3 and omega-6 on the perspective of neuroprotection in AD.

S. Banerjee · N. Nirala · P. Kumar · S. Banerjee (✉)
Department of Pharmacology and Toxicology, National institute of Pharmaceutical Education and Research, Kolkata, West Bengal, India
e-mail: sbanerjee@niperkolkata.ac.in

J. Malik et al. (eds.), *Diet and Alzheimer's Disease: Let Food be Our Medicine*, Nutritional Neurosciences, https://doi.org/10.1007/978-981-95-2736-6_10

Keywords

Alzheimer's disease · Omega-3 fatty acids · Omega-6 fatty acids · Docosahexaenoic acid (DHA) · Neuroprotection · Amyloid beta (Aβ)

10.1 Introduction

Alzheimer's disease (AD) is the most common kind of dementia. The estimated 50 million people with AD worldwide are expected to increase to 82 million by 2030 and 152 million by 2050. The clinical manifestations of AD include cognitive impairment and other behavioral abnormalities, along with anatomical abnormalities in the brain. AD is characterized neuropathologically by β-amyloid (Ab) accumulation outside cells that forms senile plaques and intraneuronal aggregation of neurofibrillary tangles. One crucial nutrient class that is essential to all living things is lipids. They are a broad set of macromolecules that have a role in many different cellular processes and tissue functions, both structurally and functionally. Lipids make up 10–12% of the brain's fresh weight and more than 50% of its dry weight, making it the body's second most lipidated organ after adipose tissue. Fatty acids, glycerolipids, phospholipids, sphingolipids, and sterols are the main lipid species found in the brain; phospholipids make up around half of the total lipid content. All lipids with different backbones, except cholesterol, have at least one fatty acid (acyl) chain. By joining a chain of fatty acids, cholesterol can also produce cholesteryl ester (CE) (esterification). Functionally, these lipids are essential parts of cellular membranes, such as those of synapses and the myelin sheath; they may also be employed as bioenergetic fuels in certain situations and transduce signals to control a variety of biological processes. Given the importance of different lipids in brain physiology, it is not unexpected that a growing body of research has identified various intricate mechanisms linking lipid metabolism to pathophysiology's associated with AD. Furthermore, different levels of efficacy against AD pathologies have been demonstrated by the development of lipid-targeting therapies (Youdim et al. 2000).

Vital fatty acids, namely α-linolenic acid (ALA) and linoleic acid (LA), are the source of ω-3 and ω-6 polyunsaturated fatty acids, respectively. A sizable amount of translational data has surfaced in the last 30 years linking polyunsaturated fatty acids (PUFA) to healthy brain development and function. The PUFA family includes omega-3 (n-3) and omega-6 (n-6) fatty acids. Since animals are unable to biosynthesize α-linolenic acid (ALA, 18:3n-3) and linoleic acid (18:2n-6), the precursors of n-3 and n-6 PUFA, respectively, they are regarded as "essential" and must be obtained from food. The n-3 PUFA docosahexaenoic acid (DHA, 22:6n-3) and the n-6 PUFA arachidonic acid (AA, 20:4n-6) are the two most prevalent PUFAs in the gray matter of mammalian brains. These fatty acids directly affect blood flow, immunological reactions, and neurotransmitter systems, as well as brain function (Singh 2005). As fatty acids directly affect the fluidity and makeup of membranes, they also play a critical role in cell membrane function. By lowering cholesterol levels in neuronal membranes, for instance, EFAs improve membrane fluidity, assisting cells in maintaining regular functions and lessening their vulnerability to harm (Fernstrom 1999). ALA, eicosapentaenoic acid, and docosapentaenoic acid are n-3 PUFA precursors of DHA that pass the blood-brain barrier. Still, because of fast oxidation, they make up less than 1% of the overall fatty acid makeup of the brain. Unesterified DHA easily diffuses from plasma into the brain and preferentially builds up in the mitochondrial membrane and synaptosomal gray matter. While dietary n-6 PUFA insufficiency results in cortical AA deficits and related drops in the AA/DHA ratio, dietary n-3 PUFA insufficiency, especially in early development, is linked to substantial cortical DHA-associated deficits and increases in the AA/DHA ratio in mice. DHA and AA have opposing impacts on inflammatory signaling pathways and synaptic signal transduction (Simopoulos 2006).

n-3 PUFAs are essential for brain growth and function, enhancing mental and physical skills development. They improve GLUT1-mediated glucose absorption and G protein-coupled signaling, which promotes hypothalamic control and lowers the risk of diseases, including diabetes and Alzheimer's (Pifferi et al. 2020). Furthermore, by strengthening SNARE protein complexes and increasing the production of lipid-transporting vesicles, n-3 PUFAs facilitate effective neurotransmitter release. By minimizing oligodendrocyte damage and axonal demyelination, they help protect the central nervous system and lessen the likelihood of multiple sclerosis (MS) (Okada et al. 2018). Additionally, n-3 PUFAs are building blocks for bioactive lipid mediators that defend against inflammation, ischemia, and damage, including protectins, resolvins, and maresins. Their benefits underscore their important role in neuroprotection and cognitive health. They may also help reduce symptoms of developmental coordination disorder (DCD) and attention-deficit hyperactivity disorder (ADHD) (Herrera Vielma et al. 2021). n-3 PUFAs are believed to exert antidepressant effects through multiple mechanisms (Bazinet and Layé 2014). Among these are the decrease of neuroinflammation, linked to the pathophysiology of depression, and the modification of the hypothalamic-pituitary-adrenal (HPA) axis, which is crucial for mood regulation and stress response. Furthermore, endocannabinoid metabolism, which impacts physiological functions including hunger, pain perception, mood, and memory, is influenced by n-3 PUFAs. When taken as a whole, these pathways demonstrate how n-3 PUFAs might help manage depressive illnesses and promote mental wellness (Sharan and Vellapandian 2024). Omega-3 polyunsaturated fatty acids (PUFAs) are components of the plasma membrane. They are involved in a number of processes, including increased synaptic development and functionality, effects on synaptic integrity and plasticity, and neuroplasticity.

Scientific data about the potential effectiveness of PUFA supplementation in neurodegenerative diseases, including Parkinson's disease (PD) and Alzheimer's disease (AD), is growing.

Dietary suggestions may be able to reduce some of the symptoms or halt the physical and cognitive deterioration of AD or PD, but they are by no means a cure. The two most significant PUFAs in the brain are omega-6, which includes linoleic acid, and arachidonic acid, which includes DHA and ARA. Omega-3 PUFAs include -linolenic acid (ALA, 18:3!-3), eicosapentaenoic acid (EPA, 20:5!-3), and docosahexaenoic acid

(DHA, 22:6!-3). These are part of the omega-6 PUFAs, which include linoleic acid (LA, 18:2!-6) and arachidonic acid (ARA, 20:4!-6). In the brain, DHA and ARA are the most significant PUFAs. Specifically, DHA accounts for 10–20% of total lipids and more than 90% of 3 PUFAs in the brain (Covington 2004). Primarily found in phosphatidylethanolamine and phosphatidylserine, it is also found in trace levels in phosphatidylcholine at endoplasmic reticula, mitochondria, and synaptic terminals. DHA can alter physiological functions and cellular characteristics, including membrane fluidity, neurotransmitter release, gene expression, myelination, neuroinflammation, and neuronal development (Kousparou et al. 2023a).

In this chapter, we explored the mechanism and function of PUFAs and their active part in neuroprotection. In this chapter, we highlighted some vital mechanistic problems related to AD and how ω3 and ω6 fatty acids are critical in acting as neuroprotective agents.

10.2 Polyunsaturated Fatty Acids (PUFAs) in the Brain: Synthesis, Metabolism Function, and Molecular Pathways

Fatty acids are carbon chains with a carboxyl group at one end and a methyl group (often called omega, ω) at the other. The first carbon after the carboxyl group is the α-carbon, and the one after that is the β-carbon. The symbol "n" is often used instead of "ω" to indicate the position of the first double bond relative to the methyl end. The position of the double bonds relative to the carboxyl group is included in the systematic nomenclature of fatty acids. This approach is used to categorize and characterize the structures of many naturally occurring fatty acid types, which are essential for many physiological functions, such as signaling, energy storage, and cell membrane structure (Müller et al. 2001; Rustan and Drevon 2005). The hydrocarbon chains of fatty acids, which are carboxylic acids, have 4–32 carbons (C4–C36). Fatty acids are hydrophobic hydrocarbon chains with a carboxyl group (-COOH) at the end, with a pKa of approximately 4.8. At physiological pH, this carboxyl group ionizes to form $-COO^-$, giving fatty acids amphipathic properties with a hydrophilic head and hydrophobic tail (Raghavendra 2024). The hydrophobic components of long-chain fatty acids (LCFAs) are very nonpolar, so they are carried in the bloodstream in association with proteins. Fatty acid esters, including triacylglycerols, cholesteryl esters, and phospholipids, constitute about 90% of the fatty acids in plasma and are found in circulating lipoprotein particles. In contrast, albumin-bound unesterified (free) fatty acids are transported throughout the circulation (Schaffer and Lodish 1995).

Alpha-linolenic acid (ALA) is the primary fatty acid of the n-3 polyunsaturated fatty acid (PUFA) class. Eicosapentaenoic acid (EPA), docosahexaenoic acid (DHA), and, to a lesser extent, docosapentaenoic acid are the major metabolic products of ALA (Padley 1994). However, EPA and DHA are primarily obtained from the diet, such as from fish oils, non-oily fish liver, and oily fish (such as sardines, mackerel, salmon, trout, and fresh tuna). Because of the restricted capacity of

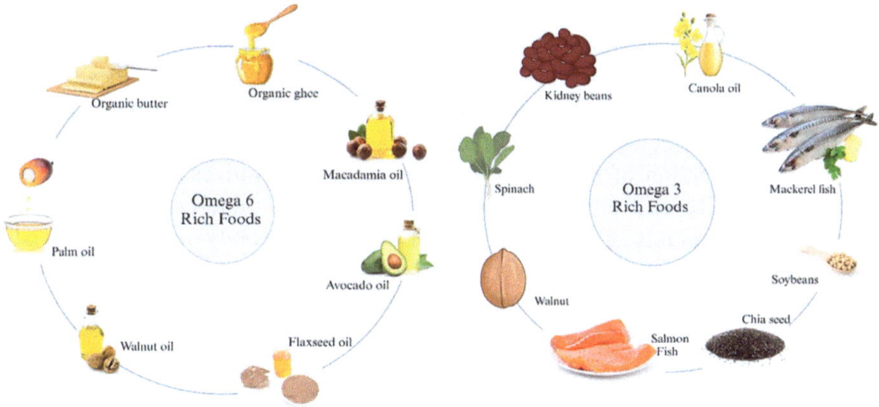

Fig. 10.1 Dietary sources of Omega 3 and Omega 6 fatty acids

humans to convert ALA into these long-chain metabolites, dietary intake is essential. Rich in the retina and other neurological tissues, DHA is crucial for signal transduction and may even control G-protein signaling (Covington 2004). The n-6 polyunsaturated fatty acid (PUFA) family is derived from linoleic acid (LA), its parent fatty acid. Its metabolites include γ-linolenic acid, dihomo-γ-linolenic acid, arachidonic acid, and, to a lesser extent adrenic acid and docosapentaenoic acid (Tvrzicka et al. 2011).

According to the length of their carbon chains, polyunsaturated fatty acids (PUFAs) may be divided into two groups: long-chain PUFAs, which have 20–24 carbon atoms, and very long-chain PUFAs, which have 26 or more carbon atoms. The "n-minus" notation is frequently used by scientists, such as nutritionists and biochemists, to characterize naturally occurring cis-unsaturated fatty acids. This scheme shows the initial double bond's location concerning the molecule's methyl end. Alpha-linolenic acid (ALA), for example, is denoted by the formula C18:3n-3, which shows that the methyl end of ALA is three carbon atoms away from the first double bond. However, the precise positions and arrangements of the remaining double bonds in the molecular structure are not specified by this naming scheme (Ratnayake and Galli 2009). Cis-unsaturated fatty acids are divided into three classes according to this categorization system: n-3 (omega-3), n-6 (omega-6), and n-9 (omega-9). Meat and other food sources naturally include the chemical structures of n-3 and n-6 PUFAs (Fig. 10.1).

10.2.1 Synthesis of Omega-3 and Omega-6 Fatty Acids

The precursor fatty acid for the n-3 (omega-3) family is alpha-linolenic acid (ALA), whereas the precursor for the n-6 (omega-6) family is linoleic acid (LA). Neither humans nor animals can synthesize ALA, one of the four most prevalent n-3 PUFAs present in animal tissues, along with eicosapentaenoic acid (EPA), docosapentaenoic acid (DPA), and docosahexaenoic acid (DHA). It is, therefore, categorized as an essential fatty acid. Animals and humans must get vital fatty acids from their diets since only plants can create them. The body then uses these important fatty acids to create longer-chain derivatives like arachidonic acid (AA), EPA, and DHA (Sinclair et al. 2002). Both human and animal research have shown that the body's conversion of ALA into its longer-chain derivatives is inefficient and ineffective, and several biological factors regulate it. It's interesting to note that dietary intake amounts of ALA and LA, rather than the body's metabolic needs, control the efficiency of this conversion. Because both fatty acids depend on the same enzyme pathways for desaturation and elongation, their competitive connection is the root cause (de Gŏmez Dumm and Brenner 1975; Emken et al. 1990).

A large dietary intake of ALA is required to ensure adequate levels of EPA and DHA in circulation and tissues since ALA is not converted into its longer-chain derivatives. Eating nutritious wild plant leaves or vegetable oils containing ALA is advised for people and animals to achieve it. Alternatively, one can avoid the desaturation and elongation processes necessary for ALA conversion by consuming diets based on marine sources, which naturally contain high quantities of EPA and DHA (Simopoulos 1999, 2004; Gibson et al. 2013). According to research, three desaturase enzymes are involved in the synthesis of 22-carbon long-chain PUFAs, including DHA (n-3) and DPAn-6 (n-6), from ALA and LA in both human and animal tissues. According to some research, the enzyme delta-4 desaturase is not involved in the conversion of ALA to DHA or LA to DPAn-6. Instead, the procedure entails a second elongation step using delta-6 desaturase, which is followed by beta-oxidation in the peripheral and circulatory tissues, producing DHA and DPAn-6 PUFAs (Gibson et al. 2013).

In both human and animal muscle tissues, phospholipids are the main source of long-chain polyunsaturated fatty acids (LCPUFAs), which are essential for the body's metabolic, functional, and physiological processes and those of organelles and tissues. Studies on animals have demonstrated that altering the dietary balance of ALA and LA can affect the amounts of n-3 and n-6 LCPUFAs in tissues. Through a sequence of desaturase and elongase-catalyzed reactions, mammals may transform ALA into LCPUFAs such as EPA, DPA, and DHA (Sprecher 2000). Unlike LA, fatty acid desaturase 1 (FADS1) and fatty acid desaturase 2 (FADS2) exhibit a greater affinity for ALA. Nevertheless, a diet rich in LA, as that of animals grown in feedlots or grain, can prevent the desaturation of ALA and 24:5n-3, a precursor of 24:6n-3, the direct precursor of DHA. ALA concentrations in tissue phospholipids and plasma are often less than 0.5%. Whether this quantity is adequate for FADS2 to successfully compete with LA, which is often more abundant in animal tissues, is yet unknown (Burdge and Calder 2005). Males metabolize ALA into EPA and DHA less effectively than females, according to studies utilizing isotope-labelled ALA. While the conversion to DHA is around 9% in females and almost 0% in men, the estimated net conversion rate of ALA to EPA is almost 21% in ladies and 8% in males. Differences in EPA and DPA levels have also been observed, with females showing higher erythrocyte phospholipid EPA but lower adipose tissue EPA and plasma DPA than men.

Furthermore, research on humans has shown a notable sex-based difference in how the body reacts to higher dietary ALA consumption. Females exhibited a much higher increase in plasma phospholipid EPA levels than men following 6 months of an ALA-rich diet (Childs et al. 2014). A comprehensive examination of genetically varied sheep grown in different production settings revealed a tiny gender variation in health-promoting fatty acid levels, with females having slightly higher levels of EPA and DHA than males. Since series-3 eicosanoids are involved in ovulation, conception, and pregnancy, female lambs may create more n-3 PUFAs as they reach reproductive maturity. Furthermore, when the father breed was Poll Dorset, lambs born to Merino dams had around 2 mg/100 g higher EPA and DHA than lambs from crossbred mothers. The PUFA-to-saturated fatty acid (SFA) ratio in meat grew gradually from second-cross Poll Dorset to first-cross Poll Dorset and then to purebred Merino, according to studies by Ponnampalam et al. Instead of a decrease in SFA levels, this increase was ascribed to an increase in PUFA content (Ponnampalam et al. 2009). Genetic factors contribute to individual differences in the conversion of LCPUFA precursors into their respective long-chain products, according to metabolic studies using stable isotope labeling, genome-wide association studies (GWAS), candidate gene single nucleotide polymorphisms (SNPs), and metabolomics. The enzymes that insert double bonds at particular points along the fatty acid chain are encoded by the fatty acid desaturase genes (FADS1 and FADS2). FADS2 is a delta-6, delta-8, and delta-4 desaturase that is selective for various fatty acid substrates, whereas FADS1 is a delta-5 desaturase (Illig et al. 2010; Park et al. 2009). FADS1 expression was considerably lower in people with the minor homozygous genotype (D/D) than in those with the major homozygous genotype (I/I). The idea that people with the D/D genotype have a poorer metabolic capacity for converting fatty acid precursors into long-chain polyunsaturated fatty acids (LCPUFAs) than those with the I/I genotype is supported by the fact that arachidonic acid (ARA) is a direct result of FADS1. According to studies, people with the I/I genotype may be more vulnerable to pro-inflammatory diseases because they can convert precursors into long-chain PUFAs. Since ARA production and the synthesis of pro-inflammatory lipoxygenase-derived metabolites have been related to differences in FADS gene single nucleotide polymorphisms (SNPs) because they efficiently convert linoleic acid (LA) into ARA.

10.2.2 Pathways of Omega-6 and Omega-3 Metabolism in the Brain

Although linoleic acid (LA) serves as a precursor fatty acid for the formation of the bioactive omega-6 PUFA, arachidonic acid, alpha-linolenic acid (ALNA) is a precursor to the bioactive omega-3 PUFAs, eicosapentaenoic acid (EPA) and docosahexaenoic acid (DHA). Both precursors rely on the same enzymes and undergo a series of desaturations and extensions to form long-chain derivatives (Saini and Keum 2018). The conversion rate of ALNA to EPA and DHA in humans is relatively low and may be further affected by a high linoleic acid diet, as it blocks the

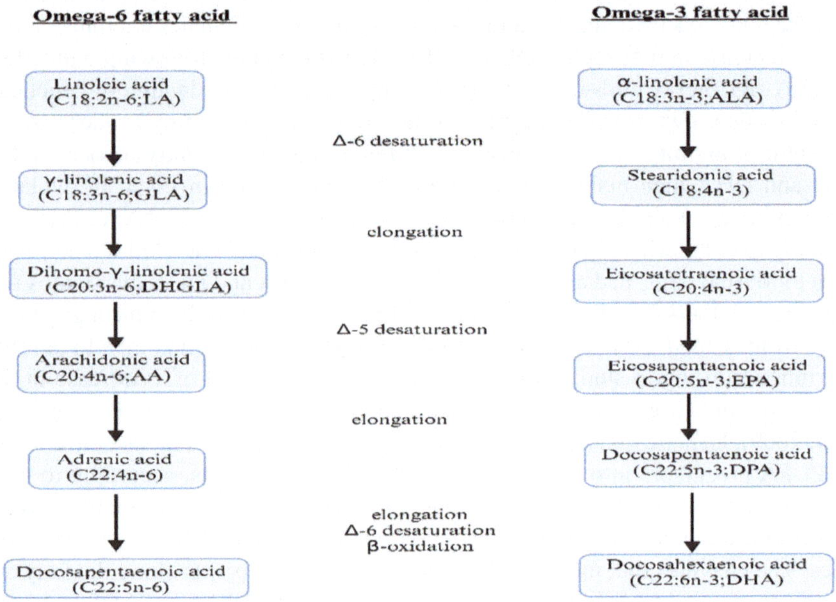

Fig. 10.2 The omega-3 and omega-6 metabolism pathways

activity of the delta-6-desaturase enzyme (Smink et al. 2012). Both eicosanoids are formed using AA and EPA/DHA as substrates. Because AA-derived eicosanoids are often more potent than EPA-derived ones, they affect the inflammatory, immunological, and circulatory systems differently. High dietary intake of EPA/DHA may remove arachidonic acid (AA) from cell membranes, causing a shift toward less effective eicosanoids of the omega-3 family. However, it is also possible that the type of PUFA present in the cell affects other cellular functions, such as the expression of regulatory protein genes unrelated to altered eicosanoid status or the phospholipid components of cell membranes (Calder 2013). Recent studies have found that a family of peroxisome proliferator-activating receptors (PPARs) function as nuclear transcription factors. These receptors can control the cellular location where fatty acids can be produced to regulate gene expression of important proteins related to lipid metabolism. PUFAs and their metabolites act as ligands for this receptor family, making this process possible (Christofides et al. 2021) (Fig. 10.2).

10.2.3 Function

High levels of DHA, arachidonic acid (AA), and palmitate but low levels of other omega-3 PUFAs, especially eicosapentaenoic acid (EPA), characterize the brain's unique fatty acid profile (Brenna and Diau 2007). EPA levels in the brain are significantly lower, 250–300 times less than DHA levels. In brain phospholipids, EPA and

AA are mainly esterified to phosphatidylinositol (PI), probably because of their comparable chain lengths. In contrast, DHA and docosapentaenoic acid (DPA) are primarily abundant in phosphatidylethanolamine (PE) and phosphatidylserine (PS) (Chen et al. 2009; Lee and Hajra 1991). Because the brain's endogenous synthesis of EPA, DPA, and DHA is restricted, it mostly depends on plasma-derived fatty acids to maintain its PUFA levels. Simple diffusion allows both EPA and DHA to pass across the blood-brain barrier, and DPA most likely does the same. Mechanisms such as β-oxidation, decreased absorption into phospholipids, elongation to DPA, and decreased recycling into membrane phospholipids cause EPA levels to stay low in the brain despite comparable uptake rates (Weiser et al. 2016; Ouellet et al. 2009). Despite the high energy cost of its production and the possibility of peroxidation, the enrichment of DHA in brain phospholipids is maintained across species. It seems to satisfy a particular demand for DHA in neuronal membranes. Although the exact causes of its preferred enrichment over EPA and DPA are still unknown, this points to DHA's unique importance in sustaining essential brain processes (Crawford et al. 1976).

Numerous chemical processes underlie the beneficial effects of DHA, which is essential for brain growth and neural cell survival. DHA (25–50 μM) stimulates the CREB pathway, improving NPC survival at different phases of differentiation, according to studies on neural progenitor cells (NPCs). In contrast, OLA did not exhibit this activation. Additionally, DHA protects neural cells, including rat cortical neurons and NGF-differentiated PC12 cells, from oxidative stress-induced apoptosis and necroptosis by inhibiting caspase, regulating the NRF2 antioxidant pathway, activating Akt, and accumulating phosphatidylserine (PS). DHA also helps non-neuronal CNS cells (Kim et al. 2022). Busting antioxidant protein production through NRF2 activation shields Schwann cells (SCs) from oxidative damage. It restores the survival of SCs that PA has damaged through Akt phosphorylation and the mTOR pathway. Conversely, high DHA concentrations (100–200 μM) harm microglial cell survival, causing morphological abnormalities and cell death. Although little information is available on astrocytes, DHA enhances neuronal survival in astrocyte-neuron co-cultures and encourages morphological development in these cells (Petermann et al. 2022).

Effective concentrations of DHA (1–30 μM) improved total neurite length, branching, and axonal growth, while other fatty acids, such as oleic acid and arachidonic acid, showed no similar effects (Kasthuri et al. 2015). Additionally, DHA-derived metabolite DEA has been identified as a potent endogenous modulator of neurite growth, acting through the GPCR GPR110. In vivo studies also confirm that diets deficient in ω-3 fatty acids, including DHA, reduce neurite lengths, highlighting DHA's critical role in maintaining neuronal structure (Calderon and Kim 2004).

FFAs and DHA are essential for synapse formation and maturation, particularly during synaptic development, where pre- and postsynaptic components are organized, and signaling pathways are orchestrated (Bazinet and Layé 2014; Innis 2008). In dendrites and axons, DHA is involved in remodeling synaptic membrane phospholipids. Studies have shown that DHA supplementation increases the expression of pre- and postsynaptic proteins such as PSD-95 and synapsin-1 as well as the levels of membrane phospholipids such as phosphatidylserine and phosphatidylinositol (Sakamoto et al. 2007). In hippocampal neurons, DHA boosts the expression of postsynaptic proteins associated with glutamatergic synapses (e.g., GluR1, GluR2, NR1, NR2A, NR2B) and increases the density and size of markers for pre- and postsynaptic sites, such as Bassoon and Homer proteins (Cao et al. 2009). These changes improve the alignment of synaptic specializations, forming more functional synapses (Carbone et al. 2020).

Essential fatty acids (EFAs) are essential for brain formation and function, especially arachidonic acid (AA) and DHA, which are particularly important during brain development. DHA is essential for developing brain structures and manufacturing neurotransmitters, whereas AA promotes brain growth in fetuses and newborns (Singh 2005). The body can convert ALA to DHA but does so inefficiently; hence, dietary DHA is necessary. As messengers, EFAs also affect the immunological response, neurotransmitter function, and signaling cascades that may encourage neuroprotection or damage. DHA-derived neuroprotectin D1 (NPD1), for instance, protects neuronal cells, fights oxidative stress, and lowers inflammation. EFA-derived prostaglandins and leukotrienes control inflammation, while omega-3 FAs such as EPA and DHA inhibit the synthesis of pro-inflammatory messengers from AA. These processes demonstrate how crucial EFAs are for preserving brain function and preventing neurological conditions like schizophrenia and Alzheimer's (Sun et al. 2018).

The functions of omega-3 and omega-6 are also vital in neuronal maintenance. Fatty acids are crucial in brain function as neuronal tissue has a very high lipid composition. The human brain holds about 60% lipids, with a 10–14% level in the grey matter and a proportionally lesser component in the white matter, DHA (Salem Jr et al. 2001; Innis 2009). About half of the total fatty acids in the phospholipids found in neuronal membranes comprise DHA and arachidonic acid (Salem Jr et al. 2001). The major functions of DHA include the size of neurons, neurogenesis, neurite growth, synapse formation and function, neuronal integrity, gene expression in the brain, glucose transport, cognitive development, and learning ability. It can easily cross the blood-brain barrier (Lassek and Gaulin 2011). A-linolenic acid, docosapentaenoic acid (n-3), and EPA make up just around 0.1% of the total fatty acids in the brain (Lin et al. 2010).

n-3 PUFAs are essential for brain growth and function, enhancing mental and physical skills development. They improve GLUT1-mediated glucose absorption and G protein-coupled signaling, which promotes hypothalamic control and lowers the risk of diseases, including diabetes and Alzheimer's (Pifferi et al. 2020). Furthermore, by strengthening SNARE protein complexes and increasing the production of lipid-transporting vesicles, n-3 PUFAs facilitate effective neurotransmitter release. By minimizing oligodendrocyte damage and axonal demyelination, they help protect the central nervous system and lessen the likelihood of multiple sclerosis (MS) (Okada et al. 2018). Additionally, n-3 PUFAs are building blocks for bioactive lipid mediators that defend against inflammation, ischemia, and damage, including protectins, resolvins, and maresins. Their benefits underscore their important role in neuroprotection and cognitive health. They may also help reduce symptoms of developmental coordination disorder (DCD) and attention-deficit hyperactivity disorder (ADHD) (Herrera Vielma et al. 2021). n-3 PUFAs are believed to exert antidepressant effects through multiple mechanisms (Bazinet and Layé 2014). Among these are the decrease of neuroinflammation, which is frequently linked to the pathophysiology of depression, and the modification of the hypothalamic-pituitary-adrenal (HPA) axis, which is crucial for mood regulation and stress response. Furthermore, endocannabinoid metabolism, which impacts physiological functions including hunger, pain perception, mood, and memory, is influenced by n-3 PUFAs. When taken as a whole, these pathways demonstrate how n-3 PUFAs might help manage depressive illnesses and promote mental wellness (Sharan and Vellapandian 2024).

10.2.4 Molecular Pathways of PUFAs in CNS

Numerous signal transduction pathways that function in neuronal membranes and, consequently, in the synaptic cleft can be modulated by PUFAs. Numerous neurotransmitters, including acetylcholine, catecholamines, and serotonin (Frazer and Hensler 1999; García-Sáinz et al. 1999; Popoli et al. 2000), interact with a class of heptahelical transmembrane receptors. G proteins transduce these receptors' signals; Salem and others have lately examined DHA's function in this process (Salem Jr et al. 2001). AC powers the cAMP messenger system. Alpha-2 and beta-adrenergic (noradrenaline and adrenaline) receptors, 5-HT1 (serotonin) receptors, and both D1 and D2 (dopamine) receptors employ this route. PUFAs have two ways to affect this pathway: they can raise both AC (Murphy 1985; Nicolas et al. 1991) and PKA activity. Conversely, PLC initiates the phosphoinositide signaling pathway, where PUFAs can influence the activity of both PLC and PKC, which take part in 5-HT2 and alpha-1 adrenergic transmission (McPhail et al. 1984). A2 and D, the other two membrane phospholipases, are likewise impacted by PUFAs (Lister et al. 1988; Shukla and Halenda 1991; Downes and Currie 1998) and play an important role in neurotransmission. Dopamine D2 receptors can activate PLA2, serotonin 5-HT2 receptors (Berg et al. 1996), muscarinic acetylcholine receptors, and glutamate. Fatty acids, liberated from the sn-2 position of phospholipids by PLA2, become the building blocks of leukotrienes, thromboxanes, lipoxins, and prostaglandins. These eicosanoids exert numerous effects on signal transduction. This difference in the type of signaling may be the reason for the different effects that these prostaglandins produce, as the prostaglandins of each family increase the level of cAMP through receptor activation of a Gs protein.

In contrast, the prostaglandins of the three other families activate a Gi protein that oppositely depresses cAMP formation. Prostaglandins of the 1 class activate phosphoinositide signaling. Ion channels, for example, Ca^{2+} and Na+, can also be regulated by PUFAs (Xiao et al. 1997; Wieland et al. 1992). Activation of Ca^{2+}-CM-PKs ushered in further events in neurotransmission and the eventual release of neurotransmitters from synaptic vesicles; PUFA effects have also been noted. Ca-ATPase in neuronal membranes maintains the thousandfold concentration gradient between extra- and intra-cellular Ca^{2+} (Wieland et al. 1992).

10.3 Effect of Omega-3 and Omega-6 in Brain and AD

10.3.1 Role of ω3 and ω6 as a Glial Cell Protector in AD

Memory loss, dementia, and cognitive impairment are the main clinical features of Alzheimer's disease (AD), which is considered one of the most complicated neurological illnesses. In addition to the large amount of research focused on neurons, there is growing evidence that glial cells—specifically, astrocytes, microglia, NG2 glia, and oligodendrocytes—may be crucial to the pathophysiology of this illness. Glial cells use growth factor production and synaptic plasticity modification to sustain and shield neurons from oxidative damage. The breakdown of Aβ in glial cells is mediated by a number of enzyme systems that have been identified. Important Aβ-degrading enzymes that lower Aβ accumulation in AD brains include neprilysin (NEP) and insulin-degrading enzyme (IDE). Furthermore, it has been demonstrated that Aβ accumulation, neurotoxicity, and inhibition of Aβ clearance are all caused by glial cell dysfunction. Furthermore, exogenous Aβ may cause glial cells to produce reactive oxygen species (ROS) and reactive nitrogen species (RNS), which in turn may contribute to the course of AD by promoting inflammation through the release of pro-inflammatory cytokines (Garcia Corrales et al. 2021). Numerous studies have shown how omega-3 polyunsaturated fatty acids (PUFAs) affect brain function and development. Nonetheless, most studies concentrated on the neuroprotective properties of docosahexaenoic acid (DHA) or eicosapentaenoic acid (EPA). Omega-3 fatty acids generated from plants have garnered attention recently because of their potential health benefits. It has been demonstrated that alpha-linolenic acid (ALA), which is concentrated in vegetable oils like perilla oil (which contains 60–70% ALA) or linseed oil (30–50% ALA), improves neurological function in the mouse brain, including neurogenesis and synapse function. In a mouse model of AD, treating with perilla oil and ALA ameliorated the cognitive impairment caused by Aβ25–35. Perilla oil, which has the highest amount of ALA, inhibited the production of inflammatory mediators in brain regions injected with Aβ25–35, including cyclooxygenase-2 (COX-2) and inducible nitric oxide synthase (iNOS). Additionally, supplementing with ALA increased the production of the enzyme IDE, which breaks down Aβ. Nevertheless, no research has documented ALA's neuroprotective capacity to alter glial cell function in response to Aβ. Using Aβ25–35-treated C6 glial cells, we evaluated the anti-inflammatory effects of ALA and examined the molecular pathways that govern the control of inflammatory response and Aβ breakdown (Lee et al. 2018).

The proper physiological operation of the brain depends on FA production. Several neurological disorders, such as multiple sclerosis (MS), Parkinson's disease (PD), and Alzheimer's disease (AD), have been linked to alterations in its metabolism. More and more evidence connects lipid metabolism to the pathogenesis of AD. Up to 70% of dementia cases are attributed to AD, which is the most common neurodegenerative disease. Two hallmarks of AD disease include Aβ plaques found in the brain's extracellular space and neurofibrillary tangles found within neurons (Murphy and LeVine 3rd 2010). The primary lipoprotein in the brain, ApoE, is isoform-dependently involved in the synthesis, aggregation, and clearance of Aβ (Wang et al. 2017). There are three main isoforms of ApoE: ApoE2, ApoE3, and ApoE4. These are expressed mainly in astrocytes and microglia. The most prevalent genetic risk factor for AD is identified to be ApoE4. The astrocytes' FA metabolism is suppressed through lipid droplet formation, reduced FA breakdown, and increased de novo FA synthesis through promoting ApoE4 (Qi et al. 2021). Patients with Alzheimer's disease (AD) have higher levels of arachidonic acid (ARA), whereas levels of linoleic acid (LA) are decreased compared with healthy controls (Zárate et al. 2017). As mentioned earlier, astrocytes are the CNS's primary source of arachidonic acid (ARA). Cascade activation of ARA increases Aβ levels, ultimately leading to functional memory impairment due to IL-1β. Transformation to a pro-inflammatory disease-associated microglia phenotype and decreased Aβ clearance are linked with ApoE4 in microglia (Norwitz et al. 2021). These pro-inflammatory microglia increase the production of fatty acids (FAs) and saturated fatty acids (SFAs). Stearic acid (C18:0) levels are significantly decreased in the frontal and temporal cortex of Alzheimer's disease (AD) patients; however, oleic acid and palmitic acid (C16:0) levels are increased in the frontal, temporal, and parietal brain. Additionally, a significant increase in monounsaturated fatty acids (MUFA) has been observed in the brain tissue of AD patients, which is directly linked to cognitive impairment and increased transcription of SCD1 (Astarita et al. 2011). SCD1 overexpression has been related to enhanced Aβ synthesis; therefore, its activity might also be diminished for the decrease in Aβ. Moreover, astrocytes in AD plaques generate higher amounts of stearic and palmitic acid, increasing β-secretase and hyperphosphorylates tau. The NPD1 and DHA levels are decreased in the AD brain. NPD1 decreases Aβ shedding through modulation of the activities of A- and b-secretases (Shah et al. 2013). Additionally, DHA can inhibit the release of Aβ peptides and is a precursor to NPD1. DHA and NPD1 increase the expression of anti-inflammatory and anti-apoptotic genes, especially those that encode Bcl-2 proteins (Lukiw et al. 2005). There is consistency in the importance of 3 PUFAs in cognitive function and loss of DHA in fractions from CSF in AD. Abnormality in the production of phosphatidic acid by phospholipase D has also been linked to AD. As AD and FA metabolism are so closely related, de novo FA synthesis may be an innovative way to modulate the pathogenesis of AD (Tanguy et al. 2019).

10.3.2 Role of DHA as a Neuroprotective Agent in AD

The main n-3 LC-PUFA in the mammalian central nervous system is DHA, which makes up around 5 g of the human brain and 15–20% of all fatty acids in the membrane lipids of the grey matter (Innis 2008). It is also prevalent in the retina's visual components. The high concentration of DHA in synaptic signaling areas, including synaptic vesicles and synaptosomal plasma membranes, underlines its crucial role in brain function and has been conserved throughout evolution. DHA's many double bonds provide great structural flexibility, supporting its many biological roles, including cognitive ones (Von Kriegstein et al. 1999). The insertion of DHA into phospholipids modifies several fundamental properties of membranes, including acyl chain organization, fluidity, elastic compression, permeability, phase behavior, speed of fusion, and flippase activity. This may affect brain functions caused by changes in membrane fatty acid composition, such as the formation of axons and dendrites, cell shape, formation of lipid rafts, G protein-coupled signaling (which regulates gene expression), neuronal polarity, plasticity, dopamine storage, vesicle formation and transport, glucose absorption, and hypothalamic control. Such changes have been linked to neurodegenerative and psychiatric disorders, like severe depression, Parkinson's disease, where there is a deterioration in motor systems, and Alzheimer's disease, where the instability in axons and dendrites characterizes (Stillwell and Wassall 2003). Beyond its direct role in brain function, DHA also gives rise to bioactive mediators such as neuroprotectins, resolvins, and maresins, which are crucial for protecting the brain against injury, ischemia, and inflammation (Serhan and Levy 2018).

DHA is a crucial molecule in cell signaling, particularly in the brain. It enhances brain glucose uptake by modulating GLUT-1 and Na+/K+ ATPase and acts as a precursor for pro-resolving mediators like protectins and resolvins, aiding in inflammation resolution. DHA-derived synaptomide influences neural signaling. DHA activates GPR40/FFAR1 receptors in the CNS, regulating pain, cognition, emotional behavior, and insulin secretion. Although DHA's effects are well understood, its molecular mechanisms in neuroprotection and neurotrophic actions remain unclear. Research into these pathways could lead to innovative therapies for CNS diseases and tumorigenesis (Petermann et al. 2022).

10.3.3 Effect of DHA in Amyloid-Beta Processing

Docosahexaenoic acid (DHA), an essential omega-3 polyunsaturated fatty acid (PUFA), has gained widespread interest for its potential role in the prevention of Alzheimer's disease (AD). It may impact this neurodegenerative disease through its ability to modulate the processing of amyloid-beta (Aβ) (Cole et al. 2009). DHA is essential for brain function and is primarily found in marine fish and algae. It supports neural transmission, plasticity, and membrane fluidity, and makes up around 60% of the fatty acids in neuronal membranes. It mainly targets synaptic membranes (Bradbury 2011). In AD patients, DHA levels are notably lower in the brain and serum, especially in critical areas, including the pons, frontal cortex, and hippocampal regions. These decreases are highly correlated with both higher Aβ deposition and cognitive decline, underscoring the critical role that DHA plays in preserving mental health as well as its potential as an AD preventative and treatment agent (Xiao et al. 2022). The buildup of Aβ peptides, pathogenic particles produced from the amyloid precursor protein (APP) by amyloidogenic processing, is a hallmark of Alzheimer's disease. In this process, β-secretase (BACE1) and γ-secretase sequentially cleave APP, leading to the formation of Aβ peptides, particularly Aβ42. These peptides bind together to form neurotoxic plaques. These plaques cause gradual cognitive deterioration by interfering with neural function (O'Brien and Wong

2011). By modifying APP processing, lowering amyloidogenic activity, and encouraging non-amyloidogenic pathways, DHA has neuroprotective benefits. DHA dramatically reduces the formation of Aβ by hindering the enzymatic activity of β- and γ-secretases.

Research shows that DHA has a dose-dependent influence on γ-secretase activity, reducing it more substantially than β-secretase activity. Remarkably, these decreases take place without changing the expression or protein levels of BACE1 and presenilin 1 (PS1), the catalytic component of γ-secretase, suggesting a direct modification of enzymatic function as opposed to an effect on protein stability or synthesis (Grimm et al. 2011). DHA's impact on cholesterol metabolism is a key mechanism behind its effects on Aβ processing. Amyloidogenic APP processing takes place on platforms made of lipid rafts, high in cholesterol and found in neuronal membranes. DHA transfers cholesterol from the raft to non-raft domains by redistributing it throughout the plasma membrane. This redistribution hinders amyloidogenic processing by upsetting the APP, BACE1, and PS1 clustering inside lipid rafts (Grimm et al. 2013). DHA also lowers cholesterol levels in neuronal membranes by inhibiting HMG-CoA reductase, the enzyme that limits the rate at which cholesterol is biosynthesized. Such alterations reduce the effectiveness of γ-secretase activity and restrict the synthesis of Aβ. Additionally, DHA modifies PS1's subcellular location, decreasing its raft domain abundance and further inhibiting amyloidogenic processing (Grimm et al. 2011).

In addition to its effects on amyloidogenic pathways, DHA stimulates neuroprotective non-amyloidogenic APP processing. This route involves α-secretase cleaving APP to produce soluble APPα (sAPPα), which supports neuronal survival and synaptic function. By stabilizing ADAM17, a crucial α-secretase enzyme, DHA increases α-secretase activity. More protein stability, not transcriptional modifications, is the cause of this stabilization. DHA improves the synthesis of neuroprotective chemicals and decreases the creation of Aβ by advancing the non-amyloidogenic pathway, which fosters a healthy neuron environment (Tang 2020). Several interconnected pathways are involved in the pleiotropic effects of DHA on Aβ processing. It changes the location of secretase components, redistributes cholesterol, boosts non-amyloidogenic processing, and directly inhibits secretase activity.

DHA is a potent regulator of AD pathogenesis because of these diverse activities, which comprehensively address the disease's fundamental processes (Grimm et al. 2011). DHA's anti-inflammatory qualities further strengthen its neuroprotective benefits. Additionally, DHA protects neurons by lowering neuroinflammation, which lessens secondary damage brought on by AD's prolonged immunological activation (Lim et al. 2005). DHA has significant medicinal implications, especially as a prophylactic. A lower incidence of AD and cognitive impairment is consistently associated with increased dietary consumption of DHA, according to epidemiological research (Cole et al. 2009). A negative association between omega-3 fatty acid consumption and AD incidence is demonstrated by observational studies, which corroborate these findings. The findings of clinical trials assessing DHA supplementation in people with existing AD, however, have been conflicting, indicating that DHA works better as a preventative measure than as a therapy for advanced illness. It emphasizes the need for early intervention as DHA can alter APP processing and lessen the buildup of Aβ before severe neuronal damage happens (Wei et al. 2023). DHA's utility as a long-term preventative agent is supported by its safety profile. DHA is a good option for dietary supplementation since it is well-tolerated, unlike pharmaceutical therapies that could have adverse side effects. Because it can alter cholesterol metabolism, it can be used as a supplement to statins, which lower cholesterol and decrease the generation of Aβ. DHA and other therapies may improve therapeutic results and provide a more comprehensive strategy for managing AD (Kelley et al. 2007). Despite its potential, typical DHA consumption in many communities is still below protective levels. For example, the usual daily intake of DHA in the American diet is around 70 mg, which is far less than the amounts found to be protective in epidemiological research (Swanson et al. 2012). This disparity emphasizes the necessity of promoting diets high in DHA, including fatty fish, supplements derived from algae, or fortified foods, to enhance brain function and lower the risk of AD. More investigation is required to improve the therapeutic effects of DHA by optimizing its dosage, formulations, and delivery systems (Bradbury 2011). In vitro studies demonstrate that DHA significantly suppresses the fibrillation of Aβ peptides, a crucial phase in AD development, supporting its function in Aβ processing. DHA prevents neurodegeneration by lowering the production of harmful Aβ aggregates (Hashimoto et al. 2009). DHA alters the shape of Aβ fibrils, resulting in amorphous aggregates instead of the usual fibrillar structures, which may lessen the toxicity of the fibrils. Moreover, DHA stabilizes Aβ oligomers, which may diminish their synaptic dysfunction while being extremely toxic, maintaining neuronal connectivity in the early stages of AD (Teng et al. 2015).

DHA's neuroprotective benefits are also greatly influenced by its antioxidant qualities. By reducing the oxidative stress associated with Aβ toxicity, DHA shields neurons from harm. In neuronal cells cultured with Aβ, it reduces cytotoxicity by lowering reactive oxygen species and lipid peroxidation (Díaz et al. 2021). By encouraging better axodendritic structures, DHA also improves cellular resilience. Mechanistically, DHA modifies the aggregation dynamics of Aβ by interacting with its hydrophobic areas. DHA also promotes the production of neuroprotective mediators, such as neuroprotectin D1 (NPD1), which has anti-inflammatory and anti-apoptotic qualities and improves neuronal survival (Lukiw et al. 2005). In vivo research supports these results, demonstrating that dietary DHA supplementation reduces amyloid load in the cortico-hippocampal areas of rats used as models for AD.

Increased cognitive performance, including improved spatial learning and memory in behavioral tasks like the Morris water maze, coincides with this decline. By destabilizing lipid rafts essential for Aβ aggregation and deposition, DHA can also change the fluidity and composition of neuronal membranes. This interference stops neurotoxic plaques from forming, highlighting DHA's complex function in AD pathophysiology. DHA supplementation has a striking effect on the model of cognitive and motor abilities (Teng et al. 2015). According to studies, DHA therapy improves motor skills, memory retention, and spatial learning. It can preserve neuronal health by lowering neurofibrillary tangles, encouraging the development of nerve fibers, and reducing amyloid-beta plaques. It is interesting to note that DHA supplementation has no discernible effect on metabolic markers, indicating that its advantages are mainly cognitive. Nonetheless, it has been shown to increase glycine levels in the AD group, offering encouraging advantages for mental performance and neuronal health without interfering with metabolic processes (Xiao et al. 2022). The importance of DHA goes beyond its function in processing Aβ. It is a marine-derived long-chain polyunsaturated fatty acid essential for developing and properly functioning the nervous system, especially the brain and retina. It has played a vital function in human evolution, especially in the formation and development of the brain. DHA is considered an essential vitamin during pregnancy and lactation since it actively contributes to the early development of the nervous system.

Additionally, neuroprotectin D1 (NPD1), a derivative of DHA, exhibits neuroprotective qualities against ischemia-reperfusion injuries, neurodegenerative disorders, and brain aging (Echeverría et al. 2017). Combating oxidative stress and neuroinflammation, two essential aspects of AD, demonstrates DHA's therapeutic promise in managing the disease. By lowering oxidative damage and controlling inflammatory reactions, DHA protects neurons. It may improve membrane fluidity and affect the location and activity of proteins involved in APP processing, emphasizing its all-encompassing neuroprotective benefits (Bazan et al. 2011).

10.3.4 Omega 3 Fatty Acid in Neuroinflammation

Omega-3 fatty acids, such as docosahexaenoic acid (DHA) and eicosapentaenoic acid (EPA), have been the subject of significant research due to their anti-inflammatory and neuroprotective properties, particularly related to nerve fiber' inflammation. Many neurological diseases, such as depression, Alzheimer's disease (AD), and other neurodegenerative diseases, are characterized by neuroinflammation, which is defined by the activation of microglia and astrocytes. Omega-3 fatty acids preserve neuronal integrity by incorporating into the cell membrane, changing inflammatory pathways, and reducing chronic inflammation (Swanson et al. 2012). The replacement of arachidonic acid (AA) in cell membranes is one of the most important mechanisms by which omega-3 fatty acids dampen inflammation, thus reducing the levels of pro-inflammatory eicosanoids synthesized through the cyclooxygenase (COX) and lipoxygenase (LOX) pathways. In addition, specific pro-resolving moieties derived from omega-3 fatty acids, such as resolvins, protectins, and maresins, play important roles in controlling inflammation and stimulating tissue regeneration. By modulating the activation of microglia and astrocytes, these biologically active lipids facilitate a shift from a pro-inflammatory phenotype to an anti-inflammatory phenotype while decreasing the generation of cytokines such as tumor necrosis factor-alpha (TNF-α), interleukin-1β (IL-1β), and interleukin-6 (IL-6) (Calder 2010). The effects of EPA and DHA on the modification of damage-associated molecular patterns (DAMPs) also include their inhibitory interaction with toll-like receptors (TLRs) and receptors for advanced glycation end products (RAGEs), such as high mobility group box 1 (HMGBI1) and S100 calcium-binding protein (S100b). Thus, fewer downstream inflammatory pathways are activated, including the nuclear factor-kappa B (NF-κB), an essential neuroinflammation regulator. Omega-3 fatty acids demonstrate neuroprotective properties. In experimental studies, for instance, omega-3 supplementation resulted in improved outcomes of chronic unexpected mild stress (CUMS) models of depression by controlling the expression of TLR4 and activity of NF-κB while reducing indicators such as neuron-specific enolase (NSE), S100β, and HMGB1 in models of traumatic brain injury (TBI) (Roh and Sohn 2018).

Omega-3 fatty acids have demonstrated variable but promising results in clinical studies. Supplementation has been associated with improvements in mood and cognitive performance, as well as reduced levels of neuroinflammatory markers such as S100β and HMGB1 in patients with neurodegenerative disorders. However, individual response heterogeneity underscores the need to identify biomarkers that predict therapy success and tailor dosage regimens. Based on new research, modifying dietary linoleic acid (LA, 18:2n-6), a precursor to AA, may augment the anti-inflammatory benefits of EPA and DHA. For example, when intake of dietary LA was lowered, and EPA and DHA supplements were increased, patients' inflammatory profiles improved, and migraine frequency was markedly reduced (Malau et al. 2024). The decrease in the levels of PUFAs, especially DHA, worsens with aging, which impacts the enzymatic pathways that synthesize and integrate PUFAs into cell membranes. Increased oxidative stress, activation of microglia, and neuroinflammation are the results of these reductions. The supplements of omega-3 fatty acids have shown the potential to reverse these changes. Fish oil-enriched diets reduced pro-inflammatory cytokine expression, decreased astrocytic activation, improved spatial memory, and restored DHA levels in aged animals. In addition, mediators derived from DHA, including neuroprotection and resolution, have shown high potency in preventing ischaemic damage and promoting inflammation resolution (Sun et al. 2018). Omega-3 fatty acids have also been shown to improve synaptic plasticity, an essential mechanism behind cognitive function. DHA has been found to enhance NMDA receptor activation and LTP through membrane fluidity changes.

Additionally, DHA elevates BDNF, a protein necessary for the survival of neurons and synaptic plasticity. DHA augments energy metabolism and brain health through pathways such as Akt and CaMKII. Such effects thus indicate the importance of preventing cognitive decline and keeping the brain healthy (Zhou et al. 2022). Recent studies emphasize the need for genotype-specific approaches to improve omega-3 fatty acid treatment. For instance, DHA effects also depend on the apolipoprotein E (APOE) genotype in that carriers of APOE4 have lower brain uptake of DHA with enhanced peripheral pro-inflammatory responses. These studies establish the promise of genotype-specific therapies designed to treat neuroinflammatory diseases with omega-3 fatty acids (Barberger-Gateau et al. 2011) (Fig. 10.3).

10.3.5 Effect of Omega-6 and Omega-3 Ratio in AD

Millions of people worldwide have Alzheimer's disease (AD), which causes gradual memory loss and cognitive decline, and the number of cases is expected to quadruple by 2050 (Brookmeyer et al. 2007). The diet's ratio of omega-6 and omega-3

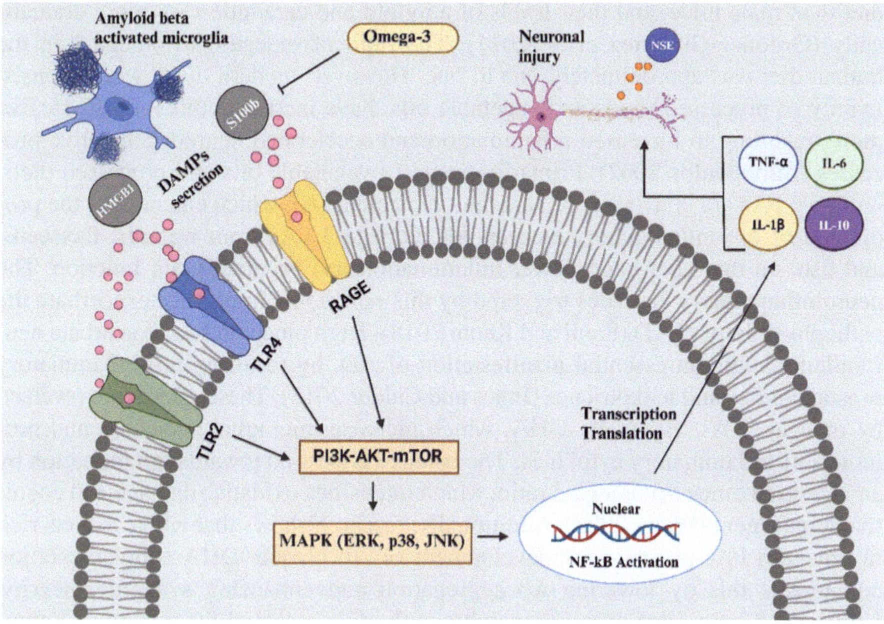

Fig. 10.3 Omega-3 fatty acids protect against AD pathology. Amyloid beta causes activation of microglia, which in turn secretes various DAMPS, causing the trigger of opening several downstream pathways, which in turn results in inflammatory cytokine secretion and neuronal death. Omega-3 inhabits this pathway and plays a vital role in neuroprotection

polyunsaturated fatty acids (PUFAs) has received considerable attention as a modifiable risk factor. Omega-3 fatty acids such as eicosapentaenoic acid (EPA), docosahexaenoic acid (DHA), and alpha-linolenic acid (ALA) have detrimental effects on inflammation and the brain. Linoleic acid (LA) is the opposite (Saini and Keum 2018). An unbalanced omega-6/omega-3 ratio, often high in the Western diet, underlies amyloid-beta (Aβ) production and other degenerative processes associated with AD. The lipid composition of the brain is negatively impacted by elevated omega-6/omega-3 ratios, which lower DHA levels necessary for synaptic integrity and neuronal function. This imbalance also changes Ceramide levels, encouraging Aβ aggregation and neuronal death (Ordóñez-Gutiérrez et al. 2021). Excessive omega-6 levels increase oxidative stress and neuroinflammation, two essential elements in the pathophysiology of AD, by activating pro-inflammatory pathways. By strengthening anti-inflammatory mechanisms, decreasing cytokine production, and fostering neuroprotection via resolvins and protectins, omega-3 FAs, on the other hand, reverse these effects (Djuricic and Calder 2021).

Research using transgenic AD animal models (e.g., APP/PS1,) demonstrates that diets rich in omega-6 FAs raise amyloid load and interfere with the metabolism of phospholipids and sphingolipids. Conversely, DHA supplementation enhances synaptic plasticity and decreases Aβ formation by lowering the omega-6/omega-3 ratio. Interestingly, female mice are more sensitive to changes in the omega-6/omega-3 diet than male mice, and their levels of amyloid and ceramide vary more dramatically (Ordóñez-Gutiérrez et al. 2021). The ratio of omega-6 to omega-3 in the human diet was approximately one to one. However, modern diets, which consist mainly of processed foods and vegetable oils, have increased this ratio to 15:1 or more, resulting in increased inflammation and accelerated neurodegenerative processes (Simopoulos 2002). Primarily found in vegetable oils and processed diets, omega-6 FAs are converted into arachidonic acid (AA), which encourages the production of pro-inflammatory eicosanoids. Omega-3 FAs from walnuts, flaxseeds, and fish, on the other hand, lower inflammation and promote brain function. The neuroinflammatory cascades triggered by this nutritional imbalance exacerbate the pathophysiology of AD (Saini and Keum 2018). High omega-6 FAs exacerbate neuroinflammation, an essential manifestation of AD, by restoring pro-inflammatory prostaglandins and leukotrienes (Innes and Calder 2018). These effects are reversed by omega-3 FAs, especially DHA, which increase microglial activation and promote anti-inflammatory cytokines. The balance is skewed towards inflammation by an increased omega-6/omega-3 ratio, which intensifies oxidative damage and cognitive impairment (Welty 2023). Additionally, research shows that whereas diets rich in omega-6 FAs promote the development of Aβ plaque, DHA supplementation counteracts this by lowering Aβ aggregation and enhancing synaptic integrity. Unbalanced ratios also decrease membrane fluidity, essential for synaptic communication, and raise oxidative stress, further compromising neuronal function (Mett 2021). Clinical data demonstrate the negative consequences of high omega-6/omega-3 ratios, including memory loss and decreased cognitive function. Furthermore, cross-sectional studies highlight the importance of dietary interventions by linking elevated plasma omega-6/omega-3 ratios to decreased DHA and EPA levels and an increased risk of AD. Increased omega-6/omega-3 ratios in brain tissues are correlated with more severe amyloid and tau pathologies, according to post-mortem examinations of AD patients. This finding emphasizes this ratio's crucial function in neurodegeneration (Sheppard and Cheatham 2013). The influence of omega-6/omega-3 ratios on AD is well supported by animal research. Amyloid plaque load, tau hyperphosphorylation, and cognitive impairments are all reduced by low dietary ratios. On the other hand, elevated ratios worsen synaptic dysfunction, neuroinflammation, and amyloidogenesis. The fact that female mice react better to omega-3 supplementation raises the possibility of sex-specific dietary approaches (Ordóñez-Gutiérrez et al. 2021) (Fig. 10.4). According to mechanistic research, omega-6 and omega-3 transpiration share enzymatic pathways. Excessive omega-6 consumption hinders omega-3 digestion, which lowers DHA production and its neuroprotective benefits (Simopoulos 2016). Omega-6/omega-3 ratios substantially impact AD; hence, dietary measures are crucial to bring them back into balance. A 4:1 to 1:1 ratio lowers inflammation and enhances cognitive function. Limiting omega-6 sources like vegetable oils and processed meals while increasing omega-3 consumption through fatty fish, flaxseeds, and supplements derived from algae is a sensible strategy. Omega-3 supplements, including DHA and EPA, may benefit high-risk people. More studies are required to clarify how omega-6/omega-3 ratios affect AD pathogenesis and identify the ideal ratios for various individuals. It

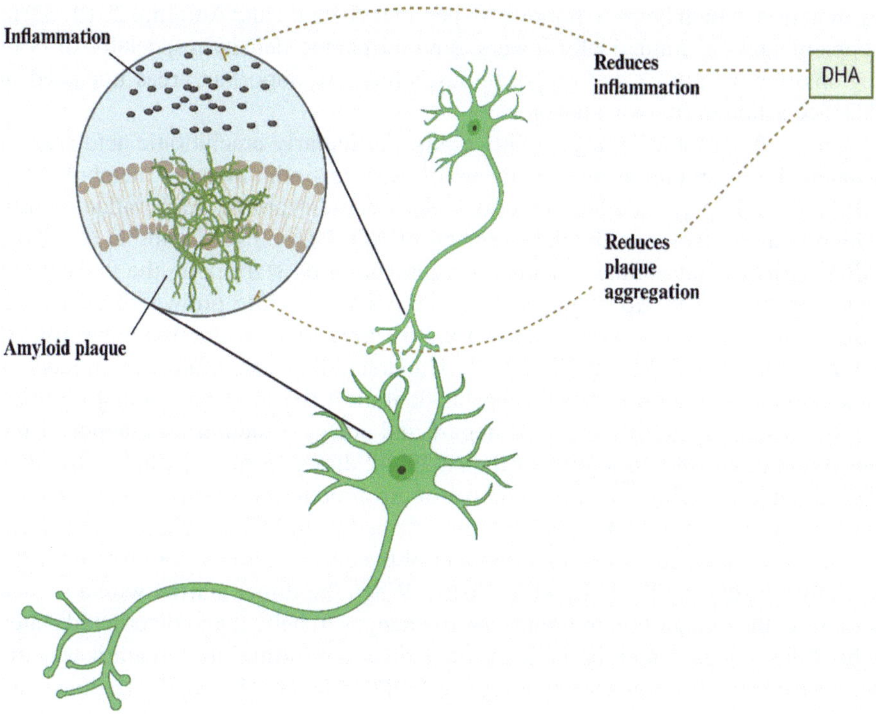

Fig. 10.4 DHA against plaque formation. DHA can provide neuroprotection by interfering with some basic pathways, which include protection against neuroinflammation, reduction of plaque aggregation

is essential to examine long-term dietary changes and how they interact with lifestyle factors like physical activity and genetic markers like APOE genotype. One possible method to reduce the incidence of AD and enhance cognitive function is to maintain a balanced omega-6/omega-3 ratio (Simopoulos 2002).

10.3.6 Effect of Omega-3 and Omega-6 in the Conformational Stability of Amyloid-Beta

It is crucial to research how omega-3 and omega-6 fatty acids affect the stability of the structure of amyloid-beta (Aβ) peptides to understand their therapeutic potential (Hjorth et al. 2013). The neuroprotective qualities of omega-3 fatty acids, especially docosahexaenoic acid (DHA), have drawn much attention. DHA's six cis-double bonds improve membrane fluidity, directly impacting the aggregation and generation of Aβ. DHA promotes the synthesis of the shorter, less hazardous Aβ40 isoform by reducing the creation of the poisonous Aβ42 isoform via modifying γ-secretase activity (Kousparou et al. 2023b). Furthermore, DHA-rich membranes promote protease activity and break the essential hydrophobic connections for β-sheet-rich aggregation, which helps to remove Aβ peptides (Fabiani and Antollini 2019). DHA supplementation dramatically lowers amyloid plaque density, especially in brain regions like the cortex and hippocampus, which are important areas impacted by AD, according to in vivo studies.

On the other hand, omega-6 fatty acids, particularly arachidonic acid (ARA), worsen the aggregation and synthesis of Aβ. Pro-inflammatory metabolites of ARA, including prostaglandins and leukotrienes, promote neuroinflammation, which is a key factor in the development of AD (Bascoul-Colombo et al. 2016). ARA-enriched membranes encourage aggregation by stabilizing the β-sheet and reducing fluidity in Aβ peptides. Because ARA lacks the protective cis-double bonds and structural flexibility of DHA, it is more likely to facilitate harmful Aβ interactions (Khondker et al. 2017). Understanding the molecular effects of omega-3 and omega-6 fatty acids has made it clear that they have a major impact on the structural stability of Aβ. Omega-6 and omega-3 membranes induce different orientations and structures of the Aβ29–42 dimer from the α-helix structure. These outcomes align with the control of Aβ production observed upon adding these fatty acids. Structural changes are obtained from MD simulations of a dimer of the 99-amino acid C-terminal segment of the amyloid precursor protein (APP), spanning residues 15–55 (Lu et al. 2019). When the dimer started with a β-sheet structure, the omega-6 membrane caused changes to helical and disordered states, whilst the omega-3 membrane kept the β-sheet conformation, though it was different from the β-sheet seen in a typical POPC bilayer. The Aβ29–42 dimer displayed a completely different conformational ensemble when combined with the omega-3, omega-6, and POPC combination (Miyashita et al. 2009). These findings elucidate the effect of fatty acid content on the structural dynamics of the Aβ peptide and show that changes in the ratio of omega-3 and omega-6 can cause significant changes in the structure of Aβ (Amtul et al. 2011). Because they are highly unsaturated, omega-3 fatty acids improve membrane fluidity, prevent Aβ from aggregating, and promote APP's non-amyloidogenic processing. Alternatively, omega-6 fatty acids stimulate amyloidogenic processes and cause membrane stiffness. This realization creates new opportunities for treatment approaches that alter lipid membranes to stop or reduce Aβ aggregation and lessen AD-related neurotoxicity (Jicha and Markesbery 2010). Making dietary adjustments or using lipid-based medications to target the mix of omega-3 and omega-6 fatty acids may be a potential way to halt the course of AD and support cognitive function (Cole et al. 2009).

10.4 Conclusion

Alzheimer's disease is a typical neurodegenerative disease that represents a challenging issue in clinical care, and over the next few decades, its burden will rise sharply. There is not yet a standard etiological therapy; most medical therapies are symptomatic. A nutritious approach, such as prophylaxis for AD, can be of immense benefit. Currently, various clinical trials are being conducted with omega-3 fatty acids as therapeutics against AD (Table 10.1).

Table 10.1 Clinical trials assessing the impact of omega-3 fatty acids supplementation in AD patients

Population characteristic	Type and dose supplementation	Exposure period	Results	References
AD (DSM-IV) MMSE 15–30 OmegAD study	1720 mg/die DHA+ 600 mg/die EPA Placebo: 4000 mg Corn oil, both groups: + 16 mg/die vitamin E	12 months	There was no significant statistical difference after 6-or 12-month treatment between groups in MMSE, ADAS-cog, CDR A subgroup with very mild cognitive dysfunction showed a reduction in decline rate	Freund-Levi et al. (2006)
AD (DSM-IV) MMSE 15–30 OmegAD study	1720 mg/die DHA+ 600 mg/die EPA Placebo: 4000 mg corn oil, both groups: + 16 mg vitamin E	12 months	Supplementation with omega-3 did not result in marked effects on neuropsychiatric symptoms except for possible positive effects on depressive symptoms (assessed by MADRS) in non-APOE E4 carriers and agitation symptoms (assessed by NPI) in APOE E4 carriers	Freund-Levi et al. (2008)
AD AD (DSM-IV) MMSE 10–26 CDR-score 1–2	720 mg/die DHA+ 1080 mg/die EPA Placebo: Olive oil Both groups: + 1.2 mg hydroquinone + 12 mg tocopherols	6 months	The treated group did not show an improvement in cognitive symptoms measured by MMSE, ADAS-cog, HDR but a relative improvement in CIBIC-plus score. In a subgroup with subjects with mild cognitive impairment (MMSE >27 e CDR 0.5–1) there was an improvement in ADAS-cog	Chiu et al. (2008)
AD MMSE 14–26 Alzheimer's disease Cooperative study (ADCS) DHA Supplementation trial USA	2000 mg/die DHA from seaweed Placebo: Corn or Soy oil	18 months	Supplementation with DHA compared with placebo did not slow the rate of cognitive and functional decline in patients with mild to moderate Alzheimer's disease assessed by MMSE, ADAS-cog, CDR, ADS-ADL, NPI	Quinn et al. (2010)
AD souvenir I Study	1700 mg/die DHA+ 600 mg/die EPA (Souvenaid) Placebo: Control drink	6 months	Supplementation with omega-3 improved delayed verbal recall. However, ADAS-cog, CIBIC-plus, NPI, ADCS-ADL, ADSC-ADL were unchanged	Scheltens et al. (2010)

(continued)

Table 10.1 (continued)

Population characteristic	Type and dose supplementation	Exposure period	Results	References
AD souvenir I Study	1700 mg/die DHA+ 600 mg/die EPA (Souvenaid) Placebo: Control drink	6 months	Souvenaid had a positive result on ADAS-cog outcome. A higher intake of Souvenaid was also associated with greater cognitive benefit	Kamphuis et al. (2011)
AD souvenir II Study	1200 mg/die DHA+ 300 mg/die EPA (Souvenaid) Placebo: Control products	6 months	In the active group, the NTB memory domain increased	Scheltens et al. (2012)
AD MMSE 14–24 Connect study	1200 mg/die DHA+ 300 mg/die EPA (Souvenaid) Placebo: Control products	6 months	Cognitive performance, as assessed by ADAS-cog, showed a decline over time in both control and active study groups, with no significant difference between study groups add-on intake of Souvenaid did not slow cognitive decline in persons treated for mild-to-moderate AD	Shah et al. (2013)
AD mild to moderate OmegAD study	1720 mg/die DHA+ 600 mg/die EPA Placebo: 4000 mg corn oil	12 months	Plasma transthyretin positively correlated with MMSE and Inversely with ADAS-cog, suggesting a potential mechanism for probable positive effects ts of omega-3 on cognition	Faxén-Irving et al. (2013)
AD MMSE 15–26 CDR 0.5–1.0 not depressed (CESD <4.0)	675 mg/die DHA+ 975 mg/die EPA Group omega-3 plus Alpha lipoic acid (LA): 675 mg/die DHA+ 975 mg/die EPA+ 600 mg/die LA Placebo: Soy oil	12 months	Active groups were no difference from the placebo group in ADAS-cog, ADL. Omega-3 + LA group showed less decline assessed by MMSE. IADL differences between placebo omega-3 and between placeboomega-3 + LA groups were observed	Shinto et al. (2014)
AD mild souvenir II study	1700 mg/die DHA+ 6 mg/die EPA (Souvenaid) Placebo: Control drink	6 months	The administration contributed to the maintenance of the Organization of brain networks in mild AD patients	de Waal et al. (2014)

(continued)

Table 10.1 (continued)

Population characteristic	Type and dose supplementation	Exposure period	Results	References
AD MMSE 16–30	625 mg/die DHA+ 600 mg/die EPA Placebo: Olive oil Both groups: + 20 Mg mixed tocopherols	4 months	The daily supplementation was associated with none or only negligible benefits on mood and cognition, assessed by MMSE, HVLT-R, BASDEC, BADLS	Phillips et al. (2015)
AD OmegAD Study	1720 mg/die DHA+ 600 mg/die EPA Placebo: 4000 mg Corn oil both groups: + 16 mg vitamin E	6 months	The daily supplementation stabilized the cognitive performance of AD subjects, assessed by ADAS-cog and MMSE scores	Eriksdotter et al. (2015)
AD OmegAD Study	1720 mg/die DHA+ 600 mg/die EPA Placebo: 4000 mg corn oil both groups: + 16 mg vitamin E	6 months	A decrease was observed in RvD1 and LXA4 production from peripheral blood mononuclear cells of AD patients who did not receive omega-3 but not in cells of AD subjects under omega-3 intake	Wang et al. (2015)
AD open label Extension study (OLE) souvenir II MMSE ≥20	1200 mg/die DHA+ 300 mg/die EPA (Souvenaid) Placebo: Control drink	6 months	The intake of Souvenaid was well tolerated with a favorable safety profile. The adherence to Souvenaid was very high reflecting its good tolerability and ease of use	Olde Rikkert et al. (2015)
AD OmegAD Study	1720 mg/die DHA+ 600 mg/die EPA Placebo: 1 g corn oil Both groups: + 16 mg vitamin E	6 months	The effect of omega-3 supplementation on MMSE and CDR appeared to be influenced by homocysteine plasma levels	Jernerén et al. (2019)

Given their anti-inflammatory and metabolic qualities, omega-3 fatty acids provide intriguing biological potential for treating various conditions. Indeed, there is positive data for Alzheimer's disease based on prospective observational research. A lower risk of developing the ailment has been linked to adopting a dietary regimen rich in omega-3 fatty acids. On the other hand, the other involvements of omega-3 fatty acids and omega-6 fatty acids in Aβ processing, metabolism, neuroinflammation, and even clearance open a new insight into new treatment strategies against AD.

References

Amtul Z et al (2011) Structural insight into the differential effects of omega-3 and omega-6 fatty acids on the production of Abeta peptides and amyloid plaques. J Biol Chem 286(8):6100–6107
Astarita G et al (2011) Elevated stearoyl-CoA desaturase in brains of patients with Alzheimer's disease. PLoS One 6(10):e24777
Barberger-Gateau P et al (2011) Dietary omega 3 polyunsaturated fatty acids and Alzheimer's disease: interaction with apolipoprotein E genotype. Curr Alzheimer Res 8(5):479–491
Bascoul-Colombo C et al (2016) Dietary DHA supplementation causes selective changes in phospholipids from different brain regions in both wild type mice and the Tg2576 mouse model of Alzheimer's disease. Biochim Biophys Acta 1861(6):524–537
Bazan NG, Molina MF, Gordon WC (2011) Docosahexaenoic acid signalolipidomics in nutrition: significance in aging, neuroinflammation, macular degeneration, Alzheimer's, and other neurodegenerative diseases. Annu Rev Nutr 31:321–351
Bazinet RP, Layé S (2014) Polyunsaturated fatty acids and their metabolites in brain function and disease. Nat Rev Neurosci 15(12):771–785
Berg KA, Maayani S, Clarke WP (1996) 5-hydroxytryptamine2C receptor activation inhibits 5-hydroxytryptamine1B-like receptor function via arachidonic acid metabolism. Mol Pharmacol 50(4):1017–1023
Bradbury J (2011) Docosahexaenoic acid (DHA): an ancient nutrient for the modern human brain. Nutrients 3(5):529–554
Brenna JT, Diau GY (2007) The influence of dietary docosahexaenoic acid and arachidonic acid on central nervous system polyunsaturated fatty acid composition. Prostaglandins Leukot Essent Fatty Acids 77(5–6):247–250
Brookmeyer R et al (2007) Forecasting the global burden of Alzheimer's disease. Alzheimers Dement 3(3):186–191
Burdge GC, Calder PC (2005) Conversion of alpha-linolenic acid to longer-chain polyunsaturated fatty acids in human adults. Reprod Nutr Dev 45(5):581–597
Calder PC (2010) Omega-3 fatty acids and inflammatory processes. Nutrients 2(3):355–374
Calder PC (2013) Omega-3 polyunsaturated fatty acids and inflammatory processes: nutrition or pharmacology? Br J Clin Pharmacol 75(3):645–662
Calderon F, Kim HY (2004) Docosahexaenoic acid promotes neurite growth in hippocampal neurons. J Neurochem 90(4):979–988
Cao D et al (2009) Docosahexaenoic acid promotes hippocampal neuronal development and synaptic function. J Neurochem 111(2):510–521
Carbone BE et al (2020) Synaptic connectivity and cortical maturation are promoted by the ω-3 fatty acid docosahexaenoic acid. Cereb Cortex 30(1):226–240
Chen CT et al (2009) Rapid beta-oxidation of eicosapentaenoic acid in mouse brain: an in situ study. Prostaglandins Leukot Essent Fatty Acids 80(2–3):157–163
Childs CE et al (2014) Increased dietary α-linolenic acid has sex-specific effects upon eicosapentaenoic acid status in humans: re-examination of data from a randomised, placebo-controlled, parallel study. Nutr J 13(1):113
Chiu CC et al (2008) The effects of omega-3 fatty acids monotherapy in Alzheimer's disease and mild cognitive impairment: a preliminary randomized double-blind placebo-controlled study. Prog Neuro-Psychopharmacol Biol Psychiatry 32(6):1538–1544
Christofides A et al (2021) The role of peroxisome proliferator-activated receptors (PPAR) in immune responses. Metabolism 114:154338
Cole GM, Ma QL, Frautschy SA (2009) Omega-3 fatty acids and dementia. Prostaglandins Leukot Essent Fatty Acids 81(2–3):213–221
Covington MB (2004) Omega-3 fatty acids. Am Fam Physician 70(1):133–140
Crawford MA, Casperd NM, Sinclair AJ (1976) The long chain metabolites of linoleic acid linolenic acids in liver and brain in herbivores and carnivores. Comp Biochem Physiol B 54(3):395–401

de Gŏmez Dumm IN, Brenner RR (1975) Oxidative desaturation of alpha-linoleic, linoleic, and stearic acids by human liver microsomes. Lipids 10(6):315–317
de Waal H et al (2014) The effect of souvenaid on functional brain network organisation in patients with mild Alzheimer's disease: a randomised controlled study. PLoS One 9(1):e86558
Díaz M, Mesa-Herrera F, Marín R (2021) DHA and its elaborated modulation of antioxidant defenses of the brain: implications in aging and AD neurodegeneration. Antioxidants (Basel) 10(6)
Djuricic I, Calder PC (2021) Beneficial outcomes of Omega-6 and Omega-3 polyunsaturated fatty acids on human health: an update for 2021. Nutrients 13(7)
Downes CP, Currie RA (1998) Lipid signalling. Curr Biol 8(24):R865–R867
Echeverría F et al (2017) Docosahexaenoic acid (DHA), a fundamental fatty acid for the brain: new dietary sources. Prostaglandins Leukot Essent Fatty Acids 124:1–10
Emken EA et al (1990) Metabolism in vivo of deuterium-labelled linolenic and linoleic acids in humans. Biochem Soc Trans 18(5):766–769
Eriksdotter M et al (2015) Plasma fatty acid profiles in relation to cognition and gender in Alzheimer's disease patients during Oral Omega-3 fatty acid supplementation: the OmegAD study. J Alzheimer's Dis 48(3):805–812
Fabiani C, Antollini SS (2019) Alzheimer's disease as a membrane disorder: spatial cross-talk among Beta-amyloid peptides, nicotinic acetylcholine receptors and lipid rafts. Front Cell Neurosci 13:309
Faxén-Irving G et al (2013) Effects on transthyretin in plasma and cerebrospinal fluid by DHA-rich n − 3 fatty acid supplementation in patients with Alzheimer's disease: the OmegAD study. J Alzheimer's Dis 36(1):1–6
Fernstrom JD (1999) Effects of dietary polyunsaturated fatty acids on neuronal function. Lipids 34(2):161–169
Frazer A, Hensler JG (1999) Serotonin involvement in physiological function and behavior. In: Basic neurochemistry: molecular, cellular and medical aspects, pp 263–292
Freund-Levi Y et al (2006) Omega-3 fatty acid treatment in 174 patients with mild to moderate Alzheimer disease: OmegAD study: a randomized double-blind trial. Arch Neurol 63(10):1402–1408
Freund-Levi Y et al (2008) Omega-3 supplementation in mild to moderate Alzheimer's disease: effects on neuropsychiatric symptoms. Int J Geriatr Psychiatry 23(2):161–169
Garcia Corrales AV et al (2021) Fatty acid synthesis in glial cells of the CNS. Int J Mol Sci 22(15)
García-Sáinz JA, Vázquez-Prado J, Villalobos-Molina R (1999) Alpha 1-adrenoceptors: subtypes, signaling, and roles in health and disease. Arch Med Res 30(6):449–458
Gibson RA et al (2013) Docosahexaenoic acid synthesis from alpha-linolenic acid is inhibited by diets high in polyunsaturated fatty acids. Prostaglandins Leukot Essent Fatty Acids 88(1):139–146
Grimm MO et al (2011) Docosahexaenoic acid reduces amyloid beta production via multiple pleiotropic mechanisms. J Biol Chem 286(16):14028–14039
Grimm MO et al (2013) The impact of cholesterol, DHA, and sphingolipids on Alzheimer's disease. Biomed Res Int 2013:814390
Hashimoto M et al (2009) Effects of docosahexaenoic acid on in vitro amyloid beta peptide 25-35 fibrillation. Biochim Biophys Acta 1791(4):289–296
Herrera Vielma F et al (2021) N-3 polyunsaturated fatty acids and their lipid mediators as a potential immune-nutritional intervention: a molecular and clinical view in hepatic disease and other non-communicable illnesses. Nutrients 13(10)
Hjorth E et al (2013) Omega-3 fatty acids enhance phagocytosis of Alzheimer's disease-related amyloid-β42 by human microglia and decrease inflammatory markers. J Alzheimer's Dis 35(4):697–713
Illig T et al (2010) A genome-wide perspective of genetic variation in human metabolism. Nat Genet 42(2):137–141
Innes JK, Calder PC (2018) Omega-6 fatty acids and inflammation. Prostaglandins Leukot Essent Fatty Acids 132:41–48

Innis SM (2008) Dietary omega 3 fatty acids and the developing brain. Brain Res 1237:35–43

Innis SM (2009) Omega-3 fatty acids and neural development to 2 years of age: do we know enough for dietary recommendations? J Pediatr Gastroenterol Nutr 48(Suppl 1):S16–S24

Jernerén F et al (2019) Homocysteine status modifies the treatment effect of Omega-3 fatty acids on cognition in a randomized clinical trial in mild to moderate Alzheimer's disease: the OmegAD study. J Alzheimer's Dis 69(1):189–197

Jicha GA, Markesbery WR (2010) Omega-3 fatty acids: potential role in the management of early Alzheimer's disease. Clin Interv Aging 5:45–61

Kamphuis PJ et al (2011) Efficacy of a medical food on cognition in Alzheimer's disease: results from secondary analyses of a randomized, controlled trial. J Nutr Health Aging 15(8):720–724

Kasthuri N et al (2015) Saturated reconstruction of a volume of neocortex. Cell 162(3):648–661

Kelley DS et al (2007) Docosahexaenoic acid supplementation improves fasting and postprandial lipid profiles in hypertriglyceridemic men. Am J Clin Nutr 86(2):324–333

Khondker A, Alsop RJ, Rheinstädter MC (2017) Membrane-accelerated amyloid-β aggregation and formation of cross-β sheets. Membranes (Basel) 7(3)

Kim HY, Huang BX, Spector AA (2022) Molecular and signaling mechanisms for docosahexaenoic acid-derived neurodevelopment and neuroprotection. Int J Mol Sci 23(9)

Kousparou C, Fyrilla M, Stephanou A, Patrikios I (2023a) DHA/EPA (Omega-3) and LA/GLA (Omega-6) as bioactive molecules in neurodegenerative diseases. Int J Mol Sci 24(13):10717

Kousparou C et al (2023b) DHA/EPA (Omega-3) and LA/GLA (Omega-6) as bioactive molecules in neurodegenerative diseases. Int J Mol Sci 24(13)

Lassek WD, Gaulin SJ (2011) Sex differences in the relationship of dietary fatty acids to cognitive measures in american children. Front Evol Neurosci 3:5

Lee CH, Hajra AK (1991) Molecular species of diacylglycerols and phosphoglycerides and the post-mortem changes in the molecular species of diacylglycerols in rat brains. J Neurochem 56(2):370–379

Lee AY et al (2018) Neuroprotective effect of alpha-linolenic acid against Aβ-mediated inflammatory responses in C6 glial cell. J Agric Food Chem 66(19):4853–4861

Lim GP et al (2005) A diet enriched with the omega-3 fatty acid docosahexaenoic acid reduces amyloid burden in an aged Alzheimer mouse model. J Neurosci 25(12):3032–3040

Lin PY, Huang SY, Su KP (2010) A meta-analytic review of polyunsaturated fatty acid compositions in patients with depression. Biol Psychiatry 68(2):140–147

Lister MD et al (1988) Kinetic analysis of the Ca2+-dependent, membrane-bound, macrophage phospholipase A2 and the effects of arachidonic acid. J Biol Chem 263(16):7506–7513

Lu Y et al (2019) Amyloid-β(29-42) dimeric conformations in membranes rich in Omega-3 and Omega-6 polyunsaturated fatty acids. J Phys Chem B 123(12):2687–2696

Lukiw WJ et al (2005) A role for docosahexaenoic acid-derived neuroprotectin D1 in neural cell survival and Alzheimer disease. J Clin Invest 115(10):2774–2783

Malau IA et al (2024) Omega-3 fatty acids and Neuroinflammation in depression: targeting damage-associated molecular patterns and neural biomarkers. Cells 13(21)

McPhail LC, Clayton CC, Snyderman R (1984) A potential second messenger role for unsaturated fatty acids: activation of Ca2+-dependent protein kinase. Science 224(4649):622–625

Mett J (2021) The impact of medium chain and polyunsaturated ω-3-fatty acids on amyloid-β deposition, oxidative stress and metabolic dysfunction associated with Alzheimer's disease. Antioxidants (Basel) 10(12)

Miyashita N, Straub JE, Thirumalai D (2009) Structures of beta-amyloid peptide 1-40, 1-42, and 1-55-the 672-726 fragment of APP-in a membrane environment with implications for interactions with gamma-secretase. J Am Chem Soc 131(49):17843–17852

Müller H, Kirkhus B, Pedersen JIJL (2001) Serum cholesterol predictive equations with special emphasis on trans and saturated fatty acids. Lipids 36(8):783–791

Murphy MG (1985) Membrane fatty acids, lipid peroxidation and adenylate cyclase activity in cultured neural cells. Biochem Biophys Res Commun 132(2):757–763

Murphy MP, LeVine H 3rd (2010) Alzheimer's disease and the amyloid-beta peptide. J Alzheimer's Dis 19(1):311–323

Nicolas C et al (1991) Dietary (n-6) polyunsaturated fatty acids affect beta-adrenergic receptor binding and adenylate cyclase activity in pig adipocyte plasma membrane. J Nutr 121(8):1179–1186
Norwitz NG et al (2021) Precision nutrition for Alzheimer's prevention in ApoE4 carriers. Nutrients 13(4)
O'Brien RJ, Wong PC (2011) Amyloid precursor protein processing and Alzheimer's disease. Annu Rev Neurosci 34:185–204
Okada L et al (2018) Omega-3 PUFA modulate lipogenesis, ER stress, and mitochondrial dysfunction markers in NASH – proteomic and lipidomic insight. Clin Nutr 37(5):1474–1484
Olde Rikkert MG et al (2015) Tolerability and safety of Souvenaid in patients with mild Alzheimer's disease: results of multi-center, 24-week, open-label extension study. J Alzheimer's Dis 44(2):471–480
Ordóñez-Gutiérrez L et al (2021) Diets with higher ω-6/ω-3 ratios show differences in ceramides and fatty acid levels accompanied by increased amyloid-Beta in the brains of male APP/PS1 transgenic mice. Int J Mol Sci 22(20)
Ouellet M et al (2009) Diffusion of docosahexaenoic and eicosapentaenoic acids through the blood-brain barrier: an in situ cerebral perfusion study. Neurochem Int 55(7):476–482
Padley FB (1994) Occurrence and characteristics of oils and fats. In: The lipid handbook, pp 47–223
Park WJ et al (2009) An alternate pathway to long-chain polyunsaturates: the FADS2 gene product Delta8-desaturates 20:2n-6 and 20:3n-3. J Lipid Res 50(6):1195–1202
Petermann AB et al (2022) Roles of the unsaturated fatty acid docosahexaenoic acid in the central nervous system: molecular and cellular insights. Int J Mol Sci 23(10)
Phillips MA et al (2015) No effect of Omega-3 fatty acid supplementation on cognition and mood in individuals with cognitive impairment and probable Alzheimer's disease: a randomised controlled trial. Int J Mol Sci 16(10):24600–24613
Pifferi F, Cunnane SC, Guesnet P (2020) Evidence of the role of Omega-3 polyunsaturated fatty acids in brain glucose metabolism. Nutrients 12(5)
Ponnampalam EN et al (2009) Polyunsaturated fats in meat from merino, first- and second-cross sheep slaughtered as yearlings. Meat Sci 83(2):314–319
Popoli M et al (2000) Second messenger-regulated protein kinases in the brain: their functional role and the action of antidepressant drugs. J Neurochem 74(1):21–33
Qi G et al (2021) ApoE4 impairs neuron-astrocyte coupling of fatty acid metabolism. Cell Rep 34(1):108572
Quinn JF et al (2010) Docosahexaenoic acid supplementation and cognitive decline in Alzheimer disease: a randomized trial. JAMA 304(17):1903–1911
Raghavendra S (2024) Biochemistry-a comprehensive guide. Academic Guru Publishing House
Ratnayake WM, Galli C (2009) Fat and fatty acid terminology, methods of analysis and fat digestion and metabolism: a background review paper. Ann Nutr Metab 55(1–3):8–43
Roh JS, Sohn DH (2018) Damage-associated molecular patterns in inflammatory diseases. Immune Netw 18(4):e27
Rustan AC, Drevon CA (2005) Fatty acids: structures and properties. Encycl Life Sciences 1:1–7
Saini RK, Keum YS (2018) Omega-3 and omega-6 polyunsaturated fatty acids: dietary sources, metabolism, and significance – a review. Life Sci 203:255–267
Sakamoto T, Cansev M, Wurtman RJ (2007) Oral supplementation with docosahexaenoic acid and uridine-5′-monophosphate increases dendritic spine density in adult gerbil hippocampus. Brain Res 1182:50–59
Salem N Jr et al (2001) Mechanisms of action of docosahexaenoic acid in the nervous system. Lipids 36(9):945–959
Schaffer JE, Lodish HF (1995) Molecular mechanism of long-chain fatty acid uptake. Trends Cardiovasc Med 5(6):218–224
Scheltens P et al (2010) Efficacy of a medical food in mild Alzheimer's disease: a randomized, controlled trial. Alzheimers Dement 6(1):1–10.e1
Scheltens P et al (2012) Efficacy of Souvenaid in mild Alzheimer's disease: results from a randomized, controlled trial. J Alzheimer's Dis 31(1):225–236

Serhan CN, Levy BD (2018) Resolvins in inflammation: emergence of the pro-resolving superfamily of mediators. J Clin Invest 128(7):2657–2669
Shah RC et al (2013) The S-connect study: results from a randomized, controlled trial of Souvenaid in mild-to-moderate Alzheimer's disease. Alzheimer's Res Ther 5(6):59
Sharan P, Vellapandian C (2024) Hypothalamic-pituitary-adrenal (HPA) Axis: unveiling the potential mechanisms involved in stress-induced Alzheimer's disease and depression. Cureus 16(8):e67595
Sheppard KW, Cheatham CL (2013) Omega-6 to omega-3 fatty acid ratio and higher-order cognitive functions in 7- to 9-y-olds: a cross-sectional study. Am J Clin Nutr 98(3):659–667
Shinto L et al (2014) A randomized placebo-controlled pilot trial of omega-3 fatty acids and alpha lipoic acid in Alzheimer's disease. J Alzheimer's Dis 38(1):111–120
Shukla SD, Halenda SP (1991) Phospholipase D in cell signalling and its relationship to phospholipase C. Life Sci 48(9):851–866
Simopoulos AP (1999) New products from the Agri-food industry: the return of n-3 fatty acids into the food supply. Lipids 34(Suppl):S297–S301
Simopoulos AP (2002) The importance of the ratio of omega-6/omega-3 essential fatty acids. Biomed Pharmacother 56(8):365–379
Simopoulos AP (2004) Omega-3 fatty acids and antioxidants in edible wild plants. Biol Res 37(2):263–277
Simopoulos AP (2006) Evolutionary aspects of diet, the omega-6/omega-3 ratio and genetic variation: nutritional implications for chronic diseases. Biomed Pharmacother 60(9):502–507
Simopoulos AP (2016) An increase in the Omega-6/Omega-3 fatty acid ratio increases the risk for obesity. Nutrients 8(3):128
Sinclair AJ, Attar-Bashi NM, Li D (2002) What is the role of alpha-linolenic acid for mammals? Lipids 37(12):1113–1123
Singh M (2005) Essential fatty acids, DHA and human brain. Indian J Pediatr 72(3):239–242
Smink W et al (2012) Linoleic and α-linolenic acid as precursor and inhibitor for the synthesis of long-chain polyunsaturated fatty acids in liver and brain of growing pigs. Animal 6(2):262–270
Sprecher H (2000) Metabolism of highly unsaturated n-3 and n-6 fatty acids. Biochim Biophys Acta 1486(2–3):219–231
Stillwell W, Wassall SR (2003) Docosahexaenoic acid: membrane properties of a unique fatty acid. Chem Phys Lipids 126(1):1–27
Sun GY et al (2018) Docosahexaenoic acid (DHA): an essential nutrient and a nutraceutical for brain health and diseases. Prostaglandins Leukot Essent Fatty Acids 136:3–13
Swanson D, Block R, Mousa SA (2012) Omega-3 fatty acids EPA and DHA: health benefits throughout life. Adv Nutr 3(1):1–7
Tang BL (2020) Enhancing α-secretase processing for Alzheimer's disease-a view on SFRP1. Brain Sci 10(2)
Tanguy E et al (2019) Phosphatidic acid: from pleiotropic functions to neuronal pathology. Front Cell Neurosci 13:2
Teng E et al (2015) Dietary DHA supplementation in an APP/PS1 transgenic rat model of AD reduces behavioral and Aβ pathology and modulates Aβ oligomerization. Neurobiol Dis 82:552–560
Tvrzicka E et al (2011) Fatty acids as biocompounds: their role in human metabolism, health and disease-a review. Part 1: classification, dietary sources and biological functions. Biomed Papers Med Faculty Palacky University Olomouc 155(2):117
Von Kriegstein K et al (1999) Distribution of synaptic vesicle proteins in the mammalian retina identifies obligatory and facultative components of ribbon synapses. Eur J Neurosci 11(4):1335–1348
Wang X et al (2015) Effects of n-3 FA supplementation on the release of proresolving lipid mediators by blood mononuclear cells: the OmegAD study. J Lipid Res 56(3):674–681
Wang J et al (2017) A systemic view of Alzheimer disease – insights from amyloid-β metabolism beyond the brain. Nat Rev Neurol 13(10):612–623

Wei BZ et al (2023) The relationship of Omega-3 fatty acids with dementia and cognitive decline: evidence from prospective cohort studies of supplementation, dietary intake, and blood markers. Am J Clin Nutr 117(6):1096–1109

Weiser MJ, Butt CM, Mohajeri MH (2016) Docosahexaenoic acid and cognition throughout the lifespan. Nutrients 8(2):99

Welty FK (2023) Omega-3 fatty acids and cognitive function. Curr Opin Lipidol 34(1):12–21

Wieland SJ, Fletcher JE, Gong QH (1992) Differential modulation of a sodium conductance in skeletal muscle by intracellular and extracellular fatty acids. Am J Phys 263(2 Pt 1):C308–C312

Xiao YF et al (1997) Suppression of voltage-gated L-type Ca2+ currents by polyunsaturated fatty acids in adult and neonatal rat ventricular myocytes. Proc Natl Acad Sci USA 94(8):4182–4187

Xiao M et al (2022) DHA ameliorates cognitive ability, reduces amyloid deposition, and nerve fiber production in Alzheimer's disease. Front Nutr 9:852433

Youdim KA, Martin A, Joseph JA (2000) Essential fatty acids and the brain: possible health implications. Int J Dev Neurosci 18(4–5):383–399

Zárate R et al (2017) Significance of long-chain polyunsaturated fatty acids in human health. Clin Transl Med 6(1):25

Zhou L et al (2022) Possible antidepressant mechanisms of omega-3 polyunsaturated fatty acids acting on the central nervous system. Front Psych 13:933704

Molecular Mechanisms and Therapeutic Potential of Ginger (*Zingiber officinale*) in Alzheimer's Disease

11

Divya Dhawal Bhandari, Jai Malik, and Prerna Sarup

Abstract

Neurodegenerative diseases represent an expanding global health burden, particularly as life expectancy increases. Alzheimer's disease (AD) is a devastating neurodegenerative disorder illustrated by loss of memory progressively and cognitive decline. The treatment includes anti-Alzheimer's drugs, but till date, there is no complete cure available with any of these drugs. Current pharmacotherapies offer only symptomatic relief and are ineffective in halting disease progression. This limitation has led to increased interest in natural agents with neuroprotective potential, thus opening newer opportunities for researchers. Herbs and herbal drugs such as allicin, ginger, curcumin, moringa, etc., are gaining much interest due to lesser side effects and higher bioavailability. Ginger (*Zingiber officinale*), a traditional medicinal plant, contains numerous bioactive compounds, such as 6-gingerol and 6-shogaol, which have exhibited the ability in modifying key pathological features of AD and has emerged as potential new drug candidates against AD. The constituents of ginger can interact with various molecular targets associated with AD treatment. Nevertheless, there is a scarcity of theoretical research in the literature exploring the potential mechanisms through which these compounds might act as anti-AD agents. A number of preclinical and clinical studies have reported the effectiveness of ginger in treating or providing symptomatic relief in case of various neurodegenerative disorders. This chapter discusses the molecular mechanisms by which ginger may exert neuroprotective effects and evaluates its therapeutic potential as an adjunct in AD treatment.

D. D. Bhandari · J. Malik · P. Sarup (✉)
University Institute of Pharmaceutical Sciences, Panjab University, Chandigarh, India

J. Malik et al. (eds.), *Diet and Alzheimer's Disease: Let Food be Our Medicine*,
Nutritional Neurosciences, https://doi.org/10.1007/978-981-95-2736-6_11

Keywords

Neurodegeneration · Alzheimer · *Zingiber officinale* · Ginger · 6-Gingerol · 6-Shogaol

Abbreviations

AD	Alzheimer's disease
NDDs	Neurodegenerative disorders
PD	Parkinson's disease
HD	Huntington's disease
ALS	Amyotrophic lateral sclerosis
NMDA	N-methyl-D-aspartate receptor
DNA	Deoxyribonucleic acid
NGF	Nerve Growth Factor
ERK	Extracellular signal regulatory kinases
CREB	cAMP response element-binding protein
LPS	Lipopolysaccharide
TNF	Tumor necrosis factor
IL	Interleukin
COX	Cyclooxygenase
MIP	Macrophage inflammatory protein
MCP	Monocyte chemoattractant protein
SOD	Superoxide dismutase
CAT	Catalase
GPx	Glutathione peroxidase
ROS	Reactive oxygen species
AChE	Acetylcholinesterase
BDNF	Brain-derived neurotrophic factor
BChE	Butyrylcholinesterase
GRE	Ginger root extract
BBB	Blood-brain barrier

11.1 Introduction

Neurodegenerative disorders (NDDs), including Alzheimer's disease (AD), Parkinson's disease (PD), Huntington's disease (HD), and amyotrophic lateral sclerosis (ALS), are progressive, debilitating conditions characterized by the gradual loss of neuronal structure and function. Despite extensive research, current pharmacotherapies offer limited efficacy, often providing only symptomatic relief and accompanied by significant side effects. This chapter explores the limitations of existing treatments and highlights the growing interest in natural compounds—particularly plant-derived phytochemicals—as potential multitargeted and safer therapeutic alternatives. We discuss their mechanisms of action (MOA), preclinical evidence, clinical evidence, and also the translational challenges that must be addressed for their successful integration into mainstream medical practice. Neurodegenerative diseases represent an expanding global health burden, particularly as life expectancy increases. These disorders involve the gradual deterioration and neuronal death, leading to cognitive and motor impairments. The major NDDs include

- *Alzheimer's Disease (AD):* Depicted by loss of memory, decline in cognition, and accumulation of amyloid-beta plaques and neurofibrillary tangles.
- *Parkinson's Disease (PD):* Involves motor dysfunction, dopaminergic neuronal loss, and presence of Lewy bodies.
- *Huntington's Disease (HD):* A genetic disorder with involuntary movements and psychiatric symptoms due to CAG trinucleotide repeat expansion.
- *Amyotrophic Lateral Sclerosis (ALS):* Features progressive muscle weakness resulting from motor neuron degeneration.

According to the World Health Organization (2021), the worldwide prevalence of these conditions is expected to double by 2050, highlighting the critical need for novel treatment strategies. Alzheimer's disease (AD), the most prevalent type of dementia, currently impacts millions of individuals around the globe. It is characterized by hallmark neuropathological features, including extracellular amyloid-beta (Aβ) plaque accumulation, intracellular neurofibrillary tangles composed of hyperphosphorylated tau, oxidative stress, neuroinflammation, and cholinergic system dysfunction. The pathogenesis of AD is multifactorial, involving genetic predispositions (e.g., APOE4), environmental factors, and aging.

Although approved treatments for Alzheimer's disease—such as acetylcholinesterase inhibitors and NMDA receptor antagonists—provide only limited symptomatic relief and do not slow the progression of the disease, additional recommendations are often made. Physicians may also suggest that AD patients take supplements like coenzyme Q10, alpha lipoic acid, G*inkgo biloba*, omega-3 fatty acids, and acetyl-L-carnitine (Bhandari et al. 2024). The development of new Alzheimer's disease medications that not only reduce side effects but also address pharmacokinetic challenges remains a highly desirable goal. Phytochemicals, which have a long history of use in human health, offer potential for treating a wide range of diseases (Jagtap et al. 2022). This has driven exploration into complementary and alternative therapies. From the time of existence of human beings in this world until today which is the era of modern medicines, people still trust natural medicines for treating many diseases. The collection of some traditionally suggested medicinal herbs, which may be used in AD, includes fennel, heena, ginseng, turmeric, ashwagandha, brahmi, shankpushpi, ginger, etc., though there is a lack of deep scientific studies. Ginger, a widely consumed rhizome in traditional medicine, is being studied for its neuroprotective and anti-inflammatory properties, particularly in age-related cognitive disorders including AD. A number of studies have reported the beneficial effects of

Ginger compounds in various computational in vitro and in vivo models of Alzheimer's disease, thus indicating that ginger compounds may play a role not only in treating the disease but also in preventing it (Schepici et al. 2021).

11.2 Phytochemistry of Ginger

Ginger (*Zingiber officinale*), a monocotyledon from Zingiberaceae family, is very known for its beneficial effects specially its tuberous rhizomes (Schepici et al. 2021). It is used all over the world as a flavoring agents or spice in various drinks and foods. *Zingiber officinale* is commercially grown in West Indies, China, and Mexico (Banerjee et al. 2011). Due to the ingredients present in ginger, it finds its applications in Traditional Chinese Medicine, Ayurvedic, and Unani-Tibb treatment systems against different diseases (Talebi et al. 2021). Some of which includes mixing ginger with milk/water for colic pain in infants and mixing with honey to treat asthmatic bronchitis, cough, hiccup, and other respiratory issues (Mahboubi 2019).

While western medicines use ginger to relieve digestive issues, Greeks use ginger as an aid to digestion and making ginger bread. Other traditional uses of ginger include its blood-purifying qualities, aphrodisiac and libido-enhancing effects, as well as its roles in relieving vomiting and nausea, stimulating appetite, reducing gas, acting as an antispasmodic, and alleviating hemorrhoids (Mahboubi 2019). The fresh ginger also finds its uses for colds, colic, nausea, asthma, swelling, indigestion, loss of appetite, rheumatism, and heart palpitations even in ancient India and China, while traditional Eastern Medicine also includes its uses as antiemetic (Kumar et al. 2012). The rhizomes of ginger are used in Iran to regulate menstrual cycle, to reduce menstrual period pain, and to increase male sexuality (Hafez 2010).

Modern research also emphasizes on studying various pharmacological activities of ginger and its associated bioactive molecules such as 6-gingerol, various polyphenols, and essential volatile oils. There are around 400 various constituents such as carbohydrates, terpenes, phenols, and lipids being found in ginger, which are further categorized into aromatic (bisabolen, borneol, camphene, cumene, pinene, and zingiberol) and pungent (capsaicin, gingerdione, gingerols, paradols, shogaols, zingerones, and zerumbone). Polyphenols acting as secondary metabolites of ginger are found in root extract and possess high antioxidant activity, which inhibits ROS (reactive oxygen species) that causes lipid peroxidation, DNA damage, coronary heart disease, and age-related health problems (Naseem et al. 2016). Lipid peroxidation of neuronal cell membranes is a key factor in the development of neurodegenerative diseases like Alzheimer's disease, which has led to interest in exploring the potential of ginger as a treatment for these disorders (Oboh et al. 2012).

Along with polyphenols, essential oils of ginger also show antibacterial activity against Gram-positive bacteria. The rhizome of *Zingiber officinale* contains 69 volatile compounds, which make up 97% of its essential oil composition. The active constituents present in essential oils of ginger include α-zingiberene (28.62%), α-curcumene (9.09%), α-farnesene (5.52%), β-bisabolene (5.40%), β-sesquiphellandrene, Camphene (9.32%), β-Phellandrene (7.97%), α-Pinene

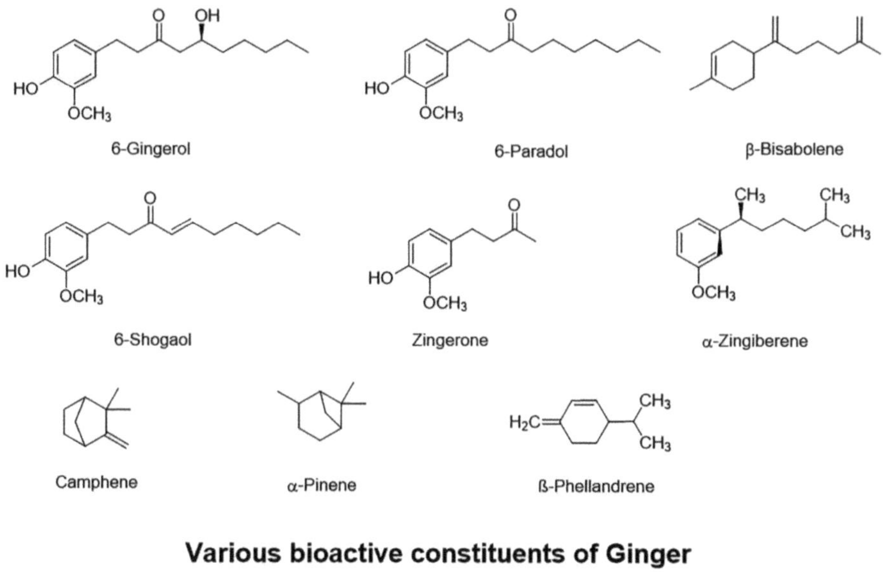

Fig. 11.1 Chemical structure of various bioactive constituents of ginger such as 6-gingerol 6-paradol, 6-shogaol, zingerone, camphene, etc

(2.57%) (Samani et al. 2013), and geranial, effective against *Staphylococcus aureus* and *Listeria monocytogenes*, while *Shigella flexneri*, *Salmonella typhimurium*, and *Escherichia coli* were resistant to the essential oil of ginger (Mahboubi 2019) (Malu et al. 2009) (Talebi et al. 2021).

Other bioactive components of ginger, which contributes toward its biological activity, include 6-gingerol, 6-shogaol, 10-gingerol, gingerdiones, gingerdiols, paradols, 6-dehydrogingerols, 12-gingerol 3,5-diacetoxy-6-gingerdioal, and 5-acetoxy-6-gingerol), among which 6-gingerol and 6-shogaol stand out as given in Fig. 11.1 (Arcusa et al. 2022) (Tanaka et al. 2015).

Oleoresins, a nonvolatile compound, are also found in rhizomes of *Zingiber officinale* and constitute around 88.6% of the total composition (Samani et al. 2013), among which Gingerols (1-(4-hydroxy-3-methoxyphenyl)-5-hydroxyalcan-3-one), Shogaols (1-(4-hydroxy-3-methoxy-phenyl)-decen-3-one), and Paradols are the most important groups. Shogaols are the more abundant components in the dried rhizome and gingerols are mainly found in the fresh rhizomes of ginger (Kiran et al. 2013) (Arcusa et al. 2022) (Schepici et al. 2021).

Recent era marks a progressive use of conventional drug therapies, but their adverse drug effects and pharmacokinetic instabilities contribute toward their inadequate efficacy, thus shifting the tables toward traditional medicinal plants. The medicinal plants are not only used for prevention of diseases but as well as a means of treatment by multiple pathological pathways, thus opening newer opportunities, which has led to their evaluation in the treatment of memory-related and neurodegenerative disorders, as it is the case of *Zingiber officinale* (Talebi et al. 2021)

(Mahdy et al. 2014). The characteristic properties, such as antioxidant and anti-inflammatory, contribute toward possible use of *Zingiber officinale* in neurodegenerative diseases, specifically Alzheimer's. The bioactive constituents present in ginger cross blood-brain barrier; thus, they can be used in the effective treatment multiple pathologies associated with neurodegenerative diseases (Simon et al. 2020). Additional clinical studies have shown that ginger supplementation increases the expression of nerve growth factor (NGF), which in turn improves memory, supports long-term hippocampal plasticity, and stimulates neurite outgrowth. Furthermore, certain preclinical studies in mice have indicated that elevated NGF levels in the hippocampus activate extracellular signal-regulated kinases (ERK) and cAMP response element-binding protein (CREB), leading to enhanced synaptogenesis (Moon et al. 2014). Additionally, ginger inhibits the expression of pro-inflammatory cytokines and chemokines in THP-1 cells, as well as adhesion and activation factors such as LPS, TNF-α, IL-1β, COX-2, MIP-1A, MCP-1, and IP-10 in mice. Research has also shown that oral administration of ethanolic ginger extract reduces neuroinflammation, leading to improvements in learning and memory, which is associated with increased expression of tight junction proteins and elevated SCFA levels. Furthermore, ginger treatment results in decreased cell death and reduced amyloid plaque accumulation (Zahedi et al. 2024).

While clinical studies in humans usually refer using mixture of herbs with ginger as one of the main ingredients among which Cognitex, a dietary supplement formulated with sage, blueberry, and *Zingiber officinale*, has been reported to show an improvement in cognitive abilities. Another placebo-controlled study using standardized ginger extracts reported improved cognitive processing abilities, with the most significant effects observed at the higher dose of 800 mg per day (Saenghong et al. 2012).

11.3 Molecular Mechanisms of Action in Alzheimer's Disease

A number of molecular mechanisms are involved in causing neuronal death in Alzheimer's disease such as Aβ aggregation, tau hyper phosphorylation, release of pro-inflammatory cytokines, and neuronal apoptosis. These various targets serve as treatment signals for anti-Alzheimer drugs against AD.

11.3.1 Antioxidant Activity

Oxidative stress is one of the earliest events in AD pathogenesis. The brain is particularly vulnerable to oxidative damage due to its high oxygen consumption and rich lipid content (Beal 1998). The oxidative imbalance generates free radicals by ROS, which in turn degenerates ageing cells and damages cellular macromolecules. The antioxidant mechanism of catalase, superoxide dismutase, and glutathione dismutase enzymes fights with these ROS (Mancuso et al. 2007a; Uttara et al. 2009). The imbalance between these enzymes and ROS level results in oxidative stress in cell followed by mitochondrial dysfunction, thus limiting the level of oxygen. The generation of O_2 and H_2O_2 during redox signaling in neurons leads to the formation of sulfenic acid (Conway and Lee 2015). Synaptic plasticity depends on calcium (Ca^{2+}) signaling. Brain cells use more oxygen than body (Halliwell 1992) and are totally dependent on ATP for its homeostasis. As energy needs increases, brain suffers oxidative stress. Reduced oxygen availability in brain cells leads to oxidative stress through the generation of reactive oxygen species (ROS), mitochondrial membrane damage, decreased cytochrome oxidase activity, impaired mitochondrial protein transport, disruption of the electron transport chain (ETC), and inhibition of ATP production, all of which contribute to neurodegeneration (Tripoli et al. 2025). ROS (reactive oxygen species) also promote Aβ aggregation, tau hyperphosphorylation, and neuronal apoptosis (Gutteridge and Halliwell 2000).

Brain cells have limited antioxidant capacity and primarily function through two mechanisms: preventing the generation of reactive oxygen species (ROS) and removing the harmful by-products resulting from oxidative stress (Minj et al. 2021). In dormant state, Keap 1 (Kelch-like ECH-associated protein) and Nrf2 (nuclear factor erythroid 2–related factor) (Dikovskaya et al. 2019) interact to regulate antioxidant effect. Under conditions of oxidative stress, Nrf2 gets dissociated from Keap1 and translocates into nucleus (Zhang et al. 2021); this increases the activity of antioxidant enzymes and reduces ROS intracellularly, thereby promoting neuroprotection, Fig. 11.2. The neurodegenerative diseases are usually associated with neurodegeneration, which is irreversible and associated with protein aggregation, neuroinflammation (Michalska and León 2020), neurotransmitters imbalance, mitochondrial dysfunction, and autophagy.

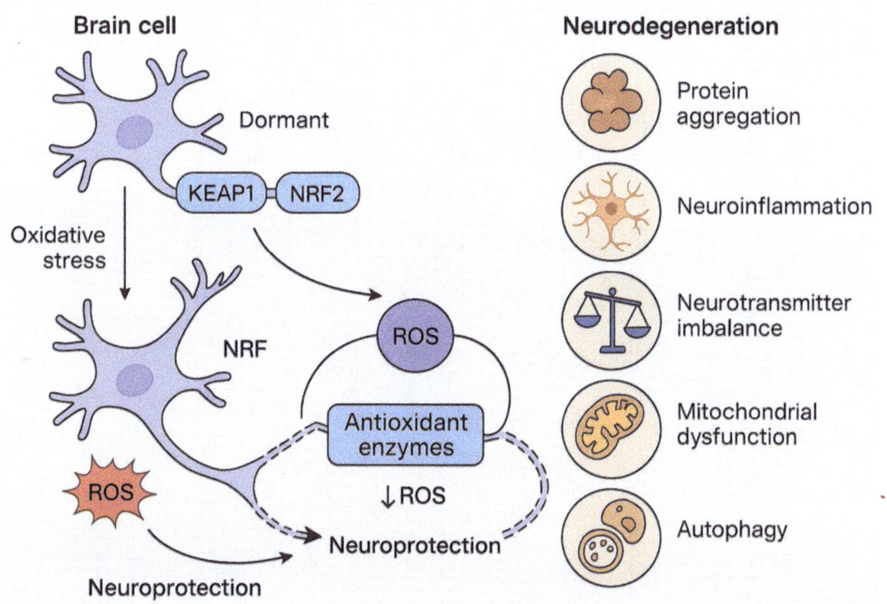

Fig. 11.2 ROS activity in brain cells under dormant state and state of oxidative stress

Ginger contains phenolic compounds, especially 6-gingerol and 6-shogaol, which possess potent free radical-scavenging properties. Research has shown that ginger boosts the expression and activity of endogenous antioxidant enzymes such as superoxide dismutase (SOD), catalase (CAT), and glutathione peroxidase (GPx), leading to decreased lipid peroxidation and reduced oxidative damage in neurons (Sahardi and Makpol 2019).

11.3.2 Anti-inflammatory Effects

Chronic neuroinflammation plays a central role in AD progression by affecting CNS (Stephenson et al. 2018). The inflammation-causing molecules are usually derived from damaged brain cells through infections, drugs, genetic mutations, or protein aggregation (Hwang et al. 2022). The formation of Aβ plaques due to ROS also induces activation of various inflammatory cells such as astrocytes and microglia. Activated microglia secrete pro-inflammatory cytokines like TNF-α, IL-1β, and IL-6, as well as chemokines, which intensify neuronal injury (Biber et al. 2008) as in Fig. 11.3. Aβ also increases activity of Nuclear Factor Kappa B (NF-κB) in brain cells, thus resulting in inflammation of neurons and their death (Friedlander 2003).

Ginger has been shown to suppress inflammation through several pathways. It inhibits the **NF-κB** signaling cascade, a central mediator of inflammatory gene transcription, and reduces microglial activation. Ginger constituents also modulate signaling molecules such as **MAPKs** and **Akt/STAT3**, leading to decreased expression of inflammatory mediators (Razak et al. 2023).

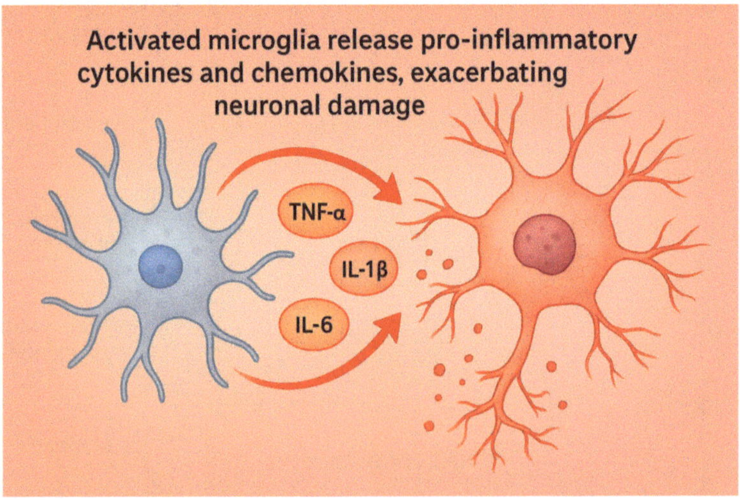

Fig. 11.3 Release of pro-inflammatory cytokines and chemokines resulting in neuroinflammation due to activated astrocytes and microglia

11.3.3 Modulation of Amyloid-Beta and Tau Pathology

Aβ aggregation and tau hyperphosphorylation are hallmark events in AD neuropathology. Alzheimer is associated with hyperphosphorylation of tau, resulting in dissociation of tau proteins, which, in turn, destabilizes microtubules from neuron. The dissociated tau proteins pair with another tau protein to form neurofibrillary tangles, thus affecting the structural integrity of neuron. This disaggregation of tau proteins from neurofibrillary tangles block transport of nutrients inside neurons, leading to tauopathy or neurodegeneration.

In a normal brain, APP, a transmembrane protein, helps in the survival, development, and repair of neurons. APP possesses two terminals as N-terminal and C-terminal (Turner et al. 2003)(Sastre et al. 2001). The C-terminal is first cleaved by the α-secretase enzyme, producing soluble APPα. This product is then further processed by γ-secretase, resulting in the formation of AICD (Amyloid Precursor Protein Intracellular Domain), which plays a role in regulating neuronal transcription and translocation (Siddappaji and Gopal 2021). In Alzheimer's disease, β-secretase (also called β-Amyloid Cleaving Enzyme-1 or BACE-1) processes amyloidogenic APP at the C-terminal end, resulting in the formation of soluble APPβ. This product subsequently activates caspase-6, which is responsible for triggering neuronal apoptosis (Selkoe 2002) and cell death as given in Fig. 11.4.

Some of the membrane-bound APP binds to γ-secretase and forms monomer fragments of Aβ. These Amyloid β peptides further aggregate to form insoluble oligomers, dense and senile Amyloid-β plaque (Ramesh et al. 2013). The so-formed plaques activate astrocytes and microglia to form cytokines and chemokines, which are responsible for the generation of ROS (Reactive Oxygen Species) (Hussain et al. 1999). The ROS generation induces mitochondrial oxidative stress and further activates an apoptotic caspase by producing p53, Bad, and Bax, resulting in lipid peroxidation, damage of membrane, and, ultimately, death of neurons (Town et al. 2002).

Ginger has demonstrated the ability to interfere with these processes. Molecular docking studies reveal that 6-gingerol binds directly to amyloidogenic sequences of Aβ, thereby preventing aggregation and fibril formation (Pan et al. 2024). Additionally, ginger compounds may downregulate **GSK-3β**, a key kinase involved in tau phosphorylation, through modulation of the **PI3K/Akt** pathway. This reduces the formation of neurofibrillary tangles, a key pathological hallmark in AD (Azam et al. 2014) as given in Fig. 11.5.

11.3.4 Cholinergic System Modulation

Cognitive symptoms in AD are partly attributed to deficits in cholinergic neurotransmission. Acetylcholinesterase (AChE) inhibitors remain the mainstay of symptomatic treatment.

Ginger extracts have shown inhibitory activity against AChE in vitro and in vivo, resulting in increased acetylcholine levels in the synaptic cleft. This may enhance

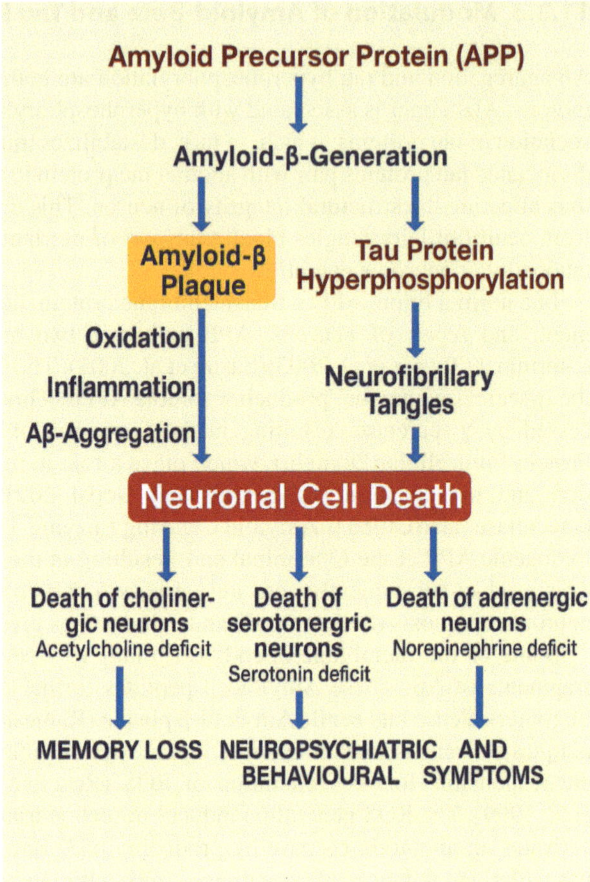

Fig. 11.4 The representation explains various molecular mechanisms responsible for the occurrence of AD due to neuronal cell death, which is dependent on APP

cognitive function and memory formation in AD models, supporting the rationale for using ginger as a natural cholinergic enhancer (Sahardi and Makpol 2019).

11.3.5 Upregulation of Neurotrophic Factors

Neurotrophic factors are the proteins produced by different cells which in turn are responsible for neurite growth and development. They belong to three different families viz neurotrophins or NGF family, glial cell line-derived neurotrophic factor (GDNF) family, and cytokine family. The NGF family comprises several neurotrophins, including brain-derived neurotrophic factor (BDNF), neurotrophin-3 (NT-3), and neurotrophin-4 (NT-4), also referred to as neurotrophin-5 (NT-5 or NT-4/5), all of which play crucial roles in synaptic plasticity, learning, and memory (Tomasello et al. 2021). Neurotrophins are being released by nervous system, gastrointestinal tract, and urinary tract (Liu 2018). They are usually released in pro-form then undergo proteolytic cleavage to form active neurotrophins (Proenca et al. 2016). The active

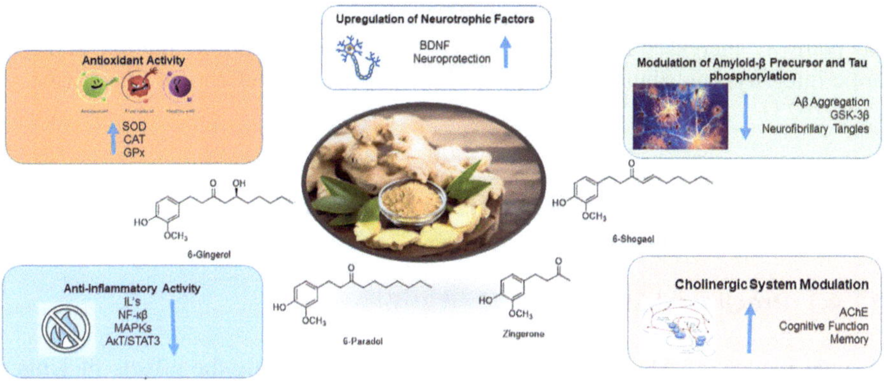

Fig. 11.5 Role of ginger and its various components in upregulation and downregulation of various pathways involved in the occurrence of AD

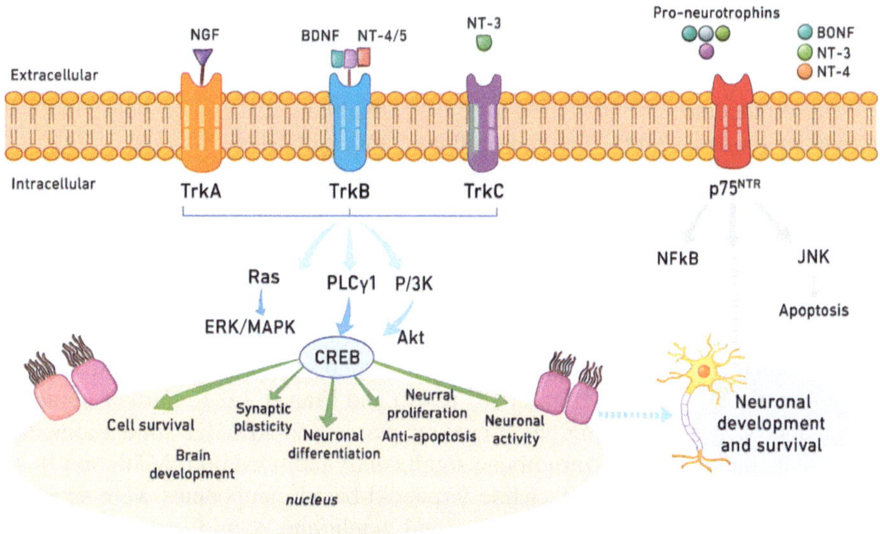

Fig. 11.6 Growth factors bind to Trk receptors, activating PI3K. PI3K activates Akt. Akt then phosphorylates and activates CREB in the nucleus. Activated CREB promotes the transcription of genes involved in cell survival, neuronal differentiation, and other neurotrophic effects

neurotrophins bind with Trk receptors strongly then p75 neurotrophin receptors. TrkA, TrkB and TrkC further acts on NGF, BDNF, NT-4/5 and NT-3 thus initiating further cascade of intracellular signaling pathways (Mohamed and El-Remessy 2015) as in Fig. 11.6. In AD, the level of neurotrophins gets changed due to polymorphism in genes (Chang et al. 2019). AD is usually associated with reduced BDNF expression which results in neuronal apoptosis (Rochat 2016).

Ginger has been shown to enhance BDNF levels through the **Akt/CREB** signaling pathway. Increased BDNF promotes neuronal survival, differentiation, and synaptic integrity, contributing to the overall neuroprotective effects of ginger (Razak et al. 2023).

11.4 Therapeutic Evidence from Preclinical and Clinical Studies

11.4.1 Preclinical Studies

Animal models of AD have been extensively used to study the effects of ginger. Findings include:

- Reduced Aβ plaque burden in the hippocampus.
- Decreased lipid peroxidation and inflammatory markers.
- Improved spatial learning and memory in Morris water maze and Y-maze tasks.
- Protection of hippocampal neurons from oxidative damage.

These outcomes are attributed to ginger's multitargeted action, particularly its impact on oxidative stress and neuroinflammatory pathways (Pan et al. 2024).

A wide array of in silico, in vitro, and in vivo models has been employed to explore the neuropharmacological potential of bioactive constituents derived from *Zingiber officinale* (ginger), as given in Table 11.1. Several molecular docking studies have revealed strong binding affinities of ginger compounds to key enzymes implicated in neurodegeneration. Specific docking analyses identified acetylcholinesterase (AChE) and butyrylcholinesterase (BChE) as principal targets, with interactions noted at critical binding sites such as Trp86 within the AChE active region (Azam et al. 2014) (Cuya et al. 2018) (Cuya and França 2020). Further computational approaches integrating target enzyme assays and ADMET simulations indicated that compounds like zerumbone significantly inhibited both AChE and BChE activities (Hwang et al. 2020), while terpenoid-based components were shown to enhance GABA-A receptor expression and ameliorate cognitive dysfunction via koniocortical network modeling (Aguiar-Furucho and Peláez 2020).

Animal model studies support these computational findings. Gingerol administered intraperitoneally (10 and 20 mg/kg) in Swiss albino mice led to a marked reduction in β-amyloid (Aβ42), β-secretase, APH1a, and COX-2 levels, while promoting α-secretase activity and preventing cognitive deficits (El Halawany et al. 2023). Similarly, methanolic extract of dried ginger exhibited potential calcium channel antagonism and cholinesterase inhibition in Sprague Dawley rats (Ghayur et al. 2008).

Combinatorial herbal formulations also demonstrated efficacy. Oral administration of a *Cyperus rotundus* and ginger extract (CP1) in Wistar rats (100–300 mg/kg) mitigated hippocampal oxidative stress, restored cholinergic tone, and reversed memory impairments (Sutalangka and Wattanathorn 2017). In C57BL/6 mice, both

Table 11.1 Various in silico, in vitro, and in vivo studies on neuropharmacological potential of ginger

S. no.	Bioactive component/ plant part providing neuroprotection	Dose	Study details
In silico studies			
1.	Various ginger-derived molecules	–	Docking studies revealed interactions at critical binding sites such as Trp86 within the AChE active region acetylcholinesterase for (AChE) and butyrylcholinesterase (BChE) as principal targets
2.	Zerumbone	–	Docking studies revealed inhibition of both AChE and BChE activities
3.	Terpenoid-based compounds	–	Docking studies revealed enhancement of GABA-A receptor expression and amelioration of cognitive dysfunction
In vitro and in vivo studies			
4.	Gingerol	10–20 mg/kg i.p.	Decreased levels of β-amyloid (Aβ42), β-secretase, APH1a, and COX-2 levels
5.	Zo.Cr extract of ginger		Calcium channel antagonism and cholinesterase inhibition
6.	Ginger + *Cyperus rotundus* extract	100–300 mg/kg	Reversal of memory impairment and restoration of cholinergic tone
7.	Fresh/dried ginger/6-gingerol	10–25 mg/kg	Improvement of behavioral performance in memory tasks and upregulation of BDNF
8.	Ginger root extract	4 g/kg	Decreased pro-inflammatory markers and enhanced cognition
9.	Ginger and peony root	50–100 mg/kg/ day	Inhibition of hippocampal Aβ aggregation, increased memory, and downregulation of neuroinflammatory markers
10.	6-Shogaol	10 mg/kg/day	Reduced astrocytic and hippocampal Aβ aggregation, enhanced cognition, and downregulation of neuroinflammatory markers
11.	Fermented ginger +6-Paradol	50–200 mg/kg	Hippocampal neuroprotection and increased memory
12.	Zerumbone	1 or 10 mg/ kg i.p. 25 mg/kg gavage	Improved memory and anxiety behavior Decreased Aβ plaque burden, improved social and nesting behaviors, and reduced inflammation
13.	Zingerone	2 or 8 mg/kg/ day	Reduced inflammation and oxidative stress
14.	Ginger essential oil	25 ppm	Neuroprotection
15.	6-Gingerol	3–10 μM	Improved SH-SY5Y cell viability and reduced Aβ25–35-induced cytotoxicity
16.	6-Gingerol	40–300 μM	Attenuated oxidative stress, inhibited Tau hyperphosphorylation, and decreased Aβ production
17.	Ginger extract	20 g/mL	Enhanced viability of PC12 and primary neuronal cells

(continued)

Table 11.1 (continued)

S. no.	Bioactive component/ plant part providing neuroprotection	Dose	Study details
18.	Dry ginger	0.02 and 0.2 mg in DMSO	Aβ aggregation and improved survival in primary rat hippocampal cells
19.	Extracts of ginger	1:20 to 1:100 ratios	Inhibition of amyloid aggregation
20.	Gingerol/Shogaol	20 μM	Reduced inflammation
21.	*Zingiber officinale, Echinacea purpurea, Centella asiatica*	150 mg/kg, 20 mg/kg, 200 mg/kg	Reduced inflammation

fresh and dried ginger (100 or 200 mg/kg), along with 6-gingerol (10 or 25 mg/kg), upregulated brain-derived neurotrophic factor (BDNF) expression and improved behavioral performance in memory tasks (Kim et al. 2018).

Dose-dependent effects of ginger root extract (GRE) were validated in Sprague Dawley rats, where higher doses (4 g/kg) significantly reduced pro-inflammatory markers and enhanced cognitive behavior (Zeng et al. 2013). In APP/PSEN1 transgenic mice, a combination of ginger and peony root (OCGP, 50 or 100 mg/kg/day) administered over 14 weeks inhibited hippocampal Aβ aggregation, improved memory, and downregulated neuroinflammatory markers (Lim et al. 2016).

6-Shogaol emerged as a potent bioactive compound, with multiple studies confirming its neuroprotective effects. In ICR mice, oral dosing (10 mg/kg/day) reduced astrocytic and hippocampal Aβ aggregation, improved memory, and downregulated neuroinflammatory markers (mg/kg in mice and 10 μM in HT22 cells), offered protection against Aβ1–42-induced cytotoxicity, improved cell viability, and suppressed amyloid aggregation (Na et al. 2016) (Na et al. 2017).

Additional studies with fermented ginger and 6-paradol combinations in ICR mice (50–200 mg/kg) showed hippocampal neuroprotection and memory enhancement, particularly at higher fermented ginger doses (Huh et al. 2018). In Sprague Dawley rats, intraperitoneal zerumbone (1 or 10 mg/kg) improved spatial memory and alleviated anxiety-like behaviors (Jafarian et al. 2019). In APP/PSEN1 mice, zerumbone (25 mg/kg gavage) and its application to microglial cultures (1–10 μg/mL) reduced Aβ plaque burden, improved social and nesting behaviors, and suppressed inflammatory cascades both in vivo and in vitro (Li et al. 2020).

Zingerone, another ginger derivative, reduced systemic inflammation and oxidative stress in YPEN-1 prostate endothelial cells and kidney cells harvested from rats pretreated with zingerone (2 or 8 mg/kg/day) (Kim et al. 2010). Ginger essential oil (25 ppm) offered neuroprotection against peroxide-induced cytotoxicity in GT1–7 cells while enhancing tamoxifen neurotoxicity (Mizuno et al. 2015).

Pretreatment with 6-gingerol (3–10 μM) improved SH-SY5Y cell viability and reduced Aβ25–35-induced cytotoxicity (Lee et al. 2011). In PC12 cells, 6-gingerol (40–300 μM) attenuated oxidative stress, inhibited tau hyperphosphorylation, and decreased Aβ production (Zeng et al. 2015). Similarly, ginger extracts enhanced viability of PC12 and primary neuronal cells at 20 g/mL concentrations (Kim et al. 2007).

Dry ginger prevented Aβ aggregation and improved survival in primary rat hippocampal cells when applied at concentrations between 0.02 and 0.2 mg in DMSO (Tung et al. 2017). Extracts of ginger from AD transgenic mouse brain tissue diluted at 1:20 to 1:100 ratios inhibited amyloid aggregation (Guo et al. 2010).

Anti-inflammatory efficacy was also validated in human THP-1 monocytic cells where a ginger and Alpinia galangal blend (255 mg pretreatment) and isolated gingerols/shogaols (20 μM) effectively reduced inflammatory responses (Grzanna et al. 2004)(Ho and Chang 2018). Additionally, exosome-like nanoparticles derived from ginger rhizomes suppressed inflammasome formation in bone-marrow-derived macrophages (Chen et al. 2019).

In a mouse model of systemic neuroinflammation, the anti-inflammatory activity was also confirmed. A combination of *Zingiber officinale* (150 mg/kg), *Echinacea purpurea* (20 mg/kg), and *Centella asiatica* (200 mg/kg) demonstrated encouraging outcomes. The studies reported downfall in C3 in prefrontal cortex of LPS-treated mice after administration of the mixture. This downfall is associated with anti-inflammatory effect and enhancement of cognitive memory (Micheli et al. 2022).

11.4.2 Clinical Studies

While human studies remain limited, preliminary trials are promising. In a double-blind, placebo-controlled study, daily supplementation with ginger extract improved working memory and attention in healthy middle-aged women. Another study in patients with mild cognitive impairment showed decreased oxidative stress biomarkers and improved cognitive scores after ginger administration (Razak et al. 2023).

Larger, multicenter trials are needed to confirm these findings and evaluate ginger's long-term efficacy and safety in AD populations.

11.4.3 Safety, Tolerability, and Bioavailability

Ginger is generally recognized as safe (GRAS) by the FDA and has a long history of culinary and medicinal use. However, limitations include:

- *Low Bioavailability:* Rapid metabolism and poor intestinal absorption reduce systemic exposure, thus reducing its availability. However, strategies, such as nanoformulations, coadministration with bioenhancers (e.g., piperine), and Liposomal and solid-lipid carriers, can increase its bioavailability.
- *Challenges in Translation*: Despite preclinical promise, clinical application faces major barriers.
- *Standardization:* Variability in plant sources and extraction processes affects reproducibility.
- *Regulatory Issues:* Classification as supplements limits regulatory oversight.

- *Drug Interactions:* Potential interactions with conventional therapies need careful evaluation.

Emerging studies on ginger nanoparticles suggest enhanced brain delivery and improved therapeutic efficacy in neurodegenerative models (Yücel et al. 2022).

11.5 Resistance and Limitations of Current Therapies

11.5.1 Limited Efficacy and Symptomatic Relief

Current treatments mainly target neurotransmitter systems without addressing the root causes of neurodegeneration such as Cholinesterase inhibitors (e.g., donepezil, rivastigmine) and NMDA receptor antagonists (e.g., memantine) offer temporary cognitive benefits without altering disease progression. Similarly, long-term use of levodopa leads to motor complications and reduced efficacy. Other approved agents like riluzole and tetrabenazine yield only modest symptomatic improvements. These drugs suffer from poor blood-brain barrier (BBB) penetration, short half-lives, and adverse side effects such as nausea, insomnia, and hallucinations (Cummings et al. 2016).

11.5.2 Mechanistic Complexity and Multifactorial Pathology

NDDs involve diverse pathogenic mechanisms including oxidative stress, mitochondrial dysfunction, protein misfolding, neuroinflammation, excitotoxicity, and autophagy impairment. Single-target interventions often fail to produce significant clinical outcomes due to disease heterogeneity and complexity.

11.5.3 Drug Resistance and Tolerance

Long-term medication use may lead to diminished therapeutic response. For instance, levodopa-induced dyskinesia is a major issue in PD, while sustained AD treatment fails to prevent cognitive decline and may provoke treatment resistance (Arbo et al. 2020).

11.6 Future Directions

Ginger is raising concerns over its use in treating AD, but researchers have to work on its low bioavailability and poor solubility. The future directions include overcoming these concerns by implementing the following strategies:

- *Advanced Delivery Systems:* Nanoformulations and liposomal carriers may improve BBB penetration and bioavailability.

- *Combination Therapies:* Coadministration with standard drugs could yield synergistic effects.
- *Systems Biology:* Integrative omics approaches may elucidate the multitarget actions of phytochemicals.
- *Personalized Medicine:* Genetic profiling (e.g., APOE status) could inform individualized treatment plans.
- *Personalized Combination Regimens:* Using precision medicine tools and multi-omics.
- *AI-Guided Drug Synergy Screening:* Identifying optimal pairs from thousands of candidates.
- *Triple Therapy Trials:* Combining antiamyloid, antitau, and anti-inflammatory agents.
- *Natural-Synthetic Synergy Studies:* Rigorous testing of nutraceutical-pharmaceutical combos.

Out of the various possible directions, combination therapy can be proved as one of the best possible approaches with earlier available drugs as an adjuvant to enhance its anti-AD potential.

11.7 The Potential for Combination Therapy in Alzheimer's Disease

Alzheimer's disease (AD) is a multifactorial neurodegenerative disorder characterized by the accumulation of amyloid-beta (Aβ) plaques, tau protein tangles, neuroinflammation, oxidative stress, mitochondrial dysfunction, and synaptic loss. Current monotherapies, including cholinesterase inhibitors and NMDA receptor antagonists, provide only modest symptomatic relief and fail to halt disease progression (Cummings et al. 2016). This limitation has fueled interest in combination therapy, which targets multiple pathogenic mechanisms simultaneously.

11.7.1 Rationale for Combination Therapy

11.7.1.1 Multifactorial Pathology Requires Multimodal Intervention

AD pathogenesis involves diverse biological pathways. Monotherapy often targets only one aspect, for example, Aβ or cholinergic transmission, while neglecting others, such as neuroinflammation or tau pathology. Combination therapy offers a more comprehensive approach by addressing several mechanisms concurrently (Tanzi 2012).

11.7.1.2 Precedent in Other Chronic Diseases

Combination therapy is a standard in chronic diseases like cancer, HIV, and hypertension, where multiple agents with different targets improve outcomes and reduce resistance (Karran and De Strooper 2016). A similar paradigm is increasingly being applied to neurodegenerative diseases, including AD.

11.7.2 Promising Combination Strategies in AD

11.7.2.1 Pharmacological Combinations

- *Donepezil + Memantine:* Approved in moderate-to-severe AD. Donepezil enhances cholinergic function, while memantine modulates glutamatergic excitotoxicity. Studies show improved cognition and functional outcomes over monotherapy (Howard et al. 2012).
- *Antiamyloid + Antitau therapies:* As new biologics emerge (e.g., lecanemab, donanemab), future regimens may combine Aβ immunotherapy with tau aggregation inhibitors.
- *NSAIDs + Antioxidants:* Agents like ibuprofen combined with vitamin E or curcumin have shown synergistic effects in reducing inflammation and oxidative damage in preclinical models.

11.7.2.2 Drug + Natural Compound Approaches

- *Rivastigmine + Curcumin:* Enhances cholinergic transmission and reduces oxidative stress and Aβ accumulation (Mancuso et al. 2007b).
- *Memantine + Resveratrol:* Provides neuroprotection through NMDA antagonism and SIRT1 activation.

11.7.2.3 Polyherbal and Nutraceutical Combinations

Formulations combining *Bacopa monnieri*, *Ginkgo biloba*, omega-3 fatty acids, and B-vitamins have been studied for memory enhancement and synaptic support. These regimens are well-tolerated and potentially synergistic (Calabrese et al. 2008)

11.7.3 Challenges in Combination Therapy

- *Pharmacokinetics and Drug Interactions:* Risk of adverse interactions or altered metabolism.
- *Clinical Trial Design:* Need for large, well-powered trials to test multiagent regimens.
- *Cost and Regulatory Hurdles:* Especially for combinations involving both drugs and supplements.
- *Patient Heterogeneity:* Requires personalized approaches based on biomarkers (e.g., APOE status).

11.8 Conclusions

Ginger exhibits promising therapeutic potential in the context of Alzheimer's disease through its multifaceted molecular actions, including antioxidant, anti-inflammatory, antiamyloidogenic, and neurotrophic mechanisms. Its ability to modulate multiple pathological processes positions it as a compelling candidate for adjunctive therapy in AD management.

However, while preclinical data are robust, clinical evidence is still emerging. Future work should focus on well-designed clinical trials, standardized dosing, improved formulations, and long-term safety assessment to validate ginger's role as a neuroprotective agent in human AD. Neurodegenerative diseases pose a substantial clinical challenge due to their multifactorial etiology and limited therapeutic options. Natural compounds, with their broad-spectrum activity, favorable safety profiles, and historical medicinal usage, present a compelling avenue for research and development. However, rigorous clinical trials, improved delivery technologies, and standardized formulations are necessary to translate their potential into effective, evidence-based therapies. Continued interdisciplinary collaboration is essential to advance natural products as integral components of NDD management.

Combination therapy holds significant promise for transforming the treatment landscape of Alzheimer's disease. By targeting multiple pathological processes simultaneously, such regimens may overcome the limitations of current monotherapies. Continued research, innovative trial designs, and a precision medicine approach will be essential to fully realize the potential of combination therapy in AD.

References

Aguiar-Furucho MA, Peláez FJR (2020) Alzheimer's disease as a result of stimulus reduction in a GABA-a-deficient brain: a Neurocomputational model. Neural Plast 2020:1–26. https://doi.org/10.1155/2020/8895369

Arbo BD et al (2020) 'Resveratrol derivatives as potential treatments for Alzheimer's and Parkinson's disease. Front Aging Neurosci 12(103):1–15. https://doi.org/10.3389/fnagi.2020.00103

Arcusa R et al (2022) Potential role of ginger (*Zingiber officinale* roscoe) in the prevention of neurodegenerative diseases. Front Nutr 9:1–14. https://doi.org/10.3389/fnut.2022.809621

Azam F et al (2014) Ginger components as new leads for the design and development of novel multi-targeted anti-Alzheimer's drugs: a computational investigation. Drug Des Devel Ther 8:2045–2059. https://doi.org/10.2147/DDDT.S67778

Banerjee S et al (2011) Zingiber officinale: a natural gold. Int J Pharm Bio Sci 2(1):283. Available at: www.ijpbs.net

Beal MF (1998) Mitochondrial dysfunction in neurodegenerative diseases. Biochim et Biophys Acta (BBA) Bioenerget 1366(1–2):211–223. https://doi.org/10.1016/S0005-2728(98)00114-5

Bhandari DD et al (2024) Neuroprotective potential of heterocyclic scaffolds vs herbal treatments: an overview. Res J Pharm Technol 17(6):2943–2948. https://doi.org/10.52711/0974-360X.2024.00460

Biber K, Vinet J, Boddeke HWGM (2008) Neuron-microglia signaling: chemokines as versatile messengers. J Neuroimmunol 198(1):69–74. https://doi.org/10.1016/j.jneuroim.2008.04.012

Calabrese C et al (2008) Effects of a standardized Bacopa monnieri extract on cognitive performance, anxiety, and depression in the elderly: a randomized, double-blind, placebo-controlled trial. J Altern Complement Med 14(6):707–713. https://doi.org/10.1089/ACM.2008.0018

Chang HM et al (2019) Neurotrophins and glial cell line-derived neurotrophic factor in the ovary: physiological and pathophysiological implications. Hum Reprod Update 25(2):224–242. https://doi.org/10.1093/humupd/dmy047

Chen X, Zhou Y, Yu J (2019) Exosome-like nanoparticles from ginger rhizomes inhibited NLRP3 Inflammasome activation. Mol Pharm 16(6):2690–2699

Conway ME, Lee C (2015) The redox switch that regulates molecular chaperones. Biomol Concepts 6(4):269–284. https://doi.org/10.1515/bmc-2015-0015

Cummings J et al (2016) Drug development in Alzheimer's disease: the path to 2025. Alzheimer's Res Ther 8(1). https://doi.org/10.1186/S13195-016-0207-9

Cuya T, França TCC (2020) A molecular modeling study of components of the ginger (*Zingiber officinale*) extract inside human butyrylcholinesterase: implications for Alzheimer disease. J Biomol Struct Dyn 38(9):2809–2815. https://doi.org/10.1080/07391102.2019.1644198

Cuya T, Baptista L et al (2018) A molecular dynamics study of components of the ginger (*Zingiber officinale*) extract inside human acetylcholinesterase: implications for Alzheimer disease. J Biomol Struct Dyn 36(14):3843–3855. https://doi.org/10.1080/07391102.2017.1401004

Dikovskaya D et al (2019) Measuring the interaction of transcription factor Nrf2 with its negative regulator Keap1 in single live cells by an improved FRET/FLIM analysis. Chem Res Toxicol 32(3):500–512. https://doi.org/10.1021/acs.chemrestox.8b00354

El Halawany AM et al (2023) Protective effects of gingerol on streptozotocin-induced sporadic Alzheimer's disease: emphasis on inhibition of β-amyloid, COX-2, alpha-, beta - secretases and APH1a. Sci Rep 13(1):9436. https://doi.org/10.1038/s41598-023-36652-w

Friedlander RM (2003) Apoptosis and caspases in neurodegenerative diseases. N Engl J Med 348(14):1365–1375. https://doi.org/10.1056/nejmra022366

Ghayur MN et al (2008) Muscarinic, Ca++ antagonist and specific butyrylcholinesterase inhibitory activity of dried ginger extract might explain its use in dementia. J Pharm Pharmacol 60(10):1375–1383. https://doi.org/10.1211/JPP/60.10.0014

Grzanna R et al (2004) Ginger extract inhibits β-amyloid peptide-induced cytokine and chemokine expression in cultured THP-1 monocytes. J Altern Complement Med 10(6):1009–1013. https://doi.org/10.1089/acm.2004.10.1009

Guo JP, Yu S, McGeer PL (2010) Simple in vitro assays to identify amyloid-β aggregation blockers for Alzheimer's disease therapy. J Alzheimer's Dis 19(4):1359–1370. https://doi.org/10.3233/JAD-2010-1331

Gutteridge JMC, Halliwell B (2000) Free radicals and antioxidants in the year 2000. A historical look to the future. Ann N Y Acad Sci 899(1):136–147. https://doi.org/10.1111/j.1749-6632.2000.tb06182.x

Hafez DA (2010) Effect of extracts of ginger Goots and cinnamon bark on fertility of male diabetic rats. J Am Sci 6(10):1–8

Halliwell B (1992) Reactive oxygen species and the central nervous system. J Neurochem 59(5):1609–1623. https://doi.org/10.1111/j.1471-4159.1992.tb10990.x

Ho SC, Chang YH (2018) Comparison of inhibitory capacities of 6-, 8- and 10-Gingerols/Shogaols on the canonical NLRP3 Inflammasome-mediated IL-1β secretion. Molecules 23(2):466. https://doi.org/10.3390/molecules23020466

Howard R et al (2012) Donepezil and Memantine for moderate-to-severe Alzheimer's disease. N Engl J Med 366(10):893–903. https://doi.org/10.1056/nejmoa1106668

Huh E et al (2018) Ginger fermented with: Schizosaccharomyces pombe alleviates memory impairment via protecting hippocampal neuronal cells in amyloid beta1-42 plaque injected mice. Food and Function 9(1):171–178. https://doi.org/10.1039/c7fo01149k

Hussain I et al (1999) Identification of a novel aspartic protease (asp 2) as β-secretase. Mol Cell Neurosci 14(6):419–427. https://doi.org/10.1006/mcne.1999.0811

Hwang J et al (2020) Biological and computational studies for dual Cholinesterases inhibitory effect of Zerumbone. Nutrients 12(5):1215. https://doi.org/10.3390/nu12051215

Hwang Y, Kim HC, Shin EJ (2022) Effect of rottlerin on astrocyte phenotype polarization after trimethyltin insult in the dentate gyrus of mice. J Neuroinflammation 19(142):1–23. https://doi.org/10.1186/s12974-022-02507-w

Jafarian S et al (2019) Effect of zerumbone on scopolamine-induced memory impairment and anxiety-like behaviours in rats. Alzheimer's and Dement Transl Res Clin Interv 22(5):637–643. https://doi.org/10.1016/j.trci.2019.09.009

Jagtap SR et al (2022) Memory enhancing activity of ginger (*Zingiber officinale*), its treatments in dementia and alzheimers disease. Int J Res Appl Sci Biotechnol 9(3):73–84. https://doi.org/10.31033/ijrasb.9.3.14

Karran E, De Strooper B (2016) The amyloid cascade hypothesis: are we poised for success or failure? J Neurochem 139:237–252. https://doi.org/10.1111/jnc.13632

Kim DSHL, Kim JY, Han YS (2007) Alzheimer's disease drug discovery from herbs: Neuroprotectivity from β-amyloid (1-42) insult. J Altern Complement Med 13(3):333–340. https://doi.org/10.1089/ACM.2006.6107

Kim MK et al (2010) Modulation of age-related NF-κB activation by dietary zingerone via MAPK pathway. Exp Gerontol 45(6):419–426. https://doi.org/10.1016/j.exger.2010.03.005

Kim CY et al (2018) Neuroprotective effect and molecular mechanism of [6]-Gingerol against scopolamine-induced amnesia in C57BL/6 mice. Evid Based Complement Alternat Med 2018(1):8941564. https://doi.org/10.1155/2018/8941564

Kiran CR et al (2013) Influence of cultivar and maturity at harvest on the essential oil composition, oleoresin and [6]-gingerol contents in fresh ginger from Northeast India. J Agric Food Chem 61(17):4145–4154. https://doi.org/10.1021/jf400095y

Kumar A, Kumar R, Kumar A (2012) Pharmacological activity of *Zingiber officinale*. Int J Pharm Chem Sci 1(3):1073–1078

Lee C et al (2011) [6]-Gingerol attenuates β-amyloid-induced oxidative cell death via fortifying cellular antioxidant defense system. Food Chem Toxicol 49(6):1261–1269. https://doi.org/10.1016/j.fct.2011.03.005

Li L et al (2020) Zerumbone ameliorates behavioral impairments and neuropathology in transgenic APP/PS1 mice by suppressing MAPK signaling. J Neuroinflammation 17(61):1–20. https://doi.org/10.1186/s12974-020-01744-1

Lim S et al (2016) An optimized combination of ginger and Peony root effectively inhibits amyloid-β accumulation and amyloid-β-mediated pathology in AβPP/PS1 double-transgenic mice. J Alzheimer's Dis 50(1):189–200. https://doi.org/10.3233/jad-150839

Liu S (2018) Neurotrophic factors in enteric physiology and pathophysiology. Neurogastroenterol Motil 30(10):e13446. https://doi.org/10.1111/nmo.13446

Mahboubi M (2019) *Zingiber officinale* Rosc. Essential oil, a review on its composition and bioactivity. Clin Phytoscience 5(1):1–12. https://doi.org/10.1186/S40816-018-0097-4

Mahdy KA et al (2014) Protective effect of ginger (*Zingiber officinale*) on Alzheimer's disease induced in rats. J Neuroinf Dis 5(2):1–10

Malu S et al (2009) Antibacterial activity and medicinal properties of ginger (Zingiber officinale). Global J Pure Appl Sci 15(3–4):365–368. https://doi.org/10.4314/gjpas.v15i3-4.48561

Mancuso C, Scapagnini G et al (2007a) Mitochondrial dysfunction, free radical generation and cellular stress response in neurodegenerative disorders. Front Biosci 12(3):1107–1123. https://doi.org/10.2741/2130/PDF

Mancuso C, Bates TE et al (2007b) Natural antioxidants in Alzheimer's disease. Expert Opin Investig Drugs 16(12):1921–1931. https://doi.org/10.1517/13543784.16.12.1921

Michalska P, Leốn R (2020) When it comes to an end: oxidative stress crosstalk with protein aggregation and Neuroinflammation induce neurodegeneration. Antioxidants 9(8):740. https://doi.org/10.3390/ANTIOX9080740

Micheli L et al (2022) Efficacy of a vegetal mixture composed of *Zingiber officinale*, *Echinacea purpurea*, and *Centella asiatica* in a mouse model of neuroinflammation: I*n vivo* and *ex vivo* analysis. Front Nutr 9:887378. https://doi.org/10.3389/FNUT.2022.887378/BIBTEX

Minj E, Upadhayay S, Mehan S (2021) Nrf2/HO-1 signaling activator Acetyl-11-keto-beta Boswellic acid (AKBA)-mediated neuroprotection in methyl mercury-induced experimental model of ALS. Neurochem Res 46(11):2867–2884. https://doi.org/10.1007/S11064-021-03366-2

Mizuno D et al (2015) An *in vitro* system comprising immortalized hypothalamic neuronal cells (GT1–7 cells) for evaluation of the neuroendocrine effects of essential oils. Evid Based Complement Alternat Med 1:343942. https://doi.org/10.1155/2015/343942

Mohamed R, El-Remessy AB (2015) Imbalance of the nerve growth factor and its precursor: implication in diabetic retinopathy. J Clin Exp Ophthalmol 6(5). https://doi.org/10.4172/2155-9570.1000483

Moon M et al (2014) 6-Shogaol, an active constituent of ginger, attenuates neuroinflammation and cognitive deficits in animal models of dementia. Biochem Biophys Res Commun 449(1):8–13. https://doi.org/10.1016/J.BBRC.2014.04.121

Na JY et al (2016) 6-Shogaol has anti-amyloidogenic activity and ameliorates Alzheimer's disease via CysLT1R-mediated inhibition of cathepsin B. Biochem Biophys Res Commun 477(1):96–102. https://doi.org/10.1016/J.BBRC.2016.06.026

Na JY et al (2017) Sortilin-related receptor 1 interacts with amyloid precursor protein and is activated by 6-shogaol, leading to inhibition of the amyloidogenic pathway. Biochem Biophys Res Commun 484(4):890–895. https://doi.org/10.1016/J.BBRC.2017.02.029

Naseem E, Shamim M, Khan NI (2016) Cardioprotective effects of herbal mixture (ginger, garlic, lemon, apple cider vinegar & honey) in experimental animal models of hyperlipidemia. Int J Biol Res 4(1):28–33

Oboh G, Ademiluyi AO, Akinyemi AJ (2012) Inhibition of acetylcholinesterase activities and some pro-oxidant induced lipid peroxidation in rat brain by two varieties of ginger (*Zingiber officinale*). Exp Toxicol Pathol 64(4):315–319. https://doi.org/10.1016/j.etp.2010.09.004

Pan Y et al (2024) Screening of active substances regulating Alzheimer's disease in ginger and visualization of the effectiveness on 6-Gingerol pathway targets. Foods 13(4). https://doi.org/10.3390/FOODS13040612

Proenca CC, Song M, Lee FS (2016) Differential effects of BDNF and neurotrophin 4 (NT4) on endocytic sorting of TrkB receptors. J Neurochem 138(3):397–406. https://doi.org/10.1111/JNC.13676

Ramesh G, Maclean AG, Philipp MT (2013) Cytokines and chemokines at the crossroads of Neuroinflammation, neurodegeneration, and neuropathic pain. Mediat Inflamm 1:480739. https://doi.org/10.1155/2013/480739

Razak AM et al (2023) Modulating effects of Zingiberaceae phenolic compounds on neurotrophic factors and their potential as Neuroprotectants in brain disorders and age-associated neurodegenerative disorders: a review. Nutrients 15(11). https://doi.org/10.3390/NU15112564

Rochat C (2016) Selective vulnerability of neuronal subtypes in als a fertile ground for the identification of therapeutic targets' update on amyotrophic lateral sclerosis. InTech. Available at: https://doi.org/10.5772/63703

Saenghong N et al (2012) *Zingiber officinale* improves cognitive function of the middle-aged healthy women. Evid Based Complement Alternat Med 1:1–9. https://doi.org/10.1155/2012/383062

Sahardi NFNM, Makpol S (2019) Ginger (*Zingiber officinale* roscoe) in the prevention of ageing and degenerative diseases: review of current evidence. Evid Based Complement Alternat Med:5054395. https://doi.org/10.1155/2019/5054395

Samani MA, Rafieian MK, Azimi N (2013) Gundelia: a systematic review of medicinal and molecular perspective. Pak J Biol Sci 16(21):1238–1247. https://doi.org/10.3923/PJBS.2013.1238.1247

Sastre M et al (2001) Presenilin-dependent γ-secretase processing of β-amyloid precursor protein at a site corresponding to the S3 cleavage of notch. EMBO Rep 2(9):835–841. https://doi.org/10.1093/EMBO-REPORTS/KVE180

Schepici G et al (2021) Ginger, a possible candidate for the treatment of dementias? Molecules 26(18). https://doi.org/10.3390/molecules26185700

Selkoe DJ (2002) Deciphering the genesis and fate of amyloid β-protein yields novel therapies for Alzheimer disease. J Clin Invest 110(10):1375–1381. https://doi.org/10.1172/JCI16783

Siddappaji KK, Gopal S (2021) Molecular mechanisms in Alzheimer's disease and the impact of physical exercise with advancements in therapeutic approaches. AIMS Neuroscience 8(3):357. https://doi.org/10.3934/NEUROSCIENCE.2021020

Simon A et al (2020) Blood-brain barrier permeability study of ginger constituents. J Pharm Biomed Anal 177. https://doi.org/10.1016/j.jpba.2019.112820

Stephenson J et al (2018) Inflammation in CNS neurodegenerative diseases. Immunology 154(2):204–219

Sutalangka C, Wattanathorn J (2017) Neuroprotective and cognitive-enhancing effects of the combined extract of *Cyperus rotundus* and *Zingiber officinale*. BMC Complement Altern Med 17(1). https://doi.org/10.1186/S12906-017-1632-4

Talebi M et al (2021) *Zingiber officinale* ameliorates Alzheimer's disease and cognitive impairments: lessons from preclinical studies. Biomed Pharmacother 133. https://doi.org/10.1016/j.biopha.2020.111088

Tanaka K et al (2015) Analysis of chemical properties of edible and medicinal ginger by metabolomics approach. Biomed Res Int 2015. https://doi.org/10.1155/2015/671058

Tanzi RE (2012) The genetics of Alzheimer disease. Cold Spring Harb Perspect Med 2(10). https://doi.org/10.1101/CSHPERSPECT.A006296

Tomasello F et al (2021) The roles of Neurotrophins in traumatic brain injury. Life 12(1):26. https://doi.org/10.3390/LIFE12010026

Town T et al (2002) p35/Cdk5 pathway mediates soluble amyloid-β peptide-induced tau phosphorylation in vitro. J Neurosci Res 69(3):362–372

Tripoli F, Ridolfi L, Scarsoglio S (2025) Assessing the cardiac function from micro-gravity to hyper-gravity conditions through a validated multiscale modelling approach. J Physiol. https://doi.org/10.1113/JP287142

Tung BT et al (2017) Antioxidant and acetylcholinesterase inhibitory activities of ginger root (*Zingiber officinale* roscoe) extract. J Complement Integr Med 14(4). https://doi.org/10.1515/jcim-2016-0116

Turner PR et al (2003) Roles of amyloid precursor protein and its fragments in regulating neural activity, plasticity and memory. Prog Neurobiol 70(1):1–32. https://doi.org/10.1016/S0301-0082(03)00089-3

Uttara B et al (2009) Oxidative stress and neurodegenerative diseases: a review of upstream and downstream antioxidant therapeutic options. Curr Neuropharmacol 7(1):65–74. https://doi.org/10.2174/157015909787602823

Yücel Ç et al (2022) Immunomodulatory and anti-inflammatory therapeutic potential of gingerols and their nanoformulations. Front Pharmacol 13:902551. https://doi.org/10.3389/FPHAR.2022.902551/XML/NLM

Zahedi E et al (2024) Ginger extract improves cognitive dysfunction via modulation of gut microbiota-derived short-chain fatty acids in D-galactose/ovariectomy-induced Alzheimer-like disease. Mol Neurobiol 62(4):5095–5108. https://doi.org/10.1007/S12035-024-04583-W/METRICS

Zeng GF et al (2013) Protective effects of ginger root extract on Alzheimer disease-induced behavioral dysfunction in rats. Rejuvenation Res 16(2):124–133. https://doi.org/10.1089/REJ.2012.1389

Zeng GF et al (2015) The role of 6-Gingerol on inhibiting amyloid β protein-induced apoptosis in PC12 cells. Rejuvenation Res 18(5):413–421. https://doi.org/10.1089/REJ.2014.1657

Zhang Q et al (2021) Activation of Nrf2/HO-1 signaling: an important molecular mechanism of herbal medicine in the treatment of atherosclerosis via the protection of vascular endothelial cells from oxidative stress. J Adv Res 34:43–63. https://doi.org/10.1016/J.JARE.2021.06.023

Carotenoids and Memory Enhancement: A Therapeutic Link to Alzheimer's Disease

12

Sunayna Choudhary, Navidha Aggarwal, and Seema Bansal

Abstract

Carotenoids, naturally occurring pigments, are found to be an effective antioxidant and anti-inflammatory component and exhibit neuroprotective and therapeutic potential in various experimental models of Alzheimer's disease (AD). Together, they provide a new and exciting nutraceutical strategy against the increasing challenge of neurodegenerative disease in an ageing population. This chapter focuses on the medicinal properties of carotenoids in memory enhancement and the prevention of AD, emphasizing mechanisms such as scavenging reactive oxygen species, regulating mitochondrial protection, modulating neuroinflammation, and acting on the gut–brain axis. Preclinical and clinical data have recently shown that carotenoids (xanthophylls and carotenes, respectively) protect against brain ageing via the regulation of key molecular pathways, including the nuclear factor erythroid 2-related factor 2 (Nrf2), the brain-derived neurotrophic factor (BDNF), and amyloid-β clearance pathways. In order to translate these exciting discoveries into clinical therapies, future approach should include organized clinical trials and new strategies for developing new drug delivery systems to improve bioavailability.

Keywords

Carotenoids · Alzheimer's disease · Memory · Gut–brain axis · Neuroinflammation · Astaxanthin · Lutein · Lycopene · BDNF · Nrf2

S. Choudhary (✉) · N. Aggarwal · S. Bansal
M. M. College of Pharmacy, Maharishi Markandeshwar (Deemed to be University), Ambala, Haryana, India

J. Malik et al. (eds.), *Diet and Alzheimer's Disease: Let Food be Our Medicine*, Nutritional Neurosciences, https://doi.org/10.1007/978-981-95-2736-6_12

Abbreviations

ACh	Acetylcholine
AChE	Acetylcholinesterase
AD	Alzheimer's disease
AMPA	α-Amino-3-hydroxy-5-methyl-4-isoxazolepropionic acid
AMPK	AMP-activated protein kinase
APP	Amyloid precursor protein
AST	Astaxanthin
Aβ	Amyloid beta
BACE1	β-site APP-cleaving enzyme 1
BBB	Blood-brain barrier
BDNF	Brain-derived neurotrophic factor
CAT	Catalase
CCDs	Carotenoid cleavage dioxygenases
CDK5	Cyclin-dependent kinase 5
ChAt	Choline acetyltransferase
CNS	Central nervous system
CoQ	Coenzyme Q
DLB	Dementia with Lewy bodies
DSM-IV	Diagnostic and statistical manual of mental disorders-IV
ERK1/2	Extracellular signal-regulated kinases 1 and 2
GPx	Glutathione peroxidase
GSH	Glutathione
GSK-3β	Glycogen synthase kinase-3β
H_2O_2	Hydrogen peroxide
HD	Huntington's disease
HO-1	Heme oxygenase-1
Iba1	Ionized calcium-binding adaptor molecule 1
IL	Interleukins
iNOS	Inducible nitric oxide synthase
Keap 1	Kelch-like ECH-associated protein 1
L-DOPA	L-Dihydroxyphenylalanine
MDA	Malondialdehyde
MPTP	1-Methyl-4-phenyl-1,2,3,6-tetrahydropyridine
mPTP	Mitochondrial permeability transition pore
mTOR	Mechanistic target of rapamycin
MWM	Morris water maze
NCEDs	9-Cis-epoxycarotenoid dioxygenases
NF-κB	Nuclear factor kappa B
NFTs	Neurofibrillary tangles
NGF	Nerve growth factor
NMDA	N-methyl-D-aspartate
NO	Nitric oxide
NQO1	NAD(P)H quinone oxidoreductase 1

Nrf2	Nuclear factor erythroid 2-related factor 2
NSAIDs	Nonsteroidal anti-inflammatory drugs
$OH^•$	Hydroxyl radicals
PD	Parkinson's disease
PI3K/Akt	Phosphoinositide 3-kinase/protein kinase B
PPARγ	Peroxisome proliferator-activated receptor gamma
PSD-95	Postsynaptic density protein-95
PSEN1	Presenilin 1
PSEN2	Presenilin 2
PSP	Progressive supranuclear palsy
ROS	Reactive oxygen species
SCFAs	Short-chain fatty acids
SOD	Superoxide dismutase
STZ	Streptozotocin
TLR4	Toll-like receptor 4
TNF-α	Tumor necrosis factor alpha
WHO	World Health Organization
6-OHDA	6-Hydroxydopamine

12.1 Introduction

Progress in medical therapies has greatly enhanced the average human longevity. The rise in life expectancy and cultural changes have elevated the number of elderly individuals and highlighted age-related issues (Mount and Downton 2006). In addition to the increase in age-related disease prevalence, geriatric diseases impose a considerable socioeconomic burden, encompassing annual expenses for nursing, home care, and medications. Furthermore, this socioeconomic load is escalating significantly and will persist in its ascent due to the continual rise in life expectancy (Abramov 2018). Among various age-related disorders, neurodegenerative disorders progress "silently" for decades before the manifestation of symptoms. The majority of approaches devised to manage these diseases are symptomatic and idiosyncratic (Castillo et al. 2019).

12.2 Neurodegenerative Diseases and Dementia

Neurodegeneration encompasses multiple levels of neurological abnormalities, such as Parkinson's disease (PD), dementia with Lewy bodies (DLB), Huntington's disease (HD), progressive supranuclear palsy (PSP), Alzheimer's disease (AD), and spinocerebellar ataxias (SCA). These disorders are characterised by various clinical manifestations, often including cognitive decline, speech, and motor abnormalities (Relja 2004). The fundamental pathogenic characteristic common to neurodegenerative illnesses is the impairment of neuronal populations in specific regions of the brain and/or spinal cord (Yacoubian 2017). Although each

Table 12.1 Key differentiating features among dementia types

Dementia subtype	Clinical presentation
Alzheimer's disease	Gradual increasing deterioration Impairment of memory in the initial phase; deficits in recalling three or five words; executive function deficits in subsequent phases
Vascular dementia	Acute or chronic onset Typically associated with cerebrovascular illness (stroke) and comorbidities (diabetes, hypertension) Minor cognitive deficits in the initial phase Potential gait impairments and falls (contingent upon the severity of the stroke)
Lewy body dementia	Cognitive fluctuations linked to Parkinson Impaired executive function and hallucinations in early stage; impairments on assessments aimed at evaluating visual perception
Frontotemporal dementia	Personality alterations and behavioral abnormalities with minimal memory impairment in the early stages

neurodegenerative condition possesses unique characteristics, these diseases may exhibit similarities in their terminal phase, with patients becoming bedridden and unable to care for themselves (Yacoubian 2017). Neurodegeneration usually arises from alterations in the intrinsic factors, which include chronic neuroinflammatory responses, accumulation or insufficient clearance of toxic proteins, and weakened endogenous antioxidant mechanisms (Jellinger 2010).

Dementia affects a significant section of the elderly population. "The American Association for Geriatric Psychiatry" defines dementia as a clinical illness characterized by diminished cognitive function that substantially impairs an individual's capacity to perform daily tasks (Phan et al. 2019). "The Diagnostic and Statistical Manual of Mental Disorders (DSM-IV)" characterizes dementia as "an acquired cognitive impairment that encompasses a decline in memory exceeding age-related expectations and at least one additional cognitive function, such as attention, visuospatial abilities, or language, or a deterioration in executive functions including planning, organisation, sequencing, or abstraction (Frota et al. 2011)." Table 12.1 outlines and summarizes the four prevalent etiologies of dementia (Duong et al. 2017; Muangpaisan 2007). Out of the various causes, AD is the predominant cause of dementia, accounting for 60% of dementia cases (Duong et al. 2017).

12.3 Alzheimer's Disease (AD)

AD is an age-related neurodegenerative condition initially identified by the renowned German pathologist, Dr. Alois Alzheimer, in 1906. Approximately 60–80% of individuals exhibit dementia attributable to AD, with the proportion rising to 90% upon reaching the age of 75 or older (Gaugler et al. 2014). Dementia linked to AD has profoundly affected the lives of both the afflicted individuals and their associates, including caregivers, family members, coworkers, and acquaintances. In 2008, the World Health Organization (WHO) identified dementia as a significant concern in developing and industrialized nations. As of 2021, about 57 million people are suffering from dementia worldwide. Among those, 60% are living in low- and middle-income countries, adding to a global dementia burden that will double every 20 years, reaching 78 million by 2030 (WHO 2021).

The primary pathologic features of AD are associated with extracellular accumulation of amyloid beta (Aβ) peptides, forming senile plaques as a consequence of abnormal cleavage of the amyloid precursor protein (APP) (O'Brien and Wong 2011). Genetic mutations in APP, presenilin 1 (PSEN1), and presenilin 2 (PSEN2) also play critical roles. The APP gene regulates the synaptic plasticity and neuronal communications, while PSEN1 and PSEN2 form an integral part of the γ-secretase complex that controls Aβ peptide generation. Mutations in these genes are associated with an increased Aβ deposition in the brain (Bekris et al. 2010; Lanoiselee et al. 2017). Another feature is hyperphosphorylation of tau proteins, which facilitates the intracellular formation of neurofibrillary tangles (NFTs) that interfere with the integrity of the neurons (Kametani and Hasegawa 2018). Furthermore, there is a decline in cholinergic neurotransmission with modifications in synaptic function in areas related to cognition (Bhushan et al. 2018; Ferreira-Vieira et al. 2016; Kumar and Nisha 2014). Beyond these central features, other contributing factors to AD pathogenesis include mitochondrial impairment (Perez Ortiz and Swerdlow 2019), alterations in calcium homeostasis (Wang et al. 2017a, b), increased oxidative stress (Huang et al. 2016), and chronic neuroinflammation (Kinney et al. 2018).

In addition to the above hypothesis, inflammation and oxidative stress play critical roles in AD, as discussed below:

12.3.1 Inflammatory Hypothesis

In the last decade, the inflammation theory has emerged as a fourth fundamental characteristic of AD, establishing a connection between Aβ plaques and the NFT hypothesis. Unlike other hypotheses, neuroinflammation results from one of the pathologies or risk factors associated with AD. Inflammation in chronic conditions causes considerable damage to multiple tissues within the body. Postmortem analyses of AD patients have shown the presence of inflammatory responses (Kinney et al. 2018). Microglia and astrocytes are the primary inflammatory elements responsible for neuronal destruction in AD. Microglia are located in the central nervous system (CNS) (Bachiller et al. 2018). Aβ catalyzes microglial activation. Initially, these microglia facilitate the degradation of amyloid plaques and the phagocytosis of Aβ, demonstrating a beneficial impact on patients suffering from AD (Baik et al. 2016; Wyss-Coray et al. 2002). However, after an extended period, these microglia expand and cannot process Aβ in the brain.

Chronically activated microglia enhance the expression of major histocompatibility complex II and increase the secretion of proinflammatory mediators, leading to damage to neurons (Bostancıklıoğlu 2019; Kinney et al. 2018; Moore and O'Banion 2002). Microglia activation has been shown to induce tau protein phosphorylation and the development of NFTs (Gorlovoy et al. 2009). Demographic investigations from the past further corroborate the inflammation theory, which shows that the use of anti-inflammatory drugs in rheumatoid arthritis has shown decreased incidence of AD (Rich et al. 1995; Szekely et al. 2004). Numerous preclinical and clinical studies have shown that various anti-inflammatory therapies, including NSAIDs, mitigate and/or avert AD pathology (Bacher et al. 2008; McGeer and McGeer 2007; Zandi et al. 2002).

12.3.2 Oxidative Stress Hypothesis

Oxidative stress advances AD development by three primary mechanisms: (i) peroxidation of macromolecules and dysfunction of mitochondria, (ii) production of reactive oxygen species (ROS), and (iii) upregulation of Aβ and tau protein synthesis (Cassidy et al. 2020). Oxidative damage has been identified in the brains of individuals with AD, accompanied by the deposition of Aβ and NFTs (Christen 2000). Free radicals and bio-metals (iron, zinc, and copper) act as catalysts and play a pivotal role in neurodegeneration (Kozlowski et al. 2009). These metals facilitate redox processes, producing extremely reactive hydroxyl radicals ($OH^{\bullet}$). The N-terminal regions of both Aβ and its precursor, APP, have a strong affinity for metal ions. Copper is a crucial mediator of $OH^{\bullet}$, leading to oxidative stress (Barnham et al. 2003). Zinc also tightly reacts with Aβ(1–40), leading to Aβ deposits in the brain. The high levels of zinc with Aβ aggregation cause oxidative damage and neurotoxicity (Cuajungco and Faget 2003).

Additionally, the brain membranes are composed of phospholipids that are targeted by free radicals (Tsaluchidu et al. 2008). The membrane phospholipids are oxidized by the enzymes lipoxygenases and cyclooxygenases, creating potent oxidized lipids, such as prostaglandins and isoprostanes. The increased oxidation of lipids and proteins in the brain causes an unusual, debilitating change, as seen in AD patients (Butterfield et al. 1997). Alteration of glutamate and creatine kinase precipitates enhanced neurotoxicity and decreased energy metabolism in AD. Additional implications include diminished action of antioxidant enzymes, Ca^{2+} dyshomeostasis, disrupting mitochondrial function and the electron transport chain. This causes the buildup of electrons in complex I and coenzyme Q (CoQ), causing free radical production, which induces cellular damage and neural dysfunction (Aliev et al. 2010; Aliev et al. 2002).

Oxidative stress also affects Aβ production by elevating APP levels or altering the β and γ secretase enzyme activities. Aβ generates a positive feedback loop on APP levels and its proteolytic route (Cheignon et al. 2018). It results in elevated quantities of Aβ oligomers, resulting in the formation of senile plaques; promoting tau protein phosphorylation; generating NFTs; and the activation of the nuclear factor kappa B (NF-κB) within neurons (Giraldo et al. 2014; Grundke-Iqbal et al. 1987; Minati et al. 2009; Praticò 2008).

12.4 Carotenoids

The carotenoids are the second most common natural component found in nature, after chlorophyll. There are almost 1181 naturally occurring carotenoids from 700 species. The discovery of carotene from carrots in 1831 by Wackenroder led to the group name (Ellison 2016). Carotenoids are lipophilic isoprenoid compounds with up to 15 conjugated double bonds in their polyene structure. They may or may not be cyclized at one or both ends, imparting particular absorption spectra; photochemical, physicochemical characteristics; and pharmacological activity (Borsarelli and Mercadante 2009). Biosynthesis of carotenoids involves the tail-to-tail linking of two C20 geranylgeranyl diphosphate molecules. All the other variants are generated from the parent skeleton of C10 carbon, which is produced in this way. Lycopene and β-carotene serve to demonstrate the fundamental structure of carotenoids. Different types of carotenoids can be obtained by (i) cyclization of this skeleton at either or both ends, (ii) by shifts in the degree of hydrogenation, and (iii) by introducing functional groups containing oxygen. Carotenoids are primarily present in linear or all-configuration and can be classified based on source, color, chemical structure, provitamin A activity, and biological activity (Jacob-Lopes et al. 2018; Ladygin 2000).

12.4.1 Source of Carotenoids

Carotenoids represent a diverse class of pigments synthesized by plants, algae, and certain bacteria. As secondary metabolites, these compounds are synthesized through enzymatic pathways and are best known for their potent antioxidant capabilities. While commonly associated with red, orange, and yellow coloration, carotenoids are also abundant in green vegetables, often masked by chlorophyll (Maoka 2020). The antioxidant activity is due to their polyene chain structures rich in conjugated double bonds, which enable effective neutralization of free radicals (Pérez-Gálvez et al. 2020). A wide variety of fruits and vegetables serve as carotenoid sources. For example, tomatoes are a rich source of lycopene, while green vegetables such as spinach, broccoli, and green beans supply α- and β-carotene. Lutein is concentrated in edible flowers like *Tropaeolum majus*, and β-cryptoxanthin is found in mandarins, corn, peas, egg yolk, and oranges. Additionally, leafy greens such as kale and spinach are notable for their high content of lutein and zeaxanthin (Priyadarshini Pradhan et al. 2022).

12.4.2 Types of Carotenoids

12.4.2.1 Based on Structure

(i) Acyclic carotenes: These compounds lack cyclic structures and possess open-chain termini at both ends. Examples include phytoene, lycopene, and phytofluene (Riaz et al. 2021; Zia-Ul-Haq et al. 2021).

(ii) Cyclic carotenes: They contain one (monocyclic) or two cyclic (bicyclic) structures, for example, α, β, γ, or δ carotene (Riaz et al. 2021; Zia-Ul-Haq et al. 2021).
(iii) Carotenols (hydroxy carotenoids): They feature one or more hydroxyl groups. For example, α-cryptoxanthin (yellow), β-cryptoxanthin (orange), lutein (yellow), and zeaxanthin (yellow–orange) (Riaz et al. 2021; Zia-Ul-Haq et al. 2021).
(iv) Epoxycarotenoid (epoxides): They are characterized by the presence of at least an epoxy (one oxygen atom linked to two carbon atoms), which may or may not connect, for example, violaxanthin, neoxanthin (Riaz et al. 2021; Zia-Ul-Haq et al. 2021).

12.4.2.2 Based on the Inclusion of Oxygen

(i) Hydrocarbon carotenoids or carotenes: They are polyunsaturated chains of hydrogen and carbon that can be either cyclic or linear and have the molecular formula $C_{40}H_{56}$. Typical colors for them include red, yellow, and orange. Carotenes are widely soluble in solvents containing hydrocarbons, for example, α-carotene, β-cryptoxanthin, lycopene, and β-carotene (Fig. 12.1). Common carotene-containing foods include oranges, peaches, watermelons, pumpkins, tomatoes, and papaya (Riaz et al. 2021; Zia-Ul-Haq et al. 2021).
(ii) Oxy-carotenoids or Xanthophylls: They contain carbon, hydrogen, and oxygen atoms arranged in a hydroxy, methoxy, epoxy, or oxy group and contain -COOH, -OH, or = O to the terminals, which makes them more water soluble. In general, xanthophylls are represented by the chemical formula $C_{40}H_{56}O_{1–4}$ and are yellow or green in color. They are abundant in leaves and indirectly contribute to photosynthesis. Xanthophylls are more polar in nature

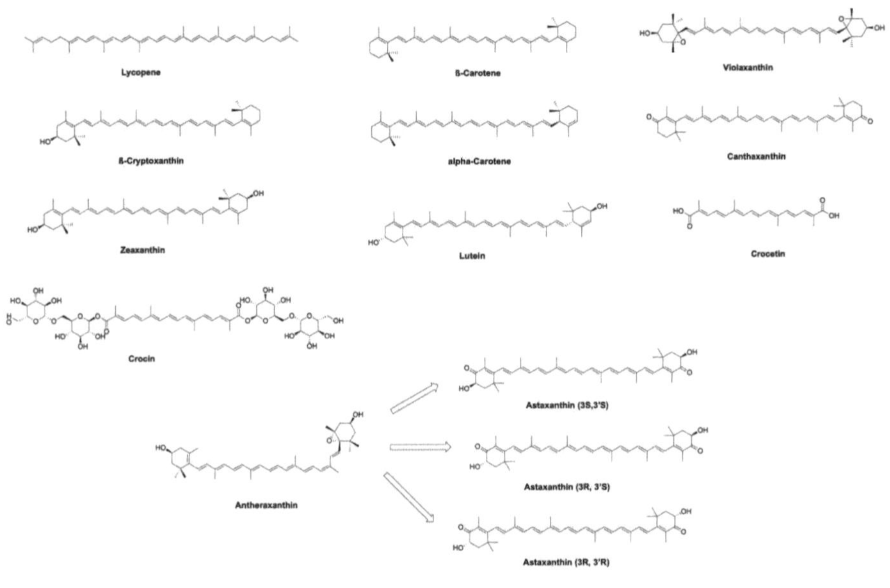

Fig. 12.1 Structures of key dietary carotenoids

and are soluble in polar solvents like alcohols. Some xanthophylls come from plants (like yellow sweet potatoes, zucchini, corn, mango, broccoli), such as lutein (Fig. 12.1), while others come from animals or marine sources (like egg yolk), such as astaxanthin, violaxanthin, cryptoxanthin, and capsanthin (Riaz et al. 2021; Zia-Ul-Haq et al. 2021). Structures of various carotenoids are shown in Fig. 12.1.

12.4.2.3 Based on Structure Modifications

(i) Allenic carotenoids: These carotenoids are characterized by the presence of a continuous double bond system, known as an allene group, for example, neoxanthin and crocoxanthin (Riaz et al. 2021; Zia-Ul-Haq et al. 2021).
(ii) Acetylenic carotenoids: This subclass features one or more triple bonds within their molecular structure, contributing to distinct chemical reactivity, for example, peridinin and fucoxanthin (Riaz et al. 2021).
(iii) Apocarotenoids/Dia-apocarotenoids: They are modified carotenoids with a carbon chain shorter than 40. Apocarotenoids (e.g., Bixin) are derived from the carotenoid skeleton due to oxidative cleavage at one end, whereas dia-apocarotenoids (e.g., Crocetin) are generated at both ends, resulting in a molecule with two distinct halves. Cleavage can be done in both enzymatic (carotenoid oxygenase) and nonenzymatic modes. The oxidative cleavage of carotenoids is primarily mediated by specific enzymes, including 9-cis-epoxycarotenoid dioxygenases (NCEDs) and carotenoid cleavage dioxygenases (CCDs). Beyond enzymatic pathways, certain carotenoids such as α-carotene, lutein, and lycopene are also susceptible to nonenzymatic degradation, which may occur during incubation, extraction, or processing procedures (Riaz et al. 2021; Zia-Ul-Haq et al. 2021).
(iv)Nor carotenoids: These are a subgroup of apocarotenoids, which have one or more carbon atoms removed from the original C_{40} carotenoid. The "nor" in the prefix reflects the loss of carbon units from oxidative cleavage by CCDs. These chemical changes often occur at the site of specific conjugation bonds of the poly chain and can lead to shorter chain derivatives as C_{13}-, C_{15}-, or C_{20}-nor-carotenoids (Riaz et al. 2021; Zia-Ul-Haq et al. 2021).
(v) Nor-carotenoids play a role in plant signaling and odor formation. β-ionone, a C_{13}-nor-carotenoid derived from β-carotene, is responsible for the characteristic scent of violets and other flowers. In humans, retinal (a C_{20}-apocarotenoid) acts as a precursor of important molecules in vision and cell signaling (Riaz et al. 2021; Zia-Ul-Haq et al. 2021).
(vi) Seco carotenoids: They are generated via the cleavage of internal C–C bonds inside the carotenoid molecule, leading to the opening of cyclic end-groups. "Seco" denotes the cleavage, resulting in linear or semi-cyclic structures with modified conjugation patterns. These alterations can markedly affect the molecule's polarity, solubility, and interaction with biological constituents. For example, bixin, a natural pigment derived from *Bixa Orellana,* is used as a food colorant (annatto) and demonstrates significant antioxidant and neuroprotective properties (Riaz et al. 2021; Zia-Ul-Haq et al. 2021).

Table 12.2 Classification of carotenoids according to various functional groups

S. no.	Functional group	Examples
1.	Acids and acid esters	Torularhodin, torularhodin methyl ester
2.	Alcohols	Alloxanthin, loraxanthin, lutein, lycoxanthin, rhodopin, zeaxanthin
3.	Aldehydes	Rhodopinal, warmingone
4.	Apocarotenoids	Apo-2-lycopenal, crocin, crocetin, paracentrone, sintaxanthin
5.	Epoxides	Citroxanthin, mutatoxanthin, neochrome
6.	Esters of alcohols	Astacein, fucoxanthin, physalein
7.	Ethers	Rhodovibrin, spheroidene
8.	Glycosides	Oscillaxanthin, phleixanthophyll
9.	Higher carotenoids	Bacterioruberin, decaprenoxanthin, Nonaprenoxanthin
10.	Hydrocarbons	Hexahydrolycopene, lycopersene, α-zeacarotene
11.	Ketones	Astacene, capsanthin, okenone, rubixanthone
12.	Nor- and seco- carotenoids	β-Carotenone, peridinin, pyrrhoxanthininol
13.	Retro and retro-apo- carotenoids	Rhodoxanthin, tangeraxanthin

(vii) Retro carotenoids: They have a unique structural alteration characterized by a displacement of double bonds (also referred to as double bond migration) inside one of the terminal rings of the carotenoid molecule. Compounds like retro-β-carotene may have potential uses in altering oxidative conditions or interacting with membrane lipids (Riaz et al. 2021; Zia-Ul-Haq et al. 2021). The rest of the functional groups in carotenoids are shown in Table 12.2.

12.4.2.4 Based on Function

(i) Primary carotenoids: They are fundamental and beneficial constituents of the photosynthetic apparatus, sometimes referred to as photosynthetic pigments, typically found in leaves, such as zeaxanthin, antheraxanthin, and neoxanthin. Lutein functions as an auxiliary pigment by transferring absorbed light energy to chlorophyll molecules, thus extending the photosynthetic light spectrum of algae (Riaz et al. 2021; Zia-Ul-Haq et al. 2021).

(ii) Secondary carotenoids: They perform important roles in various physiological and protective processes, but are not essential for the fundamental survival of plants. They are generated from primary carotenoids upon exposure to stimuli. Xanthophylls function as accessory pigments or secondary carotenoids in plants and mitigate photooxidative stress in plant tissues. Xanthophylls are commonly located in chloroplasts and are generated substantially as oily droplets. Secondary carotenoids such as canthaxanthin and astaxanthin contribute to cellular defense (Riaz et al. 2021; Zia-Ul-Haq et al. 2021).

12.4.2.5 Based on Provitamin Activity

(i) Provitamin A carotenoids: These are the carotenoids enzymatically metabolized into retinol by the body, such as α-carotene and β-cryptoxanthin (Riaz et al. 2021; Zia-Ul-Haq et al. 2021).
(ii) Non-provitamin A carotenoids: These carotenoids, such as lycopene, cannot be metabolized into retinol by the body (Riaz et al. 2021; Zia-Ul-Haq et al. 2021).

12.4.3 Mechanisms Supporting the Neuroprotective Function of Carotenoids

Carotenoids have a broad neuroprotective activity in the CNS by antioxidant, anti-inflammatory, mitochondria-stabilization, and synaptic-regulation pathways (Fig. 12.2) (Manochkumar et al. 2021; Rasmus and Kozłowska 2023). Their powerful antioxidant activity is related to their extended polyene structure that enables them to effectively scavenge ROS, thus averting lipid peroxidation of neuronal membranes, an important pathogenic hallmark in AD (Gandla et al. 2023; Park et al. 2020). Carotenoids activate the endogenous cellular antioxidant defense system, involving the translocation of "nuclear factor erythroid 2-related factor 2 (Nrf2)" that leads to the binding of Nrf2 to the antioxidant response element. This causes upregulation of antioxidant enzymes, including superoxide dismutase (SOD), glutathione peroxidase (GPx), catalase (CAT), and heme oxygenase-1 (HO-1). This transition to a redox-balanced intracellular milieu plays an important role in mitigating neuron death by oxidative stress (Ben-Dor et al. 2005; Jena et al. 2023; Terao 2023).

Simultaneously, neuroinflammation is regulated by carotenoids via the suppression of the NF-κB pathway. The IKK kinase complex (IKKα and IKKβ) constitutes the fundamental component of the NF-κB signaling cascade (Israël 2010).

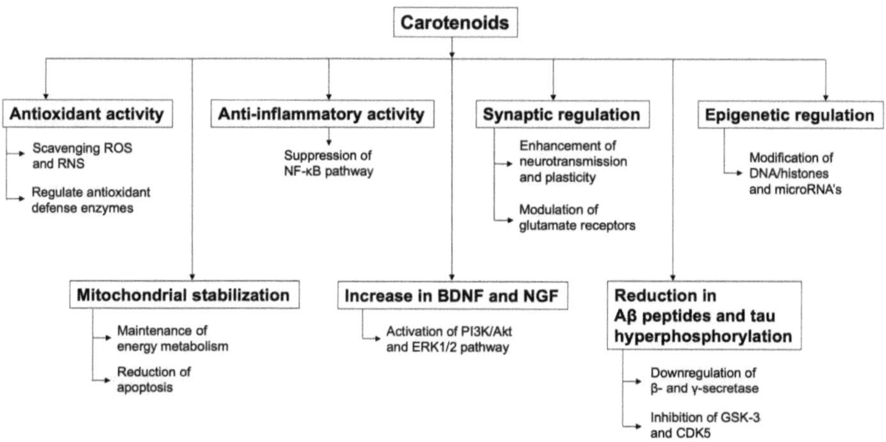

Fig. 12.2 Neuroprotective action of carotenoids

Carotenoids are recognized for their ability to block IKK activity, therefore sequestering NF-κB in the cytoplasm and regulating the production of pro-inflammatory cytokines such as TNF-α, IL-1β, IL-6, cyclooxygenase-2, and inducible nitric oxide synthase (iNOS) (Armoza et al. 2013; Karin and Delhase 2000). This results in the inhibition of microglial over-activation and astrocytic reactivity – two of the main factors underlying the neuroinflammation associated with AD (Israël 2010; Kaulmann and Bohn 2014; Linnewiel-Hermoni et al. 2014).

Furthermore, carotenoids protect mitochondrial integrity by maintaining their transmembrane potential, reducing the production of mitochondrial ROS (Ademowo et al. 2024) and the leakage of cytochrome c to the cytosol, leading to a reduction of caspase-9/caspase-3 dependent intrinsic pathway of apoptosis (Baeza-Morales et al. 2024; Bossy-Wetzel et al. 1998; Redza-Dutordoir and Averill-Bates 2016). They maintain energy metabolism in neurons by inducing ATP synthesis and blocking mitochondrial permeability transition pore (mPTP), the opening of which occurs in AD (Ademowo et al. 2024; Baburina et al. 2019; Lu et al. 2020).

Carotenoids also exert an effect on synaptic plasticity mediated through synaptophysin and postsynaptic density protein-95 (PSD-95) protein levels (Agni et al. 2024; Massaad and Klann 2011; Pietrasik et al. 2022). This recovery in synaptic structure enhances neurotransmission, especially in the cholinergic system, whereby choline acetyltransferase (ChAt) and acetylcholine (ACh) levels were increased (Mufson et al. 2008). Carotenoids influence glutamate receptor function, namely, "N-methyl-D-aspartate (NMDA)" and "α-amino-3-hydroxy-5-methyl-4-isoxazolepropionic acid (AMPA)" receptors, thereby modulating calcium influx (Le et al. 2025). Disruption of these processes can cause impairment in mitochondria and neurons (Verma et al. 2022).

Another vital aspect is that carotenoids inhibit amyloidogenic processing of APP, which reduces the production of Aβ(1–42) peptides by downregulating β- and γ-secretase (Lakey-Beitia et al. 2019; Sho et al. 2020). Finally, carotenoids inhibit the activity of kinases responsible for tau hyperphosphorylation, such as "glycogen synthase kinase-3β (GSK-3β)" and "cyclin-dependent kinase 5 (CDK5)," thus impeding the formation of NFTs (Naik et al. 2025; Rokkaku et al. 2013). In general, carotenoids boost neurotrophic signalling by inducing activation of "phosphoinositide 3-kinase/protein kinase B (PI3K/Akt)" and "extracellular signal-regulated kinases 1 and 2 (ERK1/2)" pathways, with upregulation of "brain-derived neurotrophic factor (BDNF)" and "nerve growth factor (NGF)," important for cytoprotection, synaptic plasticity, and cognitive activity (Pietrasik et al. 2022). Moreover, increasing evidence indicates that carotenoids affect the expression of genes by methylation of DNA and acetylation of histone proteins and adjust microRNAs associated with inflammation and neurodegeneration (Kim et al. 2019).

12.4.4 Carotenoids, Gut Health, and Memory Enhancement

The gut–brain axis is a two-way communication network linking the gut and the brain through biochemical molecules via neuronal, hormonal, immunological, and metabolic pathways. Its importance in mood regulation, cognition, and overall brain health is critical (Appleton 2018; Fekete et al. 2024). Studies showed that changes in the composition of gut microbiota, that is, dysbiosis, altered the permeability of the intestine, which causes systemic inflammation and the production of neurotoxic metabolites that lead to the onset of neurological diseases (Uceda et al. 2023). Carotenoids affect cognition not only due to their direct effects on the prevention of neurotransmitter loss but also due to their effects on the gut microbiome. These compounds play a role in the flourishing of the healthful microflora, especially *Lactobacillus* and *Bifidobacterium,* and the simultaneous suppression of pathogenic and pro-inflammatory bacteria (Rocha et al. 2023). This change in microflora profile supports the generation of short-chain fatty acids (SCFAs) like butyrate that maintain the integrity of the blood–brain barrier, reduce systemic inflammation, and induce the expression of neurotrophic factors (Silva et al. 2020).

Moreover, carotenoids attenuate gut-derived inflammation by downregulating "Toll-like receptor 4 (TLR4)" signaling and NF-κB activity in gut epithelial and immune cells, leading to reduced levels of circulating pro-inflammatory cytokines that may impact the brain (Coutinho-Wolino et al. 2022). This immunomodulatory effect might be particularly relevant in the elderly, in whom chronic low-level inflammation (inflammaging) is itself associated with augmented cognitive impairment. In addition to an immunomodulatory function, the antioxidant power of carotenoids is essential for maintaining the integrity of the gut epithelium by their ROS-scavenging action (Ferrucci and Fabbri 2018). This creates a healthier gut environment, which in turn promotes cognitive resilience. Effect of different carotenoids in gut is summarized in Table 12.3.

12.4.5 Carotenoids for Memory Enhancement

Some important carotenoids used to improve memory are listed:

Table 12.3 Carotenoids in regulating the gut–brain axis

Carotenoid	Gut effect	Cognitive benefit	References
Astaxanthin	Mitigates inflammation originating from the gastrointestinal tract	Augments BDNF and synaptic signaling	Bahbah et al. (2021)
Fucoxanthin	Enhances microbiota diversity	Upregulates BDNF, reduces apoptosis	Oliyaei et al. (2023)
Lutein	Boosts SCFA-producing microbes	Improves memory and cognition	Ye et al. (2024)
Lycopene	Modulates oxidative gut stress	Improves cognition, lowers inflammation	Chen et al. (2019a, b)
β-Cryptoxanthin	Restores gut barrier, lowers lipopolysaccharides	Reduces neuroinflammation	Dorey et al. (2023)

12.4.5.1 Astaxanthin

Astaxanthin (AST), a xanthophyll derivative carotenoid derived from *Haematococcus pluvialis*, has exhibited powerful neuroprotective effects in different in vitro and in vivo models of neurodegeneration. Treatment with AST (25 mg/kg) significantly ameliorated spatial learning deficits in the Morris water maze (MWM) test in APP/PS1 double transgenic mice. It reduced the Aβ(1–42) plaque load, microglial activation, and pro-inflammatory mediator production in the hippocampus and cortex. AST also recovered BDNF expression and increased activation of the PI3K/Akt signaling pathway, promoting synaptogenesis and neurogenesis (Wu et al. 2015). In a scopolamine-induced memory impairment mouse model, oral administration of AST (12 and 24 mg/kg) improved cognitive ability evaluated by MMW and Y-maze. It decreased the activity of acetylcholinesterase (AChE), restored antioxidant enzymes levels, and decreased oxidative markers levels, suggesting amelioration of oxidative damage and memory (Fassett and Coombes 2011).

AST (10 mg/kg, p.o.) effectively inhibited rotenone-induced dopaminergic neurodegeneration in PD mouse models. This neuroprotective effect was mediated by a reduction in α-synuclein aggregation, decreased mitochondria-dependent release of cytochrome c, and inhibition of caspase-3 enzymatic activation (Lotfi et al. 2024). AST also activated the Nrf2/HO-1 axis and inhibited NF-κB signaling, leading to antioxidant and anti-inflammatory effects. Verbal learning and psychomotor performance were enhanced in 96 elderly individuals who were administered 12 mg/day of AST for 12 weeks in a randomized, double-blind, placebo-controlled clinical trial. Functional MRI data showed an increased cerebral blood flow in the prefrontal cortex, due to its antioxidant properties (Nouchi et al. 2020).

12.4.5.2 β-Carotene

β-Carotene, a provitamin A carotenoid, is found in orange-colored fruits and vegetables such as carrots, pumpkins, etc. It acts as a precursor for retinoic acid, a necessary molecule in neuronal growth (Tufail et al. 2024). β-carotene disrupts Aβ(1–42) aggregation by binding to the surfaces of fibrils, which reduces the neurotoxicity (Banerjee et al. 2023). β-carotene supplementation exhibited a significant improvement in MWM performance, a decline of MDA level, and increase in the activity of SOD and CAT in the streptozotocin (STZ) induced Alzheimer's model (Hira et al. 2019). Similarly, in a model of vascular dementia, β-carotene (20 mg/kg/day) treatment in diabetic rats attenuated the cognitive deficits, reduced AChE activity, and re-established the redox balance (Lim et al. 2022).

In addition, β-carotene has shown a powerful anti-inflammatory effect in ischemic brain injury. β-carotene pretreatment significantly reduced the activity of myeloperoxidase, nitrite level, and neuronal degeneration in the rat model of occluded middle cerebral artery (Althurwi et al. 2022). It also activated the Nrf2/ARE pathway, which leads to HO-1 and glutathione S-transferase expression and subsequently decreases oxidative damage (Chen et al. 2019b). Moreover, β-carotene acts as a competitive inhibitor of AChE, as evidenced by molecular docking and in vitro enzymatic assay, reflecting its role in the retention of acetylcholine concentration at the synapse (Muthuraman et al. 2023).

Long-term intake of β-carotene improves memory in older adults (Abrego-Guandique et al. 2023). A longitudinal cohort research study showed that women in the highest percentile of β-carotene consumption had substantially superior cognitive scores after 6 years, with this link persisting even after various variables influencing the study (Engelhart et al. 2002). Another cross-sectional study with adults aged ≥60 showed that participants in the highest quartile of dietary β-carotene intake had significantly lower cognitive impairment, independent of age, education, and comorbidities (Zhong et al. 2023).

12.4.5.3 Crocin

Crocin, a water-soluble carotenoid from *Crocus sativus* (saffron), has significantly and dose-dependently improved cognitive functions and reduced neuronal death via enhancing the endogenous antioxidant defense system (Naghizadeh et al. 2013). In the d-galactose-induced and aluminum trichloride-induced neurotoxicity model, crocin (5 and 20 mg/kg) reduced the levels of oxidative stress markers, inflammatory response, and AChE activity, and increased the activity of defense enzymes and BDNF levels, thus supporting neuroplasticity (Wang et al. 2019). In vitro, crocin protected PC12 and SH-SY5Y cells against glucose deprivation (mimicking oxidative stress) and D-galactose-induced cytotoxicity by suppressing apoptotic and glycation pathways (Heidari et al. 2018; Ochiai et al. 2004).

Crocin (60 mg/kg) has been shown to provide a remarkable dopaminergic neuroprotection in "6-Hydroxydopamine (6-OHDA)"-induced PD models. It enhanced motor activity, decreased oxidative and nitrosative stress, and improved cognitive performance (Hosseini et al. 2016; Rajaei et al. 2016). In the "1-methyl-4-phenyl-1,2,3,6-tetrahydropyridine (MPTP)" mice model of PD, crocin exerted a neuroprotective effect on dopaminergic neurons of substania nigra and ventral tegmental area, by activating the "mechanistic target of rapamycin (mTOR)" pathway, which enhanced synaptic plasticity (Haeri et al. 2019; Tang et al. 2020).

In the model of ischemic brain damage, crocin administration significantly decreased infarct volume, inflammatory mediators, and oxidative stress by upregulation of "γ-glutamyl cysteinyl synthase" involved in glutathione biosynthesis (Shahbaz et al. 2022). Crocin maintained blood–brain barrier (BBB) integrity and exhibited neuronal protection in aged rats subjected to cerebral ischemia, suggesting its therapeutic significance in poststroke recovery (Zhang et al. 2017).

At the molecular level, there are numerous signaling pathways by which crocin contributes to neuroprotection. It scavenges ROS, increases antioxidant enzyme activities (Shahbaz et al. 2022), maintains cellular redox homeostasis (Wang et al. 2019), upregulates the expressions of Nrf2 and HO-1 (Chu et al. 2025), and reduces apoptosis, as indicated by downregulation of caspase-3 and Bax (Wang et al. 2019). Crocin controls mitochondrial dynamics by preventing the opening of the mitochondrial permeability transition pore (Lv et al. 2016) and upregulating mitochondrial biogenesis during neurotoxic stress, thereby preserving neuronal viability (Eteghadi et al. 2021).

12.4.5.4 Crocetin

Crocetin, a red lipophilic carotenoid extracted from saffron, has shown the ability to enhance cognitive performance, stimulate neurogenesis, and regulate signaling pathways linked to neuroprotection. Crocetin decreased Aβ buildup and neuroinflammation in transgenic AD mice, enhancing memory function by downregulating pro-inflammatory cytokines (Zhang et al. 2018). It facilitates the removal of Aβ plaques from the brain by stimulating autophagy. It activates the "STK11/LKB1-AMPK" signaling pathway, which diminishes Aβ accumulation, decreases cerebral inflammation, and enhances memory (Wani et al. 2021). In vitro investigations revealed that crocetin and its derivative trans-crocin-4 regulate the amyloidogenic pathway and impede tau protein misprocessing by inhibiting "β-site APP-cleaving enzyme 1 (BACE1), GSK-3β, and ERK1/2" (Chalatsa et al. 2019). Crocetin mitigates inflammation in RAW264.7 immune cells by inhibiting the "MEK1/JNK/NF-κB/iNOS" signaling pathway, which is frequently activated during inflammatory responses. Simultaneously, it stimulated the Nrf2/HO-1 pathway, therefore safeguarding cells from stress (Wen et al. 2021). Furthermore, administration of crocin and crocetin reduced the concentrations of inflammatory mediators, such as nitric oxide and pro-inflammatory cytokines, in brain microglial cells subjected to an inflammatory stimulus like lipopolysaccharide (Nam et al. 2010).

In PD mice, crocetin safeguarded dopaminergic neurons and motor function while diminishing oxidative damage and lipid peroxidation (Ahmad et al. 2005). Also, crocetin augmented neural stem cell proliferation, differentiation, and neurogenesis in the cerebral ischemia model (Liu et al. 2023). It facilitated neurite outgrowth and locomotor recovery during spinal cord damage (X. Wang et al., Wang et al. 2017a). In depression models, crocetin rapidly alleviated depression-like behavior via activation of the PI3K/AKT pathway (Lin et al. 2024) and also demonstrated longer-term effects by modulating MEK/ERK signaling and altering gut microbiota composition (Lin et al. 2021).

12.4.5.5 Fucoxanthin

Fucoxanthin, a carotenoid xanthophyll found exclusively in marine brown seaweeds including *Undaria pinnatifida* and *Laminaria japonica*, has shown potent neuroprotective activity by modulation of antioxidant, anti-inflammatory, and mitochondrial-regulating mechanisms. It contains an allenic bond, an epoxide ring, and a conjugated carbonyl group, which likely contribute to its ability to donate hydrogen atoms and electrons to ROS as well as its interactions with the lipid membrane (Mumu et al. 2022). In the STZ-induced model of AD in rats, fucoxanthin (50 and 100 mg/kg, p.o. for 14 days) administration significantly and dose-dependently decreased the accumulation of Aβ and the hyperphosphorylation of tau, thus improving cognitive function. It reversed the levels of oxidative stress markers (MDA and NO) while increasing antioxidant enzyme (SOD, CAT, and GPx) activities, thereby demonstrated lower oxidative stress. It also decreases pro-inflammatory cytokine levels and restores cholinergic and GABAergic neurotransmission (Dhami et al. 2021).

Additionally, fucoxanthin dose-dependently prevented hydrogen peroxide (H_2O_2)-induced oxidative damage in SH-SY5Y human neuroblastoma cells and cerebellar granule neurons. Pretreatment with fucoxanthin distinctly improved cell viability and decreased intracellular ROS and apoptotic markers. It restored the phosphorylation levels of Akt and GSK-3β while concurrently reducing ERK phosphorylation, which reduces neuronal apoptosis (Yu et al. 2017). In silico and in vitro analyses demonstrated that fucoxanthin selectively binds to D3 and D4 receptors, suggesting its potential to treat PD (Paudel et al. 2019). It protects the neuronal cells by targeting "Kelch-like ECH-associated protein 1 (Keap1)," leading to the activation of the Nrf2 antioxidant pathway, thereby reducing oxidative stress and neuronal apoptosis (Wu et al. 2021). In the l-DOPA-induced neurotoxicity model, fucoxanthin inhibits the ERK/JNK-c-Jun signaling pathway, acting as a neuroprotective agent (Paudel et al. 2019). Fucoxanthinol, a metabolite of fucoxanthin, enhances mitochondrial function and activates mitophagy pathways, thereby supporting dopaminergic differentiation (Elmorsy et al. 2025). Fucoidan, a sulfated polysaccharide, exhibited significant antioxidant activity and improved behavioral outcomes in PD models (Luo et al. 2009). Moreover, fucoxanthin has also been found to regulate neurotrophic signaling by inhibiting TNF-α, IL-1β, and IL-6 in lipopolysaccharide-induced BV-2 microglial cells. It reduces p65 nuclear translocation and degradation of IκBα. It also upregulates the expression of BDNF and phosphorylated Akt, implicating its role in neurogenesis and neuronal resilience (Zhao et al. 2017).

12.4.5.6 Lycopene

Lycopene is an open acyclic and lipophilic carotenoid containing 11 conjugated double bonds and found primarily in tomatoes, guavas, and watermelons. It has a quenching capacity for singlet oxygen 100 times that of vitamin E and β-carotene, thus demonstrating its great antioxidant potential. It is insoluble in water and ethanol and soluble in organic solvents, such as ethyl acetate and petroleum ether. It is metabolized to apo-lycopenals and other biologically active products (Bin-Jumah et al. 2022). A controlled study of 10 weeks in 20-month-old CD-1 male mice fed with lycopene-supplemented chow (0.03%, w/w) was conducted to study the effects of lycopene on cognitive ageing. Mice treated with lycopene showed enhanced spatial learning and memory in the MWM test. Biochemical tests confirmed that hippocampal MDA levels were decreased notably while antioxidant enzyme activities were increased (SOD, CAT, and GSH) and antioxidant gene expressions (HO-1, NQO1) were upregulated. The immunofluorescence staining indicated decreased microglial activation (Iba1), and ELISA showed that the level of Aβ was significantly reduced, suggesting a reduction of age-related pathological changes (Zhao et al. 2018).

Lycopene administered orally (10 mg/kg per day) for 30 days drastically reversed the rotenone-induced PD in adult Wistar rats, ameliorating cognitive and motor coordination. Lycopene treatment upregulated SOD, CAT, GSH levels; downregulated MDA content; and increased the activity of mitochondrial complex I (NADH dehydrogenase). It suppressed cytochrome c release, an important apoptotic signal, and prevented the loss of tyrosine hydroxylase-positive dopaminergic neurons in the substantia nigra. Additionally, lycopene attenuated α-synuclein accumulation and LC3B expression as a result of autophagy function restoration (Kaur et al. 2011). Lycopene has also been found to prevent oxidative stress and motor deficits induced by 3-nitropropionic acid (Sandhir et al. 2010). It also decreases cerebral infarct volume and neuro-apoptosis in ischemia-reperfusion, indicating its wider role as a neuroprotector (Wu et al. 2022). Lycopene regulates key neuroprotective signaling pathways, including activation of AMPK, PPARγ, and PI3K/Akt, and simultaneously inhibits microglial-mediated neuroinflammatory responses (Fang et al. 2020).

12.4.5.7 Zeaxanthin

A dietary xanthophyll carotenoid called zeaxanthin has shown notable neuroprotective effects in a number of experimental models. It is the most significant carotenoid in the retina, especially in the macula and lens, defending the eyes from oxidative and light-induced damage (Mrowicka et al. 2022). It reduced anxiety and depression-like behaviors by reducing hippocampal inflammation (Zhou et al. 2018) and enhanced cognitive function in diabetic rats by triggering the PI3K/AKT signaling pathway (Zhou et al. 2017). Zeaxanthin, derived from *Dunaliella salina*, reduced Aβ buildup and modulated neurotransmission to alleviate dementia in age-related neurodegenerative models (El-Baz et al. 2021). By inhibiting GSK-3β, it also reduced tau hyperphosphorylation and endoplasmic reticulum stress, suggesting that it may be used to treat AD (Zhang et al. 2022). Moreover, it decreased cerebrovascular inflammation and enhanced learning and memory in models of Aβ-induced cognitive impairment (Li et al. 2022).

Furthermore, zeaxanthin's anti-inflammatory, anti-apoptotic, and antioxidant properties protected Wistar rats against acrylamide-induced neurotoxicity (Mortazavi et al. 2025). It supported cellular defense against oxidative damage by increasing the expression of phase II antioxidant enzymes through Nrf2 activation (Zou et al. 2014). Several other bioactives have shown potential benefits in memory in preclinical and clinical studies. A concise overview is shown in Table 12.4.

12.5 Future Directions and Conclusion

Carotenoids (such as astaxanthin, lycopene, and lutein) have shown promise in ameliorating neurodegenerative disorders due to their antioxidant, anti-inflammatory, and neuroprotective effects. However, their bioavailability, metabolism, and mechanism(s) of action require further investigation. Several other factors, like formulation issues, pathophysiology, and patient-specific factors, limit their clinical application. Additionally, the absence of long-term human clinical trials and large-scale studies makes it difficult to establish standard dosage schedules. It is imperative that future studies should employ well-designed clinical trials to assess the effectiveness of carotenoids in the human population and investigate ways to increase their bioavailability through the use of innovative delivery methods.

Table 12.4 Neuroprotective effect of carotenoids in experimental and clinical models

Carotenoid	Condition/model	Effects	References
Astaxanthin	Exploratory clinical trial in patients with mild cognitive impairment	Modest improvements in reaction time, working memory, and delayed recall	ADDF (2018)
	Randomized controlled trial in recreationally active females aged 18–39	Investigating potential to mitigate cognitive decline following mental fatigue	ClinicalTrials.gov (2024)
AstaZine® (Astaxanthin)	Randomized controlled trials in healthy adults with age-related forgetfulness	Improved memory and cognitive function in older adults with mild forgetfulness	Review (2019)
Canthaxanthin	Scopolamine induced memory impairment	Improves memory	Si and Zhu (2022)
Capsanthin	Behavioral and neurochemical studies in animal models	Protects hippocampal neurons; modulates social behavior	Ang et al. (2024)
Crocetin	Brain injury in Sprague-Dawley rats	Neuroprotection, reduced apoptosis	Guo et al. (2022)
	Neurobehavioral models, opioid withdrawal	Enhances dopamine, glutamate levels; reduces withdrawal symptoms	Tavakolizadeh et al. (2024)
Crocin	Depression (Wistar rats)	Antidepressant effect	Vahdati Hassani et al. (2014)
	Clinical trial on mild-to-moderate depression	Effective treatment without side effects	Siddiqui et al. (2022)
Lutein	Randomized controlled trial in older adults	Enhances visual–spatial memory, boosts BDNF	Lindbergh et al. (2018)
Lutein and zeaxanthin	Brain and retinal oxidative stress	Protect neurons via antioxidant action	Li et al. (2023)
Lycopene	Mice with age-related cognitive deficits	Lycopene alleviated cognitive deficits via activation of liver–brain FGF21 signaling	Wang et al. (2024)
	Mice with chronic stress-induced hippocampal injury	Lycopene ameliorated hippocampal injury, suggesting neuroprotective effects	Zhang et al. (2020)
Violaxanthin	CNS inflammation models	Reduces hippocampal inflammation	Takemura et al. (2021)
α-Carotene	Population-based dietary intake analysis in older adults	Linked to slower cognitive decline	Zhong et al. (2023)

(continued)

Table 12.4 (continued)

Carotenoid	Condition/model	Effects	References
β-Carotene	Neuroblastoma BE(2)C cell line, [BALB/c mice]	Inhibits invasion, metastasis	Kim et al. (2014)
	Randomized controlled trial (physicians' health study II) in males aged ≥40 years	Long-term supplementation (mean duration 18 years) associated with modest cognitive benefits	Grodstein et al. (2007)
	Systematic review in various populations	Long-term supplementation provides cognitive benefits; short-term effects less clear	Abrego-Guandique et al. (2023)
β-Cryptoxanthin	Animal/in vitro models of neuroinflammation	Restores mitochondria, reduces neuroinflammation	León-Cépeda and Albrecht (2025)

Moreover, the interplay between carotenoids and other bioactive compounds in neurodegenerative diseases requires further investigation to understand their synergistic potential.

References

Abramov AY (2018) Different faces of neurodegeneration. Federation Eur Biochem Soc J 285:3544–3546

Abrego-Guandique DM, Bonet ML, Caroleo MC, Cannataro R, Tucci P, Ribot J, Cione E (2023) The effect of Beta-carotene on cognitive function: a systematic Review. Brain Sci 13(10). https://doi.org/10.3390/brainsci13101468

ADDF, A. s. D. D. F (2018) Cognitive vitality report: Astaxanthin. https://www.alzdiscovery.org/uploads/cognitive_vitality_media/Astaxanthin-Cognitive-Vitality-For-Researchers.pdf

Ademowo OS, Oyebode O, Edward R, Conway ME, Griffiths HR, Dias IHK (2024) Effects of carotenoids on mitochondrial dysfunction. Biochem Soc Trans 52(1):65–74. https://doi.org/10.1042/bst20230193

Agni MB, Hegde PS, Rai P, Sadananda M, K M, D. G. (2024) Astaxanthin and DHA supplementation modulates the maternal undernutrition-induced impairment of cognitive behavior and synaptic plasticity in adult life of offspring's -exploring the molecular mechanism. Mol Neurobiol 61(11):8975–8995. https://doi.org/10.1007/s12035-024-04147-y

Ahmad AS, Ansari MA, Ahmad M, Saleem S, Yousuf S, Hoda MN, Islam F (2005) Neuroprotection by crocetin in a hemi-parkinsonian rat model. Pharmacol Biochem Behav 81(4):805–813. https://doi.org/10.1016/j.pbb.2005.06.007

Aliev G, Smith MA, Seyidov D, Neal ML, Lamb BT, Nunomura A, Gasimov EK, Vinters HV, Perry G, LaManna JC, Friedland RP (2002) The role of oxidative stress in the pathophysiology of cerebrovascular lesions in Alzheimer's disease. Brain Pathol 12(1):21–35. https://doi.org/10.1111/j.1750-3639.2002.tb00419.x

Aliev G, Palacios HH, Gasimov E, Obrenovich ME, Morales L, Leszek J, Bragin V, Solís Herrera A, Gokhman D (2010) Oxidative stress induced mitochondrial failure and vascular hypoperfusion as a key initiator for the development of Alzheimer disease. Pharmaceuticals 3(1):158–187. https://doi.org/10.3390/ph3010158

Althurwi HN, Abdel-Rahman RF, Soliman GA, Ogaly HA, Alkholifi FK, Abd-Elsalam RM, Alqasoumi SI, Abdel-Kader MS (2022) Protective effect of Beta-carotene against myeloperoxidase- mediated oxidative stress and inflammation in rat ischemic brain injury. Antioxidants (Basel) 11(12). https://doi.org/10.3390/antiox11122344

Ang J-J, Low B-S, Wong P-F (2024) Dopamine and central dopaminergic circuitry in neurodegenerative diseases: roles and mechanisms of action of natural phytochemicals. Pharmacol Res Nat Products 100050:100050

Appleton J (2018) The gut-brain Axis: influence of microbiota on mood and mental health. Integr Med (Encinitas) 17(4):28–32

Armoza A, Haim Y, Basiri A, Wolak T, Paran E (2013) Tomato extract and the carotenoids lycopene and lutein improve endothelial function and attenuate inflammatory NF-κB signaling in endothelial cells. J Hypertens 31(3):521–529

Baburina Y, Krestinin R, Odinokova I, Sotnikova L, Kruglov A, Krestinina O (2019) Astaxanthin inhibits mitochondrial permeability transition pore opening in rat heart mitochondria. Antioxidants 8(12):576

Bacher M, Dodel R, Aljabari B, Keyvani K, Marambaud P, Kayed R, Glabe C, Goertz N, Hoppmann A, Sachser N, Klotsche J, Schnell S, Lewejohann L, Al-Abed Y (2008) CNI-1493 inhibits Abeta production, plaque formation, and cognitive deterioration in an animal model of Alzheimer's disease. J Exp Med 205(7):1593–1599. https://doi.org/10.1084/jem.20060467

Bachiller S, Jiménez-Ferrer I, Paulus A, Yang Y, Swanberg M, Deierborg T, Boza-Serrano A (2018) Microglia in neurological diseases: a road map to brain-disease dependent-inflammatory response [Review]. Front Cell Neurosci 12(488). https://doi.org/10.3389/fncel.2018.00488

Baeza-Morales A, Medina-García M, Martínez-Peinado P, Pascual-García S, Pujalte-Satorre C, López-Jaén AB, Martínez-Espinosa RM, Sempere-Ortells JM (2024) The antitumour mechanisms of carotenoids: a comprehensive Review. Antioxidants 13(9):1060

Bahbah EI, Ghozy S, Attia MS, Negida A, Emran TB, Mitra S, Albadrani GM, Abdel-Daim MM, Uddin MS, Simal-Gandara J (2021) Molecular mechanisms of Astaxanthin as a potential Neurotherapeutic agent. Mar Drugs 19(4). https://doi.org/10.3390/md19040201

Baik SH, Kang S, Son SM, Mook-Jung I (2016) Microglia contributes to plaque growth by cell death due to uptake of amyloid beta in the brain of Alzheimer's disease mouse model. Glia 64(12):2274–2290. https://doi.org/10.1002/glia.23074

Banerjee S, Baghel D, Pacheco de Oliveira A, Ghosh A (2023) β-Carotene, a potent amyloid aggregation inhibitor, promotes disordered aβ Fibrillar structure. Int J Mol Sci 24(6). https://doi.org/10.3390/ijms24065175

Barnham KJ, McKinstry WJ, Multhaup G, Galatis D, Morton CJ, Curtain CC, Williamson NA, White AR, Hinds MG, Norton RS, Beyreuther K, Masters CL, Parker MW, Cappai R (2003) Structure of the Alzheimer's disease amyloid precursor protein copper binding domain. A regulator of neuronal copper homeostasis. J Biol Chem 278(19):17401–17407. https://doi.org/10.1074/jbc.M300629200

Bekris LM, Yu CE, Bird TD, Tsuang DW (2010) Genetics of Alzheimer disease. J Geriatr Psychiatry Neurol 23(4):213–227. https://doi.org/10.1177/0891988710383571

Ben-Dor A, Steiner M, Gheber L, Danilenko M, Dubi N, Linnewiel K, Zick A, Sharoni Y, Levy J (2005) Carotenoids activate the antioxidant response element transcription system. Mol Cancer Ther 4(1):177–186

Bhushan I, Kour M, Kour G, Gupta S, Sharma S, Yadav AP (2018) Alzheimer's disease: causes & treatment—a review. Ann Biotechnol 1(1):1002–1010

Bin-Jumah MN, Nadeem MS, Gilani SJ, Mubeen B, Ullah I, Alzarea SI, Ghoneim MM, Alshehri S, Al-Abbasi FA, Kazmi I (2022) Lycopene: a natural arsenal in the war against oxidative stress and cardiovascular diseases. Antioxidants (Basel) 11(2). https://doi.org/10.3390/antiox11020232

Borsarelli CD, Mercadante AZ (2009) 12 thermal and photochemical degradation of carotenoids. In: Carotenoids: physical, chemical, and biological functions and properties. CRC Press Boca Raton, pp 229–250

Bossy-Wetzel E, Newmeyer DD, Green DR (1998) Mitochondrial cytochrome c release in apoptosis occurs upstream of DEVD-specific caspase activation and independently of mitochondrial transmembrane depolarization. EMBO J 17(1):37–49. https://doi.org/10.1093/emboj/17.1.37

Bostancıklıoğlu M (2019) An update on the interactions between Alzheimer's disease, autophagy and inflammation. Gene 705:157–166. https://doi.org/10.1016/j.gene.2019.04.040

Butterfield DA, Hensley K, Cole P, Subramaniam R, Aksenov M, Aksenova M, Bummer PM, Haley BE, Carney JM (1997) Oxidatively induced structural alteration of glutamine synthetase assessed by analysis of spin label incorporation kinetics: relevance to Alzheimer's disease. J Neurochem 68(6):2451–2457. https://doi.org/10.1046/j.1471-4159.1997.68062451.x

Cassidy L, Fernandez F, Johnson JB, Naiker M, Owoola AG, Broszczak DA (2020) Oxidative stress in Alzheimer's disease: a review on emergent natural polyphenolic therapeutics. Complement Ther Med 49:102294

Castillo X, Castro-Obregón S, Gutiérrez-Becker B, Gutiérrez-Ospina G, Karalis N, Khalil AA, Lopez-Noguerola JS, Rodríguez LL, Martínez-Martínez E, Perez-Cruz C, Pérez-Velázquez J, Piña AL, Rubio K, García HPS, Syeda T, Vanoye-Carlo A, Villringer A, Winek K, Zille M (2019) Re-thinking the etiological framework of neurodegeneration. Front Neurosci 13:1–25. https://doi.org/10.3389/fnins.2019.00728

Chalatsa I, Arvanitis DA, Koulakiotis NS, Giagini A, Skaltsounis AL, Papadopoulou-Daifoti Z, Tsarbopoulos A, Sanoudou D (2019) The Crocus sativus compounds trans-Crocin 4 and trans-Crocetin modulate the Amyloidogenic pathway and tau misprocessing in Alzheimer disease neuronal cell culture models. Front Neurosci 13:249. https://doi.org/10.3389/fnins.2019.00249

Cheignon C, Tomas M, Bonnefont-Rousselot D, Faller P, Hureau C, Collin F (2018) Oxidative stress and the amyloid beta peptide in Alzheimer's disease. Redox Biol 14:450–464. https://doi.org/10.1016/j.redox.2017.10.014

Chen D, Huang C, Chen Z (2019a) A review for the pharmacological effect of lycopene in central nervous system disorders. Biomed Pharmacother 111:791–801. https://doi.org/10.1016/j.biopha.2018.12.151

Chen P, Li L, Gao Y, Xie Z, Zhang Y, Pan Z, Tu Y, Wang H, Han Q, Hu X, Xin X (2019b) β-Carotene provides neuro protection after experimental traumatic brain injury via the Nrf2-ARE pathway. J Integr Neurosci 18(2):153–161. https://doi.org/10.31083/j.jin.2019.02.120

Christen Y (2000) Oxidative stress and Alzheimer disease. Am J Clin Nutr 71(2):621s–629s. https://doi.org/10.1093/ajcn/71.2.621s

Chu X, Li C, Hao Y, Shi J, Wu Z, Dong C, Xu H, Zhang S, Zhao Y, Wang H, Han X (2025) Targeting Nrf2/HO-1 signaling by crocin: role in attenuation of arsenic trioxide-induced neurotoxicity in mice. J Ethnopharmacol 337(Pt 2):118858. https://doi.org/10.1016/j.jep.2024.118858

ClinicalTrials.gov. (2024, 10 June). Impact of astaxanthin on cognition in recreationally active females. U.S. National Library of Medicine. Retrieved May 28, 2025, from https://clinicaltrials.gov/study/NCT06460181?id=NCT06460181&rank=1

Coutinho-Wolino KS, Almeida PP, Mafra D, Stockler-Pinto MB (2022) Bioactive compounds modulating toll-like 4 receptor (TLR4)-mediated inflammation: pathways involved and future perspectives. Nutr Res 107:96–116. https://doi.org/10.1016/j.nutres.2022.09.001

Cuajungco MP, Faget KY (2003) Zinc takes the center stage: its paradoxical role in Alzheimer's disease. Brain Res Brain Res Rev 41(1):44–56. https://doi.org/10.1016/s0165-0173(02)00219-9

Dhami M, Raj K, Singh S (2021) Neuroprotective effect of Fucoxanthin against Intracerebroventricular Streptozotocin (ICV-STZ) induced cognitive impairment in experimental rats. Curr Alzheimer Res 18(8):623–637. https://doi.org/10.2174/1567205018666211118144602

Dorey CK, Gierhart D, Fitch KA, Crandell I, Craft NE (2023) Low Xanthophylls, retinol, lycopene, and tocopherols in Grey and White matter of brains with Alzheimer's disease. J Alzheimer's Dis 94(1):1–17. https://doi.org/10.3233/jad-220460

Duong S, Patel T, Chang F (2017) Dementia: what pharmacists need to know. Canadian Pharm J 150(2):118–129

El-Baz FK, Abdel Jaleel GA, Hussein RA, Saleh DO (2021) Dunalialla Salina microalgea and its isolated zeaxanthin mitigate age-related dementia in rats: modulation of neurotransmission and amyloid-β protein. Toxicol Rep 8:1899–1908. https://doi.org/10.1016/j.toxrep.2021.11.021

Ellison SL (2016) Carotenoids: physiology. In: Caballero B, Finglas PM, Toldrá F (eds) Encyclopedia of food and health. Academic Press, pp 670–675. https://doi.org/10.1016/B978-0-12-384947-2.00120-3

Elmorsy EM, Al-Ghafari AB, Al Doghaither HA, Elghareeb MM, Alsaqati M (2025) Fucoxanthinol mitigates the cytotoxic effect of Chlorpyrifos and MPTP on the dopaminergic differentiation of SH-SY5Y human Neuroblastoma cells. J Mol Neurosci 75(2):46. https://doi.org/10.1007/s12031-025-02342-7

Engelhart MJ, Geerlings MI, Ruitenberg A, van Swieten JC, Hofman A, Witteman JCM, Breteler MMB (2002) Dietary intake of antioxidants and risk of Alzheimer disease. JAMA 287(24):3223–3229. https://doi.org/10.1001/jama.287.24.3223

Eteghadi MR, Nasehi M, Vaseghi S, Hesami-Tackallou S (2021) The effect of Crocin on TFAM and PGC-1α expression and catalase and superoxide dismutase activities following cholestasis-induced neuroinflammation in the striatum of male Wistar rats. Metab Brain Dis 36(7):1791–1801. https://doi.org/10.1007/s11011-021-00748-x

Fang Y, Ou S, Wu T, Zhou L, Tang H, Jiang M, Xu J, Guo K (2020) Lycopene alleviates oxidative stress via the PI3K/Akt/Nrf2pathway in a cell model of Alzheimer's disease. PeerJ 8:e9308. https://doi.org/10.7717/peerj.9308

Fassett RG, Coombes JS (2011) Astaxanthin: a potential therapeutic agent in cardiovascular disease. Mar Drugs 9(3):447–465

Fekete M, Lehoczki A, Major D, Fazekas-Pongor V, Csípő T, Tarantini S, Csizmadia Z, Varga JT (2024) Exploring the influence of gut-brain Axis modulation on cognitive health: a comprehensive review of prebiotics, probiotics, and Symbiotics. Nutrients 16(6). https://doi.org/10.3390/nu16060789

Ferreira-Vieira TH, Guimaraes IM, Silva FR, Ribeiro FM (2016) Alzheimer's disease: targeting the cholinergic system. Curr Neuropharmacol 14(1):101–115. https://doi.org/10.2174/1570159x13666150716165726

Ferrucci L, Fabbri E (2018) Inflammageing: chronic inflammation in ageing, cardiovascular disease, and frailty. Nat Rev Cardiol 15(9):505–522. https://doi.org/10.1038/s41569-018-0064-2

Frota NAF, Nitrini R, Damasceno BP, Forlenza OV, Dias-Tosta E, da Silva AB, Herrera Junior E, Magaldi RM (2011) Criteria for the diagnosis of Alzheimer's disease: recommendations of the scientific Department of Cognitive Neurology and Aging of the Brazilian academy of neurology. Dement Neuropsychol 5(3):146–152. https://doi.org/10.1590/s1980-57642011dn05030002

Gandla K, Babu AK, Unnisa A, Sharma I, Singh LP, Haque MA, Dashputre NL, Baig S, Siddiqui FA, Khandaker MU, Almujally A, Tamam N, Sulieman A, Khan SL, Emran TB (2023) Carotenoids: role in neurodegenerative diseases remediation. Brain Sci 13(3). https://doi.org/10.3390/brainsci13030457

Gaugler J, James B, Johnson T, Scholz K, Weuve J (2014) 2014 Alzheimer's disease facts and figures. A. s, Association

Giraldo E, Lloret A, Fuchsberger T, Viña J (2014) Aβ and tau toxicities in Alzheimer's are linked via oxidative stress-induced p38 activation: protective role of vitamin E. Redox Biol 2:873–877. https://doi.org/10.1016/j.redox.2014.03.002

Gorlovoy P, Larionov S, Pham TT, Neumann H (2009) Accumulation of tau induced in neurites by microglial proinflammatory mediators. FASEB J 23(8):2502–2513. https://doi.org/10.1096/fj.08-123877

Grodstein F, Kang JH, Glynn RJ, Cook NR, Gaziano JM (2007) A randomized trial of beta carotene supplementation and cognitive function in men: the physicians' health study II. Arch Intern Med 167(20):2184–2190. https://doi.org/10.1001/archinte.167.20.2184

Grundke-Iqbal I, Iqbal K, Tung Y, Quinlan M, Wisniewski H, Binder L (1987) Abnormal phosphorylation of the microtubule-associated protein?(tau) in Alzheimer cytoskeletal pathology. Alzheimer Dis Assoc Disord 1(3):202

Guo Z-L, Li M-X, Li X-L, Wang P, Wang W-G, Du W-Z, Yang Z-Q, Chen S-F, Wu D, Tian X-Y (2022) Crocetin: a systematic review. Front Pharmacol 12:745683

Haeri P, Mohammadipour A, Heidari Z, Ebrahimzadeh-Bideskan A (2019) Neuroprotective effect of crocin on substantia nigra in MPTP-induced Parkinson's disease model of mice. Anat Sci Int 94(1):119–127. https://doi.org/10.1007/s12565-018-0457-7

Heidari S, Mehri S, Shariaty V, Hosseinzadeh H (2018) Preventive effects of crocin on neuronal damages induced by D-galactose through AGEs and oxidative stress in human neuroblastoma cells (SH-SY5Y). J Pharmacopuncture 21(1):18–25. https://doi.org/10.3831/kpi.2018.21.003
Hira S, Saleem U, Anwar F, Sohail MF, Raza Z, Ahmad B (2019) B-Carotene: a natural compound improves cognitive impairment and oxidative stress in a mouse model of Streptozotocin-induced Alzheimer's disease. Biomolecules 9(9). https://doi.org/10.3390/biom9090441
Hosseini M, Rajaei Z, Alaei H, Tajadini M (2016) The effects of Crocin on 6-OHDA-induced oxidative/Nitrosative damage and motor behaviour in Hemiparkinsonian rats. Malays J Med Sci 23(6):35–43. https://doi.org/10.21315/mjms2016.23.6.4
Huang W-J, Zhang X, Chen W-W (2016) Role of oxidative stress in Alzheimer's disease. Biomed Rep 4(5):519–522. https://doi.org/10.3892/br.2016.630
Israël A (2010) The IKK complex, a central regulator of NF-kappaB activation. Cold Spring Harb Perspect Biol 2(3):a000158. https://doi.org/10.1101/cshperspect.a000158
Jacob-Lopes E, Fernandes A, Casagrande Do Nascimento T, Vera De Rosso V, Queiroz Zepka L (2018) Introductory chapter: carotenoids – a brief overview on its structure, biosynthesis, synthesis, and applications. In: Queiroz Zepka L, Jacob-Lopes E, Vera De Rosso V (eds) Progress in carotenoid research. IntechOpen. https://doi.org/10.5772/intechopen.79542
Jellinger KA (2010) Basic mechanisms of neurodegeneration: a critical update. J Cell Mol Med 14(3):457–487. https://doi.org/10.1111/j.1582-4934.2010.01010.x
Jena AB, Samal RR, Bhol NK, Duttaroy AK (2023) Cellular red-ox system in health and disease: the latest update. Biomed Pharmacother 162:114606. https://doi.org/10.1016/j.biopha.2023.114606
Kametani F, Hasegawa M (2018) Reconsideration of amyloid hypothesis and tau hypothesis in Alzheimer's disease. Front Neurosci 12(25):1–11. https://doi.org/10.3389/fnins.2018.00025
Karin M, Delhase M (2000) The IκB kinase (IKK) and NF-κB: key elements of proinflammatory signalling. Semin Immunol 12:85
Kaulmann A, Bohn T (2014) Carotenoids, inflammation, and oxidative stress – implications of cellular signaling pathways and relation to chronic disease prevention. Nutr Res 34(11):907–929. https://doi.org/10.1016/j.nutres.2014.07.010
Kaur H, Chauhan S, Sandhir R (2011) Protective effect of lycopene on oxidative stress and cognitive decline in rotenone induced model of Parkinson's disease. Neurochem Res 36(8):1435–1443. https://doi.org/10.1007/s11064-011-0469-3
Kim Y-S, Lee H-A, Lim JY, Kim Y, Jung C-H, Yoo S-H, Kim Y (2014) β-Carotene inhibits neuroblastoma cell invasion and metastasis in vitro and in vivo by decreasing level of hypoxia-inducible factor-1α. J Nutr Biochem 25(6):655–664
Kim D, Kim Y, Kim Y (2019) Effects of β-carotene on expression of selected MicroRNAs, histone acetylation, and DNA methylation in colon cancer stem cells. J Cancer Prev 24(4):224–232. https://doi.org/10.15430/jcp.2019.24.4.224
Kinney JW, Bemiller SM, Murtishaw AS, Leisgang AM, Salazar AM, Lamb BT (2018) Inflammation as a central mechanism in Alzheimer's disease. Alzheimer's Dementia (New York) 4:575–590. https://doi.org/10.1016/j.trci.2018.06.014
Kozlowski H, Janicka-Klos A, Brasun J, Gaggelli E, Valensin D, Valensin G (2009) Copper, iron, and zinc ions homeostasis and their role in neurodegenerative disorders (metal uptake, transport, distribution and regulation). Coord Chem Rev 253(21–22):2665–2685
Kumar NS, Nisha N (2014) Phytomedicines as potential inhibitors of β amyloid aggregation: significance to Alzheimer's disease. Chin J Nat Med 12(11):801–818
Ladygin V (2000) Biosynthesis of carotenoids in plastids of plants. Biochem Biokhimiia 65(10):1113–1128
Lakey-Beitia J, Kumar DJ, Hegde ML, Rao KS (2019) Carotenoids as novel therapeutic molecules against neurodegenerative disorders: chemistry and molecular docking analysis. Int J Mol Sci 20(22). https://doi.org/10.3390/ijms20225553
Lanoiselee HM, Nicolas G, Wallon D, Rovelet-Lecrux A, Lacour M, Rousseau S, Richard AC, Pasquier F, Rollin-Sillaire A, Martinaud O, Quillard-Muraine M, de la Sayette V, Boutoleau-Bretonniere C, Etcharry-Bouyx F, Chauvire V, Sarazin M, le Ber I, Epelbaum S, Jonveaux

T, Rouaud O, Ceccaldi M, Felician O, Godefroy O, Formaglio M, Croisile B, Auriacombe S, Chamard L, Vincent JL, Sauvee M, Marelli-Tosi C, Gabelle A, Ozsancak C, Pariente J, Paquet C, Hannequin D, Campion D (2017) APP, PSEN1, and PSEN2 mutations in early-onset Alzheimer disease: a genetic screening study of familial and sporadic cases. PLoS Med 14(3):e1002270. https://doi.org/10.1371/journal.pmed.1002270

Le NHT, Park SA, Kim YM, Ahn DK, Jung W, Han SK (2025) Fucoxanthin inhibits the NMDA and AMPA receptors through regulating the calcium response on substantia Gelatinosa neurons of the trigeminal subnucleus Caudalis in juvenile mice. Neural Plast 2025(1):2553040

León-Cépeda I, Albrecht SS (2025) Dietary intake of Beta Cryptoxanthin, but not other carotenoids, is associated with less frequent anxiety symptoms in US adults: a cross-sectional analysis of the National Health and nutrition examination surveys 2007 through 2012. J Acad Nutr Diet

Li X, Zhang P, Li H, Yu H, Xi Y (2022) The protective effects of Zeaxanthin on amyloid-β peptide 1-42-induced impairment of learning and memory ability in rats. Front Behav Neurosci 16:912896. https://doi.org/10.3389/fnbeh.2022.912896

Li X, Holt RR, Keen CL, Morse LS, Zivkovic AM, Yiu G, Hackman RM (2023) Potential roles of dietary zeaxanthin and lutein in macular health and function. Nutr Rev 81(6):670–683

Lim KG, Varatharajan R, Muthuraman A (2022) The attenuating effect of Beta-carotene on Streptozotocin induced diabetic vascular dementia symptoms in rats. Molecules 27(13). https://doi.org/10.3390/molecules27134293

Lin S, Li Q, Jiang S, Xu Z, Jiang Y, Liu L, Jiang J, Tong Y, Wang P (2021) Crocetin ameliorates chronic restraint stress-induced depression-like behaviors in mice by regulating MEK/ERK pathways and gut microbiota. J Ethnopharmacol 268:113608. https://doi.org/10.1016/j.jep.2020.113608

Lin S, Chen Z, Wu Z, Fei F, Xu Z, Tong Y, Sun W, Wang P (2024) Involvement of PI3K/AKT pathway in the rapid antidepressant effects of Crocetin in mice with depression-like phenotypes. Neurochem Res 49(2):477–491. https://doi.org/10.1007/s11064-023-04051-2

Lindbergh CA, Renzi-Hammond LM, Hammond BR, Terry DP, Mewborn CM, Puente AN, Miller LS (2018) Lutein and zeaxanthin influence brain function in older adults: a randomized controlled trial. J Int Neuropsychol Soc 24(1):77–90

Linnewiel-Hermoni K, Motro Y, Miller Y, Levy J, Sharoni Y (2014) Carotenoid derivatives inhibit nuclear factor kappa B activity in bone and cancer cells by targeting key thiol groups. Free Radic Biol Med 75:105–120

Liu Z, Wang Z, Zhu Z, Hong J, Cui L, Hao Y, Cheng G, Tan R (2023) Crocetin regulates functions of neural stem cells to generate new neurons for cerebral ischemia recovery. Adv Healthc Mater 12(20):e2203132. https://doi.org/10.1002/adhm.202203132

Lotfi A, Zahra A, Khazaei M (2024) Biological activities of astaxanthin in the treatment of neurodegenerative diseases. Neurodegener Dis Manag 14(6):241–256. https://doi.org/10.1080/17582024.2024.2433932

Lu P, Wong SY, Wu L, Lin D (2020) Carotenoid metabolism in mitochondrial function. Food Qual Safety 4(3):115–122. https://doi.org/10.1093/fqsafe/fyaa023

Luo D, Zhang Q, Wang H, Cui Y, Sun Z, Yang J, Zheng Y, Jia J, Yu F, Wang X, Wang X (2009) Fucoidan protects against dopaminergic neuron death in vivo and in vitro. Eur J Pharmacol 617(1–3):33–40. https://doi.org/10.1016/j.ejphar.2009.06.015

Lv B, Chen T, Xu Z, Huo F, Wei Y, Yang X (2016) Crocin protects retinal ganglion cells against H2O2-induced damage through the mitochondrial pathway and activation of NF-κB. Int J Mol Med 37(1):225–232. https://doi.org/10.3892/ijmm.2015.2418

Manochkumar J, Doss CGP, El-Seedi HR, Efferth T, Ramamoorthy S (2021) The neuroprotective potential of carotenoids in vitro and in vivo. Phytomedicine 91:153676. https://doi.org/10.1016/j.phymed.2021.153676

Maoka T (2020) Carotenoids as natural functional pigments. J Nat Med 74(1):1–16. https://doi.org/10.1007/s11418-019-01364-x

Massaad CA, Klann E (2011) Reactive oxygen species in the regulation of synaptic plasticity and memory. Antioxid Redox Signal 14(10):2013–2054. https://doi.org/10.1089/ars.2010.3208

McGeer PL, McGeer EG (2007) NSAIDs and Alzheimer disease: epidemiological, animal model and clinical studies. Neurobiol Aging 28(5):639–647. https://doi.org/10.1016/j.neurobiolaging.2006.03.013

Minati L, Edginton T, Grazia Bruzzone M, Giaccone G (2009) Reviews: current concepts in Alzheimer's disease: a multidisciplinary review. Am J Alzheimer's Disease Dementias® 24(2):95–121

Moore AH, O'Banion MK (2002) Neuroinflammation and anti-inflammatory therapy for Alzheimer's disease. Adv Drug Deliv Rev 54(12):1627–1656. https://doi.org/10.1016/s0169-409x(02)00162-x

Mortazavi Z, Ghasemzadeh Rahbardar M, Mehri S, Hosseinzadeh H (2025) Protective role of zeaxanthin on acrylamide-induced neurotoxicity in Wistar rats. Avicenna J Phytomed 15(2):1016–1029. https://doi.org/10.22038/ajp.2024.24950

Mount C, Downton C (2006) Alzheimer disease: progress or profit? Nat Med 12(7):780–784

Mrowicka M, Mrowicki J, Kucharska E, Majsterek I (2022) Lutein and Zeaxanthin and their roles in age-related macular degeneration-neurodegenerative disease. Nutrients 14(4). https://doi.org/10.3390/nu14040827

Muangpaisan W (2007) Clinical differences among four common dementia syndromes. Geriatr Aging 10(7):425–429

Mufson EJ, Counts SE, Perez SE, Ginsberg SD (2008) Cholinergic system during the progression of Alzheimer's disease: therapeutic implications. Expert Rev Neurother 8(11):1703–1718. https://doi.org/10.1586/14737175.8.11.1703

Mumu M, Das A, Emran TB, Mitra S, Islam F, Roy A, Karim MM, Das R, Park MN, Chandran D, Sharma R, Khandaker MU, Idris AM, Kim B (2022) Fucoxanthin: a promising phytochemical on diverse pharmacological targets. Front Pharmacol 13:929442. https://doi.org/10.3389/fphar.2022.929442

Muthuraman A, Ramesh M, Mustaffa F, Nadeem A, Nishat S, Paramakrishnan N, Lim KG (2023) In Silico and in vitro methods in the characterization of Beta-carotene as pharmaceutical material via acetylcholine esterase inhibitory actions. Molecules 28(11). https://doi.org/10.3390/molecules28114358

Naghizadeh B, Mansouri MT, Ghorbanzadeh B, Farbood Y, Sarkaki A (2013) Protective effects of oral crocin against intracerebroventricular streptozotocin-induced spatial memory deficit and oxidative stress in rats. Phytomedicine 20(6):537–542. https://doi.org/10.1016/j.phymed.2012.12.019

Naik RA, Rajpoot R, Koiri RK, Bhardwaj R, Aldairi AF, Johargy AK, Faidah H, Babalghith AO, Hjazi A, Alsanie WF, Alamri AS, Alhomrani M, Alsharif A, Shkodina A, Singh SK (2025) Dietary supplementation and the role of phytochemicals against the Alzheimer's disease: focus on polyphenolic compounds. J Prev Alzheimers Dis 12(1):100004. https://doi.org/10.1016/j.tjpad.2024.100004

Nam KN, Park YM, Jung HJ, Lee JY, Min BD, Park SU, Jung WS, Cho KH, Park JH, Kang I, Hong JW, Lee EH (2010) Anti-inflammatory effects of crocin and crocetin in rat brain microglial cells. Eur J Pharmacol 648(1–3):110–116. https://doi.org/10.1016/j.ejphar.2010.09.003

Nouchi R, Suiko T, Kimura E, Takenaka H, Murakoshi M, Uchiyama A, Aono M, Kawashima R (2020) Effects of lutein and Astaxanthin intake on the improvement of cognitive functions among healthy adults: a systematic review of randomized controlled trials. Nutrients 12(3):617

O'Brien RJ, Wong PC (2011) Amyloid precursor protein processing and Alzheimer's disease. Annu Rev Neurosci 34:185–204. https://doi.org/10.1146/annurev-neuro-061010-113613

Ochiai T, Soeda S, Ohno S, Tanaka H, Shoyama Y, Shimeno H (2004) Crocin prevents the death of PC-12 cells through sphingomyelinase-ceramide signaling by increasing glutathione synthesis. Neurochem Int 44(5):321–330. https://doi.org/10.1016/S0197-0186(03)00174-8

Oliyaei N, Moosavi-Nasab M, Tanideh N, Iraji A (2023) Multiple roles of fucoxanthin and astaxanthin against Alzheimer's disease: their pharmacological potential and therapeutic insights. Brain Res Bull 193:11–21. https://doi.org/10.1016/j.brainresbull.2022.11.018

Park H-A, Hayden MM, Bannerman S, Jansen J, Crowe-White KM (2020) Anti-apoptotic effects of carotenoids in neurodegeneration. Molecules 25(15):3453

Paudel P, Seong SH, Jung HA, Choi JS (2019) Characterizing fucoxanthin as a selective dopamine D(3)/D(4) receptor agonist: relevance to Parkinson's disease. Chem Biol Interact 310:108757. https://doi.org/10.1016/j.cbi.2019.108757

Perez Ortiz JM, Swerdlow RH (2019) Mitochondrial dysfunction in Alzheimer's disease: role in pathogenesis and novel therapeutic opportunities. Br J Pharmacol 176(18):3489–3507

Pérez-Gálvez A, Viera I, Roca M (2020) Carotenoids and chlorophylls as antioxidants. Antioxidants (Basel) 9(6). https://doi.org/10.3390/antiox9060505

Phan SV, Osae S, Morgan JC, Inyang M, Fagan SC (2019) Neuropsychiatric symptoms in dementia: considerations for pharmacotherapy in the USA. Drugs R D 19(2):93–115. https://doi.org/10.1007/s40268-019-0272-1

Pietrasik S, Cichon N, Bijak M, Gorniak L, Saluk-Bijak J (2022) Carotenoids from marine sources as a new approach in neuroplasticity enhancement. Int J Mol Sci 23(4):1990

Praticò D (2008) Oxidative stress hypothesis in Alzheimer's disease: a reappraisal. Trends Pharmacol Sci 29(12):609–615

Priyadarshini Pradhan S, Padhi S, Dash M et al (2022) Chapter 7 – Carotenoids. In: Kour J, Nayik GA (eds) Nutraceuticals and health care. Academic Press, pp 135–157. https://doi.org/10.1016/B978-0-323-89779-2.00006-5

Rajaei Z, Hosseini M, Alaei H (2016) Effects of crocin on brain oxidative damage and aversive memory in a 6-OHDA model of Parkinson's disease. Arq Neuropsiquiatr 74(9):723–729. https://doi.org/10.1590/0004-282x20160131

Rasmus P, Kozłowska E (2023) Antioxidant and anti-inflammatory effects of carotenoids in mood disorders: an overview. Antioxidants (Basel) 12(3). https://doi.org/10.3390/antiox12030676

Redza-Dutordoir M, Averill-Bates DA (2016) Activation of apoptosis signalling pathways by reactive oxygen species. Biochim et Biophys Acta (BBA) Mol Cell Res 1863(12):2977–2992. https://doi.org/10.1016/j.bbamcr.2016.09.012

Relja M (2004) Pathophysiology and classification of neurodegenerative diseases. Ejifcc 15(3):97–99

Review NB (2019) AstaZine Astaxanthin: two new brain health clinical studies. London, UK. Retrieved May 28, 2025, from https://nutraceuticalbusinessreview.com/astazine-astaxanthin-two-new-brain-health-clinical-studies%2D%2D160230

Riaz M, Zia-Ul-Haq M, Dou D (2021) Chemistry of carotenoids. In: Zia-Ul-Haq M, Dewanjee S, Riaz M (eds) Carotenoids: structure and function in the human body. Springer International Publishing, pp 43–76. https://doi.org/10.1007/978-3-030-46459-2_2

Rich JB, Rasmusson DX, Folstein MF, Carson KA, Kawas C, Brandt J (1995) Nonsteroidal anti-inflammatory drugs in Alzheimer's disease. Neurology 45(1):51–55. https://doi.org/10.1212/wnl.45.1.51

Rocha HR, Coelho MC, Gomes AM, Pintado ME (2023) Carotenoids diet: digestion, gut microbiota modulation, and inflammatory diseases. Nutrients 15(10). https://doi.org/10.3390/nu15102265

Rokkaku T, Kimura R, Ishikawa C, Yasumoto T, Senba M, Kanaya F, Mori N (2013) Anticancer effects of marine carotenoids, fucoxanthin and its deacetylated product, fucoxanthinol, on osteosarcoma. Int J Oncol 43(4):1176–1186

Sandhir R, Mehrotra A, Kamboj SS (2010) Lycopene prevents 3-nitropropionic acid-induced mitochondrial oxidative stress and dysfunctions in nervous system. Neurochem Int 57(5):579–587. https://doi.org/10.1016/j.neuint.2010.07.005

Shahbaz K, Chang D, Zhou X, Low M, Seto SW, Li CG (2022) Crocins for ischemic stroke: a review of current evidence. Front Pharmacol 13:825842. https://doi.org/10.3389/fphar.2022.825842

Sho M, Ichiyanagi N, Imaizumi K, Ishikawa M, Morimoto S, Watanabe H, Okano H (2020) A combinational treatment of carotenoids decreases aβ secretion in human neurons via β-secretase inhibition. Neurosci Res 158:47–55. https://doi.org/10.1016/j.neures.2019.10.006

Si P, Zhu C (2022) Biological and neurological activities of astaxanthin. Mol Med Rep 26(4):1–12

Siddiqui SA, Ali Redha A, Snoeck ER, Singh S, Simal-Gandara J, Ibrahim SA, Jafari SM (2022) Anti-depressant properties of crocin molecules in saffron. Molecules 27(7):2076

Silva YP, Bernardi A, Frozza RL (2020) The role of short-chain fatty acids from gut microbiota in gut-brain communication. Front Endocrinol (Lausanne) 11:25. https://doi.org/10.3389/fendo.2020.00025

Szekely CA, Thorne JE, Zandi PP, Ek M, Messias E, Breitner JC, Goodman SN (2004) Nonsteroidal anti-inflammatory drugs for the prevention of Alzheimer's disease: a systematic review. Neuroepidemiology 23(4):159–169. https://doi.org/10.1159/000078501

Takemura M, Sahara T, Misawa N (2021) Violaxanthin: natural function and occurrence, biosynthesis, and heterologous production. Appl Microbiol Biotechnol 105(16):6133–6142

Tang J, Lu L, Wang Q, Liu H, Xue W, Zhou T, Xu L, Wang K, Wu D, Wei F, Tao W, Chen G (2020) Crocin reverses depression-like behavior in Parkinson disease mice via VTA-mPFC pathway. Mol Neurobiol 57(7):3158–3170. https://doi.org/10.1007/s12035-020-01941-2

Tavakolizadeh M, Akbari A, Bakhtiari A, Bilondi SS, Alikhani V (2024) The impact of saffron Crocin on clinical and psychological complications of opioid substance withdrawal syndrome: a systematic Review of in vivo and clinical trial studies. J Adv Biomed Sci

Terao J (2023) Revisiting carotenoids as dietary antioxidants for human health and disease prevention. Food Funct 14(17):7799–7824

Tsaluchidu S, Cocchi M, Tonello L, Puri BK (2008) Fatty acids and oxidative stress in psychiatric disorders. BMC Psychiatry 8(Suppl 1):S5. https://doi.org/10.1186/1471-244x-8-s1-s5

Tufail T et al (2024) Nutritional benefits of lycopene and Beta-carotene: a comprehensive overview. Food Sci Nutr 12(11):8715–8741. https://doi.org/10.1002/fsn3.4502

Uceda S, Echeverry-Alzate V, Reiriz-Rojas M, Martínez-Miguel E, Pérez-Curiel A, Gómez-Senent S, Beltrán-Velasco AI (2023) Gut microbial metabolome and Dysbiosis in neurodegenerative diseases: psychobiotics and fecal microbiota transplantation as a therapeutic approach-a comprehensive narrative review. Int J Mol Sci 24(17). https://doi.org/10.3390/ijms241713294

Vahdati Hassani F, Naseri V, Razavi BM, Mehri S, Abnous K, Hosseinzadeh H (2014) Antidepressant effects of crocin and its effects on transcript and protein levels of CREB, BDNF, and VGF in rat hippocampus. DARU J Pharm Sci 22:1–9

Verma M, Lizama BN, Chu CT (2022) Excitotoxicity, calcium and mitochondria: a triad in synaptic neurodegeneration. Transl Neurodegener 11(1):3. https://doi.org/10.1186/s40035-021-00278-7

Wang X, Jiao X, Liu Z, Li Y (2017a) Crocetin potentiates neurite growth in hippocampal neurons and facilitates functional recovery in rats with spinal cord injury. Neurosci Bull 33(6):695–702. https://doi.org/10.1007/s12264-017-0157-7

Wang Y, Shi Y, Wei H (2017b) Calcium dysregulation in Alzheimer's disease: a target for new drug development. J Alzheimer's Dis Parkinsonism 7(5):374. https://doi.org/10.4172/2161-0460.1000374

Wang C, Cai X, Hu W, Li Z, Kong F, Chen X, Wang D (2019) Investigation of the neuroprotective effects of crocin via antioxidant activities in HT22 cells and in mice with Alzheimer's disease. Int J Mol Med 43(2):956–966. https://doi.org/10.3892/ijmm.2018.4032

Wang J, Li L, Li L, Shen Y, Qiu F (2024) Lycopene alleviates age-related cognitive deficit via activating liver-brain fibroblast growth factor-21 signalling. Redox Biol 77:103363. https://doi.org/10.1016/j.redox.2024.103363

Wani A, Al Rihani SB, Sharma A, Weadick B, Govindarajan R, Khan SU, Sharma PR, Dogra A, Nandi U, Reddy CN, Bharate SS, Singh G, Bharate SB, Vishwakarma RA, Kaddoumi A, Kumar A (2021) Crocetin promotes clearance of amyloid-β by inducing autophagy via the STK11/LKB1-mediated AMPK pathway. Autophagy 17(11):3813–3832. https://doi.org/10.1080/15548627.2021.1872187

Wen YL, He Z, Hou DX, Qin S (2021) Crocetin exerts its anti-inflammatory property in LPS-induced RAW264.7 cells potentially via modulation on the crosstalk between MEK1/JNK/NF-κB/iNOS pathway and Nrf2/HO-1 pathway. Oxidative Med Cell Longev 2021:6631929. https://doi.org/10.1155/2021/6631929

WHO (2021) WHO guideline on self-care interventions for health and well-being. https://iris.who.int/bitstream/handle/10665/344701/9789240033245-eng.pdf

Wu H, Niu H, Shao A, Wu C, Dixon BJ, Zhang J, Yang S, Wang Y (2015) Astaxanthin as a potential neuroprotective agent for neurological diseases. Mar Drugs 13(9):5750–5766

Wu W, Han H, Liu J, Tang M, Wu X, Cao X, Zhao T, Lu Y, Niu T, Chen J, Chen H (2021) Fucoxanthin prevents 6-OHDA-induced neurotoxicity by targeting Keap1. Oxidative Med Cell Longev 2021:6688708. https://doi.org/10.1155/2021/6688708

Wu S, Guo X, Shang J, Li Y, Dong W, Peng Q, Xie Z, Chen C (2022) Effects of lycopene attenuating injuries in ischemia and reperfusion. Oxidative Med Cell Longev 2022:9309327. https://doi.org/10.1155/2022/9309327

Wyss-Coray T, Yan F, Lin AH, Lambris JD, Alexander JJ, Quigg RJ, Masliah E (2002) Prominent neurodegeneration and increased plaque formation in complement-inhibited Alzheimer's mice. Proc Natl Acad Sci 99(16):10837–10842. https://doi.org/10.1073/pnas.162350199

Yacoubian TA (2017) Neurodegenerative disorders: why do we need new therapies? In: Adejare A (ed) Drug discovery approaches for the treatment of neurodegenerative disorders: Alzheimer's disease. Academic Press, pp 1–16

Ye J, Cheng J, Xiong R, Chen H, Huang S, Li H, Pang J, Zhang X, Zhu H (2024) Effects and mechanisms of lutein on aging and age-related diseases. Antioxidants (Basel) 13(9). https://doi.org/10.3390/antiox13091114

Yu J, Lin J-J, Yu R, He S, Wang Q-W, Cui W, Zhang J-R (2017) Fucoxanthin prevents H-induced neuronal apoptosis via concurrently activating the PI3-K/Akt cascade and inhibiting the ERK pathway. Food Nutr Res 61:1304678

Zandi PP, Anthony JC, Hayden KM, Mehta K, Mayer L, Breitner JC (2002) Reduced incidence of AD with NSAID but not H2 receptor antagonists: the Cache County study. Neurology 59(6):880–886. https://doi.org/10.1212/wnl.59.6.880

Zhang X, Fan Z, Jin T (2017) Crocin protects against cerebral- ischemia-induced damage in aged rats through maintaining the integrity of blood-brain barrier. Restor Neurol Neurosci 35(1):65–75. https://doi.org/10.3233/rnn-160696

Zhang J, Wang Y, Dong X, Liu J (2018) Crocetin attenuates inflammation and amyloid-β accumulation in APPsw transgenic mice. Immun Ageing 15:24. https://doi.org/10.1186/s12979-018-0132-9

Zhang H, Wei M, Sun Q, Yang T, Lu X, Feng X, Song M, Cui L, Fan H (2020) Lycopene ameliorates chronic stress-induced hippocampal injury and subsequent learning and memory dysfunction through inhibiting ROS/JNK signaling pathway in rats. Food Chem Toxicol 145:111688. https://doi.org/10.1016/j.fct.2020.111688

Zhang LN, Li MJ, Shang YH, Liu YR, Han-Chang H, Lao FX (2022) Zeaxanthin attenuates the vicious circle between endoplasmic reticulum stress and tau phosphorylation: involvement of GSK-3β activation. J Alzheimer's Dis 86(1):191–204. https://doi.org/10.3233/jad-215408

Zhao D, Kwon SH, Chun YS, Gu MY, Yang HO (2017) Anti-Neuroinflammatory effects of Fucoxanthin via inhibition of Akt/NF-κB and MAPKs/AP-1 pathways and activation of PKA/CREB pathway in lipopolysaccharide-activated BV-2 microglial cells. Neurochem Res 42(2):667–677. https://doi.org/10.1007/s11064-016-2123-6

Zhao B, Liu H, Wang J, Liu P, Tan X, Ren B, Liu Z, Liu X (2018) Lycopene supplementation attenuates oxidative stress, Neuroinflammation, and cognitive impairment in aged CD-1 mice. J Agric Food Chem 66(12):3127–3136. https://doi.org/10.1021/acs.jafc.7b05770

Zhong Q, Sun W, Qin Y, Xu H (2023) Association of dietary α-carotene and β-carotene intake with low cognitive performance in older adults: a cross-sectional study from the national health and nutrition examination survey. Nutrients 15(1):239

Zhou X, Wang S, Ding X, Qin L, Mao Y, Chen L, Li W, Ying C (2017) Zeaxanthin improves diabetes-induced cognitive deficit in rats through activiting PI3K/AKT signaling pathway. Brain Res Bull 132:190–198. https://doi.org/10.1016/j.brainresbull.2017.06.001

Zhou X, Gan T, Fang G, Wang S, Mao Y, Ying C (2018) Zeaxanthin improved diabetes-induced anxiety and depression through inhibiting inflammation in hippocampus. Metab Brain Dis 33(3):705–711. https://doi.org/10.1007/s11011-017-0179-x

Zia-Ul-Haq M, Dewanjee S, Riaz M (2021) Carotenoids: structure and function in the human body, vol 16. Springer

Zou X, Gao J, Zheng Y, Wang X, Chen C, Cao K, Xu J, Li Y, Lu W, Liu J, Feng Z (2014) Zeaxanthin induces Nrf2-mediated phase II enzymes in protection of cell death. Cell Death Dis 5(5):e1218–e1218. https://doi.org/10.1038/cddis.2014.190

Molecular Basis of the Role of Polyphenols in Alzheimer's Disease

13

Neha Kanojia, Nitin Verma, Komal Thapa, Chandel Prarit, and Nitin Jhangra

Abstract

Amyloid-beta (Aβ) plaques and neurofibrillary tangles build up in the brain of people with Alzheimer's disease (AD), a progressive neurodegenerative disease. Because of the limited effectiveness of current treatments, new approaches to prevention and therapy are required. The neuroprotective potential of polyphenols, a varied group of bioactive chemicals originating from plants, has attracted a lot of attention. They are potential options for treating AD because of their anti-inflammatory, antioxidant, and metal-chelating qualities, as well as their capacity to alter signaling pathways and lower Aβ aggregation. This chapter explores the molecular processes that discover the therapeutic potential of polyphenols in AD. We investigate their functions in reducing oxidative stress, preventing the development of Aβ fibrils, boosting autophagy, and adjusting gut-brain axis interactions. The chapter also covers issues related to the bioavailability of polyphenols and methods to enhance their medicinal value, such as dietary changes and delivery systems based on nanotechnology. This chapter highlights polyphenols as new modulators of Alzheimer's pathogenesis and their potential for managing neurodegenerative diseases by combining knowledge from in vitro, in vivo, and clinical research.

Keywords

Alzheimer's disease · Polyphenols · Neuroprotection · Neurodegenerative disorders

N. Kanojia · N. Verma (✉) · K. Thapa · C. Prarit · N. Jhangra
Chitkara University School of Pharmacy, Chitkara University, Baddi, Himachal Pradesh, India
e-mail: Nitin.Verma@chitkarauniversity.edu.in

J. Malik et al. (eds.), *Diet and Alzheimer's Disease: Let Food be Our Medicine*, Nutritional Neurosciences, https://doi.org/10.1007/978-981-95-2736-6_13

Abbreviations

Aβ	Amyloid-beta
AD	Alzheimer's disease
APP	Amyloid precursor protein
APOE	Apolipoprotein
UPR	Unfolded Protein Response
UPS	Ubiquitin Proteasomal System
EGCG	Epigallocatechin Gallate
BACE	(Beta-site APP cleaving enzyme 1)
GSPE	Grape Seed Polyphenolic Extract
SAMP8	Senescence-Accelerated Mice P8
PP2A	Protein Phosphatase 2A
GSK3β	Glycogen Synthase Kinase-3 beta
ROS	Reactive Oxygen Species
CREB	Cyclic AMP Response Element-Binding Protein
MAPK	Mitogen-Activated Protein Kinase
TYROBP	TYRO protein tyrosine kinase binding protein
NF-κB	Nuclear factor kappa B
STAT	Signal Transducer and Activator of Transcription
SIRT1	Sirtuin 1
5xFAD	5 familial Alzheimer's disease mutations
7,8-DHF	7,8-dihydroxyflavone
ERK	Extracellular signal-regulated kinase
CREB	Cyclic AMP response element binding protein
LTP	Long-Term Potentiation
BDNF	Brain-derived neurotrophic factor
IκB kinase	Inhibitor of kappa B kinase
NADPH	Nicotinamide Adenine Dinucleotide Phosphate
Trans-RV	Trans-Resveratrol
i.p.	Intraperitoneal

13.1 Introduction

Over a long time, the complex process of aging changes how a person normally looks (Molino et al. 2016). It involves the decline of biological processes, particularly those of the brain and cognition. Memory, thinking, orientation, comprehension, computation, learning ability, language, and judgment are all impacted when this decline exceeds what is typical of aging. People who have Alzheimer's disease (AD) experience such symptoms. Over time, the illness worsens to the point where those impacted undergo behavioral and physiological changes (Kumari et al. 2022a, b). A growing number of people are getting older, which has put AD at the forefront of contemporary study. Numerous investigations on its potential causes and risk factors have been carried out. Despite ample advancement in the studies related to its pathogenesis, the exact processes by which AD arises are still unknown, though. Since there is currently no cure, the goal of all treatment approaches is to manage behavioral changes and slow down symptoms (William Raja et al. 2023). Therefore, the majority of the study focuses on identifying risk factors that may contribute to the development of AD and all the treatments revolve around such factors. Nutrients are being extensively studied in an attempt to establish a connection between them and the neurodegeneration that develops in the brains of AD patients, as they are crucial for normal systemic processes. Numerous nutrients function as an energy source for activities in addition to being involved in metabolic events (Vauzour 2012). Because of this, changes in nutrition have a direct impact on the central nervous system, which may have consequences for neurodegeneration and the metabolism of amyloid precursor protein (APP) (Singh et al. 2008). Calorie restriction and the association between cholesterol and the apolipoprotein APOEε4 genotype are the dietary pattern factors that are deemed significant (Pasinetti et al. 2015). Furthermore, the development of AD has been connected to particular vitamins and nutrients. Numerous studies have examined the impact of nutrition on AD, emphasizing the significance of vitamins (Dhakal et al. 2019). Studies have revealed that only multivitamin combinations had safe effects, while pills containing just one type of vitamin hardly ever had any discernible effect (Caruso et al. 2022; Bukhari 2022). However, these benefits appear to be significant when a set of vitamins is taken together (Di Meo et al. 2020).

Additionally, polyphenols possess antioxidant qualities, and several animal studies have been carried out to evaluate the relationship between polyphenol-rich diets and amyloid buildup and the mitigation of oxidative stress in AD patients (Wang et al. 2025; Oladele et al. 2025). According to a study, resveratrol from grapes decreased learning impairment and hippocampus deterioration (Stankovic et al. 2025). The preventive effects of polyphenols against the beginning of AD are the main focus of this chapter (AlHayani et al. 2025; Nasiry and Khalatbary 2025). This chapter will first give a brief outline of the AD pathophysiology and its therapeutic approaches before delving deeply into the function of polyphenols.

13.2 Pathology of AD

It was initially believed that plaques and tangles were the main mediators of neurotoxicity in AD, but recent investigations have shown the significance of neuroinflammation, mitochondrial dysfunction, oxidative stress, and gut microbiome in the pathogenesis of AD. (Nguyen et al. 2025).

13.3 Therapeutic Approaches for Alzheimer's Disease

13.3.1 Metal Chelation Approach

This is supported by studies that have shown that Aβ aggregation tendency rises with oxidative stress response after association with metal ions such as Cu2+, Zn2+, and Fe2+ (Kalu and Ray 2025; Jalouli et al. 2025). A few polyphenols' protective responses have already been linked to metal chelation as one of their mechanisms (Liu et al. 2025). The impact of metal chelation on the development of Alzheimer's disease is still in its early phases, yet there is significant variation in metal ion levels among age groups.

13.3.2 Amyloid Fiber Disruption Strategy

It would be intriguing to investigate the possibility of directly targeting the mechanistic approach associated with their biogenesis (Baghcheghi et al. 2025; Islam et al. 2025a, 2025b). Lot of emphasis is now given to this approach and various agents, both natural and synthetic, have been evaluated for their antiamyloid effects whether they prevent their aggregation, increase their proteasomal degradation or clearance, or upregulation of enzymes that lead to nonpathological pathway of these proteins.

13.3.3 Antioxidant Approach

Oxidative stress has been identified as a major factor in the development of Alzheimer's disease pathogenesis (Utpal et al. 2025; Della Vedova et al. 2025). Alzheimer's disease can be effectively treated with strategies that reduce the oxidative burden (Garcia-Burgos 2025; Almutary et al. 2025).

13.3.4 Targeting Protein Homeostasis

An imbalance in the internal protein homeostasis system causes the Aβ aggregation to develop. One such reaction that is triggered in reaction to any protein problem is the unfolded protein response (UPR). Alzheimer's disease pathology can be improved with strategies that use the sudden activation of UPR (Adedara et al. 2025). Furthermore, the ubiquitin proteasomal system (UPS) mediates the breakdown of proteins that are unstructured or flawed. An aggravation of Alzheimer's pathogenic reactions is linked to defective UPS responses (Tkaczenko and Kurhaluk 2025; Islam et al. 2025a, 2025b). Thus, preserving UPS homeostasis may be a useful strategy for thwarting the pathogenic reaction associated with Alzheimer's disease.

13.3.5 Anti-inflammatory Approach

It is believed that aging-activated inflammatory cascades are crucial in regulating the aging process linked to neurodegenerative diseases (de Lima et al. 2025). It has already been proven that inflammatory cascades contribute to the amplification of

Alzheimer's disease (Gopalakrishna et al. 2025). Therefore, focusing on these inflammatory processes is a useful strategy for Alzheimer's disease treatment.

13.3.6 Dietary Approaches for Alzheimer's Disease

Numerous investigations have proposed that nutrition and diet play a therapeutic role in the development and treatment of Alzheimer's disease. Because it is associated with lower rates of morbidity and death from neurodegenerative disorders, the Mediterranean diet has become more and more popular among the general public (Tomar 2025; Li and Guo 2025).The Mediterranean diet is considered to have abundant polyphenols because of the presence of a high number of fruits and vegetables, which are naturally abundant in these compounds (Lv et al. 2025; Andishmand et al. 2025; Agarwal et al. 2025).

13.4 Polyphenols: Classification and Sources

A class of chemical molecules known as polyphenols is distinguished by the presence of several phenol structural units, which are aromatic rings with multiple hydroxyl groups. Plants that contain these phytochemicals can defend themselves against UV rays and the invasion of diseases. Diets primarily consisting of plant-based foods include them (Carrillo et al. 2025; Chowdhury et al. 2025). Flavonoids, phenolic acids, stilbenes, and lignans are the four chemical subclasses (Kelany et al. 2025; Ijaz et al. 2025). Table 13.1 summarizes the role of various polyphenols in Alzheimer's disease.

13.4.1 Flavonoids

With a 15-carbon structure, flavonoids are the most abundant and sizable classes of secondary metabolites originating from plants that have been shown to have protective effects against neurodegenerative disorders (Singh et al. 2025; Mosadegh et al. 2025). Their capacity to bind metal ions and/or scavenge free radicals is mostly due to their extremely reactive hydroxyl group (Puri et al. 2022; Chen and Sun 2025). There are various subgroups into which flavonoids can be separated: All foods contain flavonoids, with quercetin and kaempferol serving as typical molecules. Onions, broccoli, kale, blueberries, and red wine contain higher levels of flavonols (Chipofya et al. 2025). In contrast, fruits and vegetables contain far fewer flavones (luteolin and apigenin). Legumes like soy are the primary source of phytoestrogens and isoflavones (Markowska et al. 2025). The fundamental building blocks of flavonols are epicatechin and catechin, which combine to produce a variety of oligomers and polymers via C4-C6 or C4-C8 interflavan linkages. Although red wine and a variety of fruits contain flavanols, the richest sources are chocolate and green tea (Kurmi

Table 13.1 Summary of the neuroprotective effects of various polyphenols in Alzheimer's disease

Polyphenol class	Examples	Mechanism of action	Neuroprotective effects
Flavonoids	Quercetin, epigallocatechin gallate (EGCG), kaempferol	Antioxidant activity, inhibition of amyloid-beta (Aβ) aggregation, reduction of oxidative stress	Protects neurons, improves cognitive function
Phenolic acids	Ferulic acid, caffeic acid	Anti-inflammatory properties, modulation of cholinergic pathways, inhibition of Aβ toxicity	Reduces neuroinflammation, enhances synaptic plasticity
Stilbenes	Resveratrol, pterostilbene	Activates SIRT1, reduces tau phosphorylation, enhances mitochondrial function	Inhibits tau pathology, delays disease progression
Lignans	Secoisolariciresinol, matairesinol	Anti-inflammatory and antioxidant effects, modulates gut microbiota	Reduces neurotoxicity, supports brain homeostasis
Curcuminoids	Curcumin, demethoxycurcumin	Inhibits amyloid plaque formation, reduces oxidative stress and neuroinflammation	Improves memory, enhances neuronal survival
Tannins	Ellagitannins, proanthocyanidins	Chelation of metal ions, antioxidant activity, suppression of neuroinflammation	Prevents Aβ fibril formation, supports neuronal integrity

and Karati 2025). Fruits and vegetables are primarily rich in chalcone (phloretin, arbutin, etc.) and anthocyanins (cyanidin, malvidin, etc.).

13.4.2 Phenolic Acids

Benzoic and cinnamic acids are the hydroxyl derivatives of phenolic acids. Very little hydroxybenzoic acid is present in plants (Gobinath et al. 2025). Cherries and kiwis are the foods that contain the highest levels of hydroxycinnamic acids (Bojang and Varalakshmi 2025). Ferulic acid is primarily present in grains (Turer and Sanlier 2025).

13.4.3 Lignans

Lignans belong to a group of secondary metabolites found in plants. Because of their potent antioxidation, anti-inflammatory, and neuroprotective bioactivities, lignans have garnered increasing attention in recent years (El Gaamouch et al. 2022; Bukhari 2022).

13.4.4 Stilbenes

The human diet contains very little stilbene. The most well-known and researched is resveratrol, which has been proven to have anticarcinogenic properties in screening for medicinal plants. Wine contains a very small amount of resveratrol (Caruso et al. 2022).

13.5 Molecular Mechanisms of Polyphenols in AD

13.5.1 Modulation of Aβ Production, Oligomerization, and Clearance

The generation of Aβ is significantly impacted by the activities of both enzymes, which compete in APP proteolysis (Albadrani et al. 2024). Figure 13.1 represents the mechanistic approach of polyphenols in Alzheimer's disease.

Numerous studies have shown that polyphenols can either increase α-secretase activity or inhibit BACE to alter the generation of Aβ. (−)-Strong inhibitors of amyloidogenic processing include curcumin, epicatechin, epigallocatechin, and epigallocatechin-3-gallate (EGCG) (Romero-Márquez et al. 2023). In vitro tests using neuronal cell lines that expressed human APP revealed that EGCG administration dramatically reduced the generation of Aβ (Shabbir et al. 2021). In the Tg2576 animal model of AD, intraperitoneal administration of epigallocatechin-3-gallate reduced Aβ levels, reproducing these findings in vivo (Yang et al. 2024). Curcumin therapy raises α-secretase activity, according to another study (Lakey-Beitia et al. 2021). BACE activity in neural cells is inhibited by curcuminoids and epigallocatechin-3-gallate. In APP/PS1 AD mice, the BACE modulator epigallocatechin-3-gallate could inhibit lower amyloidosis and enhance cognitive performance (Xu et al. 2021; Piccialli et al. 2022). Both soluble and insoluble Aβ oligomers can be formed from Aβ monomers (Rezaee et al. 2021). It encourages the formation of smaller, amorphous, harmless protein aggregates from freshly produced oligomers (El Gaamouch et al. 2021). Another group demonstrated that (−)-epigallocatechin-3-gallate might remodel large oligomers or fibrils into less hazardous assemblies by binding to them (Wang et al. 2024). In addition to inhibiting fibril formation and destabilizing preexisting fibrils, curcumin can significantly impede the oligomerization of Aβ in a dose-dependent manner (Kabir et al. 2023; Baruah et al. 2023). The same group looked into the precise compounds that could lower Aβ (Uddin et al. 2021). Subsequent research showed that GSPE is a strong inhibitor of Aβ peptide oligomerization.

By improving Aβ elimination, polyphenols can help lessen Aβ pathogenesis. Numerous theories have been put out, such as the idea that resveratrol may encourage the intracellular breakdown of Aβ through a process involving the lysosome and autophagy (Leri et al. 2021). Furthermore, resveratrol may increase the activity of the insulin-degrading enzymes in the brain, which will break down Aβ and make it easier to remove it (Ticinesi et al. 2022).

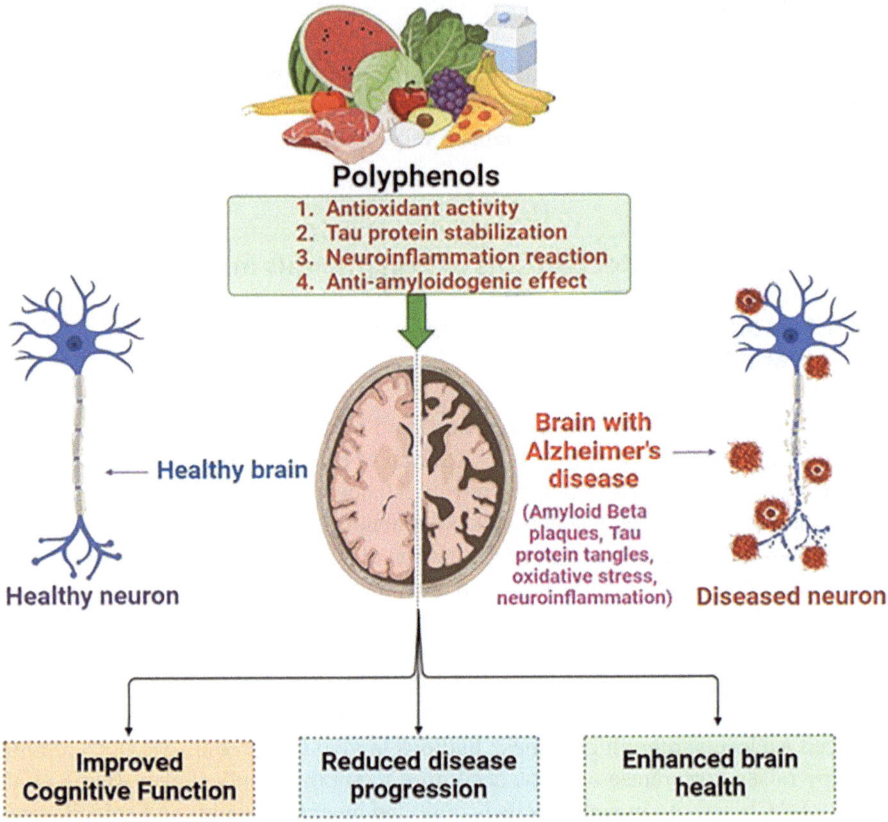

Fig. 13.1 Diagrammatic illustration of the polyphenols' molecular role in Alzheimer's disease. Key pathways that contribute to neuroprotection and cognitive improvement are depicted in the image, including antioxidant activity, prevention of amyloid-beta aggregation, control of neuroinflammation, and enhancement of synaptic function

13.5.2 Modulation of Tau Phosphorylation

Unusual microtubule-associated protein aggregation Another factor that contributes to AD pathogenesis is tau. Its capacity to bind microtubules is controlled by tau phosphorylation. The equilibrium between the phosphorylation catalyzed by PKA plus other kinases and the dephosphorylation catalyzed mostly by phosphatase 2A controls tau phosphorylation in neuronal cells (Kola et al. 2021). Through the inhibition of AD-tau kinases or the promotion of PP2A, some polyphenols can influence Tau hyperphosphorylation and the subsequent development of NFT. It has been demonstrated that resveratrol prevents Tau from being hyperphosphorylated. In addition, resveratrol prevents GSK-3β from phosphorylating Ser396 Tau in senescence-accelerated mice P8 (SAMP8) (Doğanoğlu and Erbaş 2021). By raising PP2A activity, resveratrol can also reduce Tau hyperphosphorylation and promote

Tau dephosphorylation. Similarly, curcumin therapy prevented Tau hyperphosphorylation by activating the GSK-3β pathway in an AD model that involved okadaic acid injection (Akter et al. 2021). A previous study showed in their bioavailability experiments of rats that after oral administration of GSPE, the plasma included epicatechin and methylation and glucuronidated catechin (Marianetti et al. 2022). Additionally, they found that in AD animals, only the monomeric fraction improved cognitive performance and decreased amyloid neuropathology. According to biophysical research, polyphenols can alter the structure of the Tau protein and stop it from self-association. For instance, self-assembly of Tau into β-sheet-containing filaments when arachidonic acid is present (Gan et al. 2022).

13.5.3 Modulation of AD-Associated Oxidative Stress

When these oxidants target polyunsaturated fatty acids, lipid peroxidation takes place (Oliveira and Pieniz 2024; Hügel et al. 2021). Protein carbonyl derivatives can also be produced by reactive oxygen species (ROS) attacking the side chains of amino acids or the backbones of proteins (Maqbool and Zehravi 2021). The inferior parietal lobule and hippocampal protein carbonyl content was observed to be considerably higher in AD participants than in normal controls (Calderaro et al. 2022; Maleki Dana et al. 2021). ROS-induced oxidation of structural proteins or important enzymes can seriously disrupt their physiological function, resulting in cell death and neurodegeneration. Additionally, ROS can cause DNA-protein crosslinking, single- and double-strand breakage, and nucleotide change (Zhang et al. 2022). Flavonoids have long been thought to be effective because of their strong antioxidants and inherent ability to scavenge free radicals. After 2 to 3 months of chronic nobiletin administration, AD mice's brain ROS and other oxidative stress indicators were markedly decreased. Free radicals can be effectively scavenged by quercetin, which also inhibits oxidant-induced apoptosis. Apart from its potent ability to scavenge oxygen radicals, quercetin can also bind iron and other metal ions that may be harmful to the brain and prevent lipid peroxidation. When given to an AD mouse model over an extended period, the flavone apigenin significantly reduced oxidative stress and boosted glutathione peroxidase and superoxide dismutase activity (Gea-González et al. 2022). Another potent antioxidant that can successfully stabilize ROS is curcumin. By promoting the depolarization of the mitochondria's inner membrane, it inhibits the production of reactive oxygen species. When given intravenously to rodents, it has also been demonstrated to inhibit the growth of free radicals. Resveratrol can decrease the production of ROS in vivo by increasing the activity of the superoxide dismutase enzyme. According to in vivo research, epicatechin and catechin can stop the peroxidation of lipids and the production of ROS in AD models. In rats, Aβ-mediated lipid peroxidation and ROS production in the hippocampus could be successfully inhibited by a single oral dosage of epicatechin (Islam et al. 2024). Proanthocyanidins and anthocyanins are abundant in blueberries. By blocking MAP kinase and CREB-mediated ROS signaling, blueberry extract has been demonstrated to raise redox buffer glutathione and prevent amyloid damage. Supplementing with blueberries has been demonstrated to prevent cognitive decline in persons who are at high risk of acquiring dementia (Yeung et al. 2021).

13.5.4 Polyphenols Modulate AD-Associated Inflammation

There is growing agreement that immunological disruptions play a significant role in the pathophysiology of AD. This is corroborated by genome-wide association studies that relate late-onset AD (LOAD) to genes specific to myeloid cells, including TYROBP, TREM2, and CD33. One of the key elements in the pathophysiology of AD is the process known as neuroinflammation (Li et al. 2023).

The two most researched compounds for their possible use in the treatment of AD are curcumin and resveratrol. It has been demonstrated that curcumin can inhibit the action of NF-kappa B and the related cascade of inflammation. Curcumin can also prevent the release of pro-inflammatory cytokines and chemokines that are triggered by Aβ. By inhibiting the STAT signaling pathway, resveratrol reduces Aβ-mediated microglia inflammation. Additionally, resveratrol can both in vitro and in vivo activate SIRT1. Histone deacetylase SIRT1 can epigenetically rewire inflammation. In the brains of AD mice, resveratrol therapy enhanced spatial memory, decreased neuroinflammation, and enhanced neurotrophins (Tripathi and Mazumder 2021). It has also been demonstrated that blueberry and its extract can suppress MAPK signaling, hence preventing amyloid-mediated microglia activation (Furlan and Bren 2021; Moussavi et al. 2022).

13.5.5 Modulation of Synaptic Function and Memory

In transgenic AD mice models, selective polyphenols also demonstrated encouraging results in restoring cognitive function. For instance, aged 5xFAD mice given 7,8-dihydroxyflavone (7,8-DHF) on a chronic basis showed improved performance and less synapse loss in the brain (Kumari et al. 2022a, b). Apigenin given orally improved learning and memory in 2xFAD mice and boosted ERK/CREB signaling activity (Yang et al. 2021). Fisetin-treated mice exhibited enhanced synaptic protein production, enhanced ERK/MAPK signaling activity, and enhanced cognitive performance (Jayasena et al. 2013; Emran et al. 2022). In a 1xFAD mouse model, green tea's high EGCG content can stop synaptic protein loss and cognitive decline. Similarly, anthocyanins have been shown to enhance memory performance and lessen the loss of synaptic proteins. They demonstrated in their study that cocoa extracts enhanced with epicatechin and catechin inhibited Aβ oligomerization and avoided synaptic impairments. They showed that Aβ-induced LTP deficiency may be avoided by applying cocoa extracts to mouse hippocampus slices. By activating BDNF in neural cells, cocoa has also been demonstrated to inhibit neurite degeneration caused by Aβ oligomer (Al Mamun et al. 2024; Iftikhar et al. 2022).

13.5.6 Modulation of Gut Microbiome

Their metabolites may exhibit biological actions that differ significantly from those of their parent substances. The chemical contents and neuroinflammation-related effects of the digested and undigested extracts differed (Luo et al. 2023; Grabska-Kobyłecka et al. 2023). Because of its possible connection to both health and disease, the gut microbiome is gaining more attention. Strong connections between the gut microbiota and polyphenol metabolism have been shown in recent research (Shah et al. 2024). Additionally, the gut microbiota and the central nervous system communicate in both directions. This is known as the gut-brain axis: because of its potential to impact brain function.

13.6 Implication of Various Polyphenols in Targeting Alzheimer's Disease

13.6.1 Curcumin

The South Asian continent has long used curcumin, one of the dietary polyphenols, for ages. It is also widely used in ayurvedic therapies due to its beneficial properties. The prevalent dietary use of turmeric, which has curcumin as its main ingredient, was assumed to be the cause of this (Suzuki et al. 2023; Mahboob et al. 2023). Curcumin has already been found in several scientific investigations to interfere with Aβ production and help remove its deposits, which stops Aβ aggregates from forming. As oxidation plays an important role in the onset of AD, curcumin's anti-oxidative potential may play an important role in the management of Alzheimer's disease. NF-κβ is a key transcription factor that arranges inflammatory cascades, and it coordinates the molecular effects of curcumin (Bucciantini et al. 2022). In a healthy state, IκB kinase prevents NF-κβ units from activating. On the other hand, IκB kinase-mediated regulation is loosened in response to inflammatory stimuli, which causes NF-κβ to be directed toward nuclear import and induce the production of inflammatory mediators. Conversely, it has been discovered that curcumin treatment inhibits the activation of NF-κβ, which plays an important role in the anti-inflammatory actions of curcumin. Curcumin's inhibitory activity on IκB kinase is the cause of these actions (Gentile et al. 2023).

13.6.2 Quercetin

Higher plants, fruits, and vegetables that are regularly ingested, such as onions, apples, berries, asparagus, and so on, contain quercetin, a polyphenol that falls under the flavonoid category (Tan et al. 2022). The majority of the effects observed under different situations are caused by their inflammatory and antioxidant qualities. Pharmacophores aid in the neutralization of reactive oxygen species (ROS) and account for their antioxidative potential. Inhibiting the NADPH oxidase-directed oxidative load also has an indirect effect on AMPK activity. When combined with its impact on Aβ formation, this antioxidative action is linked to neuroprotective effects in situations like Alzheimer's disease. By affecting amyloid precursor protein (APP) processing by inhibiting β-Secretase (BACE) activity, one can interfere with Aβ synthesis (Revi and Rengan 2021). Second, quercetin also modifies APP processing by inhibiting NFκβ. Since quercetin has been shown to have neuroprotective properties, the only issue impeding its clinical use is its low bioavailability, which is brought on by the gut's modification of quercetin (methylation, sulfation), which lowers its effective concentration (Ontario et al. 2022).Low doses of quercetin have been shown to have antioxidative and anti-inflammatory properties, but large doses are hazardous in in vitro experiments, even if high doses can be used to reach an effective concentration (Nian et al. 2022).The low blood–brain barrier (BBB) penetration capabilities of quercetin raise additional concerns about its potential use as a neuroprotective drug. An efficient neuro-protective concentration has been attained by directly administering quercetin via intravenous and intraperitoneal methods. As a result, the bioavailability issues surrounding the use of quercetin in clinical settings require attention.

13.6.3 Tannic Acid

Many health advantages have been linked to tannic acid, a polyphenol obtained from plants that belongs to the hydrolysable tannin class. The bioavailability and therapeutic effectiveness of tannins are determined by their levels of polymerization. Tannic acid and EGCG share structural similarities; therefore, it is anticipated that their mechanisms of action will be similar. They have been shown to affect the processing of the Amyloid Precursor Protein (APP) in Alzheimer's disease by preventing cleavage on the β-site by blocking β-secretase activity. These occurrences ultimately lead to a reduction in Aβ synthesis and are linked to better cognitive performance in Alzheimer's disease (Aluko 2021).

13.6.4 Epigallocatechin-3-Gallate (EGCG)

The polyphenolic molecule found in green tea is called EGCG, and it acts as an important phytoconstituent in targeting Alzheimer's disease because of its extraordinary capacity to operate as an Aβ fibril disruptor. Mechanistically, EGCG forms an adduct with Aβ fiber through ionic interactions based on hydrogen bonding and nonpolar interactions. Because of this compound development, these fibers are less likely to aggregate, which in turn lessens the neurodegenerative process linked to AD pathology (Nabavi et al. 2018). This indicates that EGCG's therapeutic efficacy against AD pathology appears to be mostly due to its direct influence on Aβ kinetics. Additionally, secretase activity modification triggered by EGCG decreases the production of amyloid precursor protein, which in turn inhibits Aβ aggregation.

Furthermore, EGCG may have anti-inflammatory effects by inhibiting the activation of microglia, which may influence how Alzheimer's disease progresses (Jayatunga et al. 2021).

13.6.5 Trans-Resveratrol (RV)

The ability of a cell to survive in the body depends on its ability to maintain protein homeostasis. A protein's ideal working environment is maintained by several methods. The components that are left over when disordered proteins are broken down are recycled for the subsequent protein synthesis event. In Alzheimer's disease, the advancement of the pathology is linked to an elevated UPR. It has been discovered that polyphenols such as trans-resveratrol (RV) reduce this UPR to affect the pathophysiology of Alzheimer's disease (Hu et al. 2023). It is investigated that treatment of Alzheimer's disease with RV results in a significant reduction in Aβ plaque deposition (Dubey and Singh 2023; Ferreira et al. 2024).

13.7 Conclusion

Both Tau and Aβ toxicities are pathological mechanisms implicated in the pathophysiology of Alzheimer's disease. Clinical trials continue to show that therapeutic approaches that target one or the other are ineffective. Many studies have been conducted recently to clarify the advantages of polyphenols produced from plants in the treatment of neuropathological conditions, such as AD. Any therapeutic drug must be able to pass across the blood–brain barrier to be investigated for the treatment of neurodegenerative diseases. Curcumin and resveratrol are two dietary polyphenols that have been shown to cross the blood-brain barrier, making them excellent choices for the development of neuroprotective strategies. After being taken orally, curcumin has been seen to reach its highest limit in the brain in 48 hours. The curcumin's bioavailability was successfully increased by additional injection methods, including intramuscular, oral gavage, and intraperitoneal (i.p.). Another issue with the use of curcumin is its limited shelf life. In the digestive tract and hepatic environment, curcumin has been observed to undergo glucuronidation (to generate curcumin glucuronides) and sulfate modification, which reduces its bioavailability. Resveratrol and other dietary polyphenols also have this kind of bioavailability problem. As a result, several tactics that use the suppression of the glucuronidation process will work well to raise the curcumin concentration. Curcumin's therapeutic bioavailability can be effectively increased by combining it with piperine.

The restricted brain bioavailability and poor absorption of polyphenols hinder their clinical development for AD, despite their multitargeting properties. Furthermore, after undergoing hepatic and digestive activity, the majority of polyphenol metabolite forms may not possess the same biological action as the original substance.

References

Adedara IA, Weis GC, Monteiro CS, Soares FA, Rocha JB, Schetinger MR, Emanuelli T, Aschner M (2025) Versatility of Caenorhabditis elegans as a model organism for evaluating foodborne neurotoxins and food bioactive compounds in nutritional neuroscience. Mol Neurobiol 62:1–25

Agarwal S, Kaushik S, Saha H, Paramanick D, Mazhar M, Basist P, Khan R, Alhalmi A (2025) Therapeutic potential of traditional herbal plants and their polyphenols in alleviation of mercury toxicity. Naunyn Schmiedeberg's Arch Pharmacol 398:1–27

Akter R, Chowdhury MA, Rahman MH (2021) Flavonoids and polyphenolic compounds as potential talented agents for the treatment of Alzheimer's disease and their antioxidant activities. Curr Pharm Des 27(3):345–356

Al Mamun A, Shao C, Geng P, Wang S, Xiao J (2024) Polyphenols targeting NF-κB pathway in neurological disorders: what we know so far? Int J Biol Sci 20(4):1332

Albadrani HM, Chauhan P, Ashique S, Babu MA, Iqbal D, Almutary AG, Abomughaid MM, Kamal M, Paiva-Santos AC, Alsaweed M, Hamed M (2024) Mechanistic insights into the potential role of dietary polyphenols and their nanoformulation in the management of Alzheimer's disease. Biomed Pharmacother 174:116376

AlHayani DA, Kubaev A, Uthirapathy S, Mandaliya V, Ballal S, Kalia R, Arya R, Gabble BC, Alasheqi MQ, Kadhim AJ (2025) Insights into the therapeutic potential of SIRT1-modifying compounds for Alzheimer's disease: a focus on molecular mechanisms. J Mol Neurosci 75(1):29

Almutary AG, Begum MY, Siddiqua A, Gupta S, Chauhan P, Wadhwa K, Singh G, Iqbal D, Padmapriya G, Kumar S, Kedia N (2025) Unlocking the neuroprotective potential of Silymarin: a promising ally in safeguarding the brain from Alzheimer's disease and other neurological disorders. Mol Neurobiol 62:1–23

Aluko RE (2021) Food-derived acetylcholinesterase inhibitors as potential agents against Alzheimer's disease. eFood 2(2):49–58

Andishmand H, Masoumi B, Yousefi M, Sadeghizadeh-Yazdi J et al (2025) Pharmacodynamics, pharmacokinetics, and cell signaling pathways of pomegranate peel extract. Phytochem Rev:1–24

Baghcheghi Y, Razazpour F, Mirzaee F, Dalfardi M, Pourfridoni M, Hedayati-Moghadam M (2025) Exploring the molecular mechanisms of curcumin in modulating memory impairment in neurodegenerative disorders. Mol Biol Rep 52(1):1–19

Baruah P, Moorthy H, Ramesh M, Padhi D, Govindaraju T (2023) A natural polyphenol activates and enhances GPX4 to mitigate amyloid-β induced ferroptosis in Alzheimer's disease. Chem Sci 14(35):9427–9438

Bojang KP, Varalakshmi M (2025) Effectiveness of grape-based interventions on cognitive and mental health: evidence from human trials. Curr Nutr Rep 14(1):1–12

Bucciantini M, Leri M, Scuto M, Ontario M, Salinaro AT, Calabrese EJ, Calabrese V, Stefani M (2022) Xenohormesis underlyes the anti-aging and healthy properties of olive polyphenols. Mech Ageing Dev 202:111620

Bukhari SNA (2022) Dietary polyphenols as therapeutic intervention for Alzheimer's disease: a mechanistic insight. Antioxidants 11(3):554

Calderaro A, Patanè GT, Tellone E, Barreca D, Ficarra S, Misiti F, Laganà G (2022) The neuroprotective potentiality of flavonoids on Alzheimer's disease. Int J Mol Sci 23(23):14835

Carrillo JÁ, Arcusa R, Xandri-Martínez R, Cerdá B, Zafrilla P, Marhuenda J (2025) Impact of polyphenol-rich nutraceuticals on cognitive function and neuroprotective biomarkers: a randomized, double-blind, placebo-controlled clinical trial. Nutrients 17(4):601

Caruso G, Godos J, Privitera A, Lanza G, Castellano S, Chillemi A, Bruni O, Ferri R, Caraci F, Grosso G (2022) Phenolic acids and prevention of cognitive decline: polyphenols with a neuroprotective role in cognitive disorders and Alzheimer's disease. Nutrients 14(4):819

Chen D, Sun Y (2025) Current status of plant-based bioactive compounds as therapeutics in Alzheimer's diseases. J Integr Neurosci 24(1):23090

Chipofya E, Docrat TF, Marnewick JL (2025) The neuroprotective effect of rooibos herbal tea against Alzheimer's disease: a review. Mol Nutr Food Res 69(1):e202400670

Chowdhury SR, Ghosh S, Rahaman M, Dutta S (2025) Role of polyphenols to attenuate depressive disorders and cognitive impairments associated with diabetes mellitus–a comprehensive review. Nat Prod J 15(3):E060624230792

de Lima EP, Laurindo LF, Catharin VCS, Direito R, Tanaka M et al (2025) Polyphenols, alkaloids, and Terpenoids against neurodegeneration: evaluating the neuroprotective effects of Phytocompounds through a comprehensive review of the current evidence. Meta 15(2):124

Della Vedova L, Baron G, Morazzoni P, Aldini G, Gado F (2025) The potential of polyphenols in modulating the cellular senescence process: implications and mechanism of action. Pharmaceuticals 18(2):138

Dhakal S, Kushairi N, Phan CW, Adhikari B, Sabaratnam V, Macreadie I (2019) Dietary polyphenols: a multifactorial strategy to target Alzheimer's disease. Int J Mol Sci 20(20):5090

Di Meo F, Valentino A, Petillo O, Peluso G, Filosa S, Crispi S (2020) Bioactive polyphenols and neuromodulation: molecular mechanisms in neurodegeneration. Int J Mol Sci 21(7):2564

Doğanoğlu A, Erbaş O (2021) Effects of green tea polyphenols and oxidative stress on Alzheimer's and Parkinson's diseases. J Exp Basic Med Sci 2(1):001–006

Dubey S, Singh E (2023) Antioxidants: an approach for restricting oxidative stress induced neurodegeneration in Alzheimer's disease. Inflammopharmacology 31(2):717–730

El Gaamouch F, Liu K, Lin HY, Wu C, Wang J (2021) Development of grape polyphenols as multi-targeting strategies for Alzheimer's disease. Neurochem Int 147:105046

El Gaamouch F, Chen F, Ho L, Lin HY, Yuan C, Wong J, Wang J (2022) Benefits of dietary polyphenols in Alzheimer's disease. Front Aging Neurosci 14:1019942

Emran TB, Islam F, Nath N, Sutradhar H, Das R, Mitra S, Alshahrani MM, Alhasaniah AH, Sharma R (2022) Naringin and naringenin polyphenols in neurological diseases: understandings from a therapeutic viewpoint. Life 13(1):99

Ferreira C, Vieira P, Sá H, Malva J, Castelo-Branco M, Reis F, Viana S (2024) Polyphenols: Immunonutrients tipping the balance of immunometabolism in chronic diseases. Front Immunol 15:1360065

Furlan V, Bren U (2021) Insight into inhibitory mechanism of pde4d by dietary polyphenols using molecular dynamics simulations and free energy calculations. Biomolecules 11(3):479

Gan J, Zhang X, Ma C, Sun L, Feng Y, He Z, Zhang H (2022) Purification of polyphenols from Phyllanthus emblica L. pomace using macroporous resins: antioxidant activity and potential anti-Alzheimer's effects. J Food Sci 87(3):1244–1256

Garcia-Burgos D (2025) Polyphenols as neuromodulators of learning and memory processes. In: Bioactive polyphenols for health and pathology treatment. Academic Press, pp 211–247

Gea-González A, Hernández-García S, Henarejos-Escudero P, Martínez-Rodríguez P, García-Carmona F, Gandía-Herrero F (2022) Polyphenols from traditional Chinese medicine and Mediterranean diet are effective against Aβ toxicity in vitro and in vivo in Caenorhabditis elegans. Food Funct 13(3):1206–1217

Gentile MT, Camerino I, Ciarmiello L, Woodrow P, Muscariello L, De Chiara I, Pacifico S (2023) Neuro-nutraceutical polyphenols: how far are we? Antioxidants 12(3):539

Gobinath M, Dwarampudi LP, Roychowdhury P, Yamuna K, Dutta S, Shanmugam R (2025) Traditional therapies for Alzheimer's disease. In: Multi-factorial approach as a therapeutic strategy for the management of Alzheimer's disease. Springer Nature Singapore, Singapore, pp 199–220

Gopalakrishna R, Aguilar J, Lee E, Mack WJ (2025) Neuroprotection by resveratrol-glucuronide and quercetin-glucuronide via binding to polyphenol-and glycosaminoglycan-binding sites in the laminin receptor. Neural Regen Res 20(3):819–820

Grabska-Kobyłecka I, Szpakowski P, Król A, Książek-Winiarek D, Kobyłecki A, Głąbiński A, Nowak D (2023) Polyphenols and their impact on the prevention of neurodegenerative diseases and development. Nutrients 15(15):3454

Hu D, Jin Y, Hou X, Zhu Y, Chen D, Tai J, Chen Q, Shi C, Ye J, Wu M, Zhang H (2023) Application of marine natural products against Alzheimer's disease: past, present and future. Mar Drugs 21(1):43

Hügel HM, de Silva NH, Siddiqui A, Blanch E, Lingham A (2021) Natural spirocyclic alkaloids and polyphenols as multi target dementia leads. Bioorg Med Chem 43:116270

Iftikhar A, Nausheen R, Muzaffar H, Naeem MA, Farooq M, Khurshid M, Almatroudi A, Alrumaihi F, Allemailem KS, Anwar H (2022) Potential therapeutic benefits of honey in neurological disorders: the role of polyphenols. Molecules 27(10):3297

Ijaz S, Iqbal J, Abbasi BA, Yaseen T, Ashraf Z, Iqbal R, Murtaza G, Mahmood T, Kazi M (2025) Biomedical application of green nanoparticles: insights into the anti-inflammatory molecular mechanisms. Applications and Commercialization, Expanding Nanobiotechnology, pp 182–211

Islam F, Roy S, Zehravi M, Paul S, Sutradhar H, Yaidikar L, Kumar BR, Dogiparthi LK, Prema S, Nainu F, Rab SO (2024) Polyphenols targeting MAP kinase signaling pathway in neurological diseases: understanding molecular mechanisms and therapeutic targets. Mol Neurobiol 61(5):2686–2706

Islam MR, Al-Imran MIK, Zehravi M, Sweilam SH, Mortuza MR, Gupta JK, Shanmugarajan TS, Devi K, Tummala T, Alshehri MA, Rajagopal K (2025a) Targeting signaling pathways in neurodegenerative diseases: quercetin's cellular and molecular mechanisms for neuroprotection. Anim Models Exp Med

Islam MR, Rauf A, Akter S, Akter H, Al-Imran MIK, Islam S, Nessa M, Shompa CJ, Shuvo MNR, Khan I, Al Abdulmonem W (2025b) Epigallocatechin 3-gallate-induced neuroprotection in neurodegenerative diseases: molecular mechanisms and clinical insights. Mol Cell Biochem:1–21

Jalouli M, Rahman MA, Biswas P, Rahman H, Harrath AH, Lee IS, Kang S, Choi J, Park MN, Kim B (2025) Targeting natural antioxidant polyphenols to protect neuroinflammation and neurodegenerative diseases: a comprehensive review. Front Pharmacol 16:1492517

Jayasena T, Poljak A, Smythe G, Braidy N, Muench G, Sachdev P (2013) The role of polyphenols in the modulation of sirtuins and other pathways involved in Alzheimer's disease. Ageing Res Rev 12(4):867–883

Jayatunga DPW, Hone E, Khaira H, Lunelli T, Singh H, Guillemin GJ, Fernando B, Garg ML, Verdile G, Martins RN (2021) Therapeutic potential of mitophagy-inducing microflora metabolite, urolithin a for Alzheimer's disease. Nutrients 13(11):3744

Kabir ER, Chowdhury NM, Yasmin H, Kabir MT, Akter R, Perveen A, Ashraf GM, Akter S, Rahman MH, Sweilam SH (2023) Unveiling the potential of polyphenols as anti-amyloid molecules in Alzheimer's disease. Curr Neuropharmacol 21(4):787–807

Kalu A, Ray SK (2025) Epigallocatechin-3-Gallate, quercetin, and Kaempferol for treatment of Parkinson's disease through prevention of gut Dysbiosis and attenuation of multiple molecular mechanisms of pathogenesis. Brain Sci 15(2):144

Kelany NA, Selim YA, Elshafey N, Ibrahim MA (2025) A promising strategy for treating Alzheimer's disease with Origanum majorana extraction. Egypt J Hosp Med 98:813–821

Kola A, Dudek D, Valensin D (2021) Metal complexation mechanisms of polyphenols associated to Alzheimer's disease. Curr Med Chem 28(35):7278–7294

Kumari N, Daram N, Alam MS, Verma AK (2022a) Rationalizing the use of polyphenol nano-formulations in the therapy of neurodegenerative diseases. CNS Neurol Disorders-Drug Targets 21(10):966–976

Kumari S, Goyal A, Sönmez Gürer E, Algın Yapar E, Garg M, Sood M, Sindhu RK (2022b) Bioactive loaded novel nano-formulations for targeted drug delivery and their therapeutic potential. Pharmaceutics 14(5):1091

Kurmi SPC, Karati D (2025) Inhibitory potential of polyphenolic stilbene derivatives with glycogen synthase kinase-3β (GSK-3β) for Alzheimer's disease: computational and SAR insights. Brain Disorders 17:100182

Lakey-Beitia J, Burillo AM, La Penna G, Hegde ML, Rao KS (2021) Polyphenols as potential metal chelation compounds against Alzheimer's disease. J Alzheimer's Dis 82(s1):S335–S357

Leri M, Bertolini A, Stefani M, Bucciantini M (2021) EVOO polyphenols relieve synergistically autophagy dysregulation in a cellular model of Alzheimer's disease. Int J Mol Sci 22(13):7225

Li Y, Guo B (2025) GSDMD-mediated pyroptosis: molecular mechanisms, diseases and therapeutic targets. Mol Biomed 6(1):11

Li Z, Zhao T, Shi M, Wei Y, Huang X, Shen J, Zhang X, Xie Z, Huang P, Yuan K, Li Z (2023) Polyphenols: natural food grade biomolecules for treating neurodegenerative diseases from a multi-target perspective. Front Nutr 10:1139558

Liu H, Song Y, Wang H, Zhou Y, Xu M, Xian J (2025) Deciphering the power of resveratrol in Mitophagy: from molecular mechanisms to therapeutic applications. Phytother Res

Luo M, Gan RY, Li BY, Mao QQ, Shang A, Xu XY, Li HY, Li HB (2023) Effects and mechanisms of tea on Parkinson's disease, Alzheimer's disease and depression. Food Rev Intl 39(1):278–306

Lv J, Song X, Luo Z, Huang D, Xiao L, Zou K (2025) Luteolin: exploring its therapeutic potential and molecular mechanisms in pulmonary diseases. Front Pharmacol 16:1535555

Mahboob A, Senevirathne DKL, Paul P, Nabi F, Khan RH, Chaari A (2023) An investigation into the potential action of polyphenols against human islet amyloid polypeptide aggregation in type 2 diabetes. Int J Biol Macromol 225:318–350

Maleki Dana P, Sadoughi F, Mansournia MA, Mirzaei H, Asemi Z, Yousefi B (2021) Targeting Wnt signaling pathway by polyphenols: implication for aging and age-related diseases. Biogerontology 22:479–494

Maqbool M, Zehravi M (2021) Neuroprotective role of polyphenols in treatment of neurological disorders: a review. Interven Pain Med Neuromodul 1(1)

Marianetti M, Pinna S, Venuti A, Liguri G (2022) Olive polyphenols and bioavailable glutathione: promising results in patients diagnosed with mild Alzheimer's disease. Alzheimer's Dement Transl Res Clin Interven 8(1):e12278

Markowska A, Markowska J, Stanisławiak-Rudowicz J, Kozak K, Roubinek OK, Jasińska M (2025) The role of Ferulic acid in selected malignant neoplasms. Molecules 30(5):1018

Molino S, Dossena M, Buonocore D, Ferrari F, Venturini L, Ricevuti G, Verri M (2016) Polyphenols in dementia: from molecular basis to clinical trials. Life Sci 161:69–77

Mosadegh M, Noori Goodarzi N, Erfani Y (2025) A comprehensive insight into apoptosis: molecular mechanisms, signaling pathways, and modulating therapeutics. Cancer Investig:1–26

Moussavi N, Azizullah H, Malterud KE, Inngjerdingen KT, Wangensteen H (2022) Immunomodulating polyphenols from Sideritis scardica. J Funct Foods 96:105197

Nabavi SF, Sureda A, Dehpour AR, Shirooie S, Silva AS, Devi KP, Ahmed T, Ishaq N, Hashim R, Sobarzo-Sánchez E, Daglia M (2018) Regulation of autophagy by polyphenols: paving the road for treatment of neurodegeneration. Biotechnol Adv 36(6):1768–1778

Nasiry D, Khalatbary AR (2025) Anti-ferroptotic effects of natural polyphenols in nervous system injury: a narrative literature review. Nutr Neurosci:1–16

Nguyen M, Darrel DJ, Fallot LB, Yuk SF, Limbocker R (2025) BPS2025-structural influence of polyphenols on the cytotoxicity of a pore-forming agent. Biophys J 124(3):394a

Nian Y, Zhang Y, Ruan C, Hu B (2022) Update of the interaction between polyphenols and amyloid fibrils. Curr Opin Food Sci 43:99–106

Oladele JO, Babatunde IO, Oladele OT (2025) Molecular mechanisms underpinning neuroprotection of nutritional polyphenols. In: Nanomedicine for neurodegenerative disorders: selective treatment strategies, p 331

Oliveira JT, Pieniz S (2024) Curcumin in Alzheimer's disease and depression: therapeutic potential and mechanisms of action. Braz Arch Biol Technol 67:e24220004

Ontario ML, Siracusa R, Modafferi S, Scuto M, Sciuto S, Greco V, Bertuccio MP, Salinaro AT, Crea R, Calabrese EJ, Di Paola R (2022) Potential prevention and treatment of neurodegenerative disorders by olive polyphenols and hidrox. Mech Ageing Dev 203:111637

Pasinetti GM, Wang J, Ho L, Zhao W, Dubner L (2015) Roles of resveratrol and other grape-derived polyphenols in Alzheimer's disease prevention and treatment. Biochim et Biophys Acta (BBA)-Mol Basis Dis 1852(6):1202–1208

Piccialli I, Tedeschi V, Caputo L, D'Errico S, Ciccone R, De Feo V, Secondo A, Pannaccione A (2022) Exploring the therapeutic potential of phytochemicals in Alzheimer's disease: focus on polyphenols and monoterpenes. Front Pharmacol 13:876614

Puri V, Nagpal M, Singh I, Singh M, Dhingra GA, Huanbutta K, Dheer D, Sharma A, Sangnim T (2022) A comprehensive review on nutraceuticals: therapy support and formulation challenges. Nutrients 14(21):4637

Revi N, Rengan AK (2021) Impact of dietary polyphenols on neuroinflammation-associated disorders. Neurol Sci 42:3101–3119

Rezaee N, Fernando WB, Hone E, Sohrabi HR, Johnson SK, Gunzburg S, Martins RN (2021) Potential of sorghum polyphenols to prevent and treat Alzheimer's disease: a review article. Front Aging Neurosci 13:729949

Romero-Márquez JM, Forbes-Hernández TY, Navarro-Hortal MD, Quirantes-Piné R, Grosso G, Giampieri F, Lipari V, Sánchez-González C, Battino M, Quiles JL (2023) Molecular mechanisms of the protective effects of olive leaf polyphenols against Alzheimer's disease. Int J Mol Sci 24(5):4353

Shabbir U, Tyagi A, Elahi F, Aloo SO, Oh DH (2021) The potential role of polyphenols in oxidative stress and inflammation induced by gut microbiota in alzheimer's disease. Antioxidants 10(9):1370

Shah MA, Faheem HI, Hamid A, Yousaf R, Haris M, Saleem U, Shah GM, Alhasani RH, Althobaiti NA, Alsharif I, Silva AS (2024) The entrancing role of dietary polyphenols against the most frequent aging-associated diseases. Med Res Rev 44(1):235–274

Singh M, Arseneault M, Sanderson T, Murthy V, Ramassamy C (2008) Challenges for research on polyphenols from foods in Alzheimer's disease: bioavailability, metabolism, and cellular and molecular mechanisms. J Agric Food Chem 56(13):4855–4873

Singh B, Pal S, Tiwari S (2025) Phenolic metabolites from wine and their effects on cognitive and digestive processes. Nourish Your Mind:104

Stankovic S, Mutavdzin Krneta S, Djuric D, Milosevic V, Milenkovic D (2025) Plant polyphenols as heart's best friends: from health properties, to cellular effects, to molecular mechanisms of action. Int J Mol Sci 26(3):915

Suzuki T, Ohishi T, Tanabe H, Miyoshi N, Nakamura Y (2023) Anti-inflammatory effects of dietary polyphenols through inhibitory activity against metalloproteinases. Molecules 28(14):5426

Tan W, Zheng Q, Feng K, Feng X, Zhong W, Liao C, Li S, Liu Y, Hu W (2022) Neuroprotection of Gastrodia elata polyphenols against H2O2-induced PC12 cell cytotoxicity by reducing oxidative stress. Front Pharmacol 13:1050775

Ticinesi A, Mancabelli L, Carnevali L, Nouvenne A, Meschi T, Del Rio D, Ventura M, Sgoifo A, Angelino D (2022) Interaction between diet and microbiota in the pathophysiology of Alzheimer's disease: focus on polyphenols and dietary fibers. J Alzheimer's Dis 86(3):961–982

Tkaczenko H, Kurhaluk N (2025) Antioxidant-rich functional foods and exercise: unlocking metabolic health through Nrf2 and related pathways. Int J Mol Sci 26(3):1098

Tomar VR (2025) BPS2025-the natural polyphenol myricetin as an inhibitor of SOD1 aggregation in ALS neurodegenerative disease. Biophys J 124(3):406a

Tripathi S, Mazumder PM (2021) Cellular investigations to uncover curative potentials of polyphenols-an in vitro study of apple cider vinegar (ACV) and Chrysin against Alzheimer's like pathology via down-regulation of AChE activity. IJTK 20(2):320–328

Turer BY, Sanlier N (2025) Relationship of curcumin with aging and Alzheimer and Parkinson disease, the most prevalent age-related neurodegenerative diseases: a narrative review. Nutr Rev 83(3):e1243–e1258

Uddin MS, Hasana S, Ahmad J, Hossain MF, Rahman MM, Behl T, Rauf A, Ahmad A, Hafeez A, Perveen A, Ashraf GM (2021) Anti-neuroinflammatory potential of polyphenols by inhibiting NF-κB to halt Alzheimer's disease. Curr Pharm Des 27(3):402–414

Utpal BK, Roy SC, Zehravi M, Sweilam SH, Raja AD, Haque MA, Nayak C, Balakrishnan S, Singh LP, Panigrahi S, Alshehri MA (2025) Polyphenols as Wnt/β-catenin pathway modulators: a promising strategy in clinical neurodegeneration. Anim Models Exp Med 8:266

Vauzour D (2012) Dietary polyphenols as modulators of brain functions: biological actions and molecular mechanisms underpinning their beneficial effects. Oxidative Med Cell Longev 2012(1):914273

Wang J, Zhang J, Yu ZL, Chung SK, Xu B (2024) The roles of dietary polyphenols at crosstalk between type 2 diabetes and Alzheimer's disease in ameliorating oxidative stress and mitochondrial dysfunction via PI3K/Akt signaling pathways. Ageing Res Rev 99:102416

Wang X, Zhang S, Li Y, Zhang Y (2025) The regulation of miRNAs using curcumin and other polyphenols during the prevention and treatment of Alzheimer's disease. Hum Mol Genet 34(2):117–127

William Raja TR, Duraipandiyan V, Ignacimuthu S, Janakiraman U, Packiam SM (2023) Role of polyphenols in alleviating alzheimer's disease: a review. Curr Med Chem 30(35):4032–4047

Xu H, Zhou Q, Liu B, Cheng KW, Chen F, Wang M (2021) Neuroprotective potential of mung bean (Vigna radiata L.) polyphenols in Alzheimer's disease: a review. J Agric Food Chem 69(39):11554–11571

Yang W, Cui K, Li X, Zhao J, Zeng Z, Song R, Qi X, Xu W (2021) Effect of polyphenols on cognitive function: evidence from population-based studies and clinical trials. J Nutr Health Aging 25(10):1190–1204

Yang P, Huang Q, Zhang J, Li Y, Gao H, Gu Z (2024) Natural polyphenolic nanodots for Alzheimer's disease treatment. Adv Mater 36(3):2308393

Yeung AWK, Tzvetkov NT, El-Demerdash A, Horbanczuk OK, Das N, Pirgozliev V, Lucarini M, Durazzo A, Souto EB, Santini A, Devkota HP (2021) Apple polyphenols in human and animal health. Animal Science Papers and Reports 39(2):105–118

Zhang Y, Yu W, Zhang L, Wang M, Chang W (2022) The interaction of polyphenols and the gut microbiota in neurodegenerative diseases. Nutrients 14(24):5373

Unveiling the Benefit of Maca in Preventing Alzheimer's Disease 14

Liliana Vargas-Murga

Abstract

Alzheimer's disease (AD) is a progressive brain disorder that gradually impairs memory and the ability to learn, reason, judge, communicate, and perform daily activities. In neurodegeneration models, preparations of *Lepidium meyenii*, commonly referred to as maca, have demonstrated encouraging outcomes in reducing clinical signs and symptoms. Maca preparations, particularly some of their bioactive compounds like macamide, have demonstrated promising neuroprotective effects in recent in vitro and in vivo studies. Even so, more research is needed to understand the neuroprotective mechanisms of maca preparations. Moreover, the claimed health effects are not supported by clinical trials. To maximize value perceptions, it is advised to innovate products and diversify the ways that different parts of maca are used in food and nonfood applications. In this chapter, a detailed compilation and analysis of in vitro and in vivo studies regarding the role of different maca preparations as neuroprotective agents are presented. Maca preparations could play an important role in the treatment of Alzheimer's disease and memory deficits.

Keywords

Maca · *Lepidium meyenii* · Macamides · Alzheimer's disease · Neuroprotection

L. Vargas-Murga (✉)
Department of Chemical and Agricultural Engineering and Agrifood Technology, Polytechnic School, Universitat de Girona, Girona, Catalonia, Spain
e-mail: liliana.vargas@udg.edu

J. Malik et al. (eds.), *Diet and Alzheimer's Disease: Let Food be Our Medicine*, Nutritional Neurosciences, https://doi.org/10.1007/978-981-95-2736-6_14

14.1 Introduction

Alzheimer's disease (AD) is a chronic, ongoing, neurodegenerative illness that affects significant areas of the hippocampus and cerebral cortex. It is characterized by changes in personality and sociability, cognitive decline, short- and long-term memory loss, and mental ability loss. The elderly population over 65 is the main demographic affected. Overall, dementia accounts for 50–60% of AD cases. According to estimates, over 50 million people worldwide suffer from AD, the most common type of dementia. The brains of AD patients exhibit two distinctive characteristics: (i) extracellular deposits of amyloid-beta (Aβ), a peptide produced by breaking down Aβ precursors (genetic locus 21q21–22), are found in senile plaques. Blood vessels also exhibit abnormal Aβ deposits and (ii) neurofibrillary tangles, which are dense bundles of aberrant fibers in the cytoplasm of neurons made up of a modified form of the microtubular-associated protein (Akram and Nawaz 2017; Pandey et al. 2021; Vera-López et al. 2024).

Lepidium meyenii, also known as Peruvian maca, is grown in Peru's Andes. Maca is primarily found at high elevations where it thrives in conditions of intense cold, strong winds, and sunlight. The maca is distinguished by its foliar (aboveground) and root (underground) parts. Different maca varieties are differentiated by the color of their roots, which can be white, yellow, red, purple, or black. Different maca varieties have distinct biological activities depending on their color. Maca is consumed for their nutritional and medicinal properties. Traditionally, maca was used for enhancing fertility and energy (Carvalho et al. 2024).

The in vitro neuroprotective properties of maca preparations were shown to be inhibited by several enzymes, including fatty acid amide hydrolase (FAAH), monoacylglycerol lipase (MAGL), soluble epoxide hydrolase (sEH), and cannabinoid receptors (CB1 and CB2) (Alasmari et al. 2019; Almukadi et al. 2013; Hajdu et al. 2014; Kulkarni et al. 2023; Pino-Figueroa et al. 2011; Wu et al. 2013). The in vivo neuroprotective benefits of maca preparations showed improvement in learning abilities and memory function. Due in part to its antioxidant and acetylcholinesterase (AChE) inhibitory properties, it improved experimental memory impairment brought on by ovariectomy (Rubio et al. 2006, 2007, 2011a, b; Xia et al. 2024; Yu et al. 2020).

This chapter examines the benefits of maca preparations and their most promising classes of compounds, known as macamides, in both in vitro and in vivo neuroprotective studies for the prevention of AD.

14.2 Peruvian Maca

Lepidium meyenii Walpers, maca, also known as Peruvian or Andean ginseng, is a biennial crop of the Brassicaceae family. Maca is native to the Peruvian central Andes and grows under severe environmental conditions at altitudes of 3500–4500 m. The edible portion of the plant is the colored root, classified as purple, yellow, and black varieties. It is traditionally used in preparations like juices, pudding, jam, smoothies, and infusions. Natives claimed that maca had improved fertility and nutritional qualities since the middle of the seventeenth century. Ecotypes, growth conditions, crop management, and the analysis techniques employed all have a significant impact on the chemical makeup of maca root. Minerals like iron, calcium, copper, zinc, and potassium are abundant in dried root, which also has a protein content of 13–16%, 59% hydrolysable carbohydrates, 2.2% lipids, and 8.2% soluble and insoluble fibers. Important secondary metabolites including glucosinolates and isothiocyanates, alkaloids, polysaccharides, macamides, hydantoins, thiohydantoins, and other sulfur-containing compounds, polyphenols, phytosterols, macaenes, and fatty acids are also found in maca root. Maca has historically been used to improve sexual function, fertility, and body strength. Immunomodulatory activity, enhanced spermatogenesis, memory improvement, antidepressant, and anti-inflammatory properties are among the scientific evidence supporting its biological activity. In contrast, its main class of compounds, macamides, has shown antitumoral, antifatigue, neuroprotective, and antiosteoporosis benefits, as well as improving testicular toxicity and dysfunction (Carvalho et al. 2024; Ulloa del Carpio et al. 2024).

14.3 Neuroprotective Effect

Researchers have recently become interested in maca's neuroprotective properties. From 2006 until now, 30 preclinical studies have been focused on the neuroprotection and their possible mechanism of action of different preparations of maca, mainly one of their bioactive compounds like macamides. A great variety of preparations, such as powder, extracts, isolated and synthesized macamides, crude polysaccharides, and lyophilized fermented products, were studied in 18 in vitro studies (Table 14.1) and 12 in vivo studies (Table 14.2).

14.3.1 In Vitro Studies

The in vitro neuroprotective effect of lyophilized fermented, crude polysaccharide, ethanolic extract, isolated macamides, and pentane extract was studied using H_2O_2, as an inductor of neurotoxicity and oxidative damage in 5 papers. The neuroprotective effect of lyophilized maca fermented by *Rhizopus oligosporus* through a solid-state fermentation was evaluated in H_2O_2-induced neurotoxicity in mouse hippocampal neuronal cells. The levels of bioactive substances (macamide B, polyphenols, and ergosterol) were considerably raised, and L-carnitine was also formed by the fermentation process. Higher antioxidant and α-glucosidase-inhibiting activities, as well as stronger neuroprotective and cholesterol-lowering effects, were demonstrated by fermented maca (Ryu et al. 2024). The crude polysaccharide extract (N-benzyl-hexadecanamide) protected neuronal cells from oxidative stress, suggesting a mechanism through reversing H_2O_2-induced cell morphological damage and reducing lactate dehydrogenase (LDH) leakage. The extract treatment

Table 14.1 In vitro neuroprotective assays of maca preparations

Maca preparation, Dose	Experimental model	Cellular system	Outcomes	References
Lyophilized fermented (*R. oligosporus*) through solid-state fermentation) maca (macamide B), 250 and 500 μg/mL	H_2O_2-induced neurotoxicity	Mouse hippocampal neuronal, HT-22, cell line	The neuroprotective effect was higher after 5 days of fermentation and exhibited strong correlations with the flavonoid, saponin, and macamide contents of fermented maca	Ryu et al. (2024)
5 macamides extracted by an integrated approach of deep eutectic solvent-based ultrasound-assisted extraction (DES-UAE), 5, 10, 20 μg/mL	Oxygen-glucose deprivation/ reoxygenation injury	Rat pheochromocytoma, PC12, cell	The DES-UAE extracted macamide demonstrated higher neuroprotective activity than other extracted methods	Li et al. (2023)
Hydroalcoholic extract, 100 mg/kg	sEH inhibition assay	Human and murine sEH	Showed promising inhibitory effect on sEH enzyme brain homeostasis and cognition	Kulkarni et al. (2023)
Crude polysaccharide extract (N-benzyl-hexadecanamide), 25, 50, 100 μg/mL	H_2O_2-induced oxidative impairment in SH-SY5Y cells Cell viability LDH leakage Intracellular ROS levels Assay of signaling proteins	Human neuroblastoma, SH-SY5Y, cell line	Attenuated the cell damage induced by H_2O_2. Protected neuronal cells from oxidative stress through a mechanism including a decrease in LDH leakage and reversal of H_2O_2-induced cell morphological damage. Downregulated cleaved caspase 3 and P53 protein	Zhou et al. (2022)
Yellow, purple, and black maca powder, 0.5, 1, 5 mg/mL	L-glutamic acid-induced cell damage Cell viability	Human neuroblastoma, SH-SY5Y, cell line	Efficacy against Alzheimer disease. Black maca showed significant effect on cell viability followed by yellow and purple ones	Lv et al. (2022)

Isolated 4 new macathiohydantoin derivatives (macathiohydantoin L, M, N and meyeniin D), 20 μg/L	Corticosterone-stimulated cell damage	Rat adrenal pheochromocytoma, PC12, cell line	Exhibited moderate neuroprotective activity being macathiohydantoin N the most potent neuroprotective activity	Zhang et al. (2021)
Ethanolic extract and 10 isolated macamides, EEM (12.5, 25, 50 μg/mL). Macamides (0.01, 0.1, 1,10, 100 μM)	H_2O_2-induced neurotoxicity MDA levels and GSH contents, proinflammatory cytokines TNF-α and Il-6	Rat pheochromocytoma, PC12 cell line	EEM and 10 macamides remarkably increased the cell viability, among which 4 macamides showed good neuroprotective effects. EEM inhibited the levels of proinflammatory cytokines (TNBF, II-6) and the mRNA levels of COX-2, TNF-α, and IL-6. EEM remarkably decreased MDA levels and increased GSH content, thus improving the oxidative stress state. EEM decreased the phosphorylation of IkBα and p65	Yu et al. (2020)
4 macamides: N-benzylstearamide, N-benzyloleamide, N-benzyloctadeca-9Z,12Z-dienamide, and N-benzyloctadeca-9Z,12Z,15Z-trienamide, 1, 5, 10, 50, and 100 μM	Fluorescence-based assay to assess FAAH activity		4 macamides demonstrated concentration-dependent FAAH inhibitory or modulatory activity. Double bonds appeared to have an important impact on FAAH inhibitory activity by macamides	Alasmari et al. (2019)

(continued)

Table 14.1 (continued)

Maca preparation, Dose	Experimental model	Cellular system	Outcomes	References
Ethanolic extracts and macamides M18:1 M18:2 M18:3, EEM (12.5, 25, 50 μg/mL), macamides (5, 10, 25 μM)	Corticosterone-induced neurotoxicity Cell viability, LDH release, intracellular ROS level, and MMP rates	Rat pheochromocytoma, PC12 cell line	EEM and macamides remarkably increased cell viability. EEM and macamides decreased LDH release. Macamides reduced the intracellular ROS generation. Macamides reduced the MMP decline rates and decreased the Bax/Bcl-2 ratios. M18:3 inhibited mitochondrial apoptosis, reduced the expression of pro-apoptotic proteins, namely, Bax, cytochrome C, cleaved-caspase-,3 and cleaved-PARP and increased the expression of Bcl-2. M18:3 increased the expression levels of p-Akt, p-CREB, and BNF	Yu et al. (2019)
Isolated macamide (N-benzylhexadecanamide), 12.5, 25, 50, 100, 200 μM	1-methyl-4-phenylpyridinium (MPP+)-induced neuron damage Neurotransmitter levels: Butyrulcholine, acetylcholine, choline, taurine, homovanillic acid, tyrosine, noradrenaline, serine, aspectic acid, asparagine	Mouse dopaminergic neuronal, MN9D, cell line	100 mcM of macamide significantly improved the survival rate of neurons. Increased the number of adherent cells and most of the cells had protrusive growth recovery. Increase the concentration of neurotransmitters. Effectively regulated differential lipid metabolism. Could regulated proteins related to energy metabolism	Zhou et al. (2018)

Synthesized macamides 18:1 (N-(3-methoxybenzyl)oleamide), 18:2 (N-(3-methoxybenzyl)linoleamide), 18:3 (N-(3-methoxybenzyl) linolenamide), 0.1, 0.3, 1, 3, 10, 30, 50 μM	Manganese-induced mitochondrial disruption. GSH levels, MMP, PPARγ transactivation, MAPK activation	Uppsala-87 malignant human primary glioblastoma, U-87 MG, cell line	Macamides attenuated toxicity. Macamides counteracted Mn-induced damage via a CB1 receptor-mediated mechanism. M18:3 increased GSH at a concentration of 10 and 30 μM. Macamides significantly counteracted the mitochondrial depolarization. M18:2 and M18:3 activated PPARγ receptors. M18:3 not counteracted p38 MAPK inactivation	Gugnani et al. (2018)
Total methanol extract, 2 methanol fractions (macamides), and 2 isolated macamides (N-benzylhexadecanamide and N-acetylbenzylamine), 10, 30 μg/mL	1-methyl-4-phenyl-1,2,3,6-tetrahydropyridine (MPTP)-induced zebrafish embryos. AChE and ButChE inhibition assay		80% and 100% methanol elution fractions and 2 macamides inhibited dopaminergic neuron loss in a dose-dependent manner. Increased the acetylcholine and butyrylcholine level in brain	Zhou et al. (2017)
Methanolic leaf extract, 2, 5, and 10 μg	6-hydroxydopamine (6-OHDA)-induced toxicity. LDH and SOD activity and release of NO	Rat pheochromocytoma, PC12, cell line	Increased cell viability in a dose-response manner reduced LDH activity proportional to the concentration and increased SOD activity. Decreased nitrite concentration	Rodríguez-Huamán et al. (2017)

(continued)

Table 14.1 (continued)

Maca preparation, Dose	Experimental model	Cellular system	Outcomes	References
3 isolated macamides from yellow root powder (N-(3-methoxybenzyl)-(9Z,12Z,15Z)—octadecatrienamide, (N-benzyl-(9Z,12Z,15Z)-octadecatrienamide), (N-benzyl-(9Z,12Z)-octadecadienamide), not reported	ECS targets: Arachidonoyl ethanolamide (AEA) uptake inhibition, FAAH and MAGL activity, CB1, CB2 receptor binding	-	Only N-benzyl-(9Z,12Z)-octadecadienamide showed a weak micromolar inhibition of FAAH, a significant submicromolar inhibition of AEA uptake and binding interactions with CB receptors	Hajdu et al. (2014)
Macamides and their synthetic analogs, 10 and 500 μM	FAAH inhibition	-	Macamides exhibited moderate FAAH inhibition. Synthetic (Z)-1-(Heptadec-8-en-1-yl)-3-(pyridin-3-ylmethyl) carbamate showed potent FAAH inhibition	Wu et al. (2013)
Synthetized macamide (N-3-methoxybenzyl-linoleamide), 1, 5, 10, 50, 100 μM	FAAH and MAGL inhibitory activity assays	-	Macamide showed time-dependent, mixed noncompetitive FAAH inhibitory effects. No inhibitory activity on MAGL	Almukadi et al. (2013)
Pentane extract (lipophilic compounds), 0.5, 1.5, 5 μg/mL 0.3 to 30 μg/mL (FAAH)	H_2O_2-induced oxidative damage Cell viability assay FAAH inhibitory activity assay	Rat neuroblastoma, B35, cell line	Increased viability with no concentration-dependent effects. Inhibited FAAH activity	Pino-Figueroa et al. (2011)
Pentane extract (lipophilic compounds), 0.1, 0.3, 1, 3, 10, 30 μg/mL	H_2O_2-induced neurotoxicity cell viability assay	*Orconectes limosus* (crayfish) neuronal cells	Displayed healthier dendritic cell shapes at 30 mcg/mL	Pino-Figueroa et al. (2010)

Table 14.2 In vivo neuroprotective assays of maca preparations

Maca preparation, doses, administration route, duration	Experimental model	Species	Outcomes	References
12 macamides N-benzyl-(9Z,12Z)-octadecadienamide (M 18:2), 10, 25, 50 mg/kg, IG, 28 days	Memory and learning impairment in scopolamine-induced mice Behavioral test: Morris water maze Biochemical indicators: ACh, MDA, and GSH. Activities of enzymes: AChE, BuChE, ChAT, MAO, CAT, SOD, GST, and GSH-Px Contents of proinflammation factors: IL-1β, IL-6, and TNF-α in hippocampus, serum, and colon	Male ICR mice.	M18:2 ameliorated the cognitive deficit in scopolamine-induced mice. Macamide stably binds to AChE. M18:2 inhibited the activities of AChE, BuChE, MAO and improved the content of ACh in hippocampus of scopolamine-induced mice. M18:2 improved the oxidative stress in hippocampus of scopolamine-induced mice. M18:2 reduced the positive expression rate of AChE, and increased the positive expression rates of Nrf2 and HO-1 in the hippocampal CA1 region of scopolamine-induced mice. M18:2 improved the synaptic plasticity in the hippocampus of scopolamine-induced mice. M18:2 suppressed the contents of pro-inflammatory cytokines in hippocampus, serum, and colon of scopolamine-induced mice. M18:2 did not change gut microbiota diversity and modulated the composition of gut microbiota	Xia et al. (2024)
Hydroalcoholic extract 100 mg/kg, Oral, 21 days	Chemotherapy-induced cognitive impairment Behavioral test: Morris water maze assay Oxidative stress (GSH and LPO) and inflammatory (TNF-alfa, IL-6, COX-2, BDNF) markers in the brain	Male Swiss albino mice	Was least effective	Kulkarni et al. (2023)

(continued)

Table 14.2 (continued)

Maca preparation, doses, administration route, duration	Experimental model	Species	Outcomes	References
Crude polysaccharide extract (N-benzyl-hexadecanamide), 75. 150, 300 mg/kg, oral, 8 weeks	D-gal-induced brain impairment in mice GSH-Px, MDA in brain	ICR male mice	Increased GSH-Px and reduced MDA levels	Zhou et al. (2022)
Macamide B, 30, 60, 90 mg/kg, IP	Hypoxic-ischemic brain damage-induced autophagy and apoptosis	Neonatal mice (male or female)	Significantly decrease brain damage, improve the recovery of neural function after HIBD. Also can induce the activation of PI3K/AKT signaling pathway after HIBD, enhance autophagy, and reduce hypoxic-ischemic-induced apoptosis	Yang et al. (2022)
Synthesized macamide (N-benzyl eicosapentaenamide), 250 μg/mouse, IP, 3 days	Hypoxic-ischemic brain injury	Neonatal mice	Significantly alleviates the size of cerebral infarction and improves neurobehavioral disorders after HIBD in neonatal mice. Inhibited the apoptosis of neuronal cells and improved neuronal cell survival and proliferation through the activation of phosphorylated AKT signaling. Protective property was dependent on suppression of the p53-PUMA pathway	Jiao et al. (2021)
Ethanolic extract (macamides), 100, 200, 400 mg/kg, oral, 27 days	Corticosterone-induced neurotoxicity Sucrose preference test Behavioral tests: Open field and forced swim tests Pro-inflammatory cytokines contents and BDNF levels in hippocampi of rats Neurogenesis markers DCX and BrdU	Male wistar rats	Reversed CORT-induced abnormal behaviors. Decreased the contents of TNF-alfa, II-6 in hippocampus. Increased the BDNF levels. Increased DCX and BrdU in hippocampus, promoted hippocampal neurogenesis	Yu et al. (2020)

Powder (macamides) 500 mg/kg, gavage, 5 weeks	Behavioral tests: Morris water maze test, Rotarod assessment, swimming endurance capacity test LC3, Atg7, and Benclin1 protein levels	Male ICR mice (middle-aged)	Improved cognitive function, motor coordination, and swimming endurance capacity. Increased mitochondrial respiratory function and upregulated the autophagy-related proteins in cortex	Guo et al. (2016)
Aqueous extract of black maca, 0.5 and 2 g/kg, oral, 35 days	Memory impairment induced by ovariectomy Behavioral tests: Morris water maze test, step-down avoidance test Biochemical determination of MDA, AChE, MAO brain levels	Female mice	Increased step-down latency and improved learning and memory impairment. Decreased MDA and AChE levels	Rubio et al. (2011a)
Hydroalcoholic extract of black maca (polyphenol), 0.125, 0.25, 0.5, 1 g/kg, oral, 28 days	Memory impairment induced by ethanol Behavioral performance: Open field test, water Morris maze test, escape acquisition, spatial memory test	Swiss male mice	Ameliorated the deleterious effect of ethanol during the probe trial. Improved spatial learning and memory induced by ethanol in a dose-response manner	Rubio et al. (2011b)
Pentane extract (lipophilic compounds), 3, 10, 30 mg/kg IV 24 hours	Focal ischemic stroke Infarct volume evaluation	Male Sprague-Dawley rats	Showed deleterious effects at high doses. Showed significant neuroprotective effects at the dose of 3 mg/kg	Pino-Figueroa et al. (2010)
Aqueous and hydroalcoholic extract of black maca, 0.5 and 2 g/kg 0.25 and 1 g/kg, oral 35 days	Scopolamine-induced memory impairment Behavioral tests: Water Morris maze, Step-down avoidance test AChE, MAO activities in brain	Kunming male mice	Inhibited the scopolamine-induced memory and learning impairment. Inhibited AChE activity but MAO was not affected	Rubio et al. (2007)
Aqueous extract of yellow, red and black maca, 1 g/kg, oral 21 days	Cognitive function and depression induced by ovariectomy latent learning Behavioral tests: Water finding task, Forced swimming test	Swiss female mice	Black maca showed the best response with respect to latent learning. The three varieties reduced the immobility time and increased the uterine weight	Rubio et al. (2006)

alleviated the H_2O_2-induced increases in reactive oxygen species (ROS) levels, inhibited apoptosis, relieved cell cycle arrest, and downregulated cleaved caspase 3 and P53 protein expression. The demonstrated antioxidant activity is related to its neuroprotective effects (Zhou et al. 2022). Ethanolic extract and isolated macamides remarkably increased the cell viability in H_2O_2-induced neurotoxicity. In addition, the ethanolic extract decreased the levels of proinflammatory cytokines, including tumor necrosis factor alfa (TNF-α) and interleukin 6 (IL-6), as well as their messenger RNA (mRNA) levels, suppressed nuclear factor-kappa B (NF-κB) activation, and increased the phosphorylation of IκBα and p65. Moreover, following exposure to both single and repeated dosages, the ethanolic extract showed no discernible toxicity (Yu et al. 2020). Pentane extracts, rich in lipophilic compounds, were tested for H_2O_2-induced neurotoxicity and oxidative damage. The extract reduced the H_2O_2-induced cell damage (Pino-Figueroa et al. 2010, 2011).

The enzyme FAAH is in charge of the nervous system's endocannabinoid breakdown. The inhibitory activity of FAAH was evaluated in isolated and synthesized macamides and pentane extract. Macamides (N-benzylstearamide, N-benzyloleamide, N-benzyloctadeca-9Z,12Z-dienamide, and N-benzyloctadeca-9Z,12Z,15Z-trienamide) inhibited FAAH in a concentration-dependent manner. The FAAH inhibitory activity of N-benzyloctadeca-9Z,12Z-dienamide was the highest, whereas that of N-benzylstearamide was the lowest. Furthermore, when examined after a preincubation period, N-benzylstearamide, N-benzyloleamide, and N-benzyl-octadeca-9Z,12Z-dienamide showed time-dependent inhibition, suggesting that the mechanism of inhibition for these compounds is most likely irreversible. These findings shed light on maca's mode of action and could aid in the creation of novel substances with modulatory or inhibitory effects on FAAH (Alasmari et al. 2019). Isolated macamides from yellow root powder (N-(3-methoxybenzyl)-(9Z,12Z,15Z)-octadecatrienamide), (N-benzyl-(9Z,12Z,15Z)-octadecatrienamide), and (N-benzyl-(9Z,12Z)-octadecadienamide). N-benzyl-(9Z,12Z)-octadecadienamide was investigated on different targets of the mammalian endocannabinoid system (ECS), such as FAAH, MAGL, CB1 and CB2 receptors, and N-arachidonoylethanolamide (AEA) uptake. Submicromolar and specific binding affinities for CB1 receptor were demonstrated by N-benzyl-(9Z,12Z)-octadecadienamide. It also demonstrated a strong inhibition of anandamide cellular uptake and a weak inhibition of FAAH. The pronounced ECS polypharmacology of the compound indicates that the linoleoyl group's arachidonoyl-mimicking 9Z,12Z double-bond system may be involved in the macamide's overall cannabimimetic activity (Hajdu et al. 2014). Macamides showed moderate FAAH inhibition. These amides were derived from benzylamine or 3-methoxybenzylamine as well as oleic, linoleic, and linolenic acids. Nevertheless, it was demonstrated that the synthetic carbamate, which shares structural similarities with macamides, was an irreversible FAAH inhibitor (Wu et al. 2013). N-3-methoxybenzyl-linoleamide inhibited FAAH in a time-dependent, mixed, noncompetitive manner. Also, MAGL, the second enzyme that breaks down endocannabinoids, was unaffected by the macamide (Almukadi et al. 2013). Regarding the pentane extract, it showed a significant concentration-dependent FAAH inhibitory activity at concentrations 0.1– 30 μg/mL. These findings imply that pentane extract may be used as an FAAH inhibitor. Maca lipophilic components may operate through the endocannabinoid system to shield neurons from excitotoxic damage and may also affect neurotransmission in the central nervous system (Pino-Figueroa et al. 2011).

Corticosterone (CORT) reduced cell viability and increased LDH release, intracellular ROS levels, and mitochondrial membrane potential (MMP) decline rate and induced mitochondrial apoptosis. Isolated new macathiohydantoin derivatives and ethanolic extracts of maca (EEM) and macamides M18:1, M18:2, and M18:3 were evaluated with CORT as an inductor of neurotoxicity. The macathiohydantoins L, M, N, and meyeniin D showed moderate neuroprotective activity in CORT-stimulated cell damage; however, macathiohydantoin N has the most potent neuroprotective activity (Zhang et al. 2021). Both macamides and EEM ameliorated the neurotoxicity caused by CORT. EEM improved the cell viability and decreased the LDH release. M18:1, M18:2, and M18:3 enhanced cell survival while decreasing intracellular ROS production and LDH release. M18:2 and M18:3 decreased the Bax/Bcl-2 ratios and prevented MMP reduction. M18:1 decreased intracellular ROS while leaving other variables the same. Additionally, M18:3 inhibited CORT-induced mitochondrial apoptosis, raised Bcl-2 expression levels, and decreased the expression of pro-apoptotic proteins such as cytochrome C, Bax, cleaved-caspase-3, and cleaved-poly (ADP-ribose) polymerase (cleaved-PARP). Furthermore, M18:3 enhanced protein kinase B (Akt) phosphorylation, and LY294002, a PI3K phosphorylation inhibitor, significantly decreased M18:3's capacity to guard against CORT-induced cytotoxicity. In CORT-induced PC12 cells, M18:3 also increased cyclic AMP response element-binding protein (CREB) phosphorylation and activated brain-derived neurotrophic factor (BDNF) protein levels (Yu et al. 2019).

Higher neuroprotective activity was demonstrated by macamides extracted using an integrated deep eutectic solvent-based ultrasound-assisted extraction (DES-UAE) approach in an oxygen-glucose deprivation/reoxygenation injury (Li et al. 2023). It has been demonstrated that sEH inhibition effectively reduces neuroinflammation and reverses memory impairment. According to Kulkarni et al. (2023), hydroalcoholic extracts of maca have been shown to be effective in reversing memory impairment and neuroinflammation by inhibiting sEH. L-glutamic acid can damage cells through oxidative stress, excitotoxicity, and neuronal damage and loss. A yellow maca dose of 0.5 mg/mL was unable to stop cell damage, but three types of maca root powder (yellow, purple, and black) at doses of 2 mg/mL significantly inhibited L-glutamic acid-induced toxicity after 24 hours and had a moderate effect at 1 mg/mL (Lv et al. 2022). Isolated macamide (N-benzylhexadecanamide) improved the survival rate of neurons in a 1-methyl-4-phenylpyridinium (MPP+)-induced neurodegeneration cell model. But an increase in neurotransmitter concentrations in the mouse dopaminergic neuronal (MN9D) cell line, in conjunction with integrated proteomics and lipidomics, elucidated the main neuroprotective mechanism, regulating the sphingolipid metabolism and mitochondrial function (Zhou et al. 2018). Manganese chloride exposure impairs mitochondrial function. In U-87 MG glioma cells, synthesized macamides reduced manganese (Mn)-induced neurotoxicity. Macamides inhibited Mn-induced damage via the CB1 receptor. Reduced MMP, mitochondrial disruption, and neuron loss in neurodegenerative diseases are caused by ROS, which also raise intracellular calcium and promote the opening of MPTP. Additionally, it interacted with additional targets like the human peroxisome proliferator-activated receptor gamma (PPARγ) to control inflammation, cell differentiation, metabolism, and energy homeostasis in a way similar to that of the comparable endocannabinoid AEA (Gugnani et al. 2018). Zebrafish embryos were susceptible to various neurotoxins such as 1-methyl-4-phenyl-1, 2, 3, 6-tetrahydropyridine (MPTP), which was used as an inducible model of neuronal loss. In the MPTP-induced zebrafish model, two pure compounds and three fractions exhibited robust activity. Of the fractions, the 80% methanol elution fraction demonstrated the strongest neuroprotective effects, followed by the 100% fraction. One potential mechanism for its neuroprotective action was the suppression of butyrylcholinesterase (BuChE) and AChE (Zhou et al. 2017). In rat pheochromocytoma PC12 cells, 6-hydroxydopamine (6-OHDA) causes apoptosis. 6-OHDA administration damages mitochondrial function and generates ROS. Consequently, 6-OHDA-induced oxidative stress triggers the activation of mitogen-activated protein kinase (MAPK) family members, including extracellular signal-regulated kinases 1 and 2 (ERK1/2), and encourages the death of DA neurons. In response to oxidative stress caused by 6-OHDA, methanolic extract of maca leaves demonstrated neurobiological activity of antioxidant protection, enhanced cell viability, and decreased cytotoxicity (Rodríguez-Huamán et al. 2017).

Various neurotoxicity inductors such as H_2O_2, CORT, L-glutamic acid, Mn, MPTP, MPP+, 6-OHDA, and O_2-glucose deprivation/reoxygenation were used to cause injury to the nervous cell lines. The neuroprotective properties of maca preparations, including extracts, isolated and synthesized compounds, primarily macamides, are well supported by these in vitro studies. In all cases, these preparations increased the cell viability and the survival rate of neurons. Key mechanisms involved in neuroprotection mechanisms demonstrated by these models were (i) the suppression of ROS generation, (ii) the prevention of antioxidant enzyme depletion, (iii) the reduction of inflammatory markers, (iv) the inhibition of inflammatory enzymes, (v) the inhibition of mitochondrial apoptosis, (vi) the improvement in the neurotransmitters, (vii) the inhibition of cholinergic enzymes, and (viii) the inhibition of endocannabinoide system enzymes, among others.

14.3.2 In Vivo Studies

The in vivo neuroprotective effect of macamides, powder, and aqueous and hydroalcoholic extracts was studied in seven papers using behavioral methods for the evaluation of cognitive performance in rodents (mice/rats). Some of these tests are the Water Morris maze or water-finding task, step-down avoidance, the open field, rotarod, swimming endurance capacity, escape acquisition, and forced swimming. The Water Morris maze is the most used test to assess the spatial navigational skill and memory of rodents. In the hippocampal CA1 region, macamides enhanced positive rates of nuclear factor erythroid 2-related factor 2 (Nrf2) and heme oxigenase-1 (HO-1) and prevented scopolamine-induced cognitive impairment and neurotransmitter disorders. They also enhanced synaptic plasticity by preserving synaptic morphology and boosting synapse density. Moreover, M18:2 decreased the levels of TNF-α, IL-1β, and IL-6 in the colon, serum, and hippocampal regions.

M18:2 also inhibited AChE, BuChE, and monoamine oxidase (MAO) while increasing the concentration of acetylcholine (Ach) in the hippocampus and the activity of choline acetyltransferase (ChAT). M18:2 increased MDA content, while catalase (CAT) and superoxide dismutase (SOD) activities were significantly reduced in the hippocampus of the scopolamine-treated group. Moreover, M18:2 enhanced the integrity of the colonic epithelium and partially restored the gut microbiota's composition to normal, resulting in a decrease in the genera *Clostridiales* and *Lachnospiraceae* and an increase in the genera *Muribaculaceae*, *Muribaculum*, *Alistipes*, and *Bacteroides*, which may be potential indicators of cognitive ageing (Xia et al. 2024). Aqueous and hydroalcoholic extracts of black maca improved scopolamine-induced memory impairment as measured in both the Water Morris maze and the step-down avoidance tests. While MAO activity was unaffected by black maca extracts, AChE activity was inhibited (Rubio et al. 2007). Memory, learning, and motor function losses are characteristics of a new clinical condition known as chemotherapy-induced cognitive impairment (CICI). Hydroalcoholic extract showed a slight effect on the improvement in cognitive function in CICI memory as measured in the Water Morris maze test (Kulkarni et al. 2023). Chronic administration of CORT to rodents is used to mimic chronic stress. CORT-induced neurotoxicity in a rat model was used to determine the neuroprotective effects. In rats, EEM corrected the aberrant behaviors caused by CORT, decreased TNF-α and IL-6 levels in the hippocampus, and raised the number of positive cells for doublecortin (DCX), bromodeoxyuridine (BrdU), and DCX + BrdU. EEM decreased the levels of proinflammatory cytokines such as TNF-α and IL-6, suppressed NF-κB activation, and increased the phosphorylation of IκBα and p65 (Yu et al. 2020). Using the Water Morris maze and step-down avoidance tests, an aqueous extract of black maca was used to assess learning and memory in ovariectomized (OVX) mice as well as their relationship to malonalehyde (MDA), AChE, and MAO brain levels (Rubio et al. 2011a). Black maca hydroalcoholic extract demonstrated a dose-response effect in mice given a 20% ethanol treatment as a memory impairment model. The Water Morris maze and the open field test were used to assess spatial memory and locomotor activity, respectively. AChE and MDA levels were also lowered (Rubio et al. 2011b). The Water Morris maze test, rotarod assessment, and swimming endurance capacity tests demonstrated that maca powder enhanced middle-aged mice's cognitive function, motor coordination, and endurance capacity. It also increased mitochondrial respiratory function and upregulated autophagy-related proteins in the cortex (Guo et al. 2016). Both trained and untrained mice showed reduced finding latency when given aqueous extracts of the three maca varieties—yellow, red, and black. All of the evaluated varieties decreased the amount of time that mice were immobile and increased the weight of their uteruses after ovariectomy in the forced swimming test. The variety that performed the best in the water-finding task, especially in the trained mice, was black maca (Rubio et al. 2006).

Pretreatment with macamide B can dramatically reduce brain damage and enhance neural function recovery after hypoxic-ischemic brain injury-induced autophagy and apoptosis (HIBD). In addition, macamide B pretreatment can increase autophagy, decrease hypoxic-ischemic (HI)-induced apoptosis, and trigger the activation of the PI3K/AKT signaling pathway after HIBD. Furthermore, 3-methyladenine (3-MA), an inhibitor of the PI3K/AKT signaling pathway, exacerbates HI-induced apoptosis, reverses the neuroprotective effect of macamide B on HIBD, and greatly decreases the rise in autophagy levels (Yang et al. 2022). After hypoxic-ischemic brain damage in neonatal mice, treatment with biosynthesized macamide (N-benzyl eicosapentaenamide) significantly reduces the size of the cerebral infarction and improves neurobehavioral disorders. The biosynthesized macamide prevented neuronal cells from dying after an ischemic challenge. Moreover, the macamide enhanced the survival and proliferation of neuronal cells by activating phosphorylated AKT signaling. Interestingly, the synthetic macamide's protective ability against ischemic neuronal damage was dependent on suppression of the p53 upregulated modulator of apoptosis (p53–PUMA) pathway (Jiao et al. 2021).

Finally, an ageing mouse model was induced by injecting D-galactose (D-gal) intraperitoneally, and pathological changes in the hippocampal ultrastructure were observed using transmission electron microscopy. The crude polysaccharide extract demonstrated an increase of glutathione peroxidase (GSH-Px) activity and a reduction of MDA levels. The cell structure in the low-, medium-, and high-polysaccharide extract groups tended to be normal, with large, round nuclei; a complete nuclear membrane structure; and a uniform chromatin distribution observed by electron microscopy (Zhou et al. 2022).

Cognitive impairments, motor function, and anxiety/depression were induced by scopolamine, CORT, D-gal, and conditions like OVX, CICI, and HIBD in rodents using behavioral tests. The in vivo studies revealed that maca preparations, mainly in Water Morris, the improvement in learning and memory and motor coordination was observed. Additionally, antioxidant enzymes were increased while inflammatory cytokines were reduced. Similar to in vitro models, there was a decrease in the activity of certain cholinergic enzymes and an increase in neurotransmitter levels. Furthermore, by improving the integrity of the colonic epithelium and partially reestablishing the normal composition of the gut microbiota, macamide (M18:2) supports the gut microbiota.

14.4 Conclusions

The most common neurodegenerative disease, AD, disrupts neurocognitive activity and hinders the growth of mental capacity, which leads to dementia. Notwithstanding recent developments in AD treatment, therapeutic efficacy has been modest, noncurative, and drug-sensitive. Maca root, used traditionally for enhancing fertility and other medicinal and nutritional properties, could be an alternative for AD. Although maca root has important bioactive compounds, glucosinolates, polyphenols, alkaloids, macamides, and macaenes are the major biologically active compounds. In summary, the in vitro and in vivo studies are focused on macamides, both isolated and synthesized, as the main secondary metabolites with promising neuroprotective benefit and still under continuous assessment. Clinical signs and symptoms in neurodegeneration models have been shown to improve with the aid of maca preparations. In vitro studies have mainly shown that an increase in ROS, proinflammatory cytokines, cholinergic enzymes, and neuronal apoptosis can be reduced. In addition, animal models demonstrated improvements in motor deficit, neurotransmitter levels, memory, and learning.

These results are promising for future clinical research on new therapies for treating AD. Therefore, clinical trials are required to translate experimental findings into human experience. Also, safe and effective doses of maca preparations should be determined for treating patients with neurodegenerative diseases.

References

Akram M, Nawaz A (2017) Effects of medicinal plants on alzheimer's disease and memory deficits. Neural Regen Res 12(4):660–670. https://doi.org/10.4103/1673-5374.205108

Alasmari M, Böhlke M, Kelley C, Maher T, Pino-Figueroa A (2019) Inhibition of fatty acid amide hydrolase (FAAH) by Macamides. Mol Neurobiol 56(3):1770–1781. https://doi.org/10.1007/s12035-018-1115-8

Almukadi H, Wu H, Böhlke M, Kelley CJ, Maher TJ, Pino-Figueroa A (2013) The macamide N-3-methoxybenzyl-linoleamide is a time-dependent fatty acid amide hydrolase (FAAH) inhibitor. Mol Neurobiol 48(2):333–339

Carvalho FV, da Silva VDA, Ribeiro PR (2024) Bioactive compounds and biological activities of Peruvian maca (Lepidium meyenii Walp.). In: Bioactive compounds in the storage organs of plants. Springer Nature, Cham, pp 959–980

Gugnani KS, Vu N, Rondón-Ortiz AN, Böhlke M, Maher TJ, Pino-Figueroa AJ (2018) Neuroprotective activity of macamides on manganese-induced mitochondrial disruption in U-87 MG glioblastoma cells. Toxicol Appl Pharmacol 340:67–76. https://doi.org/10.1016/j.taap.2017.12.014

Guo S-S, Gao X-F, Gu Y-R, Wan Z-X, Lu A-M, Qin Z-H, Luo L (2016) Preservation of cognitive function by Lepidium meyenii (Maca) is associated with improvement of mitochondrial activity and upregulation of autophagy-related proteins in middle-aged mouse cortex. Evid Based Complement Alternat Med 2016. https://doi.org/10.1155/2016/4394261

Hajdu Z, Nicolussi S, Rau M, Lorantfy L, Forgo P, Hohmann J, Csupor D, Gertsch J (2014) Identification of endocannabinoid system-modulating N-Alkylamides from Heliopsis helianthoides var. Scabra and Lepidium meyenii. J Nat Prod 77:1663–1669

Jiao M, Dong Q, Zhang Y, Lin M, Zhou W, Liu T, Yuan B, Yin H (2021) Neuroprotection of n-benzyl eicosapentaenamide in neonatal mice following hypoxic-ischemic brain injury. Molecules 26(11). https://doi.org/10.3390/molecules26113108

Kulkarni R, Mehta R, Goswami SK, Hammock BD, Morisseau C, Hwang SH, Mallappa O, Azeemuddin MM, Rafiq M (2023) Neuroprotective effect of herbal extracts inhibiting soluble epoxide hydrolase (sEH) and cyclooxygenase (COX) against chemotherapy-induced cognitive impairment in mice. Biochem Biophys Res Commun 667:64–72. https://doi.org/10.1016/j.bbrc.2023.05.008

Li K, Li Z, Men L, Li J, Gong X (2023) Deep eutectic solvent-based ultrasound-assisted strategy for simultaneous extraction of five Macamides from Lepidium meyenii Walp and in vitro bioactivities. Foods 12(2). https://doi.org/10.3390/foods12020248

Lv J, Li C, Zhang N, Zhang K, Gao X, Li N, Meng L, Yang Y, Zhang H, Sun J (2022) Metabolomic profiling of different Maca color types using nuclear magnetic resonance spectroscopy and multivariate data analysis. J Chem Soc Pak 44(4):356–365. https://doi.org/10.52568/001071/JCSP/44.04.2022

Pandey SN, Rangra NK, Singh S, Arora S, Gupta V (2021) Evolving role of natural products from traditional medicinal herbs in the treatment of Alzheimer's disease. ACS Chem Neurosci 12(15):2718–2728. https://doi.org/10.1021/acschemneuro.1c00206

Pino-Figueroa A, Vu H, Kelley CJ, Maher TJ (2011) Mechanism of action of Lepidium meyenii (Maca): an explanation for its neuroprotective activity. Am J Neuroprot Neuroregener 3:87–92

Pino-Figueroa A, Nguyen D, Maher TJ (2010) Neuroprotective effects of Lepidium meyenii (Maca). Ann N Y Acad Sci 1199(1):77–85

Rodríguez-Huamán Á, Casimiro-Gonzales S, Chávez-Pérez JA, Gonzales-Arimborgo C, Cisneros-Fernández R, Aguilar-Mendoza LÁ, Gonzales GF (2017) Antioxidant and neuroprotector effect of Lepidium meyenii (maca) methanol leaf extract against 6-hydroxy dopamine (6-OHDA)-induced toxicity in PC12 cells. Toxicol Mech Methods 27(4):279–285. https://doi.org/10.1080/15376516.2016.1275908

Rubio J, Caldas M, Dávila S, Gasco M, Gonzales GF (2006) Effect of three different cultivars of Lepidium meyenii (Maca) on learning and depression in ovariectomized mice. BMC Complement Altern Med 6. https://doi.org/10.1186/1472-6882-6-23

Rubio J, Dang H, Gong M, Liu X, Chen S, Gonzales GF (2007) Aqueous and hydroalcoholic extracts of black Maca (Lepidium meyenii) improve scopolamine-induced memory impairment in mice. Food Chem Toxicol 45:1882–1890

Rubio J, Qiong W, Liu X, Jiang Z, Dang H, Chen S, Gonzales GF (2011a) Aqueous extract of black Maca (Lepidium meyenii) on memory impairment induced by Ovariectomy in mice. Evid Based Complement Alternat Med 7:253958

Rubio J, Yucra S, Gasco M, Gonzales GF (2011b) Dose–response effect of black maca (Lepidium meyenii) in mice with memory impairment induced by ethanol. Toxicol Mech Methods 21(8):628–634. https://doi.org/10.3109/15376516.2011.583294

Ryu KM, Kim H, Woo J, Lim J, Kang CG, Kim SW, Kim T, Kim D (2024) Enhancement of the bioactive compounds and biological activities of maca (*Lepidium meyenii*) via solid-state fermentation with *Rhizopus oligosporus*. Food Sci Biotechnol 33(11):2585–2596. https://doi.org/10.1007/s10068-023-01508-6

Ulloa del Carpio N, Alvarado-Corella D, Quiñones-Laveriano DM, Araya-Sibaja A, Vega-Baudrit J, Monagas-Juan M, Navarro-Hoyos M, Villar-López M (2024) Exploring the chemical and pharmacological variability of Lepidium meyenii: a comprehensive review of the effects of maca. Front Pharmacol 15:1360422

Vera-López KJ, Davila-Del-Carpio G, Nieto-Montesinos R (2024) Macamides as potential therapeutic agents in neurological disorders. Neurol Int 16(6):1611–1625. https://doi.org/10.3390/neurolint16060117

Wu H, Kelley CJ, Pino-Figueroa A, Vu H, Maher TJ (2013) Macamides and their synthetic analogs: evaluation of in vitro FAAH inhibition. Bioorg Med Chem 21:5188–5197

Xia N, Xu L, Huang M, Xu D, Li Y, Wu H, Mei Z, Yu Z (2024) Neuroprotection of macamide in a mouse model of Alzheimer's disease involves Nrf2 signaling pathway and gut microbiota. Eur J Pharmacol 975. https://doi.org/10.1016/j.ejphar.2024.176638

Yang X, Wang M, Zhou Q, Bai Y, Liu J, Yang J, Li L, Li G, Luo L (2022) Macamide B pretreatment attenuates neonatal hypoxic-ischemic brain damage of mice induced apoptosis and regulates autophagy via the PI3K/AKT signaling pathway. Mol Neurobiol 59(5):2776–2798. https://doi.org/10.1007/s12035-022-02751-4

Yu Z, Jin W, Cui Y, Ao M, Liu H, Xu H, Yu L (2019) Protective effects of macamides from: Lepidium meyenii Walp. Against corticosterone-induced neurotoxicity in PC12 cells. RSC Advances 9(40):23096–23108. https://doi.org/10.1039/c9ra03268a

Yu Z, Jin W, Dong X, Ao M, Liu H, Yu L (2020) Safety evaluation and protective effects of ethanolic extract from maca (Lepidium meyenii Walp.) against corticosterone and H2O2 induced neurotoxicity. Regul Toxicol Pharmacol 111. https://doi.org/10.1016/j.yrtph.2019.104570

Zhang R, Liu J, Yan H, Peng X, Zhang L, Qiu M (2021) Macathiohydantoin L, a novel thiohydantoin bearing a thioxohexahydroimidazo [1,5-a] pyridine moiety from maca (Lepidium meyenii Walp.). Molecules 26(16). https://doi.org/10.3390/molecules26164934

Zhou Y, Li P, Brantner A, Wang H, Shu X, Yang J, Si N, Han L, Zhao H, Bian B (2017) Chemical profiling analysis of Maca using UHPLC-ESI-Orbitrap MS coupled with UHPLC-ESI-QqQ MS and the neuroprotective study on its active ingredients. Sci Rep 7. https://doi.org/10.1038/srep44660

Zhou Y, Wang H, Guo F, Si N, Brantner A, Yang J, Han L, Wei X, Zhao H, Bian B (2018) Integrated proteomics and lipidomics investigation of the mechanism underlying the neuroprotective effect of N-benzylhexadecanamide. Molecules 23(11). https://doi.org/10.3390/molecules23112929

Zhou Y, Zhu L, Li H, Xie W, Liu J, Zhang Y, Li Y, Wang C (2022) In vivo and in vitro neuroprotective effects of maca polysaccharide. Front Biosci Landmark 27(1). https://doi.org/10.31083/j.fbl2701008

Resveratrol: A Stilbene with Multifaceted Roles in Alzheimer's Disease

15

Vinod Kumar Nelson, Emani Sai Sri Jayanthi, Kiranmai Gudimetla, Vinyas Mayasa, Subhash Chandra Mandal, and Gandhodi Gopi

Abstract

Alzheimer's disease (AD) is the ultimate cause of cognitive decline in older adults, involving a steady deterioration of mental function. The main pathological hallmarks of AD include Aβ plaques (extracellular) and hyperphosphorylated tau protein (Intracellular). Several risk factors, such as genetics, aging, diet, cerebrovascular diseases, obesity, and inflammation, strongly show their impact on AD progression. On the other hand, inflammation and immune disorder also significantly impact BBB integrity, further exacerbating the disease progression. However, no specific treatment that can reverse the disease has been reported, and the current treatment strategies target only symptoms. This underscores the urgent need for an ideal drug treatment that can completely reverse the disease. Resveratrol, a bioactive polyphenol extracted from grapes and wine, has shown neuroprotective effects by enhancing the antiaging and antioxidant potentials and inhibiting inflammation in multiple models of AD. In addition, this compound also showed an impact on Aβ aggregation and clearance. However, resveratrol's poor bioavailability remains challenging, necessitating novel strategies to improve its pharmacokinetic properties. This chapter highlights the multiple

V. K. Nelson (✉) · G. Gopi
Department of Pharmaceutical Chemistry, Mahathi College of Pharmacy, Madanapalle, Andhra Pradesh, India

E. Sai Sri Jayanthi
Department of Pharmacy Practice, Pulla Reddy Institute of Pharmacy, Jawaharlal Nehru Technological University, Hyderabad, Telangana, India

K. Gudimetla · V. Mayasa
GITAM School of Pharmacy, GITAM University Hyderabad Campus, Rudraram, Telangana, India

S. C. Mandal
Department of Pharmaceutical Technology, Jadavpur University, Kolkata, West Bengal, India

J. Malik et al. (eds.), *Diet and Alzheimer's Disease: Let Food be Our Medicine*, Nutritional Neurosciences, https://doi.org/10.1007/978-981-95-2736-6_15

mechanisms of resveratrol in modulating AD pathobiology, its role in preserving BBB integrity, and its potential role in therapy.

Keywords

Alzheimer's disease · Aβ plaques · Tau aggregates · Inflammation · BBB · Resveratrol · Pharmacological effects · Clears aggregates · AD progression

15.1 Introduction

Alzheimer's disease (AD) is the ultimate cause of neurodeteriorative disease-related deaths and the leading cause of dementia in the aged population globally (Kumar Nelson et al. 2024). The main pathological A notable attribute of AD is an accumulation of protein aggregates like amyloid-beta (Aβ) plaques and tau proteins, chiefly affecting the hippocampal area (Kumar Nelson et al. 2024; Zhang et al. 2021a). These protein aggregates are essential in developing the disease symptoms (Kumar Nelson et al. 2024).

In addition, risk-related factors, such as hereditary, food habits, alcohol consumption, smoking, and age, contribute significantly to the development of the disease (Kumar Nelson et al. 2024; Silva et al. 2019). Among the array of risk elements, senescence-related factors are major contributors to the generation of vascular-based disorders of progressive neuronal loss, such as AD. Besides, aging also affects the individual immune system. It triggers the inflammation process, affecting the integrity of the BBB (Kumar Nelson et al. 2024). Disturbing the BBB structure and losing integrity is the leading cause of various neurodegenerative diseases, including AD (Kurz et al. 2022; Pugazhenthi et al. 2017).

Furthermore, the current treatment options only target the symptoms and do not modify the disease or halt the progression. Therefore, there is an immediate need for an appropriate drug candidate that can stop the AD progression. Fortunately, due to their vast chemical diversity and availability, medicinal plants are considered an ideal source for drug discovery (Chaachouay and Zidane 2024; Maghimaa et al. 2025; Priyadarshini et al. 2025; Nelson et al. 2024). The secondary metabolites like polyphenols, alkaloids, triterpenoids, flavonoids, and tannins present in plants are responsible for their health benefits against diseases like neurodegenerative disease (Bordoloi et al. 2024; Jha et al. 2024). These plant-based metabolites show various pharmacological effects like antiaging, anticholinesterase, anti-inflammatory, anti-amyloidogenic, and antioxidant to inhibit AD progression (Bordoloi et al. 2024). One such plant-derived secondary metabolite is resveratrol (RV), a polyphenol compound belonging to the class of stilbene extracted from various plant sources like soybeans, red grapes, pomegranates, Japanese knotweed, peanuts, mulberries, and blueberries. Resveratrol also exhibits different kinds of biological functions, such as antibacterial, virucidal, anticancer, angiogenesis inhibitor, immunomodulatory, antioxidant, cardioprotective, and neuroprotective effects (Zhang et al. 2021b; Salehi et al. 2018). Besides, RV and its derivatives pass the blood-brain barrier and inhibit toxin-mediated degeneration in various ND models (Breuer et al. 2006). Furthermore, RV reduces the chemical-induced seizures and symptoms associated with several CNS-related NDs, including stroke and AD, in various preclinical models by modulating multiple signaling molecules such as iNOS, COX-2, p53, FOXO, TNF-α, PGC1-α, SIRT1, Nrf2, and AMPK (Rahman et al. 2020; Chen et al. 2019). This section delves into the impact of resveratrol in diverse Alzheimer's disease models. In addition, we also elaborate on the mode of action of resveratrol in decreasing AD conditions in preclinical models. This chapter serves as a comprehensive resource for researchers working in this field.

15.2 Predisposing Factors for AD

Numerous etiological factors are involved in the advancement of AD. Among the risk factors, hereditary or genetic-based factors account for about 70% of AD cases (de Rojas et al. 2021; Parrot et al. 2023). In addition, acquired factors like cardiovascular disease, obesity, diabetes, and hypertension also promote disease progression (Petrie et al. 2018). In the section below, we describe in detail the various risk factors contributing to AD generation.

15.3 Hereditary Factors

Generally, AD is classified by onset age. Early-onset Alzheimer's disease (EOAD) arises in populations younger than 65, whereas late-onset Alzheimer's disease (LOAD) is seen in older people above 65 years. Several pieces of evidence support that genetic factors are strongly associated with AD development among the various risk factors (Mendez 2019). Specifically, APP, PSEN1, and PSEN2 gene mutations stimulate EOAD. In fact, the APP gene contains about 30 major mutations that are associated with approximately 15% of EOAD cases (Lanoiselée et al. 2017). Similarly, mutations in the PSEN1 and PSEN2 genes represent 80% and 5% of EOAD cases, respectively. Besides, mutations in the APP and PSEN1 genes promote the Aβ plaque accumulation, resulting in AD progression. Conversely, mutations in the APOE gene lead to LOAD (Lanoiselée et al. 2017). Earlier studies revealed that the APOE gene consists of three alleles (ε2, ε3, ε4) with distinct effects on AD development (Liu et al. 2013). The previous studies showed that the ε4 allele increases the disease risk, promotes the Aβ polymerization, and increases plaque deposition (Morishima-Kawashima et al. 2000). In contrast, the ε2 allele decreases the risk and enhances the clearance of Aβ plaques in the brain (Liu et al. 2013). Likewise, mutations in the TREM2 gene also improve the risk of developing AD by approximately 2.9% (Silva et al. 2019). This mutated TREM2 interacts with DAP12 and increases the production of inflammatory mediators that lead to Aβ plaque deposition (Cui et al. 2021).

15.4 Acquired Risk Factors

There are several acquired factors, such as diabetes, hypertension, obesity, cerebrovascular diseases, smoking, diet, and alcohol consumption, that elevate AD risk (Silva et al. 2019). Among the acquired risk factors, cerebrovascular diseases are the primary risk factor that promotes Alzheimer's disease progression. Previous studies revealed that cerebrovascular changes like Ischemic strokes with hemorrhage, vasculopathy, and small and large ischemic cortical infarcts enhance dementia susceptibility in individuals. In addition, the autopsy of the brain isolated from the patients suffering from these diseases showed an accumulation of $A\beta_{42}$ and tau protein aggregations (Silva et al. 2019).

In the same way, type 2 diabetes mellitus (T2DM) also augments the risk of AD. A study revealed that the increase in insulin deficiency and resistance triggers the β- and γ-secretase function that inhibits the clearance of Aβ aggregates and promotes its deposition (Rad et al. 2018). On the other hand, insulin-regulated signaling molecules such as glycogen synthase kinase 3β (GSK3β) also modulate tau phosphorylation (Beurel et al. 2015). Earlier studies also revealed that insulin resistance increases irregular activation of GSK3β, resulting in improper tau phosphorylation, leading to neurofibrillary tangles (NFTs) accumulation in the brain (Yoon et al. 2023; Hobday and Parmar 2021). Similarly, advanced glycation end products (AGEs), another signaling molecule in insulin-regulated pathways, also elevate the risk of generating AD (Kothandan et al. 2024). Previous reports disclosed that AGEs induce neuronal cell death and increase brain oxidative stress that triggers the activation of β- and γ-secretases, resulting in Aβ aggregation and ultimately promoting AD progression (Kothandan et al. 2024). Several pieces of evidence also support that obesity also leads to higher risk of dementia and, in later stages, improves Alzheimer's disease (Flores-Cordero et al. 2022). In addition smoking adds to the risk that enhances the chances of developing AD. Earlier studies revealed that smoking stimulates the production of free radicals that trigger oxidative stress (OS) and the generation of inflammatory mediators (Durazzo et al. 2014). Furthermore, smoking also promotes cerebrovascular diseases that elevate the risk of Alzheimer's disease (Durazzo et al. 2014). Likewise, diet is a key risk factor that generate AD. Previous data showed that consuming a diet with high trans fats and low antioxidants is strongly linked with developing AD (Barnard et al. 2014). High cholesterol also increases the risk of developing AD. Previous studies reported that patients with high cholesterol increased the risk of developing atherosclerosis as well as AD. Besides, cholesterol also damages the BBB integrity, which enhances the chance of developing neurodegenerative diseases, including AD (Ransmayr 2003).

15.5 Pathophysiology of Alzheimer's Disease

Despite enormous advancements in research in the area of NDs like Alzheimer's disease, the exact etiology and pathology of the disease remain unclear. This ambiguity imposes various challenges in developing an ideal drug candidate to reverse the disease. However, the piling of Aβ plaques and generation of neurofibrillary tangles are still the main pathological hallmarks of AD (Murphy and LeVine 3rd. 2010). In addition, mitochondrial dysfunction, oxidative stress, and increased metal ion deposition also drive AD progression (Misrani et al. 2021). In the section below, we describe in detail the pathogenesis mechanism of AD.

15.6 Aβ Cascade

Aggregation of Aβ proteins, the primary element in the amyloid plaques, is the main characteristic lesion of the disease. This Aβ protein is a proteolytic fragment synthesized from amyloid precursor protein (APP), a type I transmembrane protein located in dendritic cells and axons of neurons (O'brien and Wong 2011). This amyloid precursor protein (APP) was cleaved by two main pathways: a nonamyloidogenic α-secretase-dependent pathway and an amyloidogenic β- and γ-secretase-dependent pathway. In a nonamyloidogenic pathway, the APP breaks and produces soluble APPα molecules that are likely nontoxic and provide a neuroprotective effect against excitotoxicity (Habib et al. 2017). In addition, sAPPα also improves the survivability of various neurons and increases its plasticity. On the other hand, in the amyloidogenic pathway, the APP is cleaved by amyloidogenic β (predominately BACE1 enzyme) and γ-secretase enzymes (Cole and Vassar RJMn. 2007). In this pathway, initially, APP is cleaved by β-secretase or BACE1 enzyme and produces a soluble molecule called sAPPβ, which promotes neuronal cell death and a carboxy-terminal complex that helps in cell membrane formation (Chow et al. 2010). This sAPPβ later also cleaved by an γ-secretase complex that contains four proteins (presenilin 1 or 2, nicastrin, APH-1, and PEN-2 and produces Aβ peptides (Hur 2022). Each Aβ peptide comprises at least 38–43 amino acids. Among these Aβ peptides, Aβ40 and Aβ42 are predominant and pathogenic. These Aβ40 and Aβ42 promote the pathogenesis of Alzheimer's disease through producing various inflammatory mediators.

15.7 Resveratrol and Its Effect on AD

AD is the most death-causing ND, with a significant decline in cognitive abilities. Besides, this is the main reason for dementia that affects millions of people, leading to disability and also decreasing the quality of life of the aged population. Furthermore, over 50 million people are affected yearly, and the number could increase to 152 million by 2050 (Kumar Nelson et al. 2024). No specific treatment can yet cure the disease. Besides, currently recommended drugs only manage the symptoms often linked to side effects (Kumar Nelson et al. 2024). Thus, there is a massive requirement for an ideal drug candidate that can significantly affect the disease's progression.

Natural products, herbal sources play a key role in identifying novel drugs for various dreadful diseases, including NDs. One plant-sourced polyphenol,

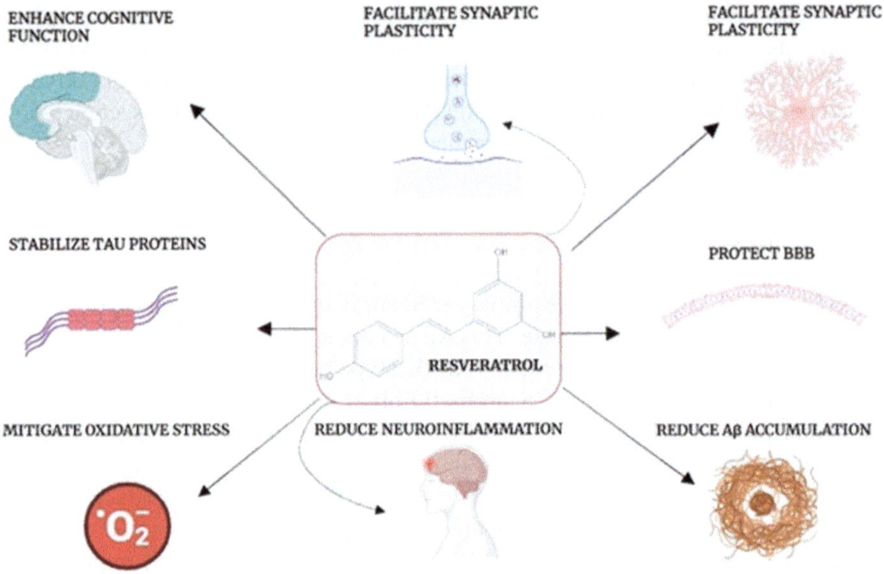

Fig. 15.1 Showing various pharmacological effects of resveratrol that inhibits the progression of AD

resveratrol, comes chiefly from grapefruits. This compound showed various pharmacological effects, including neuroprotective effects shown in Fig. 15.1 (Zhang et al. 2021b; Rahman et al. 2020). Earlier in vitro and in vivo evidence also concluded that resveratrol was found effective in AD models and inhibited the pathogenesis of Alzheimer's disease by modulating various signaling molecules such as ERK, PGC-1α, SIRT1, HO-1, TNF-α, p53, FOXO (Rahman et al. 2020). Previous studies showed that resveratrol mainly targets the Aβ cascade pathway and inhibits Aβ accumulation. In addition, RV also clears Aβ aggregation and decreases neuronal damage shown in Fig. 15.1 (Rahman et al. 2020). Hence, in the sections below, we describe in detail the effect of resveratrol on Alzheimer's disease. It offers a standalone guide for researchers in this field.

15.8 Resveratrol Modulates Aβ Production

15.8.1 Resveratrol Inhibits β-Secretase or BACE-1 Enzyme

BACE-1 or β-secretase predominately manifests in the neurons of the brain and promotes the cleavage of APP, which plays a significant role in the pathogenesis of the disease by inducing Aβ accumulation. Besides, Aβ triggers the generation of stress markers like 4-hydroxynonenal in the brain cells, enhances expression of BACE-1, and promotes Aβ aggregation and neurotoxicity. Therefore, modulating the activity of BACE-1 decreases the Aβ accumulation and its associated toxicity. A study showed that RV suppresses BACE-1 activity, Aβ aggregation, and its associated disease pathogenesis in a cell model. Another study also demonstrated that RV and its structural analogs that contain at least two electron-donating groups at the ortho position to 4'-OH significantly inhibit the activity of BACE-1 (Jia et al. 2017a). In another experiment, RV was found to inhibit the γ-secretase function and elevate the α-secretase role in neuro2a, which improved the AD conditions. On the other hand, several compounds like LY2886721, E2609, and AZD3293 were in various stages of clinical trials for Alzheimer's disease and were reported to exhibit multiple adverse effects. Therefore, aiming BACE-1 using plant-derived compounds like RV may provide a promising treatment strategy for AD (Jia et al. 2017a).

15.8.2 Resveratrol Activates Autophagy and Inhibits Aβ Production

Two main defense pathways, Ubiquitin Proteasome System (UPS) and autophagy, clear protein aggregates, pathogens, and dysfunctional organelles and regulating protein homeostasis (Li et al. 2022). However, UPS is ineffective in eliminating debris like protein aggregates and pathogens in neuronal and glial cells (Jansen et al. 2014). Hence, autophagy is the primary defensive mechanism in the brain cells to maintain protein homeostasis (Gómez-Virgilio et al. 2022). On the other hand, autophagy inhibits the activity of PSEN1, thereby decreasing the function of the γ-secretase enzyme, which upregulates the Aβ production. In a similar manner, resveratrol also activates the autophagy function and inhibits Aβ production in HEK 293 cells by decreasing PSEN1 and γ-secretase activity. Likewise, in another study, RV enhances autophagy activity by triggering the TyrRS-PARP1-SIRT1 pathway and inhibiting the Aβ accumulation in PC-12 cells (Jia et al. 2017b).

15.9 Resveratrol Inhibits Aβ Aggregation

Aβ-aggregation triggers microglial and astrocyte activation, induces an inflammatory response, and contributes to AD progression (Deng et al. 2024). Aβ inhibition may serve as an Alzheimer's therapy. Resveratrol was also reported to inhibit the Aβ-aggregation in various experimental models. A study reported that RV binds to the Aβ peptide, ultimately interfering with the Aβ aggregation process (Al-Edresi et al. 2020). In this way, resveratrol inhibits the Aβ aggregation and halts the disease progression.

15.10 Resveratrol Clears the Aβ Aggregation

Accumulation of Aβ aggregates triggers various signaling molecules that induce toxicity in neuronal cells and ultimately contribute to disease progression. Hence, targeting signaling molecules that can clear Aβ aggregates could be a potential drug candidate to treat Alzheimer's disease. Previously documented evidence also suggested that activating the signals involved in initiating Aβ-degrading enzymes (ADEs) and BBB transporter proteins promotes microglial phagocytosis and clears Aβ protein aggregates (Hampel et al. 2021). Hence, activating signals such as ADEs and BBB transporter proteins by using external agents like plant-derived molecules could be potential drug candidates for AD. One such compound is resveratrol; this compound upregulates ADE function and clears the Aβ aggregates by enhancing the activity of neprilysin (NEP) and angiotensin-converting enzyme (ACE) in SK-N-SH and HEK293 cells (Jia et al. 2017b). In addition, RV also activates plasminogen activators like t-PA and u-PA that contribute to eliminating Aβ-aggregates and its associated neurotoxicity in HUVEC cells (Zhou et al. 2019; Cho et al. 2014). Likewise, RV also restores BBB integrity by enhancing the activity of transporter proteins like P-gp, LRP1, and RAGE (Jia et al. 2017b). These BBB transporter proteins interact with Aβ protein and restrict its entry. Another study also reported that RV elevates the activity of LRP1 in SH-SY5Y cells and promotes brain Aβ clearance in transgenic AD models (Jia et al. 2017b).

15.11 Resveratrol Effect on Tau Protein Phosphorylation

Tau is a phosphoprotein and a main microtubule-associated protein (MAP) of a fully grown neuron. This tau protein majorly contributes to the stabilization of microtubules that are integrated into the axons and dendrites (Rahman et al. 2020; Rawat et al. 2022). The tau usually consists of 2–3 moles of phosphate groups in healthy adult brains. However, in the pathological state, tau gets hyperphosphorylated (three- to fourfold more hyperphosphorylated) and forms neurofibrillary tangles, which is the major pathological hallmark of the disease. This hyperphosphorylated tau loses its original function, results in neuronal damage and synaptic impairment, and contributes to AD progression (Iqbal et al. 2010). On the other hand, protein phosphatase 2A is the leading enzyme that dephosphorylates tau at AD-related sites. Therefore, increasing the activity of PP2A could be an AD drug prospect. Phytocompounds like resveratrol enhance the function of PP2A and remove the phosphate groups at specific sites of tau by reducing the activity of MID1 protein, which suppresses the activity of PP2A in primary cortical neurons (Schweiger et al. 2017).

15.12 Resveratrol Antioxidant and Inflammation

Oxidative stress promotes abnormal APP proteolysis through activating β- and γ-secretase enzymes, accumulating Aβ aggregates in the cortex and hippocampal region (Tamagno et al. 2021). This increase in Aβ plaque accumulation further increases ROS production, accelerating the AD progression (Tamagno et al. 2021). Besides, Aβ aggregates also induce inflammation by activating microglia and astrocytes that further release various inflammatory signaling molecules such as cytokines NO and TNF-α. These inflammatory signaling molecules amplify the Aβ-aggregation, leading to neuronal dysfunction and death (Tamagno et al. 2021). Hence, modulating oxidative stress and corresponding inflammatory response by using external agents could develop a potent therapeutic tool for AD. Resveratrol is one such compound that mitigates ROS and its corresponding inflammatory response. A study reported that RV elevated its antioxidant function by inhibiting iNOS and lipid peroxidation production (Rahman et al. 2020). In addition, this compound was also found to activate the heme oxygenase (HO-1) enzyme. According to earlier studies, it was found that AD and DM co-occur, and both diseases exhibit similar kinds of pathogenesis like oxidative stress, inflammation, and cell death. In addition, both AD and DM are interlinked, and one disease can induce the other as well (Rahman et al. 2020). The previous study results also revealed that resveratrol simultaneously inhibited the pathogenesis of both diseases by altering the signaling molecules in oxidative stress and inflammation (Ma et al. 2019). Similarly, regular administration of red wine to Tg2576 mice showed a significant improvement in AD condition by improving cognitive abilities (Pasinetti et al. 2015). The same study found that treating wine to the mice inhibited the accumulation of Aβ aggregates. Likewise, treating 10 and 20 mg of resveratrol in dementia rats under intracerebroventricular (ICV) streptozotocin (STZ) reduced the level of MDA by increasing glutathione levels (Sharma and Gupta 2002).

15.13 Conclusion and Future Directions

AD ranks highest in dementia-related deaths among the elderly. Aβ and tau accumulation in key brain regions defines the pathology. These protein aggregates trigger inflammatory responses and damage BBB integrity, which promotes disease progression. On the other hand, the current drug treatments recommended for AD only provide symptomatic relief but show no effect on the disease course. Given the severity of the disease and the limitations of available treatment, there is an urgent need for an ideal drug candidate. Phytocompound resveratrol is one such compound that is gaining massive attention for its beneficial effects in various neurodegenerative disease models. The preclinical results obtained from multiple experiments suggested that RV potentially inhibited AD progression by modulating the signaling molecules involved in enhancing Aβ clearance, preserving BBB integrity, and inflammation. This multi-beneficial effect makes resveratrol a qualified drug candidate for AD therapy. However, due to its poor bioavailability, resveratrol concentration is insufficient to prove its effect in humans with AD. Hence, novel drug delivery techniques are essential to improve its therapeutic efficacy. Therefore, future research should focus on stabilizing RV pharmacokinetics by nanoformulations, structural modifications, and combination therapies. In conclusion, RV can provide hope to develop a better therapeutic tool to treat AD. In this chapter, we discuss in detail the mode of action of resveratrol in various AD models, a guide for ongoing studies.

References

Al-Edresi S, Alsalahat I, Freeman S, Aojula H, Penny J (2020) Resveratrol-mediated cleavage of amyloid β(1-42) peptide: potential relevance to Alzheimer's disease. Neurobiol Aging 94:24–33

Barnard ND, Bunner AE, Agarwal U (2014) Saturated and trans fats and dementia: a systematic review. Neurobiol Aging 35(Suppl 2):S65–S73

Beurel E, Grieco SF, Jope RS (2015) Glycogen synthase kinase-3 (GSK3): regulation, actions, and diseases. Pharmacol Ther 148:114–131

Bordoloi S, Pathak K, Devi M, Saikia R, Das J, Kashyap VH, Das D, Ahmad MZ, Abdel-Wahab BA (2024) Some promising medicinal plants used in Alzheimer's disease: an ethnopharmacological perspective. Discover Appl Sci 6(5):215

Breuer C, Wolf G, Andrabi SA, Lorenz P, Horn TF (2006) Blood-brain barrier permeability to the neuroprotectant oxyresveratrol. Neurosci Lett 393(2–3):113–118

Chaachouay N, Zidane L (2024) Plant-derived natural products: a source for drug discovery and development. Drugs Drug Candidates 3(1):184–207

Chen JY, Zhu Q, Zhang S, OuYang D, Lu JH (2019) Resveratrol in experimental Alzheimer's disease models: a systematic review of preclinical studies. Pharmacol Res 150:104476

Cho KS, Lee EJ, Kwon KJ, Gonzales EL, Kim YB, Cheong JH et al (2014) Resveratrol down-regulates a glutamate-induced tissue plasminogen activator via Erk and AMPK/mTOR pathways in rat primary cortical neurons. Food Funct 5(5):951–960

Chow VW, Mattson MP, Wong PC, Gleichmann M (2010) An overview of APP processing enzymes and products. NeuroMolecular Med 12(1):1–12

Cole SL, Vassar RJMn. (2007) The Alzheimer's disease β-secretase enzyme. BACE1 2:1–25

Cui X, Qiao J, Liu S, Wu M, Gu W (2021) Mechanism of TREM2/DAP12 complex affecting β-amyloid plaque deposition in Alzheimer's disease modeled mice through mediating inflammatory response. Brain Res Bull 166:21–28

de Rojas I, Moreno-Grau S, Tesi N, Grenier-Boley B, Andrade V, Jansen IE et al (2021) Common variants in Alzheimer's disease and risk stratification by polygenic risk scores. Nat Commun 12(1):3417

Deng Q, Wu C, Parker E, Liu TC, Duan R, Yang L (2024) Microglia and astrocytes in Alzheimer's disease: significance and summary of recent advances. Aging Dis 15(4):1537–1564

Durazzo TC, Mattsson N, Weiner MW (2014) Smoking and increased Alzheimer's disease risk: a review of potential mechanisms. Alzheimers Dement 10(3 Suppl):S122–S145

Flores-Cordero JA, Pérez-Pérez A, Jiménez-Cortegana C, Alba G, Flores-Barragán A, Sánchez-Margalet V (2022) Obesity as a risk factor for dementia and Alzheimer's disease: the role of leptin. Int J Mol Sci 23(9)

Gómez-Virgilio L, Silva-Lucero MD, Flores-Morelos DS, Gallardo-Nieto J, Lopez-Toledo G, Abarca-Fernandez AM et al (2022) Autophagy: a key regulator of homeostasis and disease: an overview of molecular mechanisms and modulators. Cells 11(15)

Habib A, Sawmiller D, Tan J (2017) Restoring soluble amyloid precursor protein α functions as a potential treatment for Alzheimer's disease. J Neurosci Res 95(4):973–991

Hampel H, Hardy J, Blennow K, Chen C, Perry G, Kim SH, Villemagne VL, Aisen P, Vendruscolo M, Iwatsubo T, Masters CL (2021) The amyloid-β pathway in Alzheimer's disease. Mol Psychiatry 26(10):5481–5503

Hobday AL, Parmar MS (2021) The link between diabetes mellitus and tau hyperphosphorylation: implications for risk of Alzheimer's disease. Cureus 13(9):e18362

Hur JY (2022) γ-Secretase in Alzheimer's disease. Exp Mol Med 54(4):433–446

Iqbal K, Liu F, Gong CX, Grundke-Iqbal I (2010) Tau in Alzheimer disease and related tauopathies. Curr Alzheimer Res 7(8):656–664

Jansen AH, Reits EA, Hol EM (2014) The ubiquitin proteasome system in glia and its role in neurodegenerative diseases. Front Mol Neurosci 7:73

Jha SK, Nelson VK, Suryadevara PR, Panda SP, Pullaiah CP, Nuli MV et al (2024) Cannabidiol and neurodegeneration: from molecular mechanisms to clinical benefits. Ageing Res Rev 100:102386

Jia Y, Wang N, Liu X (2017a) Resveratrol and amyloid-beta: mechanistic insights. Nutrients 9(10):1122

Jia Y, Wang N, Liu X (2017b) Resveratrol and amyloid-Beta: mechanistic insights. Nutrients 9:10

Kothandan D, Singh DS, Yerrakula G et al (2024) Advanced glycation end products-induced Alzheimer's disease and its novel therapeutic approaches: a comprehensive review. Cureus. 16(5):e61373

Kumar Nelson V, Jha NK, Nuli MV, Gupta S, Kanna S, Gahtani RM et al (2024) Unveiling the impact of aging on BBB and Alzheimer's disease: factors and therapeutic implications. Ageing Res Rev 98:102224

Kurz C, Walker L, Rauchmann BS, Perneczky R (2022) Dysfunction of the blood-brain barrier in Alzheimer's disease: evidence from human studies. Neuropathol Appl Neurobiol 48(3):e12782

Lanoiselée HM, Nicolas G, Wallon D, Rovelet-Lecrux A, Lacour M, Rousseau S et al (2017) APP, PSEN1, and PSEN2 mutations in early-onset Alzheimer disease: a genetic screening study of familial and sporadic cases. PLoS Med 14(3):e1002270

Li Y, Li S, Wu H (2022) Ubiquitination-proteasome system (UPS) and autophagy two Main protein degradation machineries in response to cell stress. Cells 11(5)

Liu CC, Liu CC, Kanekiyo T, Xu H, Bu G (2013) Apolipoprotein E and Alzheimer disease: risk, mechanisms and therapy. Nat Rev Neurol 9(2):106–118

Ma X, Sun Z, Han X, Li S, Jiang X, Chen S et al (2019) Neuroprotective effect of resveratrol via activation of Sirt1 signaling in a rat model of combined diabetes and Alzheimer's disease. Front Neurosci 13:1400

Maghimaa M, Sagadevan S, Suryadevara PR, Sudhan HH, Burle GSR, Ruokolainen J, Nelson VK, Kesari KK (2025) Cytotoxicity and targeted drug delivery of green synthesized metallic nanoparticles against oral cancer: a review. Inorg Chem Commun 173:113806

Mendez MF (2019) Early-onset Alzheimer disease and its variants. Continuum (Minneapolis, Minn) 25(1):34–51

Misrani A, Tabassum S, Yang L (2021) Mitochondrial dysfunction and oxidative stress in Alzheimer's disease. Front Aging Neurosci 13:617588

Morishima-Kawashima M, Oshima N, Ogata H, Yamaguchi H, Yoshimura M, Sugihara S et al (2000) Effect of apolipoprotein E allele epsilon4 on the initial phase of amyloid beta-protein accumulation in the human brain. Am J Pathol 157(6):2093–2099

Murphy MP, LeVine H 3rd. (2010) Alzheimer's disease and the amyloid-beta peptide. J Alzheimer's Dis 19(1):311–323

Nelson VK, Nuli MV, Ausali S, Gupta S, Sanga V, Mishra R et al (2024) Dietary anti-inflammatory and anti-bacterial medicinal plants and its compounds in bovine mastitis associated impact on human life. Microb Pathog 192:106687

O'brien RJ, Wong PC (2011) Amyloid precursor protein processing and Alzheimer's disease. Annu Rev Neurosci 34(1):185–204

Parrot A, Barral M, Amiot X, Bachmeyer C, Wagner I, Eyries M et al (2023) Hereditary hemorrhagic telangiectasia. Rev Mal Respir 40(5):391–405

Pasinetti GM, Wang J, Ho L, Zhao W, Dubner L (2015) Roles of resveratrol and other grape-derived polyphenols in Alzheimer's disease prevention and treatment. Biochim Biophys Acta 1852(6):1202–1208

Petrie JR, Guzik TJ, Touyz RM (2018) Diabetes, hypertension, and cardiovascular disease: clinical insights and vascular mechanisms. Can J Cardiol 34(5):575–584

Priyadarshini S, Goyal K, Gupta S, Roy A, Biswas R, Patra S, Chauhan P, Wadhwa K, Singh G, Kamal M, Iqbal D (2025) Polypharmacology and neuroprotective effects of gingerol in Alzheimer's disease. Mol Neurobiol:1–21

Pugazhenthi S, Qin L, Reddy PH (2017) Common neurodegenerative pathways in obesity, diabetes, and Alzheimer's disease. Biochim Biophys Acta Mol basis Dis 1863(5):1037–1045

Rad SK, Arya A, Karimian H, Madhavan P, Rizwan F, Koshy S et al (2018) Mechanism involved in insulin resistance via accumulation of β-amyloid and neurofibrillary tangles: link between type 2 diabetes and Alzheimer's disease. Drug Des Devel Ther 12:3999–4021

Rahman MH, Akter R, Bhattacharya T, Abdel-Daim MM, Alkahtani S, Arafah MW et al (2020) Resveratrol and neuroprotection: impact and its therapeutic potential in Alzheimer's disease. Front Pharmacol 11:619024

Ransmayr G (2003) Cholesterol and statins in Alzheimer disease. Wien Med Wochenschr 153(11–12):258–259

Rawat P, Sehar U, Bisht J, Selman A, Culberson J, Reddy PH (2022) Phosphorylated tau in Alzheimer's disease and other Tauopathies. Int J Mol Sci 23(21)

Salehi B, Mishra AP, Nigam M, Sener B, Kilic M, Sharifi-Rad M et al (2018) Resveratrol: a double-edged sword in health benefits. Biomedicine 6(3)

Schweiger S, Matthes F, Posey K, Kickstein E, Weber S, Hettich MM, Pfurtscheller S, Ehninger D, Schneider R, Krauß S (2017) Resveratrol induces dephosphorylation of tau by interfering with the MID1-PP2A complex. Sci Rep 7(1):13753

Sharma M, Gupta YK (2002) Chronic treatment with trans resveratrol prevents intracerebroventricular streptozotocin induced cognitive impairment and oxidative stress in rats. Life Sci 71(21):2489–2498

Silva MVF, Loures CMG, Alves LCV, de Souza LC, Borges KBG, Carvalho MDG (2019) Alzheimer's disease: risk factors and potentially protective measures. J Biomed Sci 26(1):33

Tamagno E, Guglielmotto M, Vasciaveo V, Tabaton M (2021) Oxidative stress and beta amyloid in Alzheimer's disease. Which comes first: the chicken or the egg? Antioxidants 10(9):1479

Yoon JH, Hwang J, Son SU, Choi J, You SW, Park H et al (2023) How can insulin resistance cause Alzheimer's disease? Int J Mol Sci 24(4)

Zhang XX, Tian Y, Wang ZT, Ma YH, Tan L, Yu JT (2021a) The epidemiology of Alzheimer's disease modifiable risk factors and prevention. J Prev Alzheimers Dis 8(3):313–321

Zhang LX, Li CX, Kakar MU, Khan MS, Wu PF, Amir RM et al (2021b) Resveratrol (RV): a pharmacological review and call for further research. Biomed Pharmacother 143:112164

Zhou X, Yang J, Zhou M, Zhang Y, Liu Y, Hou P et al (2019) Resveratrol attenuates endothelial oxidative injury by inducing autophagy via the activation of transcription factor EB. Nutr Metabol 16:42

Mushrooms and Their Protective Role in Alzheimer's Disease

16

Richa Shri

Abstract

Alzheimer's disease (AD) involves slow, progressive, irreversible neurodegeneration resulting in gradual memory loss and decline in cognitive abilities. This has a profound impact on the well-being of the patients and caregivers. Numerous factors—both modifiable as well as nonmodifiable—lead to this complex disorder. One modifiable cause is lifestyle that includes food and exercise. Dietary habits have a strong correlation with gut health as well as cognitive performance. A healthy gut promotes brain well-being and vice versa. Food components that effect the gut-brain axis can prove beneficial for the prevention and management of AD. One such food is mushrooms. Mushrooms or macro-fungi are called "superfood," because they contain nutritional components and bioactive myco-constituents that display a wide range of health benefits. The purpose of this chapter was to explore the importance of incorporating mushrooms in diet for promoting brain health and functions. Mushrooms are reported to boost both brain health and gut functions. Experimental studies with mushroom extracts and constituents strongly corroborate their neuroprotective effects via modulation of diverse mechanisms. There are a number of human studies—epidemiological and interventional, on effect of dietary intake of mushrooms on cognitive health. Many studies support the neuroprotective and cognition enhancing effects of mushrooms in diet of elderly adults. Though systematic human trials with more mushroom species and evaluation of simultaneous impact on brain and gut biomarkers are required to generate irrefutable evidence of their neuroprotective effects, it may be concluded that regular intake of mushrooms in diet can promote cognitive abilities.

R. Shri (✉)
Department of Pharmaceutical Sciences and Drug Research, Punjabi University, Patiala, Punjab, India
e-mail: richashri@pbi.ac.in

J. Malik et al. (eds.), *Diet and Alzheimer's Disease: Let Food be Our Medicine*, Nutritional Neurosciences, https://doi.org/10.1007/978-981-95-2736-6_16

Keywords

Alzheimer's disease · Dietary mushroom · Dual pathway · Gut-brain health · Human studies

Abbreviations

AD	Alzheimer's disease
DASH diet	Dietary Approaches to Stop Hypertension
GI system	Gastrointestinal system
MIND diet	Mediterranean—DASH Intervention for Neurodegenerative Delay
MITO diet	Mitochondrial health diet
NGF	Nerve growth factor

16.1 Introduction

Oscar Wilde said "Memory is the diary we all carry about us." Alzheimer's disease (AD) deprives the patient of this "diary!" AD involves slow, progressive, irreversible degeneration of the structure and functions of neurons especially of the cortex and hippocampus resulting in memory loss, decline in cognitive abilities, gradual inability to carry out daily activities, and ultimately mortality (Craig and Stitzel 2004; Polis and Samson 2019). These debilitating changes are life-altering for the patient as well as their caregivers. Hence, it is important to understand the causes and management strategies of AD. Figure 16.1 summarizes the risk factors,

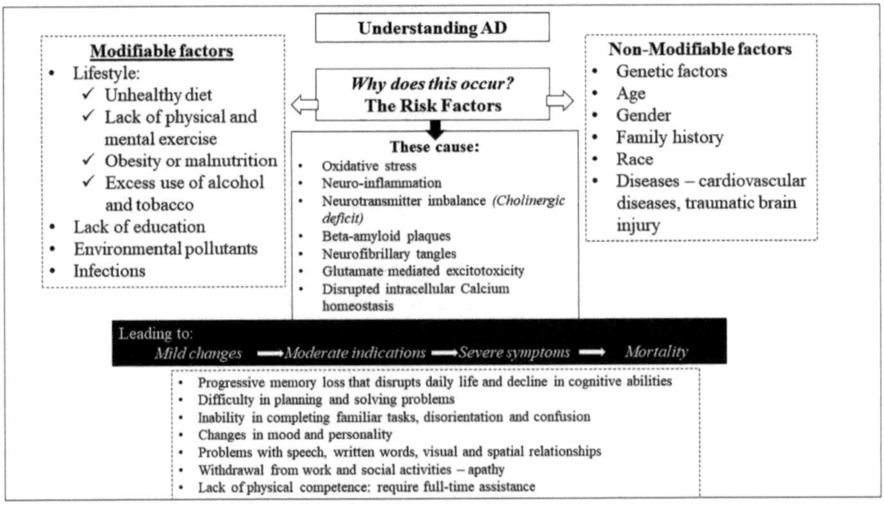

Fig. 16.1 Risk factors, pathophysiological changes, and the symptoms of AD

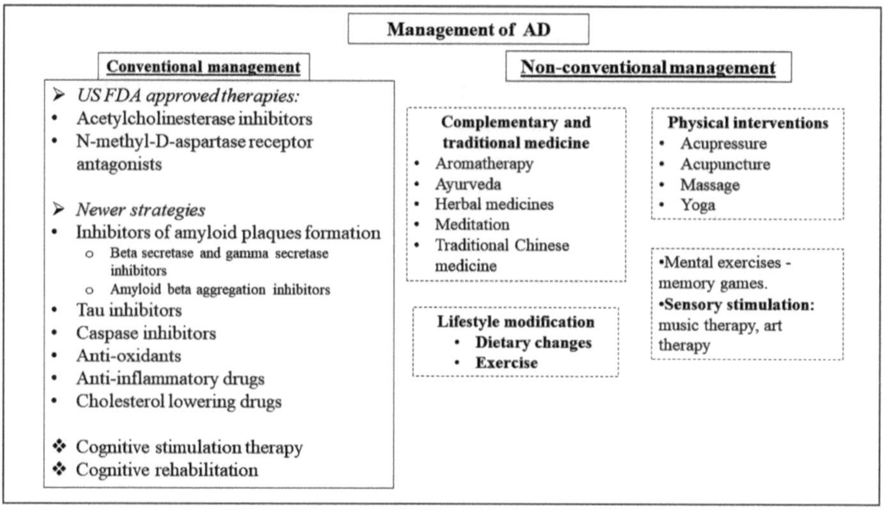

Fig. 16.2 Management of Alzheimer's disease

pathophysiological changes, and the symptoms of AD (Katzung 2012; Alzheimer's Association 2025; Alzheimer's Society 2025).

AD leads to immeasurable morbidity and results in enormous social and economic burden. Hence, numerous management strategies are available. These include conventional as well as alternative and traditional options. These approaches bring relief from the symptoms and can slow disease progression; however, the neurodegenerative changes cannot be reversed and the side effects of the conventional drug therapy can be extremely draining. Since the management has to be long term—as long as the patient is alive, options have to be generated that may provide neuroprotection, prevent the onset of neurodegeneration, retard the disease progression, and offer relief from symptoms. Figure 16.2 summarizes some of the available management strategies (Dhikav and Anand 2012; Katzung 2012; Parums 2021; Beversdorf et al. 2023; Cummings et al. 2023).

Poor lifestyle practices (which include poor dietary habits and inadequate exercise) are a major modifiable factor that contributes to the occurrence as well as progression of AD. Hence, one possible way to prevent as well as slow down the neurodegenerative changes is to modify lifestyle! This includes intake of a neuroprotective diet (Gómez-Pinilla 2008; Duplantier and Gardner 2021).

16.2 Diet and Brain Health

Diet or food that one consumes provides nourishment and building blocks for growth and development as well as energy, aids metabolism, helps in normal tissue repair, and protects against diseases by boosting immunity. There is no single magic potion against AD. It is imperative to understand which foods are protective and

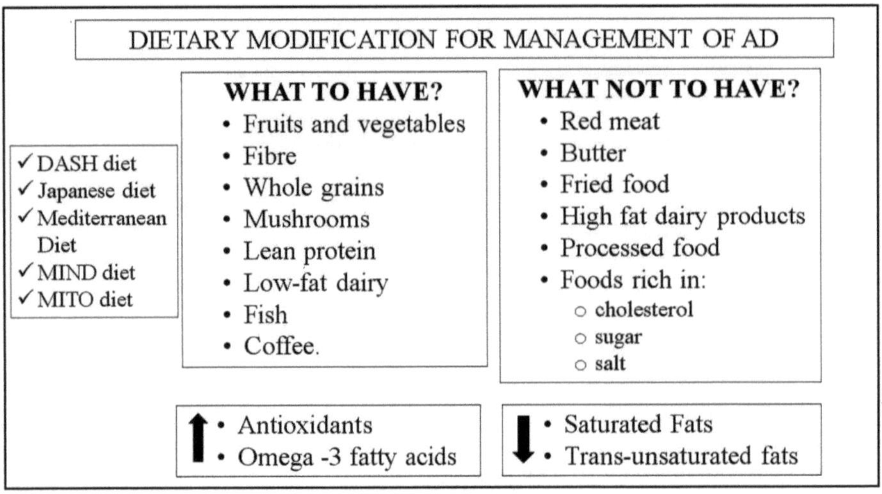

Fig. 16.3 Dietary changes for management of AD(DASH diet: Dietary Approaches to Stop Hypertension; MIND diet: Mediterranean-DASH Intervention for Neurodegenerative Delay; MITO diet: mitochondrial health diet)

which foods must be avoided to prevent and manage AD (Fig. 16.3; nia.nih.gov 2023) and then alter the diet.

The food that is consumed impacts every part of the body. **Most significantly, our diet influences the gut and the brain**! Every dietary component passes through the gastrointestinal (GI) system and thus has the most marked impact there (Bull and Plummer 2014). The brain requires much larger amounts of energy as compared to other parts of the body. Energy transfer from food has a direct bearing on the brain functions. Food components like omega-3-fatty acids, polyphenols, vitamins, selenium, etc., help in maintaining brain health and cognitive performance (Hu et al. 2013; Fortune et al. 2019; American Heart Association 2021).

In traditional systems of medicine such as Ayurveda, diet, digestion, and lifestyle are the foundation for holistic health. Hippocrates also said that health and disease begin in the digestive system. Contemporary researchers also recognize the interdependence of gut health and brain well-being. This bidirectional interaction, mediated via the vagus nerve, between the brain and the GI system is referred to as the gut-brain axis (Wallace 2020). Stress and cognitive impairments can lead to GI disorders, and GI disturbances can influence mood and cognitive processes (Kelly et al. 2015). What we eat influences the brain and GIT and brain health impacts the gut and gut health impacts the brain (Gómez-Pinilla 2008; Carabotti et al. 2015). Hence, it may be stated that ***food affects the brain directly and indirectly*** (Fig. 16.3; McEvoy et al. 2017; Cha et al. 2024).

- Direct influence of food is by providing nutrients for brain development and functioning, for meeting the energy requirements along with dietary components

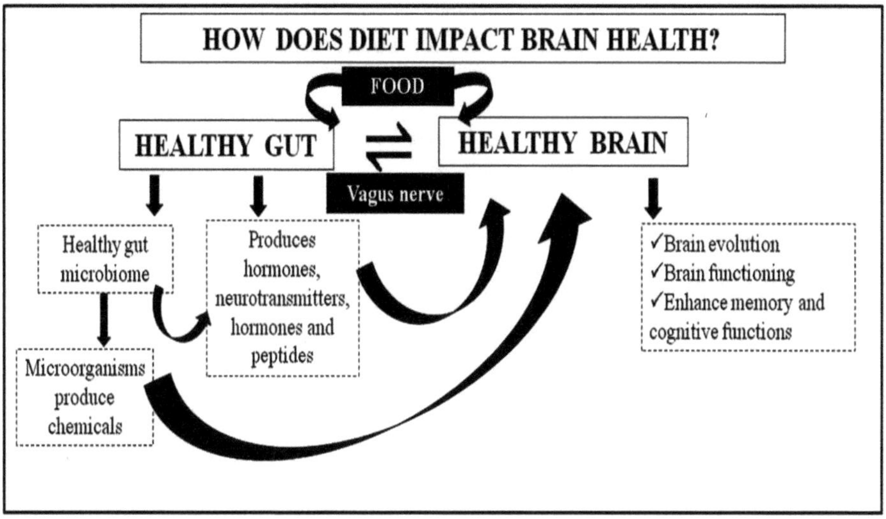

Fig. 16.4 Impact of food on the brain

that counter oxidative stress and neuro-inflammation, help to maintain neurotransmitter balance, and increase synaptic plasticity. These are brain foods!

- Indirect effect of food on the brain is through the GI system. The gut produces hormones, neurotransmitters, and peptides that are transferred to the brain and influence memory and cognitive functions. Food also helps to maintain the microbiome of the gut, and this aids in the production of neurotransmitters by the gut. The chemicals produced by the microorganisms also influence mood and brain functions (Fig. 16.4).

An ideal dietary modification should involve incorporation of foods, which have neuroprotective effect and also enhance gut health. One such dietary component may be mushrooms.

16.3 Importance of Dietary Mushrooms

Mushrooms or macro-fungi have attracted human beings since ancient times because of their distinctive appearance. These are a rich source of dietary components, contain a variety of bioactive mycoconstituents, and display a wide range of health benefits and medicinal activities. Mushrooms help to maintain ecological balance, are valuable for human health, and are commercially profitable. Globally, mushrooms occupy a place of pride in the kitchen—valued for their distinct flavor and nutrient value (Anusiya et al. 2021; Chang and Buswell 2023). These are of greatly valued commercially as they are consumed as food, soups, teas, dietary supplements, nutraceuticals and are used for production of bio-pesticides, cosmaceuticals and "mushroom pharmaceuticals" (Wasser 2010; Yadav and Negi 2021). Some of the most popular edible mushrooms include the following (Anusiya et al. 2021):

- White button mushrooms (*Agaricus bisporus*).
- Honey mushrooms (*Armillaria mellea*).
- Caterpillar mushrooms (*Cordyceps sinensis*).
- Reishi mushrooms (*Ganoderma lucidum*).
- Maitake mushrooms (*Glifolia frondosa*).
- Lion's mane mushrooms (*Hericium erinaceus*).
- Chaga mushroom (*Inonotus obliquus*).
- Shiitake mushroom (*Lentinula edodes*).
- King oyster mushroom (*Pleurotus eryngii*).
- Oyster mushroom (*Pleurotus florida*).
- Turkey tail mushroom (*Coriolus versicolor*).

Mushrooms are "nutrient dense" as these contain vitamins (A, B2, B3, Folate, B5, C, and D) and minerals (calcium, phosphorus, iron, copper, chlorine, sodium, zinc, manganese, bromine, potassium, selenium). Mushrooms contain essential fatty acids (Sande et al. 2019). They primarily consist of moisture (80–90%), fiber (8–10%), antioxidants, and beta-glucans (Bernaś et al. 2006; Cerletti et al. 2021; Chatterjee and Samajpati 2021). These are low calorie, low fat, low salt, and zero cholesterol foods that are safe for all age groups. (A word of caution—some people are allergic to mushrooms and should not eat them). Along with these, mushrooms are rich in immunostimulatory oligosaccharides, lipopolysaccharides,

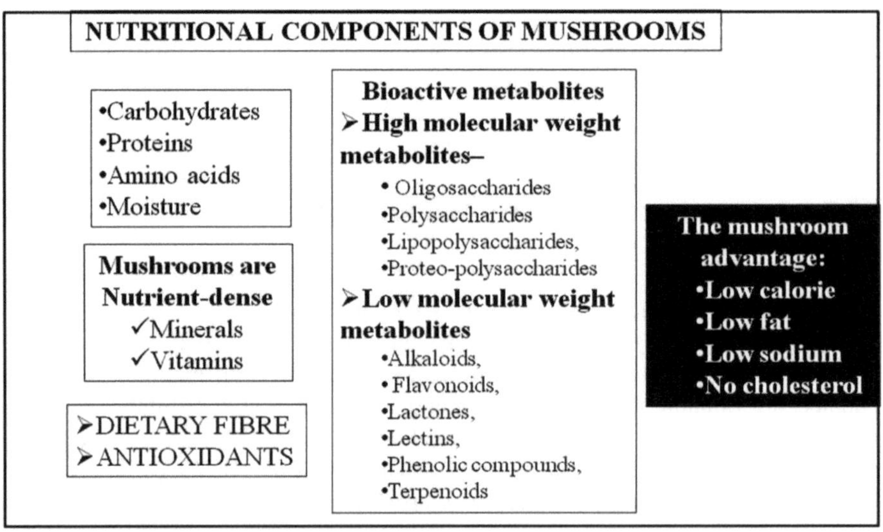

Fig. 16.5 Various components of mushrooms

proteo-polysaccharides and secondary metabolites like alkaloids, flavonoids, lactones, lectins, phenolic compounds, terpenoids, antioxidants like glutathione and L-ergothioneine (Fig. 16.5), which are responsible for the wide array of pharmacological activities (e.g., antianxiety, anticancer, antidepressant, antidiabetic, antihypercholesterolemia, antiobesity, antioxidant, antimicrobial, antiviral, immunity enhancing, and neuroprotection) demonstrated by numerous mushrooms in in vitro and in vivo models as well as in human trials (Chang and Wasser 2012; Anusiya et al. 2021).

16.4 Role of Dietary Mushrooms in the Management of AD

Literature shows that many culinary mushrooms such as *Antrodia camphorata, Coriolus versicolor, Ganoderma lucidum, Grifola frondosa, Hericium erinaceus, Lignosus rhinocerotis, Pleurotus species, Sarcodon* species, and *Trimella fuciformis* have demonstrated marked memory-enhancing effects in experimental models (Sabaratnam et al. 2013; Phan et al. 2015; Kaur et al. 2017; Phan et al. 2017; Kaur et al. 2019; Randhawa et al. 2021). Scientific reports show that mushroom extracts or constituents reduce neurofibrillary tangles and β-amyloid plaques; reverse cholinergic deficit, stimulate synthesis of nerve growth factor (NGF); counter free radicals and oxidative stress, and neuro-inflammation in experimental animals (Wong et al. 2017; Li et al. 2023; Silva et al. 2023).

But the most prevalent mode of consuming mushrooms by humans is in diet. Hence, it is prudent to investigate if mushrooms when taken regularly in diet have a protective as well as disease-mitigating effect for AD patients.

As described above, mushrooms have diverse constituents and these act by various mechanisms to impart brain health. It is important to mention that mushrooms may protect brain well-being by dual routes:

- By acting directly on the brain (Uffelman et al. 2024).
- By maintaining gut health and providing indirect protection to the brain (Jayachandran et al. 2017).

Figure 16.6 summarizes the dual path by which mushrooms influence and promote brain health.

Numerous observational/ experiential reports emphasize the importance of mushrooms in diet for improving memory, prevention of AD as well as delaying the progression of the disease. Many clinical studies have been carried out to investigate the cognition-protective capacity of dietary mushrooms. Results of some investigations in the last 15 years are presented below.

- A double-blind, parallel group, placebo-controlled trial by Mori et al. (2009) revealed that use of dry powder of *Hericium erinaceus* fruiting bodies (as 250 mg tablets with 96% dry powder of mushroom, three times a day for 16 weeks) to 50–80-year-old Japanese patients with mild cognitive decline resulted in

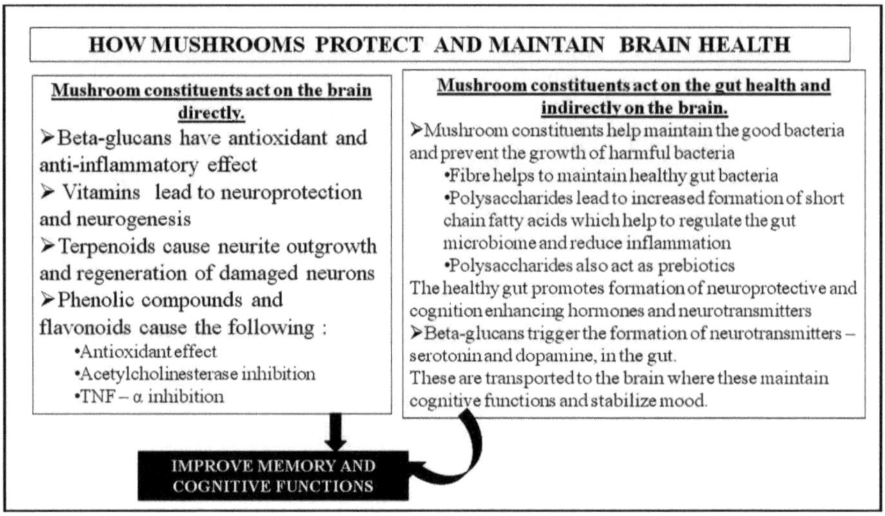

Fig. 16.6 Dual path of memory-enhancing effect of mushrooms

improvement in cognitive function scores and with no adverse effect. In 2020, Li et al. carried out a double-blind randomized placebo-controlled trial on patients with mild dementia. The patients were given three capsules daily [these 350 mg capsules contained the active constituent of *Hericium erinaceus*—erinacine (5 mg/g)] for 49 weeks. A significant improvement in cognitive functions and daily activities was observed.

- Zhang et al. (2017) carried out Ohsaki cohort study, which included 13,230 Japanese individuals (of 65 years or more age group). The results of the study support that regular mushroom intake reduces the risk of dementia.
- Ban et al. (2018) reported that dietary supplement of *Tremella fuciformis* (600 and 1200 mg/day) in a randomized, placebo-controlled, double-blind trial improved scores in subjective memory complaint questionnaire and short-term memory of the patients in comparison to the placebo group.
- Feng et al. (2019) carried out a cross-sectional study on the effect of mushroom consumption and cognitive deficit in 663 patients in Singapore. The researchers concluded that participants who had more than two portions of mushrooms in a week had improved cognitive performance.
- Ba et al. (2022) investigated data of 2840 older adults aged 60 years and more. Data was collected from the 2011–2014 National Health and Nutrition Examination Survey. Higher consumption of mushrooms leads to improvement in certain cognitive functions. The authors concluded that a regular intake of mushrooms could diminish the decline in cognitive abilities. In 2021, the same research group (Ba et al. 2021) carried out a prospective cohort study on 30,378 patients and a meta-analysis of five studies involving 601,893 patients. They concluded that regular mushroom diet led to increase in antioxidant defense and reduced the risk of "all-cause" mortality.

- Yanshree et al. (2022) presented a critical review on six preclinical and three clinical investigations of effect of *Hericium erinaceus* (extracts/constituents) on AD. They concluded that *H. erinaceus* has distinct promise for improving memory in AD patients, but more clinical studies are required.
- Gou et al. (2023) examined the association between dietary patterns and cognitive functions of 884 participants. They concluded that intake of vegetables and mushrooms in our diet prevents decline in cognitive abilities.
- Aoki et al. (2024) did a prospective study involving 3750 Japanese people (40–64 years). They observed that in women, mushroom diet prevented disabling dementia.
- Cha et al. (2024) conducted an exhaustive narrative systematic review with an objective to examine the role of dietary mushrooms on maintaining neurocognition and mood. They examined a total of 34 human studies—these included 24 epidemiological and ten interventional studies.
  - The epidemiological studies included cross-sectional or case-controlled or prospective studies, which were mainly on Asian population (Japanese, Chinese and Korean). Mushrooms were an integral part of the diet. There was improvement in mood and sleep and a reduction in anxiety and depression. Diet without mushroom increased risk of decline, while diets with mushrooms led to better cognition and verbal memory. Moreover, studies on cognitively healthy older adults showed that they consumed larger quantities of mushrooms in their diet. One study emphasized that physical exercise had more impact on cognitive functions when accompanied with dietary modification (Wei et al. 2022).
  - The interventional studies included nine randomized clinical trials mainly on Japanese and Chinese populations who had intervention with mushrooms like Reishi, Lion's Mane, Caterpillar, Shiitake, or vitamin D-enriched button mushrooms. There was no consistent pattern in the impact of mushroom intake and neurocognition. Some studies showed an improvement in mood and cognitive function following mushroom intervention. Most significant effects were observed with Lion's mane as compared to other mushrooms.
- Uffelman et al. (2024) carried out a randomized controlled feeding trial to study the impact of a healthy U.S. Mediterranean-style dietary pattern (MED) with or without mushrooms. The group with MED and mushrooms were given 84 g/d of mushrooms (4 days/week white button and 3 days/week oyster). The results showed that there was improvement in activity and immediate memory in middle-aged and older adult who were taking MED with as well as without mushroom supplementation.
- Yang et al. (2024) studied the association between edible mushrooms and algae intake with decline in cognitive functions in individuals aged 65 years and above in China. They concluded that diet with mushrooms and algae reduced the risk of cognitive impairment.

Literature shows that mushrooms have been consumed as a part of human diet since the beginning of life. Mushrooms, their extract, and mycoconstituents have demonstrated neuroprotective potential in experimental models. Human studies conducted to examine the effect of mushroom in diet have also indicated their neuroprotective and neuro-cognition enhancing potential. However, these investigations are not exhaustive and sometimes, the results are not conclusive.

16.4.1 So, What Are the Future Directions?

- More randomized placebo-controlled trials.
- More edible mushroom species should also be investigated.
- Systematic trials that include intervention with mushrooms—pre and post designs as well as compared with placebo.
- Trials that include determination of quantification of mushroom diet— how much to eat and how often.
- Trials should include evaluation of behavior, psychological parameters along with determination of biomarkers—both in the brain and in the gut and brain imaging. This would facilitate elucidation of the mechanisms by which dietary mushrooms lead to neuroprotection and memory enhancing effect.

16.5 Conclusion

The objective of this chapter was to highlight the importance of mushrooms in diet. It is evident that mushrooms with an array of constituents and nutrients are like "Superfoods," which nourish, help to fight different diseases, and impart health and well-being. Regular consumption of mushrooms in our diet prevents neurodegenerative changes, enhances neuro-cognition as well as neurogenesis and delays the progression of AD. Mushrooms act on the brain—improve neurotransmitter balance, reduce attack by free radicals, prevent Tau protein formation, decrease neuroinflammation, and enhance neurogenesis. Mushrooms also regulate gut health, which also promotes brain health indirectly. Thus, incorporation of mushrooms in diet may impart neuroprotection by these dual pathways and bring relief to the millions suffering from AD and improve their quality of life.

References

Alzheimer's Society (2025) Official website. Registered office 43–44, Crutched Friars, London, EC3N2AE

Alzheimer's Association (2025) Official website. 225 N Michigan Ave. Floor 17 Chicago, IL 60601. Official website

American Heart Association (2021) What is Brain Health? https://www.heart.org/en/health-topics/brain-health

Anusiya G, Gowthama Prabu U, Yamini NV, Sivarajasekar N, Rambabu K, Bharath G, Banat F (2021) A review of the therapeutic and biological effects of edible and wild mushrooms. Bioengineered 12(2):11239–11268

Aoki S, Yamagishi K, Maruyama K, Ikeda A, Nagao M, Noda H, Umesawa M, Hayama-Terada M, Muraki I, Okada C, Tanaka M, Kishida R, Kihara T, Takada M, Shimizu Y, Ohira T, Imano H, Sankai T, Okada T, Tanigawa T, Kitamura A, Kiyama M, Iso H (2024) Mushroom intake and risk of incident disabling dementia: the Circulatory Risk in Communities Study (CIRCS). Br J Nutr 131(9):1641–1647

Ba DM, Gao X, Al-Shaar L, Muscat J, Chinchilli VM, Ssentongo P, Zhang X, Liu G, Beelman RB, Richie JP Jr (2021) Prospective study of dietary mushroom intake and risk of mortality: results from continuous National Health and Nutrition Examination Survey (NHANES) 2003-2014 and a meta-analysis. Nutr J 20(1):80

Ba DM, Gao X, Al-Shaar L, Muscat J, Chinchilli VM, Ssentongo P, Beelman RB, Richie J (2022) Mushroom intake and cognitive performance among US older adults: the national health and nutrition examination survey, 2011–2014. Br J Nutr 128:2241–2248

Ban S, Lee SL, Jeong HS, Lim SM, Park S, Hong YS, Kim JE (2018) Efficacy and safety of *Tremella fuciformis* in individuals with subjective cognitive impairment: a randomized controlled trial. J Med Food 21:400–407

Bernaś E, Grażyna J, Zofia L (2006) Edible mushrooms as a source of valuable nutritive constituents. Acta Sci Pol Technol Aliment 5(1):5–20

Beversdorf DQ, Crosby HW, Shenker JI (2023) Complementary and alternative medicine approaches in Alzheimer disease and other neurocognitive disorders. Mo Med 120(1):70–78. PMID: 36860601; PMCID: PMC9970340

Bull MJ, Plummer NT (2014) Part 1: The human gut microbiome in health and disease. Integr Med (Encinitas) 13(6):17–22. PMID: 26770121; PMCID: PMC4566439

Carabotti M, Scirocco A, Maselli MA, Severi C (2015) The gut-brain axis: interactions between enteric microbiota, central and enteric nervous systems. Ann Gastroenterol 28(2):203–209. PMID: 25830558; PMCID: PMC4367209

Cerletti C, Esposito S, Iacoviello L (2021) Edible mushrooms and beta-glucans: impact on human health. Nutrients 13(7):2195

Cha S, Bell L, Shukitt-Hale B, Williams CM (2024) A review of the effects of mushrooms on mood and neurocognitive health across the lifespan. Neurosci Biobehav Rev 158:105548

Chang S, Buswell J (2023) Medicinal mushrooms: past, present and future. Adv Biochem Eng Biotechnol 184:1–27. https://doi.org/10.1007/10_2021_197. PMID: 35220455

Chang ST, Wasser SP (2012) The role of culinary-medicinal mushrooms on human welfare with a pyramid model for human health. Int J Med Mushrooms 14(2):95–134

Chatterjee A, Samajpati JN (2021) Nutritional value of edible mushrooms. J Mycopathol Res 59(4):357–362

Craig CR, Stitzel RE (2004) Modern pharmacology with clinical applications, 6th edn. Lippincott Williams & Wilkins

Cummings J, Zhou Y, Lee G, Zhong K, Fonseca J, Cheng F (2023) Alzheimer's disease drug development pipeline. Alzheimers Dement (N. Y) 9(2):e12385

Dhikav V, Anand KS (2012) Complementary and alternative medicine usage among Alzheimer's disease patients. Int Psychogeriatr 24(8):1361–1362

Duplantier SC, Gardner CD (2021) A critical review of the study of neuroprotective diets to reduce cognitive decline. Nutrients 13:2264

Feng L, Cheah IK-M, Ng MM-X, Li J, Chan SM, Lim SL, Mahendran R, Kua E-H, Halliwell B (2019) The association between mushroom consumption and mild cognitive impairment: a community-based cross-sectional study in Singapore. J Alzheimer's Dis 68:197–203

Fortune NC, Harville EW, Guralnik JM et al (2019) Dietary intake and cognitive function: evidence from the Bogalusa Heart Study. Am J Clin Nutr 109(6):1656–1663

Gómez-Pinilla F (2008) Brain foods: the effects of nutrients on brain function. Nat Rev Neurosci 9(7):568–578

Gou R, Qin J, Pang W, Cai J, Luo T, He K, Xiao S, Tang X, Zhang Z, Li Y (2023) Correlation between dietary patterns and cognitive function in older Chinese adults: a representative cross-sectional study. Front Nutr 10:1093456

Hu N, Yu JT, Tan L, Wang YL, Sun L, Tan L (2013) Nutrition and the risk of Alzheimer's disease. Biomed Res Int 2013:Article ID 524820

Jayachandran M, Xiao J, Xu B (2017) A critical review on health promoting benefits of edible mushrooms through gut microbiota. Int J Mol Sci 18(9):1934

Katzung BG (2012) Special aspects of geriatric pharmacology in basic & clinical pharmacology edited by Bertram G, 12th edn. The McGraw-Hill Companies, Inc., Katzung, MD. ISBN: 978-0-07-176402-5

Kaur R, Singh V, Shri R (2017) Anti-amnesic effects of *Ganoderma* species: a possible cholinergic and antioxidant mechanism. Biomed Pharmacother 92:1055–1061

Kaur A, Randhawa K, Singh V, Shri R (2019) Bioactivity-guided isolation of acetylcholinesterase inhibitor from *Ganoderma mediosinense* (Agaricomycetes). Int J Med Mushrooms 21(8):755–763

Kelly JR, Kennedy PJ, Cryan JF, Dinan TG, Clarke G, Hyland NP (2015) Breaking down the barriers: the gut microbiome, intestinal permeability and stress-related psychiatric disorders. Front Cell Neurosci 2015(9):392

Li IC, Chang HH, Lin CH, Chen WP, Lu TH, Lee LY, Chen YW, Chen YP, Chen CC, Lin DP (2020) Prevention of early Alzheimer's disease by Erinacine A-enriched hericium erinaceus mycelia pilot double-blind placebo-controlled study. Front Aging Neurosci 12:155

Li N, Li H, Liu Z et al (2023) Unveiling the therapeutic potentials of mushroom bioactive compounds in Alzheimer's disease. Foods 12:2972

McEvoy CT, Guyer H, Langa KM, Yaffe K (2017) Neuroprotective diets are associated with better cognitive function: the health and retirement study. J Am Geriatr Soc 65(8):1857–1862

Mori K, Inatomi S, Ouchi K, Azumi Y, Tuchida T (2009) Improving effects of the mushroom Yamabushitake (Hericium erinaceus) on mild cognitive impairment: a double-blind placebo-controlled clinical trial. Phytother Res 23(3):367–372

nia.nih.gov (2023) An official website of National Institute of Health. What Do We Know about Diet and Prevention of Alzheimer's disease? National Institute of Ageing 2023

Parums DV (2021) Editorial: targets for disease-modifying therapies in Alzheimer's disease, including amyloid b and tau protein. Med Sci Monit 27:e934077

Phan CW, David P, Naidu M, Wong KH, Sabaratnam V (2015) Therapeutic potential of culinary-medicinal mushrooms for the management of neurodegenerative diseases: diversity, metabolite, and mechanism. Crit Rev Biotechnol 35(3):355–368

Phan CW, David P, Sabaratnam V, Kim KS, Zhao TT, Shin KS, Park HJ, Cho YJ, Lee KE, Kim SH (2017) Edible and medicinal mushrooms: emerging brain food for the mitigation of neurodegenerative diseases. J Med Food 20(1):1–10

Polis B, Samson AO (2019) A new perspective on Alzheimer's disease as a brain expression of a complex metabolic disorder. In: Wisniewski T (ed) Alzheimer's disease [Internet]. Codon Publications, Brisbane (AU)

Randhawa K, Singh V, Kaur S, Kaur R, Kumar S, Shri R (2021) Isolation of Pleurotus Florida derived acetylcholinesterase inhibitor for the treatment of cognitive dysfunction in mice. Food Sci Human Wellness 10:490–496

Sabaratnam V, Kah-Hui W, Naidu M, Rosie DP (2013) Neuronal health–can culinary and medicinal mushrooms help? J Trad Complement Med 3(1):62–68

Sande D, Oliveira GP, Moura MAFE, Martins BA, Lima MTNS, Takahashi JA (2019) Edible mushrooms as a ubiquitous source of essential fatty acids. Food Res Int 125:108524

Silva AM, Preto M, Grosso C, Vieira M, Delerue-Matos C, Vasconcelos V, Reis M, Barros L, Martins R (2023) Tracing the path between mushrooms and Alzheimer's disease—a literature review. Molecules 28:5614

Uffelman CN, Harold R, Hodson ES, Chan NI, Foti D, Campbell WW (2024) Effects of consuming white button and oyster mushrooms within a healthy Mediterranean-style dietary pattern on changes in subjective indexes of brain health or cognitive function in healthy middle-aged and older adults. Foods 13:2319

Wallace RK (2020) The microbiome in health and disease from the perspective of modern medicine and Ayurveda. Medicina (Kaunas) 56(9):462

Wasser SP (2010) Medicinal mushroom science: history, current status, future trends, and unsolved problems. Int J Med Mushrooms 12(1):1

Wei K, Yang J, Lin S, Mei Y, An N, Cao X, Jiang L, Liu C, Li C (2022) Dietary habits modify the association of physical exercise with cognitive impairment in community-dwelling older adults. J Clin Med 11(17):5122

Wong KH, Ng CC, Kanagasabapathy G, Yow YY, Sabaratnam V (2017) An overview of culinary and medicinal mushrooms in neurodegeneration and neurotrauma research. Int J Med Mushrooms 19(3):191–202

Yadav D, Negi PS (2021) Bioactive components of mushrooms: processing effects and health benefits. Food Res Int 148:110599

Yang Y, Zhu D, Qi R, Chen Y, Sheng B, Zhang X (2024) Association between intake of edible mushrooms and algae and the risk of cognitive impairment in Chinese older adults. Nutrients 16:637

Yanshree YWS, Fung ML, Lee CW, Lim LW, Wong KH (2022) The monkey head mushroom and memory enhancement in Alzheimer's disease. Cells 11:2284

Zhang S, Tomata Y, Sugiyama K, Sugawara Y, Tsuji I (2017) Mushroom consumption and incident dementia in elderly Japanese: the Ohsaki Cohort 2006 study. J Am Geriatr Soc 2017(65):1462–1469

Alzheimer's Disease Management Through the Gut-Brain Axis

17

Poonam Negi, Nishtha Singh, and Pranshul Sethi

Abstract

Alzheimer's disease (AD) is a neurodegenerative disorder that primarily affects the elderly population. Clinically, it is characterized by memory loss, cognitive decline, personality changes, and impaired executive functions. Amyloid-beta (Aβ) plaque accumulation, hyperphosphorylation of tau protein, neuroinflammation, oxidative stress, and widespread neuronal loss, particularly in the hippocampus and cortex, are the main pathological features of AD. Despite extensive research, effective disease-modifying treatments remain elusive, warranting the urgent need for novel therapeutic approaches. Gut-brain axis (GBA), a bidirectional communication network between the CNS and the gastrointestinal tract, involves neural pathways (such as the vagus nerve), endocrine signals, immune modulation, and microbial metabolites. The gut microbiota residing in the human intestine is responsible for maintaining gut integrity, regulating immune responses, producing neuroactive compounds, and modulating brain function. Gut dysbiosis, which is a disruption in the composition and function of gut microbiota, has been increasingly attributed to the development and progression of neurodegenerative diseases, including AD.

Gut dysbiosis contributes to AD pathology through different mechanisms. An altered microbial profile enhances intestinal permeability ("leaky gut"), allowing the translocation of microbial products (such as LPS) into systemic circulation. These pro-inflammatory molecules can cross the blood-brain barrier (BBB), triggering neuroinflammation and exacerbating Alzheimer's disease (AD) pathol-

P. Negi (✉)
Chitkara College of Pharmacy, Chitkara University, Rajpura, India

N. Singh · P. Sethi
Department of Pharmacology, School of Pharmaceutical Sciences, Shoolini University, Solan, Himachal Pradesh, India

J. Malik et al. (eds.), *Diet and Alzheimer's Disease: Let Food be Our Medicine*, Nutritional Neurosciences, https://doi.org/10.1007/978-981-95-2736-6_17

ogy, including Aβ deposition and tau hyperphosphorylation. Additionally, dysbiosis can impair the production of beneficial microbial metabolites (such as SCFAs), which have neuroprotective and anti-inflammatory properties. This altered microbial signaling may influence cognitive and emotional functions by changing the synthesis and regulation of neurotransmitters, such as serotonin, dopamine, and gamma-aminobutyric acid (GABA). Modulation of the compelling link between the gut microbiota and brain health, that is, the gut-brain axis, has emerged as a potential therapeutic strategy for AD. Several microbiota-targeted interventions are being explored to restore microbial balance and improve neurological outcomes. These include the administration of probiotics (beneficial bacteria), prebiotics (nondigestible food components that promote the growth of beneficial microbes), and synbiotics (a combination of both). Dietary modifications, particularly adherence to anti-inflammatory and fiber-rich diets such as the Mediterranean or MIND diets, have also shown potential in promoting gut and brain health. Another emerging intervention is fecal microbiota transplantation (FMT), where stool from a healthy donor is transferred to an AD patient to repopulate the gut with a healthy microbial community. The gut-brain axis represents a novel and promising frontier in understanding and managing Alzheimer's disease. A gut microbiota-targeted approach to modulate systemic inflammation can augment neuroprotection and reduce cognitive decline. In the future, research efforts should focus on elucidating precise microbial products associated with AD and developing personalized microbiomes, thereby improving clinical outcomes and the quality of life of AD patients.

Keywords

Prebiotics · Gut dysbiosis · Lipopolysaccharides · Synbiotics · Mediterranean diet · Probiotics · Gut microbiota · Neurodegeneration · Dementia

Abbreviations

AD	Alzheimer's Disease
Aβ	Amyloid Beta
Aβ40 / Aβ42	Amyloid-beta isoforms (40 amino acid/42 amino acid variants)
APP	Amyloid Precursor Protein
BACE1	Beta-site Amyloid Precursor Protein-Cleaving Enzyme 1
BDNF	Brain-Derived Neurotrophic Factor
CNS	Central Nervous System
BBB	Blood-Brain Barrier
GBA	Gut-Brain Axis
ENS	Enteric Nervous System
ANS	Autonomic Nervous System
HPA	Hypothalamic-Pituitary-Adrenal (Axis)
NFT	Neurofibrillary Tangles

LPS	Lipopolysaccharide
MAMPs	Microbial-Associated Molecular Patterns
TNF-α	Tumor Necrosis Factor-Alpha
IL-1β	Interleukin-1 Beta
IL-6	Interleukin-6
IL-10	Interleukin-10
GALT	Gut-Associated Lymphoid Tissue
SCFAs	Short-Chain Fatty Acids (Butyrate, Acetate, Propionate)
NO	Nitric Oxide
ROS	Reactive Oxygen Species
SOD	Superoxide Dismutase
MDA	Malondialdehyde
GSH	Glutathione
GSK-3β	Glycogen Synthase Kinase-3 Beta
CDK5	Cyclin-Dependent Kinase 5
FXR	Farnesoid X Receptor
TGR5	Takeda G-Protein-Coupled Receptor 5
FMT	Fecal Microbiota Transplantation
GI	Gastrointestinal
H_2S	Hydrogen Sulfide
PYY	Peptide YY
GLP-1	Glucagon-Like Peptide-1
GABA	Gamma-Aminobutyric Acid
Ach	Acetylcholine
5-HT	5-Hydroxytryptamine (Serotonin)

17.1 Introduction

Alzheimer's disease (AD) is a progressive neurodegenerative disorder and the most prevalent form of dementia, accounting for approximately 60–70% of cases globally. As of 2025, over 55 million individuals are living with dementia worldwide, with projections indicating a rise to 78 million by 2030 and 139 million by 2050, primarily driven by aging populations. AD is characterized by a gradual decline in cognitive functions, including memory, reasoning, language, and executive skills, ultimately leading to complete dependence and death (Nichols et al. 2019).

The neuropathological hallmarks of AD include the accumulation of extracellular amyloid-beta (Aβ) plaques and intracellular neurofibrillary tangles composed of hyperphosphorylated tau protein. These pathological features are accompanied by synaptic dysfunction, neuronal loss, gliosis, and chronic neuroinflammation. Despite extensive research, the precise etiology of AD remains elusive, and current therapeutic strategies offer only symptomatic relief without altering disease progression (Kumar et al. 2025).

17.1.1 Limitations of Traditional Therapeutic Approaches

Conventional treatment strategies for AD have mostly targeted the amyloid cascade theory, which holds that the buildup of Aβ peptides starts a series of neurodegenerative events. Several pharmaceutical treatments, including β-secretase inhibitors, γ-secretase modulators, and monoclonal antibodies targeting Aβ, have been developed to reduce Aβ generation, prevent aggregation, or enhance its clearance (WU et al. 2022). Nevertheless, there has been limited clinical success with these therapies, and numerous large-scale clinical trials have failed to demonstrate any discernible cognitive benefits. Likewise, treatments targeting neurotransmitter regulation, neuroinflammation, and tau pathology have not yielded the expected results (Congdon et al. 2023). Given that AD is a complex condition combining genetic, environmental, metabolic, and lifestyle factors, a single therapy target might not be enough. As a result, there is increasing interest in investigating complementary and alternative methods that take into account the intricate interactions between systemic elements that contribute to the pathophysiology of AD (Chen et al. 2021).

17.1.2 Emergence of the Gut-Brain Axis in Neurodegenerative Research

In the field of neuroscience and neurodegenerative research, the gut-brain axis (GBA) has garnered considerable attention recently. The gastrointestinal (GI) tract and the central nervous system (CNS) have a bidirectional communication network known as the GBA, which includes metabolic, immunological, hormonal, and neurological pathways (Lu et al. 2024). This complex system facilitates ongoing communication between the gut microbiota and the brain, influencing a range of physiological functions, including mood, immunological responses, and cognition. An intricate and ever-changing community of bacteria, known as the gut microbiota, is found in the human gut and is essential for maintaining homeostasis. Metabolic disorders, autoimmune diseases, and neuropsychiatric illnesses are among the many conditions associated with dysbiosis, or alterations in the composition and function of the gut microbiota. Recent research indicates that gut dysbiosis may have a role in the onset and advancement of neurodegenerative illnesses, such as AD, by modifying neurochemical pathways, disrupting the blood-brain barrier (BBB), and causing systemic inflammation (Carabotti et al. 2015).

17.1.3 Mechanisms Linking Gut Dysbiosis to Alzheimer's Disease

Increased intestinal permeability caused by dysbiosis may allow bacterial endotoxins, such as lipopolysaccharides (LPS), to enter the bloodstream. A major characteristic of AD is persistent neuroinflammation, which is caused by these endotoxins' ability to penetrate the blood-brain barrier and activate astrocytes and microglia (Rosendo-Silva et al. 2023).

Microbial metabolites and inflammatory cytokines can influence tau phosphorylation and aggregation, potentially contributing to the development of neurofibrillary tangles (Rawat et al. 2022).

Important for cognitive function, the gut microbiota produces and regulates neurotransmitters like dopamine, serotonin, and gamma-aminobutyric acid (GABA). Dysbiosis may interfere with these processes, resulting in behavioral and cognitive deficits. Gut microbes influence the balance of lipids and glucose, as well as host metabolism. Because metabolic disorders increase oxidative stress and insulin resistance, they can worsen AD pathogenesis (Mroke et al. 2024).

17.1.4 Evidence Supporting the Gut-Brain Axis in Alzheimer's Disease

The impact of the gut microbiota on the central nervous system is demonstrated by changes in behavior, cognition, and brain disease in germ-free and antibiotic-treated mice (Fig. 17.1). It has been demonstrated that fecal microbiota transplantation (FMT) from AD patients to healthy mice causes AD-like pathologies and cognitive impairments (Grabrucker et al. 2023). The gut microbiota makeup of AD patients and healthy controls differs, according to comparative investigations. These

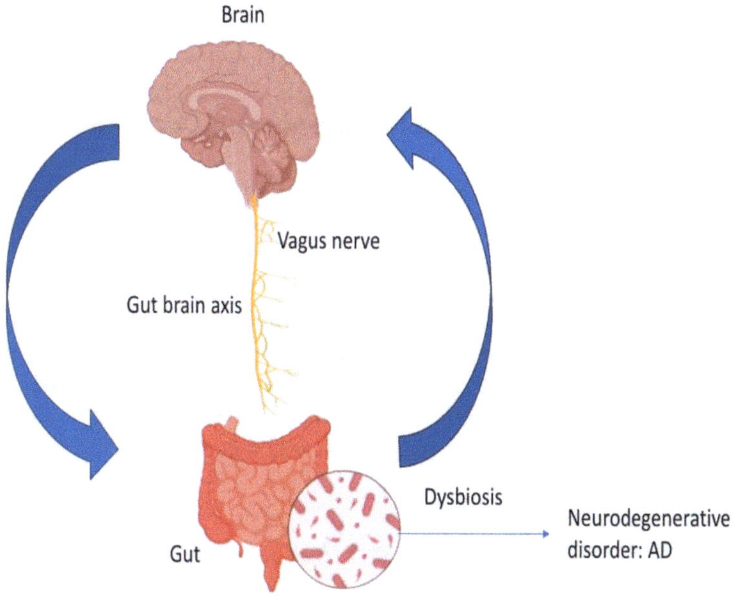

Fig. 17.1 The picture illustrates the gut-brain axis, emphasizing the two-way communication that occurs between the gut and the brain through the immune system, gut bacteria, the vagus nerve, and circulatory routes. It highlights the dynamic interplay between the brain and gut activities, demonstrating how neurotransmitters (e.g., serotonin, GABA) and microbial metabolites affect brain function. (This image is created with Biorender.com)

alterations include a decrease in microbial diversity and a change in the abundance of particular bacterial taxa. Microbial profiles have been found to correlate with indicators of AD pathology, including tau and Aβ levels (Ping Liu et al. 2019).

According to encouraging findings from early clinical trials examining the impact of probiotics, prebiotics, and nutritional treatments on cognitive performance in AD patients, altering the gut flora may have therapeutic benefits (Li et al. 2024).

17.1.5 Potential Microbiota-Targeted Therapeutic Strategies

Probiotics: Beneficial bacterial strains are administered to boost gut health and microbial balance. Prebiotics are indigestible dietary fibers that encourage the development of good gut flora. Symbiotics: A blend of prebiotics and probiotics that work together to improve the makeup of the gut microbiota (Chen Li et al. 2020).

Dietary Changes: To promote a healthy gut microbiome, adopt a diet high in fiber, polyphenols, and omega-3 fatty acids, such as the Mediterranean diet (Merra et al. 2020).

Fecal Microbiota Transplantation (FMT): restoring microbial diversity and function in patients by transferring fecal matter from healthy donors (Zikou et al. 2024).

Pharmacological agents: creation of medications that target signaling pathways or microbial metabolites implicated in the GBA (Brown and Hazen 2017).

17.2 The Gut-Brain Axis: Components and Communication

The central nervous system (CNS) and the gastrointestinal (GI) tract can interact dynamically thanks to the gut-brain axis (GBA), a complex, two-way communication network. Numerous physiological, metabolic, cognitive, and emotional processes are coordinated because of this connection (Aljeradat et al. 2024). Neural, hormonal, immunological, and microbial signaling pathways are among the many interrelated components and signaling pathways that comprise the GBA, rather than a single system. Knowing these elements and how they work is essential to understanding how gut health might affect brain function, particularly in neurodegenerative conditions like Alzheimer's disease (AD) (Zhou et al. 2023).

17.2.1 Central Nervous System (CNS)

Processing and reacting to signals from the body, including those originating from the gut, is primarily controlled by the central nervous system (CNS), which comprises the brain and spinal cord. It processes information, receives sensory input, and produces behavioral and motor reactions. Within the framework of the GBA, the central nervous system (CNS) deciphers signals from the GI tract, including microbial cues, pain, inflammation, and satiety. These signals' dysregulation may be a factor in neuropsychiatric and neurodegenerative diseases (Thau et al. 2025).

Crucially, neuroimaging research has shown that the composition of the gut microbiota can influence the structure and function of the brain, including areas related to mood, pain perception, and cognition. In AD, signals from a dysbiosis gut microbiome may worsen central nervous system disease, including amyloid plaque buildup and neuroinflammation (Kowalski and Mulak 2019).

17.2.2 Enteric Nervous System (ENS)

A vast network of neurons embedded in the walls of the GI tract, the Enteric Nervous System (ENS), is often referred to as the "second brain." It controls semiautonomous digestive functions such peristalsis, the release of enzymes, and the absorption of nutrients. Via the vagus nerve and spinal afferents, the ENS extensively connects with the central nervous system (CNS) and contains more neurons than the spinal cord, with over 100 million (Chaudhry et al. 2023).

Additionally, gut bacteria influence the ENS. Some microbial metabolites have the ability to directly affect enteric neurons, altering barrier integrity, gastrointestinal motility, and even behavior and mood. In AD, ENS failure may impact gut barrier permeability and systemic inflammation, which could lead to neuroinflammatory processes in the brain (Shengwen Lu et al. 2024).

17.2.3 Autonomic Nervous System (ANS) and the Vagus Nerve

The Autonomic Nervous System (ANS) controls involuntary physiological processes such as parasympathetic channel connecting the brain and gut within the GBA. Signals from the gut can travel directly to the brainstem and higher brain areas, such as the cortex and hypothalamus, through the vagus nerve. Gastrointestinal hormones, inflammatory signals, and microbiological contaminants can trigger it. In addition to modifying mood and cognition, vagal stimulation has been demonstrated to lower systemic inflammation. On the other hand, this communication may be hampered by vagus nerve injury or malfunction, which could exacerbate neurodegenerative diseases (Waxenbaum et al. 2025).

In AD, systemic inflammation or compromised microbiota may alter vagus nerve transmission, which in turn impairs the brain's ability to regulate inflammation and maintain homeostasis (Navid Faraji et al. 2025).

17.2.4 Hypothalamic-Pituitary-Adrenal (HPA) Axis

One important neuroendocrine mechanism that regulates the body's reaction to stress is the HPA axis. Adrenocorticotropic hormone (ACTH) is released by the pituitary gland in response to the hypothalamus' secretion of corticotropin-releasing hormone (CRH), which in turn causes the adrenal glands to release cortisol (Herman et al. 2016).

Chronic stress and HPA axis dysregulation are associated with increased gut permeability, altered microbiota composition, and heightened neuroinflammation. Elevated cortisol levels can also impair hippocampal function and memory, both of which are critical in AD pathology (Rusch et al. 2023).

Additionally, the HPA axis can be influenced by the microbiota; for instance, germ-free mice exhibit heightened stress reactions that subside when specific probiotic strains are administered. This demonstrates how GBA communication is reciprocal and relevant to AD (Rusch et al. 2023).

17.2.5 Immune System and Gut-Associated Lymphoid Tissue (GALT)

An integral aspect of the GBA is the immune system, specifically the gut-associated lymphoid tissue (GALT). Up to 70% of the body's immune cells are found in the gut, making it the biggest organ in the body. To maintain immune system balance, it continuously monitors antigens.Microbial imbalance in the context of gut dysbiosis can lead to the immune system becoming overactive, resulting in the secretion of pro-inflammatory cytokines such as TNF-α, IL-1β, and IL-6. These cytokines can cross the blood-brain barrier, enter the systemic circulation, and induce neuroinflammation, a defining feature of AD (Ruth and Field 2013). Chronic gut inflammation can also increase intestinal permeability, also known as "leaky gut," which allows microbial compounds, such as lipopolysaccharides (LPS), to enter the bloodstream and fuel central and systemic inflammation (Brown and Heneka 2024).

17.2.6 Microbial Metabolites and Their Role in Brain Function

Human physiology, including brain function, is significantly impacted by the diverse range of bioactive compounds produced by the gut bacteria. Among the important microbial metabolites in GBA communication are:

The microbial fermentation of dietary fibre produces short-chain fatty acids (SCFAs), such as butyrate, acetate, and propionate, which promote the integrity of the gut barrier, regulate immunological responses, and can penetrate the blood-brain barrier to affect neuroinflammation and neurogenesis (Tran and Mohajeri 2021).

Tryptophan metabolites: Tryptophan is a necessary amino acid and a precursor to serotonin, and its metabolism is regulated by gut flora. When tryptophan metabolism is shifted toward the kynurenine pathway due to dysbiosis, neurotoxic metabolites are produced, which may contribute to cognitive loss in AD (Agus et al. 2018). Gut bacteria can alter secondary bile acids, which then interact with nuclear receptors including FXR and TGR5 to affect inflammation and metabolism. In people with AD, altered bile acid profiles have been noted.

Derived from microbial metabolism and dietary polyphenols, phenolic compounds and polyamines are molecules with anti-inflammatory and antioxidant properties that may help prevent the development of dementia (Kan Gao et al. 2020).

17.2.7 Signaling Modalities in the Gut-Brain Axis

Neural signaling: Mechanical, chemical, and immunological signals are among the sensory data that are sent from the gut to the brain via afferent fibers of the spinal and vagus nerves.

Endocrine signaling: Hunger, mood, and stress responses are modulated by gut-derived hormones, including ghrelin, peptide YY (PYY), glucagon-like peptide-1 (GLP-1), and cortisol. These hormones also affect brain function and may alter the risk of AD.

Immune signaling: Gut-resident immune cells release cytokines and chemokines that can enter the bloodstream or interact with the blood-brain barrier to modulate neuroinflammatory responses relevant to AD (Browning et al. 2017).

Microbial metabolite signaling: As previously mentioned, bile acids, SCFAs, and substances that resemble neurotransmitters function as molecular messengers that affect the central nervous system, the ENS, and systemic immunity (Chen et al. 2021).

17.3 Gut Microbiota Alterations in Alzheimer's Disease

The gut microbiota, a large population of bacteria, viruses, fungi, and archaea that reside in the gastrointestinal tract, is essential to host health because it regulates the immune system, metabolism, and brain communication. Dysbiosis, or disruptions in the gut microbial environment, may be strongly related to the pathogenesis of Alzheimer's disease (AD), according to new research (Colella et al. 2023).

Recent research has shown that AD patients and age-matched healthy individuals differ significantly in the composition and function of their gut microbiota, both in human and animal models. These changes may support systemic inflammation, disruption of the blood-brain barrier (BBB), and neuroinflammation, which may contribute to the start and progression of AD in addition to reflecting disease-associated changes (Hou et al. 2022).

17.3.1 Reduced Microbial Diversity

The reduction in gut microbial diversity in AD patients is one of the most consistent results from several studies. A healthy microbiome is characterized by microbial variety, which is linked to improved immunological regulation, increased resistance to infections, and a more robust metabolic capability (Chandra et al. 2023).

It has been noted that AD patients have reduced alpha diversity (species richness and evenness within a sample), which may indicate a damaged microbiome. Less diversity hinders the ecosystem's ability to produce vital metabolites, such as short-chain fatty acids (SCFAs), which maintain intestinal integrity and regulate inflammation (Murray et al. 2022).

Therefore, immunological modulation and gut-brain connection may be compromised by the loss of microbial richness, which would exacerbate cognitive impairment (Safarchi et al. 2025).

17.3.2 Decreased Abundance of Beneficial Bacteria

A characteristic of dysbiosis associated with AD is the loss of good gut bacteria, especially those with neuroprotective and anti-inflammatory qualities:

Bifidobacterium spp.: These Gram-positive anaerobic bacteria are among the first to colonize the infant gut and continue to provide essential health benefits into adulthood. They are known to produce SCFAs (especially acetate), enhance mucosal barrier function, and regulate immune responses. Reduced *Bifidobacterium* levels in AD have been associated with impaired barrier function, which facilitates the systemic translocation of microbial products, such as lipopolysaccharide (LPS) and other endotoxins.

Lactobacillus species: By generating lactic acid, regulating the immune system, and affecting the synthesis of neurotransmitters like gamma-aminobutyric acid (GABA), these bacteria help to maintain gut homeostasis. Their decrease in AD patients may lead to decreased amounts of healthy neuroactive substances, impaired intestinal barrier function, and diminished anti-inflammatory signaling. In animal models, certain *Lactobacillus* strains have even been demonstrated to enhance memory and lessen amyloid buildup.

Collectively, the decline of these probiotic genera may encourage a gut environment that is pro-inflammatory, which in turn fuels neuroinflammation and accelerates neurodegenerative processes (Kandpal et al. 2022).

17.3.3 Increased Abundance of Pro-inflammatory Genera

In addition to the loss of helpful microorganisms, AD patients have an excess of potentially dangerous bacteria, especially those linked to inflammation:

Escherichia/Shigella spp.: Patients with AD often have higher levels of these Gram-negative bacteria. Lipopolysaccharide (LPS), a strong endotoxin that triggers the innate immune system through Toll-like receptor 4 (TLR4), is something they are known to produce. LPS has the ability to break down tight connections in the blood-brain barrier and the gut, allowing systemic inflammation to enter the brain and promoting tau hyperphosphorylation, amyloid genesis, and microglial activation (Yadav et al. 2022).

Desulfovibrio spp.: These sulphate-reducing bacteria can harm epithelial cells and jeopardize the integrity of mucosal membranes by producing hydrogen sulfide and other harmful compounds. Increased *Desulfovibrio* levels may contribute to neuroinflammation in AD and have also been connected to Parkinson's disease (Munteanu et al. 2024).

Proteobacteria: Often raised in dysbiosis conditions, this phylum contains a large number of harmful bacteria. Its proliferation is thought to be an indication of pro-inflammatory potential and microbial imbalance.

As long-term immune stimulants, these pro-inflammatory taxa may cause low-grade inflammation, which eventually leads to neuronal loss and dysfunction—two key aspects of AD pathophysiology (Sowmiya et al. 2024).

17.3.4 Role of Microbial Products in AD Pathogenesis

Increased microbial-derived pro-inflammatory chemical synthesis and systemic circulation are among the most important features of gut dysbiosis in AD. These include:

- Lipopolysaccharides (LPS): *Escherichia coli* and other Gram-negative bacteria contain LPS in their outer membrane. Patients with AD have been found to have elevated amounts of LPS in their blood and cerebrospinal fluid (CSF). After passing through the weakened intestinal barrier and entering the bloodstream, LPS eventually reaches the brain, where it activates microglia, leading to long-term neuroinflammation and promoting the accumulation of amyloid-β (Aβ) (Farhana and Khan 2025).
- Bacterial amyloids: Although they differ physically from human Aβ, certain gut bacteria make amyloid-like proteins that can cross-seed with host proteins to speed up the development of amyloid fibrils in the brain, which causes amyloid plaque (Sampson et al. 2020).
- Cytokines: Dysbiosis-induced gut inflammation can trigger immune cells to release pro-inflammatory cytokines like IL-6, TNF-α, and IL-1β, which can cross the blood-brain barrier and worsen neuroinflammatory cascades in the brain. Neuronal apoptosis and cognitive impairment are linked to persistently elevated levels of these cytokines (Min'an Zhao et al. 2023).

17.3.5 Mechanisms Linking Gut Dysbiosis to Neurodegeneration

There are multiple ways to conceive how changed gut microbiota and neurodegeneration in AD are related:

Increased Gut Permeability ("Leaky Gut"): When intestinal epithelial cells' tight connections are broken by dysbiosis, inflammatory mediators and microbial metabolites can move into the bloodstream (Fig. 17.2).

Blood-Brain Barrier Disruption: Peripheral cytokines, LPS, and microbial metabolites can enter the central nervous system and cause neurodegeneration when the BBB is weakened by systemic inflammation and circulating toxins.

Microglial Activation: In response to peripheral inflammatory signals, the brain's resident immune cells, known as microglia, become activated and release complement proteins, cytokines, and reactive oxygen species (ROS), which harm synapses and neurons.

Amyloid genesis: Plaque is created when amyloid-β peptides are produced in excess and are not cleared well due to inflammatory signaling.

Tau Pathology: The development of neurofibrillary tangles may be facilitated by oxidative stress and inflammation, which can also increase tau phosphorylation.

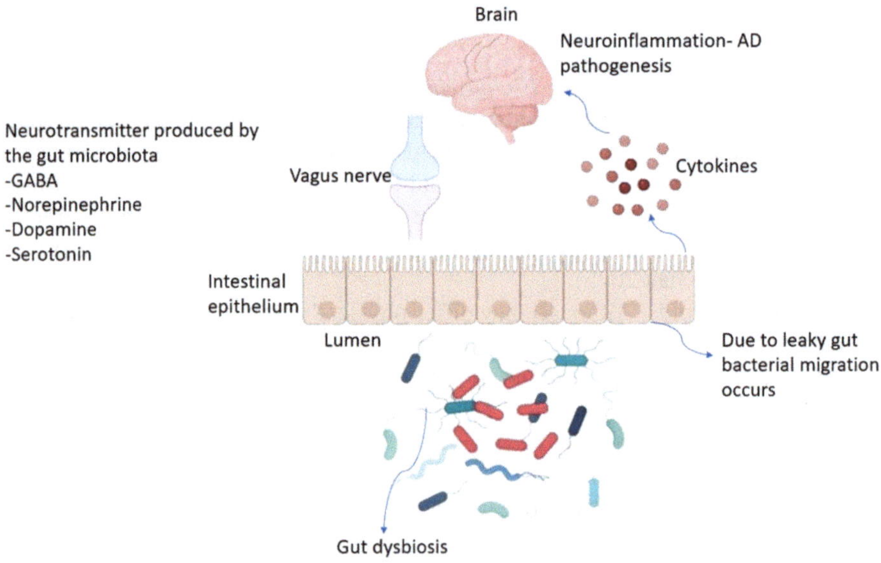

Fig. 17.2 This picture demonstrates the gut-brain axis in Alzheimer's disease by showing how a leaky gut releases microbial compounds into the bloodstream, which in turn causes proinflammatory cytokines. These cause neuroinflammation and microglial activation when they enter the bloodstream and travel through the vagus nerve to the brain. Further impacting brain function and AD progression, it also demonstrates how the gut bacteria communicate with the brain by generating neurotransmitters (such as GABA and serotonin). (This image is created with Biorender.com)

Neurotransmitter Imbalance: In AD patients, dysbiosis can change the availability and production of neurotransmitters such as GABA, serotonin, and dopamine, which can affect behavior, mood, and thought processes (Mou et al. 2022).

17.4 Mechanistic Links Between the Gut Microbiota and Alzheimer's Disease

Developing microbiome-targeted therapies requires an understanding of the molecular mechanisms linking the gut microbiota to Alzheimer's disease. There are several physiological and molecular explanations for how gut dysbiosis may impact AD neurodegenerative processes. These include oxidative stress, mitochondrial dysfunction, tau and amyloid diseases, chronic neuroinflammation, and compromised neurotransmitter signaling (Yang et al. 2024).

17.4.1 Neuroinflammation

Alzheimer's disease is characterized by neuroinflammation, which is becoming more widely acknowledged as a key factor in the development and course of the illness. The regulation of peripheral and central immune responses is significantly influenced by the gut flora.

The intestinal epithelium serves as a selective barrier in a healthy state, keeping pathogenic microorganisms out of the bloodstream. On the other hand, dysbiosis causes "leaky gut," or increased intestinal permeability, which makes it possible for microbial-associated molecular patterns (MAMPs) such peptidoglycans and lipopolysaccharides (LPS) to move over the gut barrier. After being in use, these elements can exert the following effects:

Immune cells (such as monocytes and macrophages) can be activated to produce proinflammatory cytokines (IL-6, TNF-α, and IL-1β). Transcend the blood-brain barrier (BBB), especially if inflammation or aging has already weakened it.

Cause the brain's indigenous immune cells, known as microglia, to become active.

The prolonged release of reactive oxygen species (ROS), nitric oxide (NO), and pro-inflammatory cytokines brought on by chronic microglial activation creates a toxic environment that exacerbates synapse loss, neuronal dysfunction, and cognitive decline.

Furthermore, data from AD mice models indicates that antibiotic-induced gut microbiota depletion enhances cognitive performance and decreases microglial activation, hence confirming a gut-driven mechanism in neuroinflammation (Kinney et al. 2018).

17.4.2 Amyloid and Tau Pathology

The accumulation of extracellular amyloid-β (Aβ) plaques and intracellular neurofibrillary tangles composed of hyperphosphorylated tau proteins are classical pathological hallmarks of AD. Recent findings suggest that the gut microbiota can influence these proteinopathies through several routes:

17.4.2.1 Bacterial Amyloids and Cross-Seeding

Functional amyloids, like curli fibers, are produced by some bacteria, including *Salmonella* and *Escherichia coli*, and they structurally resemble human amyloid proteins. These are amyloids from bacteria: Acting as molecular mimics, they can facilitate cross-seeding, which is the misfolding and aggregation of host Aβ proteins. Activate Toll-like receptor 2 (TLR2) and other innate immune receptors to promote neuroinflammatory signaling (Bloom 2014).

17.4.2.2 Lipopolysaccharide-Induced Amyloid Genesis

It has been demonstrated that LPS, a strong endotoxin produced by Gram-negative bacteria, increases the expression of amyloid precursor protein (APP) and causes neuronal cells to cleave it into Aβ peptides. Additionally, LPS can encourage:

Disruption of microglial phagocytosis and other Aβ clearance mechanisms.

increased tau phosphorylation, possibly as a result of kinases like GSK-3β and CDK5 being activated by cytokines (Brown and Heneka 2024).

17.4.2.3 Bile Acid Metabolism and Tau Pathways

The biotransformation of bile acids, which function as signaling molecules in the central nervous system, depends heavily on gut bacteria. The brain's bile acid receptors, including FXR and TGR5, may be modulated by altered bile acid profiles seen in AD patients, which may impact tau phosphorylation pathways, inflammation, and neuronal energy consumption (Romero-Ramírez and Mey 2024).

17.4.3 Mitochondrial Dysfunction and Oxidative Stress

The maintenance of brain function depends on mitochondrial health, and AD is strongly linked to its degradation. Through a number of pathways, gut dysbiosis can exacerbate mitochondrial dysfunction:

Decreased SCFA production: Butyrate, a crucial colonocyte energy source, influences oxidative phosphorylation and gene expression, which in turn impacts neuronal mitochondrial function. Depletion of butyrate may affect mitochondrial bioenergetics, and it has been demonstrated to have neuroprotective and anti-inflammatory properties.

Increased oxidative stress: Inflammation brought on by dysbiosis produces too many reactive oxygen species (ROS), which harm proteins, lipids, and mitochondrial DNA (mtDNA). This may worsen neuronal stress, hinder ATP synthesis, and encourage apoptotic cell death.

Microbial-derived toxins: Certain dysbiosis bacteria, including Desulfovibrio, produce excess amounts of hydrogen sulfide (H_2S), which can also interfere with mitochondrial activity and redox balance in cerebral tissue. When combined, these disruptions reduce neuronal viability and may hasten the process of neurodegeneration (Das and Ganesh 2023).

17.4.4 Neurotransmitter Regulation

The gut microbiota has the amazing ability to produce, alter, and break down a wide range of neurotransmitters and neuromodulators, which can affect behavior and brain function. These consist of:

17.4.4.1 Gamma-Aminobutyric Acid (GABA)

The main inhibitory neurotransmitter in the brain, GABA, is produced by the bacteria *Lactobacillus* and *Bifidobacterium*. Increased neuronal excitability, anxiety, and cognitive impairment are linked to decreased GABA levels. GABA production may be hindered by a dysbiosis microbiota, which could exacerbate neuropsychiatric symptoms associated with AD (Mhanna et al. 2024).

17.4.4.2 Serotonin (5-HT)

More than 90% of serotonin is produced in the gut, but signals from the microbiota are crucial for its cerebral effects. Tryptophan, the precursor of serotonin, is influenced by the gut flora, which also affects the kynurenine pathway, which is associated to neurotoxicity and cognitive impairment when dysregulated (Emilia Nunzi et al. 2025).

17.4.4.3 Dopamine and Norepinephrine

These catecholamines play a role in executive function, motivation, and attention. Dopamine can be produced by gut microorganisms like *Bacillus* and *Escherichia* species, while norepinephrine synthesis is impacted by microbial manipulation of tyrosine metabolism. Apathy and executive dysfunction in AD may be exacerbated by disruptions in these pathways (Hamamah et al. 2022).

17.4.4.4 Acetylcholine

A vital component of memory and learning, acetylcholine is severely reduced in AD. Acetylcholinesterase, the enzyme that breaks down acetylcholine, may be impacted by microbial byproducts, and certain gut bacteria may have an impact on choline metabolism (Chen et al. 2021).

17.5 Microbiota-Targeted Interventions for Alzheimer's Disease

17.5.1 Probiotics

"Live microorganisms, which, when administered in adequate amounts, confer a health benefit on the host" is how the World Health Organization defines probiotics. Because probiotics can alter the gut-brain axis, there has been a recent spike in interest in employing them as a therapeutic approach for neurological diseases, including Alzheimer's disease (AD). Probiotics may influence a number of pathways implicated in AD pathogenesis, such as neuroinflammation, oxidative stress, and synaptic function, by modifying the makeup of the gut microbiota and reestablishing intestinal homeostasis (Maftei et al. 2024).

Mechanisms of Action Relevant to AD

1. Anti-inflammatory Effects
   Modifying inflammatory responses is one of the most important ways by which probiotics exert their neuroprotective effects.
   Pro-inflammatory cytokine production is decreased by several strains, including *Lactobacillus rhamnosus* and *Bifidobacterium longum* (e.g., IL-6, TNF-α, IL-1β). They reduce microglial activation and inflammation of the central nervous system (CNS) by upregulating the expression of anti-inflammatory cytokines such as IL-10. Probiotics can lessen systemic inflammation and its effects on the central nervous system by strengthening the gut barrier and blocking the translocation of endotoxins like lipopolysaccharide (LPS) (Filidou et al. 2024).
2. Antioxidant Defense
   One of the main causes of AD-related neurodegeneration is oxidative stress. Probiotics:
   Increase the activity of natural antioxidant enzymes such as glutathione peroxidase, catalase, and superoxide dismutase (SOD). Decrease lipid peroxidation indicators such as malondialdehyde (MDA). Preserve the integrity of neurons by mitigating mitochondrial dysfunction through redox balance regulation (Niedzielska et al. 2016).
3. Cognitive Function and Synaptic Plasticity
   Improving synaptic proteins including synapsis and brain-derived neurotrophic factor (BDNF). Decreasing neuronal loss and increasing hippocampus neurogenesis in AD mice models. Adjusting the levels of neurotransmitters that are essential for memory and behavior, such as acetylcholine, GABA, and serotonin (Mostafa et al. 2025).

Preclinical Evidence
Probiotic supplementation has continuously demonstrated positive outcomes in animal models of AD:

In transgenic APP/PS1 mice, *Lactobacillus plantarum* was shown to enhance spatial learning and memory while lowering microglial activation and amyloid-β deposition.

By lowering inflammatory cytokines and raising BDNF expression, *Bifidobacterium breve* was discovered to enhance learning capacity in elderly rats. These benefits were further enhanced by combined probiotics (symbiotics and prebiotic fibers), which increased the development of good bacteria and the synthesis of short-chain fatty acids (SCFAs) like butyrate, which improve neuroprotection (Hu et al. 2024).

Decreased SCFA production: Butyrate, a crucial colonocyte energy source, influences oxidative phosphorylation and gene expression, which in turn impacts neuronal mitochondrial function. Depletion of butyrate may affect mitochondrial bioenergetics, and it has been demonstrated to have neuroprotective and anti-inflammatory properties (Rasoul Mirzaei et al. 2022).

17.5.2 Prebiotics and Synbiotics

Nondigestible fibers known as prebiotics specifically nourish good gut flora. Fructooligosaccharides (FOS), galactooligosaccharides (GOS), and inulin are typical examples. Synbiotics are a combination of probiotics and prebiotics that work in concert to improve the generation of SCFAs, lower inflammation, and maintain the integrity of the gut barrier (Davani-Davari et al. 2019).

17.5.3 Dietary Interventions

A healthy microbiome and a lower risk of AD have been linked to the Mediterranean diet, which is high in polyphenols, omega-3 fatty acids, fiber, and fermented foods.

- By altering metabolism and modifying the makeup of the microbiota, ketogenic diets may help lessen the buildup of Aβ.
- Animal models have demonstrated that intermittent fasting has neuroprotective effects and positively changes gut microbial patterns (Dissanayaka et al. 2024).

17.5.4 Fecal Microbiota Transplantation (FMT)

The purpose of fecal microbiota transplantation (FMT), a therapeutic procedure, is to restore a diverse and balanced gut microbiome by infusing a processed feces sample from a healthy donor into the recipient's gastrointestinal tract. Since gut dysbiosis has been linked more and more to the pathogenesis of complicated neurological illnesses like Alzheimer's disease (AD), FMT may be useful in treating these infections, even though it is most frequently used to treat recurrent *Clostridium difficile* infections (Gupta et al. 2016).

Rationale Behind FMT in AD

The idea behind FMT in AD is based on the mounting data that connects changed gut microbiota to:

Chronic systemic and neuroinflammation; increased permeability of the gut and blood-brain barrier; and dysregulated microbial metabolite synthesis, including bile acids, lipopolysaccharides (LPS), and short-chain fatty acids (SCFAs) (Biazzo and Deidda 2022).

- Aggregation of amyloid-beta (Aβ) and hyperphosphorylation of tau, by substituting a healthy, varied microbial population for a dysbiosis microbiota, FMT seeks to:
- Restore immunological and intestinal homeostasis.
- Modify neuroimmune signaling.
- Improve gut-brain connection and maybe lessen AD-related neuropathology by normalizing microbial metabolite patterns (Kim et al. 2020).

Evidence from Animal Studies

FMT has the ability to alter neuropathology and cognitive function in animal models of Alzheimer's disease, according to several preclinical studies:

1. **Cognitive Improvement**
   FMT from healthy mice into transgenic AD mouse models (e.g., APP/PS1 mice) has resulted in notable enhancements in:
   - Memory and spatial learning (as measured by the Y-maze and Morris Water Maze tests).
   - The plasticity of hippocampal synapses.
   - Behavioral indicators like anxiety and feelings of depression (Sun et al. 2019).
2. **Reduction in Amyloid Pathology**
   FMT has been linked to decreased amyloid-beta plaque burden and a lower Aβ42/Aβ40 ratio in brain tissues in AD models.
   - Reduced expression of β-secretase (BACE1) and amyloid precursor protein (APP).
3. **Attenuation of Neuroinflammation**
   FMT has immunomodulatory effects as well:
   TNF-α, IL-1β, and IL-6 are among the pro-inflammatory cytokines that are downregulated.
   - Promotes the expression of anti-inflammatory cytokines (such as IL-10).
   - Prevents the activation of microglia and astrocytes, which are important mediators of neuroinflammation.
4. **Restoration of Microbial Diversity**
   FMT from young or healthy donors reduces dangerous pro-inflammatory bacteria like *Escherichia/Shigella* while increasing alpha diversity and beneficial microbial taxa (like *Lactobacillus* and *Bifidobacterium*) (Zhiyan Zou et al. 2025).

17.5.5 Antibiotic Stewardship

Broad-spectrum antibiotic use can aggravate dysbiosis and deplete beneficial species, even if some medications can temporarily lower harmful microorganisms.

- Although research on humans is still unclear, targeted, brief use of antibiotics in animal models has demonstrated decreased Aβ deposition (Patangia et al. 2022).

17.6 Challenges and Future Perspectives

Although the gut-brain axis has a lot of intriguing potential, there are a few issues that need to be resolved:

- Treatment generalization is made more difficult by individual differences in microbiota makeup.
- Standardized procedures for probiotic supplementation and FMT are lacking.
- Lack of clarity regarding causality: it's not apparent if gut dysbiosis causes or results from AD.
- Regulatory obstacles and ethical issues with microbiota-based treatments (Franco et al. 2024).

17.7 Conclusion

An innovative paradigm for comprehending and treating Alzheimer's disease is the gut-brain axis. Changing the gut microbiota with food, probiotics, prebiotics, and new treatments like FMT may open the door to new, multifaceted therapy modalities. In the upcoming decades, tailored gut-brain therapies could provide more efficient, comprehensive, and long-lasting treatment of AD as our knowledge of the microbiome grows.

References

Agus A, Planchais J, Sokol H (2018) Gut microbiota regulation of tryptophan metabolism in health and disease. Cell Host Microbe 23(6):716–724. https://doi.org/10.1016/j.chom.2018.05.003

Aljeradat B, Kumar D, Abdulmuizz S, Kundu M, Almealawy YF, Batarseh DR, Atallah O, Ennabe M, Alsarafandi M, Alan A, Weinand M (2024) Neuromodulation and the gut–brain axis: therapeutic mechanisms and implications for gastrointestinal and neurological disorders. Pathophysiology 31(2):244–268. https://doi.org/10.3390/pathophysiology31020019

Białoń MN, Górka DHNOZD, Górka MM (2024) The brain-gut axis: communication mechanisms and the role of the microbiome as a neuroprotective factor in the development of neurodegenerative diseases: a literature overview. AIMS Neurosci 11(3):289–311. https://doi.org/10.3934/Neuroscience.2024019

Biazzo M, Deidda G (2022) Fecal microbiota transplantation as new therapeutic avenue for human diseases. J Clin Med 11(14):4119. https://doi.org/10.3390/jcm11144119

Bloom GS (2014) Amyloid-β and tau: the trigger and bullet in Alzheimer disease pathogenesis. JAMA Neurol 71(4):505–508. https://doi.org/10.1001/jamaneurol.2013.5847

Brown JM, Hazen SL (2017) Targeting of microbe-derived metabolites to improve human health: the next frontier for drug discovery. J Biol Chem 292(21):8560–8568. https://doi.org/10.1074/jbc.R116.765388

Brown GC, Heneka MT (2024) The endotoxin hypothesis of Alzheimer's disease. Mol Neurodegener 19(1):30. https://doi.org/10.1186/s13024-024-00722-y

Browning KN, Verheijden S, Boeckxstaens GE (2017) The vagus nerve in appetite regulation, mood, and intestinal inflammation. Gastroenterology 152(4):730–744. https://doi.org/10.1053/j.gastro.2016.10.046

Carabotti M, Scirocco A, Maselli MA, Severi C (2015) The gut-brain axis: interactions between enteric microbiota, central and enteric nervous systems. Ann Gastroenterol 28(2):203–209. https://www.ncbi.nlm.nih.gov/pmc/articles/PMC4367209/

Chandra S, Sisodia SS, Vassar RJ (2023) The gut microbiome in Alzheimer's disease: what we know and what remains to be explored. Mol Neurodegener 18(1):9. https://doi.org/10.1186/s13024-023-00595-7

Chaudhry TS, Senapati SG, Gadam S, Mannam HPSS, Voruganti HV, Abbasi Z, Abhinav T, Challa AB, Pallipamu N, Bheemisetty N, Arunachalam SP (2023) The impact of microbiota on the gut-brain axis: examining the complex interplay and implications. J Clin Med 12(16):5231. https://doi.org/10.3390/jcm12165231

Chen X, Drew J, Berney W, Lei W (2021) Neuroprotective natural products for Alzheimer's disease. Cells 10(6):1309. https://doi.org/10.3390/cells10061309

Colella M, Charitos IA, Ballini A, Cafiero C, Topi S, Palmirotta R, Santacroce L (2023) Microbiota revolution: how gut microbes regulate our lives. World J Gastroenterol 29(28):4368–4383. https://doi.org/10.3748/wjg.v29.i28.4368

Congdon EE, Ji C, Tetlow AM, Jiang Y, Sigurdsson EM (2023) Tau-targeting therapies for Alzheimer disease: current status and future directions. Nat Rev Neurol 19(12):715–736. https://doi.org/10.1038/s41582-023-00883-2

Das TK, Ganesh BP (2023) Interlink between the gut microbiota and inflammation in the context of oxidative stress in Alzheimer's disease progression. Gut Microbes 15(1):2206504. https://doi.org/10.1080/19490976.2023.2206504

Davani-Davari D, Negahdaripour M, Karimzadeh I, Seifan M, Mohkam M, Masoumi SJ, Berenjian A, Ghasemi Y (2019) Prebiotics: definition, types, sources, mechanisms, and clinical applications. Foods (Basel, Switzerland) 8(3):92. https://doi.org/10.3390/foods8030092

Dissanayaka DMS, Jayasena V, Rainey-Smith SR, Martins RN, Fernando WMADB (2024) The role of diet and gut microbiota in Alzheimer's disease. Nutrients 16(3):412. https://doi.org/10.3390/nu16030412

Faraji N, Payami B, Ebadpour N, Gorji A (2025) Vagus nerve stimulation and gut microbiota interactions: A novel therapeutic avenue for neuropsychiatric disorders, 105990., ISSN 0149-7634. Neurosci Biobehav Rev 169. https://doi.org/10.1016/j.neubiorev.2024.105990

Farhana A, Khan YS (2025) Biochemistry, lipopolysaccharide [Updated 2023 Apr 17]. In: StatPearls [Internet]. StatPearls Publishing, Treasure Island (FL). Available from: https://www.ncbi.nlm.nih.gov/books/NBK554414/

Filidou E, Kandilogiannakis L, Shrewsbury A, Kolios G, Kotzampassi K (2024) Probiotics: shaping the gut immunological responses. World J Gastroenterol 30(15):2096–2108. https://doi.org/10.3748/wjg.v30.i15.2096

Franco CD, Sagar RS, Bokhari SFH (2024) From microbes to memories: challenges and future perspectives regarding the gut-brain Axis for improved cognitive health in Alzheimer's. Cureus 16(1):e52795. https://doi.org/10.7759/cureus.52795

Gao K, Mu CL, Farzi A, Zhu WY (2020) Tryptophan metabolism: a link between the gut microbiota and brain. Adv. Nutr. (Bethesda, Md) 11(3):709–723. https://doi.org/10.1093/advances/nmz127

Grabrucker S, Marizzoni M, Silajdžić E, Lopizzo N, Mombelli E, Nicolas S, Dohm-Hansen S, Scassellati C, Moretti DV, Rosa M, Hoffmann K, Cryan JF, O'Leary OF, English JA, Lavelle

A, O'Neill C, Thuret S, Cattaneo A, Nolan YM (2023) Microbiota from Alzheimer's patients induce deficits in cognition and hippocampal neurogenesis. Brain 146(12):4916–4934. https://doi.org/10.1093/brain/awad303

Gupta S, Allen-Vercoe E, Petrof EO (2016) Fecal microbiota transplantation: in perspective. Ther Adv Gastroenterol 9(2):229–239. https://doi.org/10.1177/1756283X15607414

Hamamah S, Aghazarian A, Nazaryan A, Hajnal A, Covasa M (2022) Role of microbiota-gut-brain Axis in regulating dopaminergic signaling. Biomedicine 10(2):436. https://doi.org/10.3390/biomedicines10020436

Herman JP, McKlveen JM, Ghosal S, Kopp B, Wulsin A, Makinson R, Scheimann J, Myers B (2016) Regulation of the hypothalamic-pituitary-adrenocortical stress response. Compr Physiol 6(2):603–621. https://doi.org/10.1002/cphy.c150015

Hou K, Wu ZX, Chen XY, Wang JQ, Zhang D, Xiao C, Zhu D, Koya JB, Wei L, Li J, Chen ZS (2022) Microbiota in health and diseases. Signal Transduct Target Ther 7(1):135. https://doi.org/10.1038/s41392-022-00974-4

Hu F, Gao Q, Zheng C et al (2024) Encapsulated *lactiplantibacillus plantarum* improves Alzheimer's symptoms in APP/PS1 mice. J Nanobiotechnol 22:582. https://doi.org/10.1186/s12951-024-02862-1

Kandpal M, Indari O, Baral B, Jakhmola S, Tiwari D, Bhandari V, Pandey RK, Bala K, Sonawane A, Jha HC (2022) Dysbiosis of gut microbiota from the perspective of the gut-brain axis: role in the provocation of neurological disorders. Meta 12(11):1064. https://doi.org/10.3390/metabo12111064

Kim MS, Kim Y, Choi H, Kim W, Park S, Lee D, Kim DK, Kim HJ, Choi H, Hyun DW, Lee JY, Choi EY, Lee DS, Bae JW, Mook-Jung I (2020) Transfer of a healthy microbiota reduces amyloid and tau pathology in an Alzheimer's disease animal model. Gut 69(2):283–294. https://doi.org/10.1136/gutjnl-2018-317431

Kinney JW, Bemiller SM, Murtishaw AS, Leisgang AM, Salazar AM, Lamb BT (2018) Inflammation as a central mechanism in Alzheimer's disease. Alzheimers Dement (New York, N. Y.) 4:575–590. https://doi.org/10.1016/j.trci.2018.06.014

Kowalski K, Mulak A (2019) Brain-gut-microbiota Axis in Alzheimer's disease. J Neurogastroenterol Motil 25(1):48–60. https://doi.org/10.5056/jnm18087

Kumar A, Sidhu J, Lui F et al (2025) Alzheimer Disease. [Updated 2024 Feb 12]. In: StatPearls [Internet]. StatPearls Publishing, Treasure Island (FL). Available from: https://www.ncbi.nlm.nih.gov/books/NBK499922/?

GBD 2019 Dementia Forecasting Collaborators(2022) Estimation of the global prevalence of dementia in 2019 and forecasted prevalence in 2050: an analysis for the global burden of disease study 2019. Lancet Public Health 7(2):e105–e125. https://doi.org/10.1016/S2468-2667(21)00249-8

Li C, Niu Z, Zou M, Liu S, Wang M, Gu X, Lu H, Tian H, Jha R (2020) Probiotics, prebiotics, and synbiotics regulate the intestinal microbiota differentially and restore the relative abundance of specific gut microorganisms. J Dairy Sci 103(7):5816–5829. https://doi.org/10.3168/jds.2019-18003

Li H, Cui X, Lin Y, Huang F, Tian A, Zhang R (2024) Gut microbiota changes in patients with Alzheimer's disease spectrum based on *16S rRNA* sequencing: a systematic review and meta-analysis. Front Aging Neurosci 16:1422350. https://doi.org/10.3389/fnagi.2024.1422350

Liu P, Wu L, Peng G, Han Y, Tang R, Ge J, Zhang L, Jia L, Yue S, Zhou K, Li L, Luo B, Wang B (2019) Altered microbiomes distinguish Alzheimer's disease from amnestic mild cognitive impairment and health in a Chinese cohort. Brain Behav Immun 80:633–643. https://doi.org/10.1016/j.bbi.2019.05.008

Lu S, Zhao Q, Guan Y, Sun Z, Li W, Guo S, Zhang A (2024) The communication mechanism of the gut-brain axis and its effect on central nervous system diseases: a systematic review. Biomed Pharmacothe 178:117207. https://doi.org/10.1016/j.biopha.2024.117207

Maftei NM, Raileanu CR, Balta AA, Ambrose L, Boev M, Marin DB, Lisa EL (2024) The potential impact of probiotics on human health: an update on their health-promoting properties. Microorganisms 12(2):234. https://doi.org/10.3390/microorganisms12020234

Merra G, Noce A, Marrone G, Cintoni M, Tarsitano MG, Capacci A, De Lorenzo A (2020) Influence of Mediterranean diet on human gut microbiota. Nutrients 13(1):7. https://doi.org/10.3390/nu13010007

Mhanna A, Martini N, Hmaydoosh G, Hamwi G, Jarjanazi M, Zaifah G, Kazzazo R, Haji Mohamad A, Alshehabi Z (2024) The correlation between gut microbiota and both neurotransmitters and mental disorders: a narrative review. Medicine 103(5):e37114. https://doi.org/10.1097/MD.0000000000037114

Mirzaei R, Dehkhodaie E, Bouzari B, Rahimi M, Gholestani A, Hosseini-Fard SR, Keyvani H, Teimoori A, Karampoor S (2022) Dual role of microbiota-derived short-chain fatty acids on host and pathogen. Biomed Pharmacother 145:112352. https://doi.org/10.1016/j.biopha.2021.112352

Mostafa M, Disouky A, Lazarov O (2025) Therapeutic modulation of neurogenesis to improve hippocampal plasticity and cognition in aging and Alzheimer's disease. Neurotherapeutics 22(3):e00580. https://doi.org/10.1016/j.neurot.2025.e00580

Mou Y, Du Y, Zhou L, Yue J, Hu X, Liu Y, Chen S, Lin X, Zhang G, Xiao H, Dong B (2022) Gut microbiota interact with the brain through systemic chronic inflammation: implications on neuroinflammation, neurodegeneration, and aging. Front Immunol 13:796288. https://doi.org/10.3389/fimmu.2022.796288

Mroke P, Goit R, Rizwan M, Tariq S, Rizwan AW, Umer M, Nassar FF, Torijano Sarria AJ, Singh D, Baig I (2024) Implications of the gut microbiome in Alzheimer's disease: a narrative review. Cureus 16(11):e73681. https://doi.org/10.7759/cureus.73681

Munteanu C, Onose G, Rotariu M, Poştaru M, Turnea M, Galaction AI (2024) Role of microbiota-derived hydrogen sulfide (H_2S) in modulating the gut–brain axis: implications for Alzheimer's and Parkinson's disease pathogenesis. Biomedicine 12(12):2670. https://doi.org/10.3390/biomedicines12122670

Murray ER, Kemp M, Nguyen TT (2022) The microbiota-gut-brain axis in Alzheimer's disease: a review of taxonomic alterations and potential avenues for interventions. Arch Clin Neuropsychol 37(3):595–607. https://doi.org/10.1093/arclin/acac008

Nichols E, Szoeke CEI, Vollset SE, Abbasi N, Abd-Allah F, Abdela J, Murray CJL (2019) Global, regional, and national burden of Alzheimer's disease and other dementias, 1990–2016: a systematic analysis for the global Burden of Disease Study 2016. Lancet Neurol 18(1):88–106. https://doi.org/10.1016/S1474-4422(18)30403-4

Niedzielska E, Smaga I, Gawlik M et al (2016) Oxidative stress in neurodegenerative diseases. Mol Neurobiol 53:4094–4125. https://doi.org/10.1007/s12035-015-9337-5

Nunzi E, Pariano M, Costantini C, Garaci E, Puccetti P, Romani L (2025) Host-microbe serotonin metabolism. Trends Endocrinol Metab 36(1):83–95. https://doi.org/10.1016/j.tem.2024.07.014

Patangia DV, Anthony Ryan C, Dempsey E, Paul Ross R, Stanton C (2022) Impact of antibiotics on the human microbiome and consequences for host health. Microbiol Open 11(1):e1260. https://doi.org/10.1002/mbo3.1260

Rawat P, Sehar U, Bisht J, Selman A, Culberson J, Reddy PH (2022) Phosphorylated tau in Alzheimer's disease and other tauopathies. Int J Mol Sci 23(21):12841. https://doi.org/10.3390/ijms232112841

Romero-Ramírez L, Mey J (2024) Emerging roles of bile acids and TGR5 in the central nervous system: molecular functions and therapeutic implications. Int J Mol Sci 25(17):9279. https://doi.org/10.3390/ijms25179279

Rosendo-Silva D, Viana S, Carvalho E, Reis F, Matafome P (2023) Are gut dysbiosis, barrier disruption, and endotoxemia related to adipose tissue dysfunction in metabolic disorders? Overview of the mechanisms involved. Intern Emerg Med 18(5):1287–1302. https://doi.org/10.1007/s11739-023-03262-3

Rusch JA, Layden BT, Dugas LR (2023) Signalling cognition: the gut microbiota and hypothalamic-pituitary-adrenal axis. Front Endocrinol 14:1130689. https://doi.org/10.3389/fendo.2023.1130689

Ruth MR, Field CJ (2013) The immune modifying effects of amino acids on gut-associated lymphoid tissue. J Anim Sci Biotechnol 4:27. https://doi.org/10.1186/2049-1891-4-27

Safarchi A, Al-Qadami G, Tran CD, Conlon M (2025) Understanding dysbiosis and resilience in the human gut microbiome: biomarkers, interventions, and challenges. Front Microbiol 16:1559521. https://doi.org/10.3389/fmicb.2025.1559521

Sampson TR, Challis C, Jain N, Moiseyenko A, Ladinsky MS, Shastri GG, Thron T, Needham BD, Horvath I, Debelius JW, Janssen S, Knight R, Wittung-Stafshede P, Gradinaru V, Chapman M, Mazmanian SK (2020) A gut bacterial amyloid promotes α-synuclein aggregation and motor impairment in mice. *eLife* 9:e53111. https://doi.org/10.7554/eLife.53111

Sowmiya S, Dhivya LS, Rajendran P, Harikrishnan N, Ankul Singh S (2024) Exploring the potential of probiotics in Alzheimer's disease and gut dysbiosis. IBRO Neurosci Rep 17:441–455. https://doi.org/10.1016/j.ibneur.2024.11.004

Sun J, Xu J, Ling Y et al (2019) Fecal microbiota transplantation alleviated Alzheimer's disease-like pathogenesis in APP/PS1 transgenic mice. Transl Psychiatry 9:189. https://doi.org/10.1038/s41398-019-0525-3

Thau L, Reddy V, Singh P (2025) Anatomy, central nervous system [Updated 2022 Oct 10]. In: StatPearls [Internet]. Treasure Island (FL), StatPearls Publishing. Available from: https://www.ncbi.nlm.nih.gov/books/NBK542179/

Tran SM, Mohajeri MH (2021) The role of gut bacterial metabolites in brain development, aging and disease. Nutrients 13(3):732. https://doi.org/10.3390/nu13030732

Waxenbaum JA, Reddy V, Varacallo MA (2025) Anatomy, autonomic nervous system. [Updated 2023 July 24]. In: StatPearls [Internet], Treasure Island (FL), StatPearls Publishing. Available from: https://www.ncbi.nlm.nih.gov/books/NBK539845/

Wu T, Lin D, Cheng Y, Jiang S, Riaz MW, Fu N, Mou C, Ye M, Zheng Y (2022) Amyloid cascade hypothesis for the treatment of Alzheimer's disease: progress and challenges. Aging Dis 13(6):1745–1758. https://doi.org/10.14336/AD.2022.0412

Yadav P, Lee YH, Panday H, Kant S, Bajwa N, Parashar R, Jha SK, Jha NK, Nand P, Lee SS, Jha AK (2022) Implications of microorganisms in Alzheimer's disease. Curr Issues Mol Biol 44(10):4584–4615. https://doi.org/10.3390/cimb44100314

Yang J, Liang J, Hu N, He N, Liu B, Liu G, Qin Y (2024) The gut microbiota modulates neuroinflammation in Alzheimer's disease: elucidating crucial factors and mechanistic underpinnings. CNS Neurosci Ther 30(10):e70091. https://doi.org/10.1111/cns.70091

Zhao M, Chu J, Feng S, Guo C, Xue B, He K, Li L (2023) Immunological mechanisms of inflammatory diseases caused by gut microbiota dysbiosis: a review. Biomed Pharmacother 164:114985. https://doi.org/10.1016/j.biopha.2023.114985

Zhou SY, Guo ZN, Yang Y, Qu Y, Jin H (2023) Gut-brain axis: mechanisms and potential therapeutic strategies for ischemic stroke through immune functions. Front Neurosci 17:1081347. https://doi.org/10.3389/fnins.2023.1081347

Zikou E, Koliaki C, Makrilakis K (2024) The role of fecal microbiota transplantation (FMT) in the management of metabolic diseases in humans: a narrative review. Biomedicine 12(8):1871. https://doi.org/10.3390/biomedicines12081871

Zou Z, Lei D, Yin Y, Xu R, Luo H, Chen T, Liu M, Li X (2025) Frequent fecal microbiota transplantation improves cognitive impairment and pathological changes in Alzheimer's disease FAD4T mice via the microbiota-gut-brain axis. Heliyon 11:e42925. https://doi.org/10.1016/j.heliyon.2025.e42925. https://www.researchgate.net/publication/389328876_Frequent_fecal_microbiota_transplantation_improves_cognitive_impairment_and_pathological_changes_in_Alzheimer's_disease_FAD4T_mice_via_the_microbiota-gut-brain_axis

Exploring the Neuroprotective Potential of Extra Virgin Olive Oil: A Dietary Approach to Mitigating Alzheimer's Disease

18

Monika Sachdeva

Abstract

A promising dietary intervention for supporting brain function and maybe slowing the course of Alzheimer's disease (AD) is extra virgin olive oil (EVOO). Packed with monounsaturated fatty acids, especially oleic acid, EVOO supports neuronal function and communication by preserving the integrity of brain cell membranes. In addition to its beneficial fat content, EVOO has potent polyphenolic chemicals with high anti-inflammatory and antioxidant qualities, including oleeuropein, hydroxytyrosol, and oleocanthal. By neutralizing reactive oxygen species and reducing inflammatory responses in the brain, these bioactives fight oxidative stress and chronic neuroinflammation, two major factors that contribute to AD pathogenesis. According to research, EVOO helps the brain's glymphatic system remove amyloid-beta, which may decrease the buildup of plaque, a defining characteristic of AD. It also seems to control the phosphorylation of tau proteins, which lessens the development of neurofibrillary tangles that disrupt brain transmission and synaptic function. Additionally, EVOO promotes neurogenesis and cognitive resilience by increasing brain-derived neurotrophic factor (BDNF) levels and synaptic plasticity. Existing research shows that adding high-quality EVOO to a balanced diet, especially the Mediterranean dietary pattern, offers a useful, natural strategy for long-term brain health, even if extensive human trials are still being conducted. When combined with other healthy lifestyle choices like exercise and mental stimulation, EVOO might be a convenient and successful dietary approach to lower the risk of AD and promote cognitive function. This study provides insights into how dietary choices might affect brain health by highlighting EVOO's distinct nutritional profile, neuroprotective processes, and possible involvement in preventing and delaying the course of

M. Sachdeva (✉)
Fatima College of Health Sciences, Alain, UAE
e-mail: Monika.sachdeva@actvet.gov.ae

J. Malik et al. (eds.), *Diet and Alzheimer's Disease: Let Food be Our Medicine*,
Nutritional Neurosciences, https://doi.org/10.1007/978-981-95-2736-6_18

AD. Knowing how the bioactive ingredients in EVOO work together may lead to novel dietary treatments for neurodegenerative illnesses, making EVOO an essential part of brain-healthy eating plans.

Keywords

Alzheimer's disease · Antioxidants · Brain health · Extra virgin olive oil · Neuroprotection · Polyphenols

18.1 Introduction

The progressive neurodegenerative condition known as Alzheimer's disease (AD) affects a person's memory, thinking, and behavior, which eventually affects their capacity to do daily tasks. Alzheimer's is becoming more common as life expectancy rises worldwide, making it a serious public health concern. Although pharmaceutical therapies are the subject of much investigation, dietary changes that may help prevent or reduce the course of illness are receiving more attention (Leri et al. 2023). Extra virgin olive oil (EVOO), which is well known for its health advantages, particularly in the context of the Mediterranean diet, is one such potential dietary component. Polyphenolic chemicals and monounsaturated fatty acids (MUFAs), which are abundant in olive oil, have anti-inflammatory, antioxidant, and neuroprotective qualities. These features have raised interest among scientists in olive oil's ability to counteract the main symptoms of Alzheimer's disease, such as the buildup of amyloid-beta plaque, oxidative stress, tau protein tangles, and neuroinflammation. Additionally, people following a Mediterranean diet, in which olive oil is the main source of fat, have demonstrated reduced incidence of Alzheimer's disease and cognitive decline. Examining the molecular makeup, mechanisms of action, and new data from preclinical and clinical research, this study investigates the complex link between olive oil and Alzheimer's disease (Guruvaiya and Subramaniam 2023). By understanding how olive oil supports brain health, we can better assess its potential as a natural, accessible strategy for preventing or mitigating AD.

18.1.1 Nutritional Composition of Olive Oil

Extra virgin olive oil (EVOO), in particular, has a special and strong nutritional profile that adds to its many health advantages, including the possibility that it might prevent Alzheimer's disease. Monounsaturated fatty acids (MUFAs) make up the majority of olive oil, with oleic acid being the most prevalent kind. By lowering inflammation and improving cell membrane integrity, oleic acid, a heart-healthy lipid, not only improves cardiovascular function but also fosters brain health (Qusa et al. 2019; Lauretti et al. 2020). For neurons, which depend on stable, fluid membranes for effective communication and signal transmission, this is especially crucial. Furthermore, MUFAs have been linked to decreased oxidative stress and enhanced insulin sensitivity, two characteristics that are becoming more widely acknowledged as being pertinent to neurodegenerative illnesses like Alzheimer's. EVOO is abundant in bioactive substances, especially polyphenols like oleeuropein, hydroxytyrosol, and oleocanthal, in addition to its fatty acid content. As strong antioxidants, these polyphenols aid in scavenging free radicals that lead to oxidative damage, a major factor in brain damage or Alzheimer's disease pathogenesis. Additionally, they have anti-inflammatory properties that lessen the low-grade, chronic inflammation that is frequently seen in Alzheimer's patients. Interestingly, studies have demonstrated that oleocanthal possesses ibuprofen-like qualities, providing natural anti-inflammatory advantages without the negative effects of medications.

The capacity of these polyphenols to penetrate the blood-brain barrier, a selective membrane that protects the brain from dangerous chemicals while permitting vital nutrients to pass through them, makes them particularly fascinating. Because of this capacity, the protective chemicals found in olive oil can interact directly with brain cells, perhaps preventing the development of amyloid-beta plaque, decreasing the aggregation of tau proteins, and enhancing neuronal survival in general (Xu et al. 2018). Olive oil is becoming known as a natural ally in the battle against AD because of its mix of MUFAs and polyphenols, which make it a nutrient-dense, brain-supporting diet.

18.1.2 Extra-Virgin Olive Oil (EVOO) and Its Phenolic Compounds: A Natural Ally Against Alzheimer's Disease

Beyond its flavor, extra-virgin olive oil (EVOO) contains a rich blend of bioactive phenolic chemicals that are becoming more extensively acknowledged for their possible therapeutic benefits in Alzheimer's disease (AD). This wide range of chemicals changes as the olive ages and the oil is extracted, which has a direct impact on the health-promoting qualities of EVOO. Hydroxytyrosol (HT), Oleuropein (OLE), as well as tyrosol compounds, all of which have strong antioxidant along with neuroprotective properties, are important phenolics. Oleuropein aglycone (OLG), a derivative of OLE, which is found in large quantities in olive fruit, has potent anti-inflammatory and anti-amyloid properties (Abdallah et al. 2023; Batarseh and Kaddoumi 2018). Renowned for its remarkable antioxidant potency, HT aids in combating oxidative stress, a primary cause of brain damage in AD. Along with phenolic alcohols like tyrosol and HT, EVOO also includes lesser-known but no less important minor chemicals, such as secoiridoids including oleocanthal (OC) and oleacein (OH-OC). Together, these substances can lessen oxidative damage and persistent neuroinflammation, two major factors in the development of AD. Along with other phenolics including phenolic acids, flavonoids, as well as lignans (such as pinoresinol and acetoxypinoresinol), variations like deacetoxy OLE and ligstroside aglycone further strengthen EVOO's antioxidant repertoire. In the end, this complex web of phenolics helps to halt the neurodegenerative processes typical of AD by actively neutralizing reactive oxygen species (ROS), unstable chemicals that cause

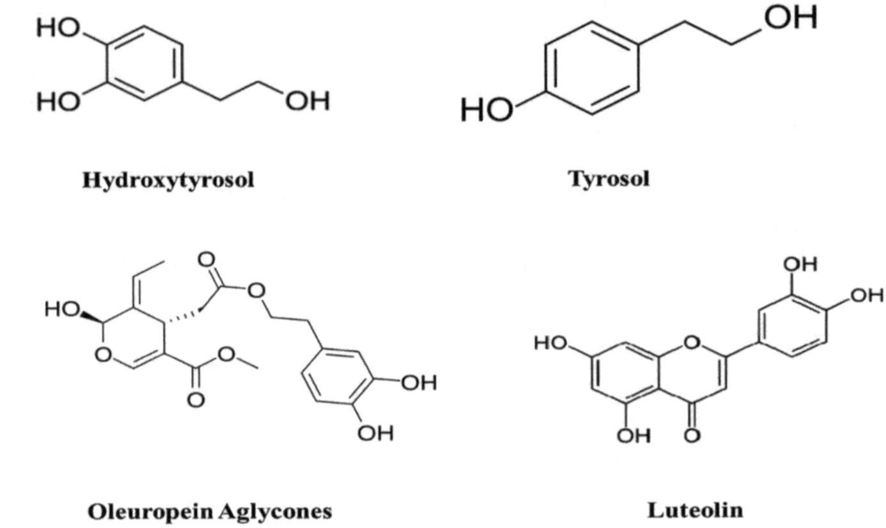

Fig. 18.1 The molecular structures of key phenolic compounds found in EVOO

cellular damage, and shielding neuronal cells from oxidative stress (Casamenti and Stefani 2017). The potential of EVOO as a natural, multifunctional medicinal agent is highlighted by the phenolic profile's extreme diversity and complexity. The function of EVOO in neuroprotection as well as Alzheimer's disease prevention is still an intriguing and developing field of research as scientists continue to understand the complex interactions between these substances.

The molecular structures of the main phenolic compounds found in EVOO are depicted in Fig. 18.1, emphasizing the variety of chemical configurations that give EVOO its neuroprotective qualities. Key ingredients include hydroxytyrosol (HT), tyrosol, and oleuropein (OLE), each of which has a unique function in preventing oxidative stress and neuroinflammation, two important processes in the development of Alzheimer's disease. Secoiridoids that have anti-inflammatory properties similar to those of ibuprofen, such as oleocanthal (OC) along with hydroxyl-oleocanthal (OH-OC), are also shown. The antioxidant potential of EVOO is further increased by phenolic acids, lignans, flavonoids such as pinoresinol and acetoxypinoresinol, or minor phenolic elements like deacetoxy oleuropein aglycone as well as ligstroside aglycone. These compounds' structural variety reflects their multifunctional properties, which include scavenging reactive oxygen species (ROS), blocking amyloid-beta plaque formation, and suppressing tau protein hyperphosphorylation, which is a hallmark of Alzheimer's pathogenesis. This comprehensive illustration highlights the phenolic profile of EVOO's biochemical complexity, offering insight into how these components work in concert to enhance its potential as a natural therapeutic agent in the treatment of neurological diseases.

18.1.3 Effect of Extra-Virgin Olive Oil and Its Phenolics on Alzheimer's Disease Markers

Through several interrelated processes, extra-virgin olive oil (EVOO) as well as its bioactive phenolic components have shown exceptional promise in preventing and reducing the main pathological hallmarks of Alzheimer's disease (AD). Disrupting the aggregation of amyloid-beta peptides, the harmful protein fragments that create plaques in the brains of people with AD, is one of the main ways by which EVOO exerts it protective effects. These substances not only stop amyloid-beta from initially clumping together, but they also improve the brain's natural clearance mechanisms, such as the glymphatic system, to more effectively eliminate these dangerous proteins. This lessens the accumulation of plaques that might otherwise cause neuronal damage and cognitive deterioration. Furthermore, the hyperphosphorylation of tau proteins, which results in the development of neurofibrillary tangles, is inhibited by the phenolic compounds found in EVOO, especially oleuropein and oleocanthal. These tangles hinder communication and ultimately cause cell death by interfering with the vital transport mechanisms in neurons. EVOO supports healthy synaptic function, preserves cognitive functions, and helps to maintain the structural integrity of neurons by controlling tau phosphorylation.

Additionally, EVOO supports the blood-brain barrier, an essential barrier that keeps dangerous chemicals out of the brain. By making this barrier stronger, fewer neurotoxic substances may enter, which would otherwise worsen neurodegeneration. The neuroprotective advantages of EVOO are further enhanced by the polyphenols' strong anti-inflammatory as well as antioxidant qualities. By neutralizing reactive oxygen species (ROS), which lead to cellular degradation, they combat oxidative stress, a significant factor in brain injury. At the same time, they reduce chronic neuroinflammation, which speeds up the development of AD and is frequently brought on by amyloid-beta deposits (Rita et al. 2023; Siddique et al. 2019). Over time, EVOO protects the structure and function of the brain by reducing inflammation and oxidation, which makes the brain less prone to AD progression.

18.1.4 Anti-inflammatory and Antioxidant Effects

Oxidative stress and chronic inflammation are major factors in the development of Alzheimer's disease (AD). Amyloid-beta plaques along with neurofibrillary tangles, two pathological characteristics that impair neuronal transmission and cause brain cell death, are fueled by these processes. Bioactive substances with potent anti-inflammatory and antioxidant qualities found in olive oil, especially EVOO, may be able to reverse these harmful processes. Because of their neuroprotective properties, polyphenols including oleuropein, hydroxytyrosol, as well as oleocanthal are some of the most researched ingredients in olive oil. These substances counteract reactive oxygen species (ROS), which are unstable chemicals that oxidatively damage DNA, proteins, and lipids in cells. By reducing oxidative stress, olive oil helps protect neurons from early degeneration and supports mitochondrial function, which is essential for cellular energy production and survival.

Particularly noteworthy for its distinct anti-inflammatory qualities is oleocanthal. By blocking cyclooxygenase (COX) enzymes, oleocanthal functions similarly to non-steroidal anti-inflammatory medications (NSAIDs), such as ibuprofen, according to studies. By lowering the synthesis of pro-inflammatory chemicals, this inhibition lessens neuroinflammation. Oleocanthal's capacity to regulate inflammation is particularly beneficial, because chronic neuroinflammation is a major component contributing to the acceleration of cognitive loss and the advancement of AD (Abuznait et al. 2013; Al Rihani et al. 2019). Additionally, the polyphenols in olive oil may affect microglial cells, which are the immune cells that exist in the brain and are in charge of removing amyloid-beta plaques. These cells can more successfully carry out their preventive, plaque-clearing roles by lowering inflammation, which may halt the progression of illness. Olive oil's anti-inflammatory and antioxidant properties work together to provide a neuroprotective environment that may boost brain resistance against age-related damage, postpone the onset of Alzheimer's disease, and maintain cognitive performance. This dual-action mechanism underscores why olive oil is a cornerstone of brain-healthy diets, like the Mediterranean diet, which has been linked to lower rates of Alzheimer's disease.

18.1.5 Amyloid-Beta Clearance and Tau Protein Regulation

The accumulation of harmful protein clumps called amyloid-beta plaques, which impair neuronal function and cause extensive inflammation in the brain, is one of the clinical characteristics that distinguish Alzheimer's disease (AD). According to new research, eating olive oil helps the brain's natural process of eliminating these toxic proteins. EVOO has been shown in animal experiments to increase the functioning of the brain's specialized waste elimination network, the lymphatic system. This mechanism, which removes amyloid-beta plaques and other metabolic waste, works best while you sleep. It is thought that the polyphenols in olive oil, especially oleocanthal, improve glymphatic function, speeding up the removal of amyloid-beta and lowering its buildup.

Olive oil appears to regulate tau protein, another major cause of Alzheimer's disease, in addition to amyloid-beta. Normally, tau helps neurons maintain their structural integrity by facilitating the formation of microtubules, which carry messages and nutrients across cells. When tau is inappropriately hyperphosphorylated in Alzheimer's disease, it separates from microtubules and forms tangles that suffocate neurons and ultimately cause cell death. According to research, the beneficial components in olive oil may alter tau phosphorylation, so avoiding the formation of these damaging tangles (Tsolaki et al. 2020; Klimova et al. 2019). Olive oil's ability to treat several neurodegenerative pathways at once is highlighted by its dual-action impact of increasing amyloid-beta clearance and controlling tau.

18.1.6 Supporting Synaptic Function and Neurogenesis

Synaptic plasticity, the brain's capacity to adapt by creating and fortifying connections between neurons, is crucial to preserve normal brain function. Monounsaturated fats found in olive oil, especially oleic acid, support the fluidity and structural integrity of neuronal cell membranes, both of which are necessary for effective signal transmission. Learning, memory, and general cognitive function are supported by neurons that can interact more efficiently, thanks to a flexible, well-maintained membrane.

The polyphenols in olive oil, particularly oleuropein, have been connected to elevated levels of BDNF, a protein essential for neurogenesis, the process by which the brain produces new neurons in addition to preserve cellular integrity. For cognitive resilience, memory formation, as well as post-injury healing processes, BDNF promotes the survival, development, and differentiation of neurons. Cognitive decline and a higher risk of neurodegenerative disorders like Alzheimer's are linked to low BDNF levels. Olive oil may assist maintain neurogenesis along with synaptic function by promoting the creation of BDNF, which would reverse the brain degradation observed in Alzheimer's patients (Siddique et al. 2020). When taken as a whole, these outcomes imply that olive oil does more than simply shield the brain from harm; it also actively promotes the brain's capacity to repair, adapt, and remain resilient in the face of Alzheimer's disease's increasing obstacles. This regenerative potential is a key reason why olive oil continues to attract attention in neuroscience and nutritional research.

18.1.7 The Mediterranean Diet and Cognitive Health

The Mediterranean diet, which is often praised for fostering general health and longevity, is based on olive oil. This diet has attracted a lot of interest in the context of brain health, since it may offer protection against Alzheimer's disease (AD) and cognitive decline. The Mediterranean diet, which is high in fruits, vegetables, whole grains, nuts, fish, and most importantly EVOO, provides a potent blend of anti-inflammatory, healthy fat, and antioxidant chemicals that all work together to enhance brain function. The study examined the benefits of a Mediterranean diet supplemented with nuts or EVOO versus a low-fat diet control group. Results demonstrated that participants who consumed the olive-oil-enhanced Mediterranean diet showed significant improvements in memory, executive function, and cognitive flexibility, key components of brain performance. Moreover, they exhibited a reduced risk of developing mild cognitive impairment (MCI), a common early indicator of Alzheimer's.

It is believed that the many synergistic effects of the diet contribute to its cognitive advantages. While polyphenols fight against oxidative stress and neuroinflammation, two factors strongly linked to Alzheimer's disease, the monounsaturated fats in olive oil enhance vascular health and guarantee sufficient blood flow to the brain. The Mediterranean diet also promotes healthy cholesterol and blood sugar levels, which lower risk factors that eventually lead to cognitive deterioration. One of the most successful, scientifically supported dietary interventions for maintaining cognitive health and possibly postponing the onset of Alzheimer's disease is the Mediterranean diet, which is fueled by the potent bioactive compounds in olive oil and promotes the consumption of nutrients that are good for the brain while guarding against Alzheimer's disease.

18.1.8 Human and Animal Studies

Both animal and human studies provide compelling evidence supporting olive oil's role in protecting against Alzheimer's disease (AD). In animal models of AD, diets supplemented with EVOO consistently show improvements in cognitive performance, alongside a marked reduction in the buildup of amyloid-beta plaques, a hallmark of Alzheimer's pathology. These studies demonstrate the enhanced function of the glymphatic system (the brain's waste-clearance system), improved synaptic integrity, and reduced neuroinflammation in animals fed with EVOO-enriched diets. For example, a study involving transgenic mice prone to developing Alzheimer's-like symptoms found that animals consuming an olive oil-rich diet displayed improved spatial memory and learning abilities compared to control groups. Brain tissue analysis further revealed lower tau protein tangles and better synaptic function, suggesting that olive oil's bioactive components directly support neuronal health and resilience.

Human trials, though more complex due to genetic and environmental variables, are beginning to yield similarly encouraging results. These individuals demonstrated significant cognitive improvements particularly in memory recall and processing speed compared to those on a standard diet. Moreover, biomarker analysis indicated reduced levels of phosphorylated tau proteins and inflammatory markers associated with neurodegeneration, suggesting that olive oil's protective effects extend beyond cognition to the underlying biological processes driving Alzheimer's progression (Cordero et al. 2018; Kaddoumi et al. 2022). While larger, long-term human studies are needed to solidify these findings, the growing body of evidence from both animal models and clinical research underscores olive oil's potential as a powerful, accessible dietary intervention to slow or prevent Alzheimer's disease, making it an essential component in brain health strategies.

18.2 Discussion

18.2.1 Role of Vitamins in Alzheimer's Disease

18.2.1.1 Antioxidant Vitamins and AD

Vitamins that include antioxidants protect the body from "free radicals," which damage cells and can cause heart disease, cancer, as well as Alzheimer's disease. There are several antioxidants, such as resveratrol, beta-carotene, and vitamins C and E. They are present in plant-based foods including greens, berries, along with bell peppers. Resveratrol can be found in peanuts, grapes, red wine, and some types of dark chocolate. According to scientific research, it may have antiaging properties and lower the risk of contracting specific diseases. Researchers thought for a long that resveratrol might contain compounds able to protect the brain against AD symptoms. Daily resveratrol doses did slow the advancement of the illness, according to new research. Free radicals commonly build up in nerve cells as people age. Increasing oxidative stress is one of the factors contributing to the onset and advancement of neurodegenerative processes. Studying Alzheimer's sufferers' brains has revealed that the body has been exhibiting signs of oxidative stress to stop free radical damage. Consequently, it would appear that increasing antioxidant intake would be advantageous.

Although experts are unsure which antioxidants are superior to others, eating meals is probably a better approach to obtaining antioxidants than taking tablets. The probable association between dietary antioxidants as well as AD may be explained by several metabolic mechanisms. Age-related oxidative stress on brain tissue as well as lymphocytes, increased free radical buildup, and weakened antioxidant systems have all been shown. In the first place, antioxidants may lessen the level of oxidative stress in the brain. Antioxidants may thereby reduce the amount of amyloid buildup, neuronal cell death, and DNA degradation in the brain (Rigacci 2015). Second, since atherosclerosis and oxidative processes are linked to atherogenesis and AD is linked to cardiovascular risk factors, a high antioxidant intake may lower the risk of dementia by lowering the risk of atherosclerosis. Dementia risk may be decreased by preventing the onset of these events, which are all significant neuropathological characteristics of AD. The pleiotropic actions of antioxidants, which include decreasing oxidized lipid membranes, avoiding nucleic acid damage, and inhibiting protein carbonylation, make it difficult to identify a single mechanism that is disturbed and leads to the development of AD. Positive outcomes have been shown in the prevention and treatment of AD with a class of vitamins.

18.2.2 Categories of Vitamins Having a Role in Alzheimer's Disease (AD)

Figure 18.2 illustrates the key categories of vitamins that play a crucial role in Alzheimer's disease (AD) prevention and management. These include fat-soluble vitamins like vitamin E, known for its potent antioxidant properties that help protect neurons from oxidative damage, and vitamin D, which supports neuroprotection and reduces inflammation. Water-soluble vitamins, particularly B vitamins such as B6, B9 (folate), and B12, are essential for regulating homocysteine levels an amino acid linked to cognitive decline when elevated. Vitamin C, another powerful water-soluble antioxidant, further aids in reducing oxidative stress and supporting the immune system's role in brain health. Together, these vitamin categories contribute to maintaining neuronal integrity, supporting cognitive function, and mitigating the

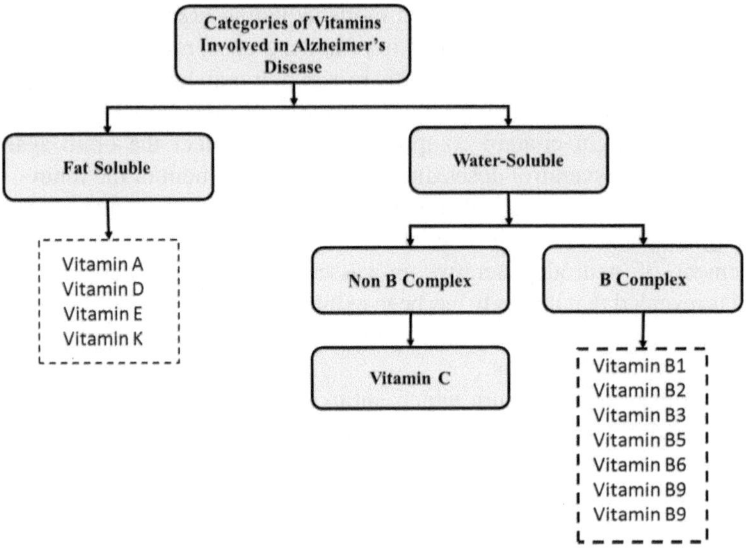

Fig. 18.2 Several categories of vitamins are recognized for their potential in alleviating AD

pathological processes associated with AD, underscoring their importance in both dietary strategies and potential therapeutic approaches.

Vitamins are necessary to keep the brain healthy, and a higher risk of AD is increasingly linked to vitamin deficiencies. Numerous vitamin types help maintain cognitive function, safeguard neurons, and fight against the underlying processes that cause AD, including neuroinflammation, oxidative stress, as well as the accumulation of amyloid-beta plaques. Lowering excessive levels of homocysteine, an amino acid linked to an increased risk of AD, requires B vitamins, particularly B9, B6, (folate), as well as B12. They also promote the production of neurotransmitters and healthy neuron activity, both of which enhance memory and cognitive abilities. Because of its potent antioxidant qualities, vitamin E helps scavenge dangerous free radicals and shield brain cells from oxidative damage, a major cause of neurodegeneration (Silva et al. 2023). Vitamin D has emerged as a vital player in neuroprotection, with studies suggesting it supports immune function in the brain, reduces inflammation, and promotes clearance of amyloid plaques. Additionally, vitamin C enhances antioxidant defenses and supports vascular health, which is critical for maintaining adequate blood flow to the brain. Together, these vitamins form a protective network that supports brain resilience. Long-term brain health may be promoted by ensuring sufficient consumption of these vital nutrients through food or supplements, which may help moderate cognitive decline as well as lower the risk of AD.

18.2.2.1 B Vitamins (B1, B6, B9, B12)

Because of their function in controlling the metabolism of homocysteine, an amino acid that, when high, is associated with neurotoxicity as well as an increased risk of AD, B vitamins are crucial for preserving brain health. Every B vitamin has a distinct role in promoting cognitive performance and preventing neuronal damage. Thiamine, or vitamin B1, is essential for neurons to produce energy. Given the high energy requirements of brain cells, it facilitates the conversion of glucose into adenosine triphosphate (ATP), the main source of energy for cells. This mechanism is disturbed by thiamine deficit, which results in neuronal energy failure, memory loss, and other hallmarks of AD. In fact, research has shown that thiamine deficiencies are commonly observed in AD patients, contributing to cognitive decline and brain atrophy.

Vitamin B6 (Pyridoxine) is vital for regulating homocysteine levels, preventing vascular damage that can impair blood flow to the brain. It also supports the production of key neurotransmitters such as serotonin, dopamine, and gamma-aminobutyric acid (GABA), which influence mood, behavior, and cognitive stability. A deficiency in B6 can lead to imbalanced neurotransmitter levels, contributing to mood disturbances, poor concentration, and cognitive deterioration factors often observed in AD. Vitamin B9 (folate) is essential for DNA synthesis, repair, and methylation processes that maintain healthy brain cells and promote normal cellular function. Folate deficiency has been linked to increased homocysteine levels, leading to oxidative stress, neuronal damage, and brain shrinkage. In older adults, low folate levels are associated with cognitive decline and an increased risk of dementia, making it a critical nutrient for brain protection. Moreover, folate supports the production of neurotransmitters and helps regulate inflammatory responses in the brain, both of which are vital for maintaining cognitive health.

Vitamin B12 (Cobalamin) is indispensable for nerve cell maintenance and the formation of the myelin sheath a protective layer that insulates nerve fibers and ensures efficient signal transmission between neurons. Low B12 levels are commonly observed in individuals with cognitive impairment and AD, often accompanied by brain volume loss. A deficiency can lead to nerve degeneration, memory loss, and confusion, which are frequently misattributed to aging or dementia. B12 also works synergistically with folate to regulate homocysteine levels, further contributing to neuroprotection. Collectively, these B vitamins are vital for preserving brain function, protecting against neurodegenerative processes, and potentially slowing the progression of Alzheimer's disease.

18.2.2.2 Vitamin D

Vitamin D plays a vital role in maintaining brain health, thanks to its neuroprotective and anti-inflammatory properties. It supports cognitive function through multiple mechanisms, including regulating calcium balance a crucial factor for proper neuronal signaling and survival. Vitamin D also promotes the expression of nerve growth factor (NGF), which supports the growth, maintenance, and repair of neurons, ensuring the brain's structural and functional integrity. Additionally, it helps to modulate the immune response, reducing chronic inflammation that contributes to neurodegeneration. Emerging research highlights a strong link between vitamin D deficiency and Alzheimer's disease (AD). Low vitamin D levels are associated with increased beta-amyloid accumulation, a hallmark of AD that forms toxic plaques in the brain, disrupting neuronal communication and triggering cell death. Furthermore, vitamin D deficiency exacerbates neuronal inflammation, promoting the release of pro-inflammatory cytokines that intensify brain damage and accelerate cognitive decline. Studies suggest that individuals with insufficient vitamin D are more likely to experience memory loss, impaired cognitive function, and a higher risk of developing AD. Ensuring adequate vitamin D intake through sunlight exposure, diet, or supplementation may offer a protective strategy against AD, potentially slowing disease progression and preserving cognitive health.

18.2.2.3 Vitamin E (Tocopherols and Tocotrienols)

Vitamin E is a powerful fat-soluble antioxidant that plays a critical role in protecting brain cells from oxidative damage caused by free radical's unstable molecules that contribute to cellular deterioration. In AD, oxidative stress is a major factor driving neurodegeneration. It accelerates harmful processes like amyloid-beta plaque aggregation and tau protein hyperphosphorylation, both of which are key pathological hallmarks of the disease. By neutralizing free radicals, vitamin E helps safeguard neurons from this destructive chain reaction, supporting brain health and cognitive function.

Alpha-tocopherol, the most biologically active and widely studied form of vitamin E, is essential for maintaining the stability and integrity of cell membranes. It shields neurons from oxidative injury and has demonstrated promising potential in slowing cognitive decline in AD patients. Clinical studies have indicated that alpha-tocopherol supplementation may help delay the progression of functional impairment, promoting better daily living abilities in individuals with mild-to-moderate AD. Beyond its antioxidant capacity, alpha-tocopherol also supports immune function and helps modulate inflammatory responses in the brain, contributing to overall neuroprotection. Tocotrienols, though less explored, are gaining attention for their unique neuroprotective properties (Angeloni et al. 2017). They possess potent anti-inflammatory and antioxidant capabilities that may complement alpha-tocopherol's effects. Emerging research suggests that tocotrienols can help protect against neuroinflammation, mitochondrial dysfunction, and neuronal apoptosis processes that exacerbate AD progression. Their distinct chemical structure allows tocotrienols to penetrate deeper into brain cell membranes, potentially offering enhanced protection against lipid peroxidation, a damaging consequence of oxidative stress. Together, the tocopherols and tocotrienols within vitamin E form a powerful defense against oxidative damage and inflammation associated with Alzheimer's disease, making them an important area of ongoing research in neuroprotection and cognitive longevity.

18.2.2.4 Vitamin C (Ascorbic Acid)

Vitamin C, also known as ascorbic acid, is a vital antioxidant that plays a multifaceted role in supporting brain health and combating Alzheimer's disease (AD). One of its key functions is enhancing the regenerative capacity of vitamin E, working synergistically to extend vitamin E's neuroprotective effects. When vitamin E neutralizes free radicals, it becomes oxidized and vitamin C helps restore it to its active form, allowing it to continue shielding brain cells from oxidative damage. This antioxidant partnership is crucial for mitigating the harmful effects of reactive oxygen species (ROS), which are elevated in AD and contribute to neuronal injury, amyloid-beta accumulation, and tau protein hyperphosphorylation. Beyond its antioxidant properties, vitamin C plays a pivotal role in neurotransmitter synthesis. It supports the production of dopamine and norepinephrine, neurotransmitters essential for regulating mood, attention, and cognitive function. Dopamine is critical for motivation and reward processing, while norepinephrine aids in alertness and memory retrieval. Deficiencies in these neurotransmitters are often observed in AD and other neurodegenerative diseases, contributing to cognitive decline and mood disturbances. By facilitating their synthesis, vitamin C helps maintain neural communication and emotional stability, positioning it as a key nutrient in promoting cognitive resilience and potentially slowing Alzheimer's progression.

18.2.2.5 Vitamin K

Vitamin K, though often overlooked, plays a crucial role in maintaining brain health and may offer protective benefits against AD. One of its primary functions in the brain is supporting the synthesis of sphingolipids: vital fat molecules that contribute to the structure and function of brain cell membranes. Sphingolipids are essential for cell signaling, neuronal survival, and myelin sheath integrity, all of which are critical for maintaining cognitive function. Disruptions in sphingolipid metabolism have been associated with neurodegenerative diseases, including AD, highlighting the importance of vitamin K in preserving neural structure and communication pathways.

Moreover, vitamin K possesses anti-inflammatory and antioxidant properties that further enhance its neuroprotective potential. Chronic inflammation and oxidative stress are key drivers of AD pathology, contributing to neuronal damage, amyloid-beta accumulation, and tau protein hyperphosphorylation. Vitamin K's ability to modulate inflammatory responses and neutralize reactive oxygen species may help to counteract these processes, reducing the progression of neurodegeneration. Emerging research has linked vitamin K deficiency to cognitive impairment and an increased risk of developing neurodegenerative disorders. This growing body of evidence underscores the need to explore vitamin K's therapeutic potential further, positioning it as an essential, though underappreciated, player in promoting cognitive health and resilience against AD. Each vitamin category supports brain function through unique, yet interconnected mechanisms from regulating oxidative stress and inflammation to maintaining nerve cell integrity and neurotransmitter balance. Deficiencies in these vitamins can contribute to AD progression, highlighting the importance of a nutrient-rich diet or supplementation, particularly in older adults at risk of cognitive decline. Future research should continue to explore the optimal vitamin intake levels and synergistic effects of these nutrients in preventing and managing Alzheimer's disease.

Equally vital is the role of nutrition in mitigating AD risk and supporting brain health. Deficiencies in essential vitamins, commonly observed in dementia patients, may exacerbate cognitive decline. A nutrient-rich diet, particularly one abundant in fruits and vegetables, can provide key vitamins that contribute to neuroprotection.

Vitamins such as B-complex (B1, B6, B9, B12), C, E, and K are crucial for maintaining neuronal function, reducing oxidative stress, and preventing inflammation, all factors that are implicated in AD progression. Increased awareness about these nutritional strategies empowers patients and caregivers to adopt informed, proactive measures that may help manage symptoms and enhance quality of life.

18.2.3 Olive Oil: Role of Nutrition in Maintaining Brain Health

Olive oil, particularly EVOO, stands out not only for its culinary appeal but also for its powerful role in supporting brain health. Packed with monounsaturated fats and bioactive phenolic compounds such as hydroxytyrosol, oleuropein, and tyrosol olive oil exerts significant neuroprotective effects. These compounds help combat oxidative stress, a major contributor to neuronal damage and cognitive decline. The brain, being highly metabolically active, produces an abundance of reactive oxygen species (ROS). When left unchecked, these ROS lead to lipid peroxidation, protein damage, and DNA fragmentation, key processes that accelerate neurodegeneration. Olive oil's antioxidants neutralize these harmful free radicals, reducing oxidative damage and preserving brain cell integrity.

Olive oil has strong anti-inflammatory qualities in addition to its antioxidant capability. It is becoming more well-acknowledged that neurodegenerative illnesses such as Alzheimer's are characterized by persistent low-grade inflammation in the brain. Like ibuprofen, EVOO's phenolic components, especially oleocanthal, have anti-inflammatory properties that help reduce neuroinflammatory reactions. This is important, because chronic inflammation leads to neuronal death, synaptic dysfunction, and the accumulation of toxic proteins like tau tangles and amyloid-beta, which are important indicators of Alzheimer's disease pathogenesis. Additionally, olive oil promotes vascular health, which is critical for preserving cognitive function. It supports healthier blood arteries and improves cerebral blood flow by lowering LDL cholesterol and raising HDL cholesterol in blood lipid profiles. Enhanced blood flow ensures that the brain receives sufficient oxygen and nutrients, supporting synaptic activity and overall neuronal performance. This vascular benefit also lowers the risk of stroke and ischemic injuries, both of which are linked to cognitive impairment and dementia.

Moreover, research highlights olive oil's role in promoting autophagy: the brain's natural waste clearance process. Autophagy helps clear out misfolded proteins, including amyloid-beta plaques, which accumulate in Alzheimer's disease. By enhancing this cellular housekeeping, olive oil contributes to maintaining a cleaner, healthier neural environment. Another critical aspect is olive oil's influence on the blood-brain barrier (BBB), a selective barrier that shields the brain from harmful substances. EVOO supports the integrity of this barrier, preventing neurotoxins and inflammatory molecules from crossing into brain tissue. A compromised BBB is often observed in Alzheimer's patients, leading to increased vulnerability to toxins and pathogens. By preserving this barrier, olive oil helps maintain brain homeostasis and protects against further damage.

When integrated into a balanced, nutrient-rich diet, such as the Mediterranean diet olive oil, this works synergistically with other brain-boosting foods like fruits, vegetables, whole grains, nuts, and fatty fish. This combination provides essential vitamins, minerals, and omega-3 fatty acids, all of which contribute to improved cognitive function, reduced risk of neurodegenerative diseases, and better overall brain health. Olive oil's multifaceted benefits ranging from antioxidant and anti-inflammatory effects to enhanced blood flow, autophagy, and BBB protection position it as a key nutritional ally for brain health. Embracing olive oil as part of a healthy lifestyle may not only support cognitive resilience but also reduce the risk of neurodegenerative diseases like Alzheimer's, highlighting the profound connection between nutrition and lifelong brain vitality.

18.3 Conclusion

Extra virgin olive oil (EVOO) has gained recognition as a valuable dietary component in supporting brain health and potentially reducing the risk of Alzheimer's disease. Rich in monounsaturated fatty acids, particularly oleic acid, EVOO helps maintain the structure and function of brain cell membranes, ensuring proper neuronal communication. Additionally, it contains a wealth of polyphenolic compounds, such as oleuropein, hydroxytyrosol, and oleocanthal, which possess potent antioxidant and anti-inflammatory properties. These bioactive substances are essential for reducing chronic inflammation as well as neutralizing dangerous reactive oxygen species, two major mechanisms that accelerate the development of Alzheimer's disease. According to recent studies, EVOO supports the brain's glymphatic system, which aids in the removal of amyloid-beta plaques, a hallmark of Alzheimer's disease. Olive oil has the potential to delay plaque accumulation and lessen neuronal damage by promoting this natural waste-removal mechanism. Additionally, it seems to control the phosphorylation of tau proteins, reducing the development of neurofibrillary tangles that impair synaptic connections and cognitive function. In addition to its protective properties, EVOO promotes synaptic plasticity and increases BDNF, a protein necessary for neurogenesis and cognitive resilience. This not only helps maintain existing neural networks but also supports the growth of new brain cells, promoting long-term cognitive health. While large-scale clinical trials continue to explore the full extent of EVOO's neuroprotective benefits, incorporating high-quality olive oil into a balanced diet, especially within the Mediterranean dietary pattern, offers a practical, natural strategy for promoting brain health. When combined with other healthy lifestyle habits, such as regular physical activity and cognitive engagement, EVOO may serve as a powerful, accessible tool in supporting cognitive function and reducing the risk of Alzheimer's disease.

References

Abdallah IM, Al-Shami KM, Alkhalifa AE, Al-Ghraiybah NF, Guillaume C, Kaddoumi A (2023) Comparison of oleocanthal-low EVOO and oleocanthal against amyloid-β and related pathology in a mouse model of Alzheimer's disease. Molecules. https://doi.org/10.3390/molecules28031249

Abuznait AH, Qosa H, Busnena BA, El Sayed KA, Kaddoumi A (2013) Olive-oil-derived oleocanthal enhances β-amyloid clearance as a potential neuroprotective mechanism against Alzheimer's disease: in vitro and in vivo studies. ACS Chem Neurosci. https://doi.org/10.1021/cn400024q

Al Rihani SB, Darakjian LI, Kaddoumi A (2019) Oleocanthal-rich extra-virgin olive oil restores the blood-brain barrier function through NLRP3 inflammasome inhibition simultaneously with autophagy induction in TgSwDI mice. ACS Chem Neurosci. https://doi.org/10.1021/acschemneuro.9b00175

Angeloni C, Malaguti M, Barbalace MC, Hrelia S (2017) Bioactivity of olive oil phenols in neuroprotection. https://doi.org/10.3390/ijms18112230

Batarseh YS, Kaddoumi A (2018) Oleocanthal-rich extra-virgin olive oil enhances donepezil effect by reducing amyloid-β load and related toxicity in a mouse model of Alzheimer's disease. J Nutr Biochem. https://doi.org/10.1016/j.jnutbio.2017.12.006

Casamenti F, Stefani M (2017) Olive polyphenols: new promising agents to combat aging-associated neurodegeneration. https://doi.org/10.1080/14737175.2017.1245617

Cordero JG, García-Escudero R, Avila J, Gargini R, García-Escudero V (2018) Benefit of oleuropein aglycone for Alzheimer's disease by promoting autophagy. https://doi.org/10.1155/2018/5010741

Guruvaiya P, Subramaniam K (2023) Therapeutic potential of extra-virgin olive oil against Alzheimer's disease progression. https://doi.org/10.1007/978-981-99-0677-2_3

Kaddoumi A et al (2022) Extra-Virgin Olive Oil Enhances the Blood–Brain Barrier Function in Mild Cognitive Impairment: A Randomized Controlled Trial. Nutrients. https://doi.org/10.3390/nu14235102

Klimova B, Novotný M, Kuca K, Valis M (2019) Effect of an extra-virgin olive oil intake on the delay of cognitive decline: role of secoiridoid oleuropein? https://doi.org/10.2147/NDT.S218238

Lauretti E, Nenov M, Dincer O, Iuliano L, Praticò D (2020) Extra virgin olive oil improves synaptic activity, short-term plasticity, memory, and neuropathology in a tauopathy model. Aging Cell. https://doi.org/10.1111/acel.13076

Leri M et al (2023) EVOO polyphenols exert anti-inflammatory effects on the microglia cell through TREM2 signaling pathway. Pharmaceuticals. https://doi.org/10.3390/ph16070933

Qusa MH, Siddique AB, Nazzal S, El Sayed KA (2019) Novel olive oil phenolic (−)-oleocanthal (+)-xylitol-based solid dispersion formulations with potent oral anti-breast cancer activities. Int J Pharm. https://doi.org/10.1016/j.ijpharm.2019.118596

Rigacci S (2015) Olive oil phenols as promising multi-targeting agents against Alzheimer's disease. Adv Exp Med Biol. https://doi.org/10.1007/978-3-319-18365-7_1

Rita L et al (2023) Alzheimer's disease: using gene/protein network machine learning for molecule discovery in olive oil. Hum Genomics. https://doi.org/10.1186/s40246-023-00503-6

Siddique AB et al (2019) Novel liquid-liquid extraction and self-emulsion methods for simplified isolation of extra-virgin olive oil phenolics with emphasis on (−)-oleocanthal and its oral anti-breast cancer activity. PLoS One. https://doi.org/10.1371/journal.pone.0214798

Siddique AB, King JA, Meyer SA, Abdelwahed K, Busnena B, El Sayed K (2020) Safety evaluations of single dose of the olive Secoiridoid S-(−)-Oleocanthal in Swiss albino mice. Nutrients. https://doi.org/10.3390/nu12020314

Silva P, Rodríguez-Pérez M, Gómez-Torres Ó, Burgos-Ramos E (2023) Olive oil and wine as source of multi-target agents in the prevention of Alzheimer disease. https://doi.org/10.1017/S095442242100041X

Tsolaki M et al (2020) A randomized clinical trial of Greek high phenolic early harvest extra virgin olive oil in mild cognitive impairment: the MICOIL pilot study. J Alzheimer's Dis. https://doi.org/10.3233/JAD-200405
Xu F, Li Y, Zheng M, Xi X, Zhang X, Han C (2018) Structure properties, acquisition protocols, and biological activities of oleuropein aglycone. https://doi.org/10.3389/fchem.2018.00239

Vitamins and Their Role in Alzheimer's Disease

19

Dipesh Joshi

Abstract

Alzheimer's disease (AD) is a progressive neurological condition characterized by degeneration of brain cells that leads to decline in memory and cognitive functioning. There is a progressive decline of memory and cognitive function in people with AD. The histopathological characteristics include amyloid plaques, neurofibrillary tangles, and depletion of neuronal synapses in the brain. Multiple factors, including nutritional aspects, have been associated with AD. There is increasing evidence that diet and nutrition, as key modifiable lifestyle factors, are linked to the risk of AD. A deficiency or an excess of dietary factors such as vitamins has been reported to contribute to AD development. Vitamins are involved in DNA methylation, homocysteine metabolism, reduction in β-amyloid-induced lipid peroxidation, and oxidative stress in key brain areas involved in memory and cognition. Therefore, dietary interventions, which can alter gene expression and influence the human genome, might contribute to the prevention of Alzheimer's disease. Current treatment options are predominantly limited to delay disease progression and hence have a limited beneficial effect. Preventive measures against dementia are therefore in focus and being explored. Vitamin supplementation may positively impact delaying and preventing cognitive decline and the clinical condition of patients with Alzheimer's disease. This chapter will examine the existing evidence regarding the use of vitamins as a treatment option for Alzheimer's disease.

Keywords

Alzheimer's disease · Diet · Dementia · Memory · Neurodegenerative · Oxidative stress · Vitamin

D. Joshi (✉)
Novotech (Australia) Pty Ltd., Sydney, NSW, Australia

J. Malik et al. (eds.), *Diet and Alzheimer's Disease: Let Food be Our Medicine*, Nutritional Neurosciences, https://doi.org/10.1007/978-981-95-2736-6_19

Abbreviations

Aβ	Amyloid-beta
AD	Alzheimer's Disease
APP	Amyloid Precursor Protein
CSF	Cerebrospinal Fluid
DNA	Deoxyribonucleic acid
NMDA	N-methyl-D-aspartate
RDA	Recommended Dietary Allowance
ROS	Reactive Oxygen Species

19.1 Introduction

Alzheimer's disease (AD) is an advancing degenerative disease that affects memory and cognitive function. It is the most prevalent type of dementia among older adults (Tarawneh and Holtzman 2012; Lane et al. 2018), with a number of cases on the rise worldwide. The hallmark features of AD, that is, "plaques" and "tangles," were initially documented by Alois Alzheimer in 1907. The neurodegeneration that occurs in AD is influenced by inflammation, immune responses, and the impact of free radicals (Cummings and Cole 2002). Numerous factors have been linked to AD development, and these include old age, past brain injuries, family history, diabetes, obesity, and hypertension (Hersi et al. 2017). Moreover, lifestyle factors like diet, exercise, training, smoking, and alcohol consumption are thought to contribute to the development of AD (Garre-Olmo 2018; Mielech et al. 2020).

People with AD face the twin challenge of delayed diagnosis and the absence of neuroprotective or therapeutic drug treatments. AD is recognized much later downstream when cognitive symptoms manifest. Unfortunately, existing treatments offer limited and short-term relief from symptoms such as memory loss. In AD, the presymptomatic phase lasts 1–2 decades before the onset of symptoms; this adds to the treatment complexity and capability. During this presymptomatic phase, the brain undergoes through many biochemical changes that results in cognitive impairment. Introducing changes to the diet, and intake of nutritional supplements, may have a preventive impact on AD. Epidemiological studies suggest that dietary habits can influence the onset and progression of AD (Gillette-Guyonnet et al. 2013).

For AD, currently, there is no effective prevention nor cure, and hence the existing research efforts focus on preventive intervention (Śliwińska and Jeziorek 2021). Acetylcholinesterase inhibitors (e.g., rivastigmine and donepezil) and N-methyl-D-aspartate (NMDA) receptor antagonists (e.g., memantine) are some of the commonly used drugs for AD. Unfortunately, these drugs are limited to decelerating AD progression but cannot stop the disease (Brunton et al. 2007). Given the limited relief provided by current medications available in the market, researchers are exploring diet modification and physical training as key treatment strategies (Botchway et al. 2018). This chapter will focus specifically on the vitamins and how the deficiency and an excess may impact development of AD (Mielech et al. 2020; An et al. 2019; Mecocci and Polidori 2012).

19.2 Vitamin A

Vitamin A (retinol, retinoic acid) has a vital role in the optimal functioning of the brain (Lane and Bailey 2005; Mey and McCaffery 2004; Ransom et al. 2014; Stoney and McCaffery 2016). Vitamin A is processed from retinol to retinoic acid (active metabolite) in the body. Vitamin A is transformed into its bioactive forms retinoic acid, which acts as a powerful signaling molecule on entering the brain (Fragoso et al. 2012; Olson and Mello 2010). Retinoic acid receptors are expressed in striatum, cortex, and hippocampus, which are the key brain areas that are relevant to cognition (Kennedy and Haskell 2011). Retinoic acid binds to Retinoic Acid Receptors (RARs) and Retinoid X Receptors (RXRs) in the nucleus (Blomhoff and Blomhoff 2006; Lane and Bailey 2005; Ransom et al. 2014) to regulate the transcription of genes involved in cognition, learning, and memory processes (Lane and Bailey 2005; Olson and Mello 2010; Shearer et al. 2012). It plays a critical role in the growth, differentiation, survival, and plasticity of the neurons (Olson and Mello 2010; Sun and Kawaguchi 2011), including secretion of brain neurotransmitters (Sodhi and Singh 2014; Lane and Bailey 2005). Vitamin A has a significant role in membrane stability, hippocampal synaptic plasticity, and dopaminergic function. Alterations in vitamin A level can lead to associated impairments in memory and learning behaviors. The precise regulation of retinoic acid levels is therefore critical to cognition, learning, and memory processes (Olson and Mello 2010).

In the aging population, insufficient vitamin A levels pose a significant risk for AD and represent a major concern (Shahar et al. 2013). Vitamin A helps regulate multiple genes associated with AD within the body's natural defenses (Carter 2011). In AD, the transport of retinoid from the intestine to target tissues, such as the brain, has been noted to be altered. Dietary deficiency of vitamin A disrupts retinoid signaling pathway, leading to amyloid β deposition in the brain (Corcoran et al. 2004) and in turn contributing to the onset of amyloid pathology in the central nervous system (Goncalves et al. 2013). Preclinical studies have shown that aging rodents exhibit reduced serum retinol levels and decreased expression of brain vitamin A receptors (RARs and RXRs), which are linked to memory changes (Enderlin et al. 1997; Etchamendy et al. 2001; Mingaud et al. 2008; Touyarot et al. 2013). These observations are consistent with the clinical evidence available from studies focusing on AD pathology. In elderly individuals, reduced serum retinol levels (Leotsinidis et al. 2000; Succari et al. 1991) have been linked to cognitive decline, as demonstrated by clinical assessments like the Mini-Mental State Examination (MMSE) (Huang et al. 2018; Zeng et al. 2017). Furthermore, the expression of vitamin A receptors (RARα and RXRβ) in peripheral blood mononuclear cells (lymphocytes and monocytes) is decreased in elderly individuals (65 years old) compared to younger individuals (25 years old) (Brtko et al. 2007; Feart et al. 2005b). Clinical AD studies report that in comparison to age-matched controls, patients with AD exhibit lower levels of serum retinol (Bourdel-Marchasson et al. 2001; Lopes da Silva et al. 2014; Rinaldi et al. 2003) and diminished expression of retinoic acid receptors (RARs) and the enzyme responsible for retinoic acid synthesis (retinal dehydrogenase) in the neocortex (Corcoran et al. 2004).

In nonprimate studies, administering pharmacological retinoic acid or dietary vitamin A enhances synaptic plasticity, boosts hippocampal neurogenesis, and alleviates memory deficits in aged animals (Bonhomme et al. 2014; Etchamendy et al. 2001; Feart et al. 2005a; Mingaud et al. 2008; Touyarot et al. 2013). Multiple lines of evidence from clinical studies suggest that retinoids may offer a promising therapeutic approach for AD (Goodman 2006; Goodman and Pardee 2003; Lee et al. 2009; Lerner et al. 2012; Ono and Yamada 2012), as they modulate various targets implicated in the disease, including amyloid β (Aβ) production by stimulation of retinoid signaling (Ding et al. 2008; Koryakina et al. 2009; Tippmann et al. 2009; Chakrabarti et al. 2016; Muñoz-Cabrera et al. 2019; Tousi 2015).

19.3 B Vitamins

The B vitamins (also called B-complex vitamins) are a collection of eight water-soluble vitamins (B1: thiamin, B2: riboflavin, B3: niacin, B5: pantothenic acid, B6: pyridoxine, B7: biotin, B9: folic acid or folate, and B12: cyanocobalamin). The B vitamins act as cofactors/coenzymes in multiple enzymatic processes (including DNA methylation) that support key functions within the central nervous system (Hanna et al. 2022; Robinson et al. 2018). The B vitamins are crucial for neuronal health and are vital for brain development and neuronal activity. Among the B vitamins, B6, B12, and folic acid (B9) specifically have a role in neurodegenerative processes and AD (Wang et al. 2022). These B vitamins are involved in regulating homocysteine levels, which are linked to cognitive function (Soumare et al. 2006; Behrouzi et al. 2000) by facilitating homocysteine conversion to methionine and cysteine (Bottiglieri 2005).

Several studies have reported decreased DNA methylation, increased homocysteine levels, and reduced vitamin B levels in AD (Liu et al. 2022; Van Dam and Van Gool 2009). The B vitamins and homocysteine levels have an inverse relationship, and the homocysteine levels in plasma can be modified by dietary B vitamin supplementation (Selhub et al. 1993). Therefore, elevated homocysteine levels are an adjustable risk factor for cognitive impairment and a significant predictor of AD incidence (Beydoun et al. 2014). Increased homocysteine and reduced B vitamin levels are both linked to a higher risk of AD (Shen and Ji 2015; Wang et al. 2021). The B vitamins have a protective effect against cognitive impairment and decline (Kado et al. 2005; Duthie et al. 2002; Tucker et al. 2005).

In AD, DNA methylation is decreased, and homocysteine levels are increased (Meng et al. 2019). Both observations are associated with a heightened risk of AD. Vitamin B, particularly B12, plays a crucial role in the methionine cycle. It helps convert homocysteine into methionine, with the assistance of vitamin B6 and folic acid as cofactors. This transformation reaction induces DNA methylation (Mohajeri et al. 2015; Athanasopoulos et al. 2016). Oulhaj et al. (2016) reported that treating participants with mild cognitive impairment using vitamin B led to a decrease in total homocysteine levels, which subsequently slowed the rate of gray matter volume reduction/cognitive decline. Increased levels of folic acid and vitamin B12 have protective impact against cognitive impairment (Meng et al. 2019). An et al. (2019) further corroborated this by finding that inadequate vitamin B12 intake is associated with cognitive dysfunction. Adequate levels of B vitamins (such as folic acid, vitamin B6, and vitamin B12) had a protective effect on cognition.

Current research presents conflicting findings on the effectiveness of B vitamin supplementation in enhancing cognitive function and preventing AD (Tardy et al. 2020; Wang et al. 2022; Gil Martínez et al. 2022; Zhou et al. 2022). While some studies report potential benefits in reducing or delaying cognitive impairment with B vitamin supplementation (Gil Martínez et al. 2022), there are other reports that have yielded inconclusive results regarding the impact of B-vitamin supplementation on reducing homocysteine levels and enhancing cognition and memory (Tan et al. 2023; Dangour et al. 2015). Future research targeting longitudinal data collection is necessary to develop a better understanding for the relationship between the B vitamins (particularly B6, B12, and folic acid), homocysteine and AD.

19.4 Vitamin C

Vitamin C (ascorbic acid) is a powerful antioxidant and a vital nutrient for optimal brain function. It plays a crucial role in brain development, regulating neurotransmitters, facilitating glutamate-driven nerve signaling, and preserving oxidative balance (He et al. 2015). It inhibits lipid peroxidation in biological systems and serves as a key defense against free radicals in both whole blood and plasma (Padayatty et al. 2003; Naidu 2003). The regions of the human brain that have the highest concentrations of vitamin C are the cerebral cortex, hippocampus, and amygdala (Travica et al. 2017). Increasing evidence indicates that vitamin C may help mitigate factors associated with AD development (Yao et al. 2004). Vitamin C's neuroprotective effects are primarily due to its ability to modulate neuroinflammation, the scavenging activity against reactive oxygen species (ROS), DNA demethylation, the binding of iron, copper, and zinc, and the inhibition of the fibrillation of amyloid-beta (Aβ) peptide (Choudhry et al. 2012; Kocot et al. 2017; Huff et al. 2021).

A number of clinical trials and large population studies have explored the connection between AD and vitamin C (Monacelli et al. 2017). Individuals with cognitive impairment and AD exhibit lower plasma vitamin C levels (Charlton et al. 2004; Rinaldi et al. 2003; Bhatti et al. 2016). Research investigations point toward a link between cognitive impairment and low antioxidant status. Morris and colleagues (1998) found that vitamins C and E supplementation synergistically lowered the risk of AD (Zandi et al. 2004). A subsequent large population-based study conducted by Engelhart et al. (2002) reported that a high intake of vitamin C and vitamin E was linked to a reduced risk of AD (Engelhart et al. 2002). The association was stronger among individuals with the lowest vitamin C levels, such as smokers. Multiple studies have explored the relationship between plasma vitamin C levels and cognitive decline, including AD. The main findings emanating from these studies indicated a notable link between vitamin C deficiency (less than 20 μM) and mild cognitive impairment (Goodwin et al. 1983). A mean cerebrospinal fluid

(CSF)-to-plasma vitamin C ratio was significantly lower in individuals with AD compared to controls (Quinn et al. 2003). Furthermore, a higher CSF-to-plasma vitamin C ratio was linked to a slower progression of AD (Harrison et al. 2014).

As increased levels of ROS are a hallmark of AD and accelerate its progression, dietary antioxidants such as vitamin C are vital to negate the harmful effects of oxidative stress. It has been observed that neurodegenerative diseases characterized by high oxidative stress have been observed to have reduced vitamin C levels due to oxidation in the brain. When ROS levels are elevated, vitamin C becomes unavailable to regulate neuronal metabolism. Consequently, oxidative stress and increased ROS production are crucial factors in the progression of neuronal deterioration in AD. Vitamin C, as a vital antioxidant, alleviates the damage caused by ROS by lowering their levels, thereby delaying disease advancement. Although researchers are still determining the optimal dosage, a diet rich in vitamin C may benefit individuals with AD, as demonstrated by various observational studies and clinical trials (Polidori and Nelles 2014). Vitamin C supplementation contributes to the proteolytic processing of amyloid precursor protein (APP), resulting in a reduction of Aβ peptides (Costanzi et al. 2008). Animal model studies yielded identical results, showing that increased vitamin C supplementation reduced the Aβ plaque burden in the hippocampus and cortical regions of mice (Kook et al. 2014). Researchers have reported that even a mild deficiency of vitamin C may accelerate amyloid pathogenesis via modulation of oxidative stress mechanism leading to impaired cognition (Dixit et al. 2015). Vitamin C solution treatment for a period of 6 months in an AD mouse model demonstrated a significant decline in ROS, Aβ peptides, and synaptophysin. These findings underscore the significance of vitamin C as a beneficial functional nutrient in preventing AD (Murakami et al. 2011).

It is crucial to note that preventing vitamin C deficiency through a healthy diet is likely more advantageous for protecting against age-related decline in cognitive function and AD, than relying on supplements. This is because the body's ability to transport ingested vitamin C from the intestines into the bloodstream is limited by saturable sodium-dependent vitamin C transporters, leading to an overestimation of the benefits of supplements (Covarrubias-Pinto et al. 2015; Reddy and Beal 2008; Bowman 2012).

19.5 Vitamin D

Insufficient vitamin D levels in the blood are linked to a higher risk of AD, cognitive issues, and worsening cognitive function in individuals with dementia (Littlejohns et al. 2014; Aguilar-Navarro et al. 2019; Llewellyn 2010; Gezen-Ak et al. 2014). Given the high prevalence of vitamin D deficiency in older adults (Holick 2007), this remains a significant cause for concern. The hippocampus and dentate gyrus, two important brain regions involved in memory, are among the many brain regions that express vitamin D receptors (Eyles et al. 2005). 1,25 dihydroxy-vitamin D3 (1,25-D3), the active form of vitamin D, controls the expression of neurotrophins like neurotrophin 3 (nerve growth factor) and glial-derived neurotrophic factor

(Annweiler et al. 2013). It also influences the survival, development, and function of neural cells (Fernandes de Abreu et al. 2009). Amyloid plaques are a key feature of AD. Vitamin D activates macrophages, which helps in clearing these plaques (Masoumi et al. 2009; Mizwicki et al. 2012). Vitamin D also decreases amyloid-induced cytotoxicity and programmed cell death in primary cortical neurons (Dursun et al. 2011, 2013). A significant link has been observed between the over-expression of vitamin D receptors or vitamin D supplementation and the reduction of amyloid-β protein precursor (AβPP) (Moneim 2015; Bhatti et al. 2016; Littlejohns et al. 2014; Darwish et al. 2015).

Vitamin D supplementation with existing AD treatment options has demonstrated a synergistic neuroprotective effect with improvement in cognition in AD. Annweiler et al. (2011, 2014) in their research work demonstrated that vitamin D3 supplementation enhanced cognition and memory in individuals with moderate AD who were also taking memantine, likely due to the combined neuroprotective effects (reduced neuronal loss) of both treatments. This evidence brings to light a new multitarget therapeutic class for AD treatment (Annweiler et al. 2011). Besides, vitamin D affects several mechanisms of AD pathogenesis including production, clearance, phagocytosis, and enzymatic degradation of Aβ peptides as well as tau phosphorylation (Balion et al. 2012). Supplementation with vitamin D has beneficial effect and improves cognitive deficit in AD. In mild AD, high vitamin D levels have been shown to improve cognition (Shen and Ji 2015). Furthermore, preclinical studies have shown that vitamin D supplementation enhances learning and memory in aged rats, mitigating age-related cognitive decline (Briones and Darwish 2012). Vitamin D supplements, therefore, may help delay or even prevent the onset of dementia and AD in older adults (Littlejohns et al. 2014). Future research into the relationship between gene–environment interactions and their impact on AD development, along with metabolic and endocrine factors, could uncover new treatment options for this progressive debilitating condition in aging population (Annweiler et al. 2011; Balion et al. 2012; Banerjee et al. 2015).

Lai et al. (2022) have found contrasting results, indicating that vitamin D supplementation may exacerbate the progression of AD (Lai et al. 2022). The results of their study on mice suggest that vitamin D deficiency might be a result of AD rather than a contributing factor. Their population-based longitudinal study results aligned with their findings in mice, warning against the extended use of vitamin D in individuals with AD. These observations warrant more research in this domain to unravel the role of vitamin D in AD pathogenesis.

19.6 Vitamin E

Vitamin E has potent antioxidant, neuro-protective, and anti-inflammatory properties. It influences gene expression and mechanisms that could contribute to the development of AD (Jiang 2014). Vitamin E comprises of a family of eight naturally occurring homologues: four tocopherols and four tocotrienols that have α, β, γ, and δ isoform. These homologues are unevenly distributed across various food sources, including vegetable oils, grains, and different types of nuts and seeds (Bieri 1984). Among the tocopherols, α-tocopherol is the main isoform present in vitamin E supplements (Grimm et al. 2016).

Numerous studies have highlighted the positive impact of vitamin E supplementation on inflammatory stress markers, cellular signaling, and immune function in humans, as well as its role in Alzheimer's disease-related pathology (Lee and Wan 2000; De la Fuente et al. 2008). Furthermore, preclinical studies have found a link between lower vitamin E levels and increased expression of genes related to AD progression, including those involved in apoptosis regulation, neurotransmission, and Aβ metabolism (Rota et al. 2005).

The meta-analysis conducted by two different research groups found comparable results, noting lower vitamin E levels in AD patients compared to cognitively normal individuals (Mullan et al. 2018; Dong et al. 2018). Another meta-analysis performed by De Wilde et al. (2017) found significantly lower vitamin E levels in the brains of individuals with AD. They concluded that the nutrient status of the brain mirrors that of the systemic circulation (De Wilde et al. 2017). A 2017 study found that vitamin E supplementation was linked to a reduced risk of cognitive decline in a group of 560 AD patients (Basambombo et al. 2017). Additional studies support these observations and have confirmed that higher vitamin E intake from dietary sources is associated with a reduced risk of developing AD (Morris et al. 2002; Engelhart et al. 2002).

Researchers have carried out randomized trials to explore the effectiveness of vitamin E as a potential treatment for AD. These studies concluded that vitamin E supplementation slowed the progression of moderate AD, both on its own and when used in combination with other therapies, compared to a placebo (Sano et al. 1997; Dysken et al. 2014). Current evidence in support of vitamin E providing clinical benefit in people with AD remains inconsistent and inconclusive. Despite growing evidence of heightened oxidative stress in AD and lower levels of circulating vitamin E, clinical trials exploring vitamin E as a treatment have not yet met the anticipated outcomes. Some of the limitations contributing to this unmatched expectation could be due to differences in study design, choice of intervention (with most clinical trials focusing solely on the α-tocopherol isoform), inherent weaknesses in studies examining oxidative stress, and the selection of the most appropriate oxidative stress markers (Winterbourn 2014; Zielonka et al. 2017; Frijhoff et al. 2015).

19.7 Summary and Future Directions

Based on the available research data, it is evident that alteration in vitamin levels plays a role in AD and dementia. While the focus may be on developing new drugs for AD treatment, monitoring vitamin levels and making appropriate dietary and supplementary adjustments may complement the existing or new drug treatments being explored. It is important to note that changed levels of vitamins can impact nervous system per se and in turn lead to brain-related abnormalities and disorders such as AD and other neuropsychiatric disorders. Monitoring vitamin levels early, particularly in individuals with a genetic predisposition to AD and other neuropsychiatric disorders, could enable timely interventions to treat or slow cognitive decline in those with AD and dementia. The evidence available is currently inadequate to either support or refute the idea that vitamins are an effective clinical intervention for slowing or preventing the onset of AD.

Some of the inconsistent results observed with vitamin administration in AD and dementia may be attributed to various limitations that affect the interpretation of clinical outcomes. Observational studies on vitamin intake, plasma levels, and the positive effects of vitamins on aging and cognition show significant inconsistency. It is important to recognize that these studies often excluded elderly individuals with polypharmacy or comorbidities. Such exclusion criteria considerably weaken the validity of the results by omitting the populations most vulnerable to vitamin deficiency. The classification of groups according to vitamin level status differs greatly across studies, thereby introducing an element of variability. Inconsistent and inaccurate study designs and methodologies, the absence of standardization between single nutrient intake and multivitamins, and the lack of systematic monitoring of plasma vitamin levels can also impact the reliability and accuracy of the results.

Finally, although researchers hypothesize that increased vitamin or dietary intake might reduce the risk of developing AD, it remains uncertain whether the oxidative stress linked to AD causes the decrease in antioxidants, or if the reduced antioxidants contribute to the disease's progression. Further research is needed to clarify the precise mechanistic role of vitamins in pathophysiology and prevention of AD.

Acknowledgments No sources of funding were used to assist in the writing of this chapter.

Disclosure The author declares no conflict of interest related to this work. The views expressed in this chapter are those of the author and do not reflect the views of the affiliated organization. The affiliated organization neither endorses nor was involved in the publication of this work.

References

Aguilar-Navarro SG, Mimenza-Alvarado AJ, Jiménez-Castillo GA, Bracho-Vela LA, Yeverino-Castro SG, Ávila-Funes JA (2019) Association of Vitamin D with mild cognitive impairment and Alzheimer's dementia in older mexicanadults. Rev Invest Clin 71(6):381–386

An Y, Feng L, Zhang X, Wang Y, Wang Y, Tao L, Qin Z, Xiao R (2019) Dietary intakes and biomarker patterns of folate, vitamin B6, and vitamin B12 can be associated with cognitive impairment by hypermethylation of redox-related genes NUDT15 and TXNRD1. Clin Epigenetics 11:139

Annweiler C, Fantino B, Parot-Schinkel E, Thiery S, Gautier J, Beauchet O (2011) Alzheimer's disease—input of vitamin D with memantine assay (AD-IDEA trial): study protocol for a randomized controlled trial. Trials 12:230

Annweiler C, Montero-Odasso M, Hachinski V, Seshadri S, Bartha R, Beauchet O (2013) Vitamin D concentration and lateral cerebral ventricle volume in older adults. Mol Nutr Food Res 57:267–276

Annweiler C, Karras SN,Anagnostis P, Beauchet O. Vitamin D (2014) supplements: a novel therapeutic approach for Alzheimer patients. FrontPharmacol. Jan 28;5:6

Athanasopoulos D, Karagiannis G, Tsolaki M (2016) Recent findings in Alzheimer disease and nutrition focusing on epigenetics. Adv Nutr 7(5):917–927

Balion C, Griffith L, Strifler L, Henderson M, Patterson C, Hecknman G, Llewellyn D, Raina P (2012) Vitamin D, cognition, and dementia: a systematic review and meta-analysis. Neurology 79:1397–1405

Banerjee A, Khemka V, Ganguly A, Roy D, Ganguly U, Chakrabarti S (2015) Vitamin D and Alzheimer's disease: neurocognition to therapeutics. Int J Alzheimer's Dis 192747:1

Basambombo L, Carmichael P, Côté S, Laurin D (2017) Use of vitamin E and C supplements for the prevention of cognitive decline. Ann Pharmacother 51(2):118–124

Behrouzi P, Grootswagers P, Keizer P, Smeets E, Feskens E, de Groot L, van Eeuwijk F (2000) Dietary intakes of vegetable protein, folate, and vitamins B-6 and B-12 are partially correlated with physical functioning of Dutch older adults using copula graphical models. J Nutr 150(3):634–643

Beydoun M, Beydoun H, Gamaldo A, Teel A, Zonderman A, Wang Y (2014) Epidemiologic studies of modifiable factors associated with cognition and dementia: systematic review and meta-analysis. BMC Public Health 14:643

Bhatti A, Usman M, Ali F, Satti S (2016) Vitamin supplementation as an adjuvant treatment for Alzheimer's disease. J Clin Diagn Res 10:OE07–OE11

Bieri J (1984) Sources and consumption of antioxidants in the diet. J Am Oil Chem Soc 61:1917–1918

Blomhoff R, Blomhoff HK (2006) Overview of retinoid metabolism and function. J Neurobiol 66:606–630

Bonhomme D, Pallet V, Dominguez G, Servant L, Henkous N, Lafenêtre P, Higueret P, Béracochéa D, Touyarot K (2014) Retinoic acid modulates intrahippocampal levels of corticosterone in middle-aged mice: consequences on hippocampal plasticity and contextual memory. Front Aging Neurosci 7(6):6

Botchway B, Moore M, Akinleye F, Iyer I, Fang M (2018) Nutrition: review on the possible treatment for Alzheimer's disease. J Alzheimer's Dis 61(3):867–883

Bottiglieri T (2005) Homocysteine and folate metabolism in depression. Prog Neuro-Psychopharmacol Biol Psychiatry 29:1103–1112

Bourdel-Marchasson I, Delmas-Beauvieux M, Peuchant E, Richard-Harston S, Decamps A, Reignier B, Emeriau J, Rainfray M (2001) Antioxidant defences and oxidative stress markers in erythrocytes and plasma from normally nourished elderly Alzheimer patients. Age Ageing 30:235–241

Bowman G (2012) Ascorbic acid, cognitive function, and Alzheimer's disease: a current review and future direction. Biofactors 38:114–122

Briones T, Darwish H (2012) Vitamin D mitigates age-related cognitive decline through the modulation of pro-inflammatory state and decrease in amyloid burden. J Neuroinflammation 9:244

Brtko J, Rock E, Nezbedova P, Krizanova O, Dvorcakova M, Minet-Quinard R, Farges M, Ribalta J, Winklhofer-Roob B, Vasson M, Macejova D (2007) Age-related change in the retinoid X receptor beta gene expression in peripheral blood mononuclear cells of healthy volunteers: effect of 13-cis retinoic acid supplementation. Mech Ageing Dev 128(11–12):594–600

Brunton L, Lazo J, Parker K (2007) Goodman and Gilman's the pharmacological, 1st edn. McGraw-Hill Professional Czelej, Lublin. red. Polish edition Buczko

Carter C (2011) The fox and the rabbits—environmental variables and population genetics, (1) replication problems in association studies and the untapped power of GWAS (2) vitamin a deficiency, herpes simplex reactivation and other causes of Alzheimer's disease. ISRN Neurol 2011:394678

Chakrabarti M, McDonald A, Will Reed J, Moss M, Das B, Ray S (2016) Molecular signaling mechanisms of natural and synthetic retinoids for inhibition of pathogenesis in Alzheimer's disease. J Alzheimer's Dis 50:335–352

Charlton K, Rabinowitz T, Geffen L, Dhansay M (2004) Lowered plasma vitamin C, but not vitamin E, concentrations in dementia patients. J Nutr Health Aging 8:99–107
Choudhry F, Howlett D, Richardson J, Francis P, Williams R (2012) Pro-oxidant diet enhances beta/gamma secretase-mediated APP processing in APP/PS1 transgenic mice. Neurobiol Aging 33:960–968
Corcoran J, So P, Maden M (2004) Disruption of the retinoid signalling pathway causes a deposition of amyloid beta in the adult rat brain. Eur J Neurosci 20:896–902
Costanzi E, Martino S, Persichetti E, Tiribuzi R, Massini C, Bernardi G, Orlacchio A, Orlacchio A (2008) Effects of vitamin C on fibroblasts from sporadic Alzheimer's disease patients. Neurochem Res 33(12):2510–2515
Covarrubias-Pinto A, Moll P, Solis-Maldonado M, Acuna A, Riveros A, Miro M, Papic E, Beltran F, Cepeda C, Concha II, Brauchi S, Castro M (2015) Beyond the redox imbalance: oxidative stress contributes to an impaired GLUT3 modulation in Huntington's disease. Free Radic Biol Med 89:1085–1096
Cummings JL, Cole G (2002) Alzheimer disease. JAMA 287(18):2335–2338
Dangour A, Allen E, Clarke R, Elbourne D, Fletcher A, Letley L, Richards M, Whyte K, Uauy R, Mills K (2015) Effects of vitamin B-12 supplementation on neurologic and cognitive function in older people: a randomized controlled trial. Am J Clin Nutr 102(3):639–647
Darwish H, Zeinoun P, Ghusn H, Khoury B, Tamim H, Khoury S (2015) Serum 25-hydroxyvitamin D predicts cognitive performance in adults. Neuropsychiatr Dis Treat 11:2217–2223
De la Fuente M, Hernanz A, Guayerbas N, Manuel Victor V, Amalich F (2008) Vitamin E ingestion improves several immune functions in elderly men and women. Free Radic Res 42(3):272–280
De Wilde M, Vellas B, Girault E, Yavuz A, Sijben J (2017) Lower brain and blood nutrient status in Alzheimer's disease: results from meta-analyses. Alzheimers Dement 3(3):416–431
Ding Y, Qiao A, Wang Z, Goodwin J, Lee E, Block M, Allsbrook M, McDonald M, Fan G (2008) Retinoic acid attenuates beta-amyloid deposition and rescues memory deficits in an Alzheimer's disease transgenic mouse model. J Neurosci 28:11622–11634
Dixit S, Bernardo A, Walker J, Kennard J, Kim G, Kessler E, Harrison F (2015) Vitamin C deficiency in the brain impairs cognition, increases amyloid accumulation and deposition, and oxidative stress in APP/PSEN1 and normally aging mice. ACS Chem Neurosci 6(4):570–581
Dong Y, Chen X, Liu Y, Shu Y, Chen T, Xu L, Li M, Guan X (2018) Do low-serum vitamin E levels increase the risk of Alzheimer disease in older people? Evidence from a meta-analysis of case-control studies. Int J Geriatr Psychiatry 33(2):e257–e263
Dursun E, Gezen-Ak D, Yilmazer S (2011) A novel perspective for Alzheimer's disease: vitamin D receptor suppression by amyloid-β and preventing the amyloid-β induced alterations by vitamin D in cortical neurons. J Alzheimer's Dis 23:207–219
Dursun E, Gezen-Ak D, Yilmazer S (2013) A new mechanism for amyloid-β induction of iNOS: vitamin D-VDR pathway disruption. J Alzheimer's Dis 36:459–474
Duthie S, Whalley L, Collins A, Leaper S, Berger K, Deary I (2002) Homocysteine, B vitamin status, and cognitive function in the elderly. Am J Clin Nutr 75:908–913
Dysken M, Sano M, Asthana S, Vertrees J, Pallaki M, Llorente M, Love S, Schellenberg G, McCarten J, Malphurs J, Prieto S, Chen P, Loreck D, Trapp G, Bakshi R, Mintzer J, Heidebrink J, Vidal-Cardona A, Arroyo L, Cruz A, Zachariah S, Kowall N, Chopra M, Craft S, Thielke S, Turvey C, Woodman C, Monnell K, Gordon K, Tomaska J, Segal Y, Peduzzi P, Guarino P (2014) Effect of vitamin E and memantine on functional decline in Alzheimer disease: the TEAM-AD VA cooperative randomized trial. JAMA 311(1):33–44
Enderlin V, Alfos S, Pallet V, Garcin H, Azais-Braesco V, Jaffard R, Higueret P (1997) Aging decreases the abundance of retinoic acid (RAR) and triiodothyronine (TR) nuclear receptor mRNA in rat brain: effect of the administration of retinoids. FEBS Lett 412:629–632
Engelhart M, Geerlings M, Ruitenberg A, van Swieten J, Hofman A, Witteman J, Breteler M (2002) Dietary intake of antioxidants and risk of Alzheimer disease. JAMA 287(24):3223–3229
Etchamendy N, Enderlin V, Marighetto A, Vouimba RM, Pallet V, Jaffard R, Higueret P (2001) Alleviation of a selective age-related relational memory deficit in mice by pharmacologically induced normalization of brain retinoid signaling. J Neurosci 21:6423–6429

Eyles D, Smith S, Kinobe R, Hewison M, McGrath J (2005) Distribution of the vitamin D receptor and 1 alpha-hydroxylase in human brain. J Chem Neuroanat 29:21–30

Feart C, Mingaud F, Enderlin V, Husson M, Alfos S, Higueret P, Pallet V (2005a) Differential effect of retinoic acid and triiodothyronine on the age-related hypo-expression of neurogranin in rat. Neurobiol Aging 26:729–738

Feart C, Pallet V, Boucheron C, Higueret D, Alfos S, Letenneur L, Dartigues J, Higueret P (2005b) Aging affects the retinoic acid and the triiodothyronine nuclear receptor mRNA expression in human peripheral blood mononuclear cells. Eur J Endocrinol 152:449–458

Fernandes de Abreu DA, Eyles D, Féron F (2009) Vitamin D, a neuro-immunomodulator: implications for neurodegenerative and autoimmune diseases. Psychoneuroendocrinology 34(suppl 1):S265–S277

Fragoso Y, Campos NS, Tenrreiro B, Guillen F (2012) Systematic review of the literature on vitamin a and memory. Dement Neuropsychol 6(4):219–222

Frijhoff J, Winyard PG, Zarkovic N, Davies SS, Stocker R, Cheng D (2015) Clinical relevance of biomarkers of oxidative stress. Antioxid Redox Signal 23(14):1144–1170

Garre-Olmo J (2018) Epidemiology of Alzheimer's disease and other dementias. Rev Neurol 66:377–386

Gezen-Ak D, Yilmazer S, Dursun E (2014) Why vitamin D in Alzheimer's disease? The hypothesis. J Alzheimer's Dis 40:257–269

Gil Martínez V, Avedillo Salas A, Santander Ballestín S (2022) Vitamin supplementation and dementia: a systematic review. Nutrients 14(5):1033

Gillette-Guyonnet S, Secher M, Vellas B (2013) Nutrition and neurodegeneration: epidemiological evidence and challenges for future research. Br J Clin Pharmacol 75(3):738–755

Goncalves M, Clarke E, Hobbs C, Malmqvist T, Deacon R, Jack J, Corcoran J (2013) Amyloid beta inhibits retinoic acid synthesis exacerbating Alzheimer disease pathology which can be attenuated by an retinoic acid receptor alpha agonist. Eur J Neurosci 37:1182–1192

Goodman A (2006) Retinoid receptors, transporters, and metabolizers as therapeutic targets in late onset Alzheimer disease. J Cell Physiol 209:598–603

Goodman A, Pardee A (2003) Evidence for defective retinoid transport and function in late onset Alzheimer's disease. Proc Natl Acad Sci USA100:2901–2905

Goodwin J, Goodwin J, Garry P (1983) Association between nutritional status and cognitive functioning in a healthy elderly population. JAMA 249:2917–2921

Grimm M, Mett J, Hartmann T (2016) The impact of vitamin E and other fat-soluble vitamins on Alzheimer's disease. Int J Mol Sci 17(11):1785

Hanna M, Jaqua E, Nguyen V, Clay J (2022) B vitamins: functions and uses in medicine. Perm J 26(2):89–97

Harrison F, Bowman G, Polidori M (2014) Ascorbic acid and the brain: rationale for the use against cognitive decline. Nutrients 6:1752–1781

He X, Kim M, Kim S, Yi S, Rhee Y, Kim T, Lee E, Park C, Dixit S, Harrison F, Lee S (2015) Vitamin C facilitates dopamine neuron differentiation in fetal midbrain through TET1- and JMJD3-dependent epigenetic control manner. Stem Cells 33(4):1320–1332

Hersi M, Irvine B, Gupta P, Gomes J, Birkett N, Krewski D (2017) Risk factors associated with the onset and progression of Alzheimer's disease: a systematic review of the evidence. Neurotoxicology 61:143–187

Holick MF (2007) Vitamin D deficiency. N Engl J Med 357:266–281

Huang X, Zhang H, Zhen J, Dong S, Guo Y, Van Halm-Lutterodt N, Yuan L (2018) Diminished circulating retinol and elevated α-TOH/retinol ratio predict an increased risk of cognitive decline in aging Chinese adults, especially in subjects with ApoE2 or ApoE4 genotype. Aging (Albany NY) 10:4066–4083

Huff T, Sant D, Camarena V, Van Booven D, Andrade N, Mustafi S, Monje P, Wang G (2021) Vitamin C regulates Schwann cell myelination by promoting DNA demethylation of pro-myelinating genes. J Neurochem 157(6):1759–1773

Jiang Q (2014) Natural forms of vitamin E: metabolism, antioxidant, and anti-inflammatory activities and their role in disease prevention and therapy. Free Radic Biol Med 72:76–90

Kado D, Karlamangla A, Huang M, Troen A, Rowe J, Selhub J, Seeman T (2005) Homocysteine versus the vitamins folate, B6, and B12 as predictors of cognitive function and decline in older high-functioning adults: MacArthur studies of successful aging. Am J Med 118:161–167

Kennedy D, Haskell C (2011) Vitamins and cognition: what is the evidence? Drugs 71:1957–1971

Kocot J, Luchowska-Kocot D, Kiełczykowska M, Musik I, Kurzepa J (2017) Does vitamin C influence neurodegenerative diseases and psychiatric disorders? Nutrients 9(7):659

Kook S, Lee K, Kim Y, Cha M, Kang S, Baik S, Lee H, Park R, Mook-Jung I (2014) High-dose of vitamin C supplementation reduces amyloid plaque burden and ameliorates pathological changes in the brain of 5XFAD mice. Cell Death Dis 5(2):e1083

Koryakina A, Aeberhard J, Kiefer S, Hamburger M, Kuenzi P (2009) Regulation of secretases by all-trans-retinoic acid. FEBS J 276:2645–2655

Lai R, Hsu C, Yu B, Lo Y, Hsu Y, Chen M, Juang J (2022) Vitamin D supplementation worsens Alzheimer's progression: animal model and human cohort studies. Aging Cell 21(8):e13670

Lane M, Bailey S (2005) Role of retinoid signalling in the adult brain. Prog Neurobiol 75:275–293

Lane C, Hardy J, Schott J (2018) Alzheimer's disease. Eur J Neurol 25(1):59–70

Lee C, Wan F (2000) Vitamin E supplementation improves cell-mediated immunity and oxidative stress of Asian men and women. J Nutr 130(12):2932–2937

Lee H-P, Casadesus G, Zhu X, Lee H-G, Perry G, Smith M, Gustaw-Rothenberg K, Lerner A (2009) All-trans retinoic acid as a novel therapeutic strategy for Alzheimer's disease. Expert Rev Neurother 9:1615–1621

Leotsinidis M, Alexopoulos A, Schinas V, Kardara M, Kondakis X (2000) Plasma retinol and tocopherol levels in Greek elderly population from an urban and a rural area: associations with the dietary habits. Eur J Epidemiol 16:1009–1016

Lerner AJ, Gustaw-Rothenberg K, Smyth S, Casadesus G (2012) Retinoids for treatment of Alzheimer's disease. Biofactors 38:84–89

Littlejohns T, Henley WE, Lang I, Annweiler C, Beauchet O, Chaves P, Fried L, Kestenbaum B, Kuller L, Langa K, Lopez O, Kos K, Soni M, Llewellyn D (2014) Vitamin D and the risk of dementia and Alzheimer disease. Neurology 83(10):920–928

Liu X, Wu P, Liu Y, Jiang Y, Wan M, Xiao X, Yang Q, Jiao B, Liao X, Wang J, Liu S, Zhang X, Shen L (2022) Association between serum vitamins and the risk of Alzheimer's disease in Chinese population. J Alzheimer's Dis 85(2):829–836

Llewellyn DJ, Lang IA, Langa KM, Muniz-Terrera G, Phillips CL, Cherubini A, Ferrucci L, Melzer D (2010) Vitamin D andrisk of cognitive decline in elderly persons. Arch Intern Med 170(13):1135–1141

Lopes da Silva S, Vellas B, Elemans S, Luchsinger J, Kamphuis P, Yaffe K, Sijben J, Groenendijk M, Stijnen T (2014) Plasma nutrient status of patients with Alzheimer's disease: systematic review and meta-analysis. Alzheimers Dement 10:485–502

Masoumi A, Goldenson B, Ghirmai S, Avagyan H, Zaghi J, Abel K, Zheng X, Espinosa-Jeffrey A, Mahanian M, Liu P, Hewison M, Mizwickie M, Cashman J, Fiala M (2009) 1alpha,25-dihydroxyvitamin D3 interacts with curcuminoids to stimulate amyloid-beta clearance by macrophages of Alzheimer's disease patients. J Alzheimer's Dis 17(3):703–717

Mecocci P, Polidori M (2012) Antioxidant clinical trials in mild cognitive impairment and Alzheimer's disease. Biochim Biophys Acta (BBA) Mol Basis Dis 1822:631–638

Meng H, Li Y, Zhang W, Zhao Y, Niu X, Guo J (2019) The relationship between cognitive impairment and homocysteine in a B12 and folate deficient population in China: a cross-sectional study. Medicine (Baltimore) 98(47):e17970

Mey J, McCaffery P (2004) Retinoic acid signaling in the nervous system of adult vertebrates. Neuroscientist 10:409–421

Mielech A, Puścion-Jakubik A, Markiewicz-Żukowska R, Socha K (2020) Vitamins in Alzheimer's disease-review of the latest reports. Nutrients 12(11):3458

Mingaud F, Mormede C, Etchamendy N, Mons N, Niedergang B, Wietrzych M, Pallet V, Jaffard R, Krezel W, Higueret P, Marighetto A (2008) Retinoid hyposignaling contributes to aging-related decline in hippocampal function in short-term/working memory organization and long-term declarative memory encoding in mice. J Neurosci 28:279–291

Mizwicki M, Menegaz D, Zhang J, Barrientos-Durán A, Tse S, Cashman J, Griffin P, Fiala M (2012) Genomic and nongenomic signaling induced by 1α,25(OH)2-vitamin D3 promotes the recovery of amyloid-β phagocytosis by Alzheimer's disease macrophages. J Alzheimer's Dis 29(1):51–62
Mohajeri M, Troesch B, Weber P (2015) Inadequate supply of vitamins and DHA in the elderly: implications for brain aging and Alzheimer-type dementia. Nutrition 31(2):261–275
Monacelli F, Acquarone E, Giannotti C, Borghi R, Nencioni A (2017) Vitamin C, aging and Alzheimer's disease. Nutrients 9(7):670
Moneim AE (2015) Oxidant/antioxidant imbalance and the risk of Alzheimer's disease. Curr Alzheimer Res 12:335–349
Morris M, Evans D, Bienias J, Tangney C, Bennett D, Aggarwal N, Wilson R, Scherr P (2002) Dietary intake of antioxidant nutrients and the risk of incident Alzheimer disease in a biracial community study. JAMA 287(24):3230–3237
Mullan K, Cardwell C, McGuinness B, Woodside J, McKay G (2018) Plasma antioxidant status in patients with Alzheimer's disease and cognitively intact elderly: a meta-analysis of case-control studies. J Alzheimer's Dis 62(1):305–317
Muñoz-Cabrera J, Sandoval-Hernández A, Niño A, Báez T, Bustos-Rangel A, Cardona-Gómez G, Múnera A, Arboleda G (2019) Bexarotene therapy ameliorates behavioral deficits and induces functional and molecular changes in very-old triple transgenic mice model of Alzheimer's disease. PLoS One 14:e0223578
Murakami K, Murata N, Ozawa Y, Kinoshita N, Irie K, Shirasawa T, Shimizu T (2011) Vitamin C restores behavioral deficits and amyloid-β oligomerization without affecting plaque formation in a mouse model of Alzheimer's disease. J Alzheimer's Dis 26(1):7–18
Naidu K (2003) Vitamin C in human health and disease is still a mystery? An overview. Nutr J 2:7
Olson C, Mello C (2010) Significance of vitamin a to brain function, behavior and learning. Mol Nutr Food Res 54:489–495
Ono K, Yamada M (2012) Vitamin a and Alzheimer's disease. Geriatr Gerontol Int 12:180–188
Oulhaj A, Jernerén F, Refsum H, Smith AD, de Jager CA (2016) Omega-3 fatty acid status enhances the prevention of cognitive decline by B vitamins in mild cognitive impairment. J Alzheimers Dis 50(2):547–557
Padayatty S, Katz A, Wang Y, Eck P, Kwon O, Lee J, Chen S, Corpe C, Dutta A, Dutta S et al (2003) Vitamin C as an antioxidant: evaluation of its role in disease prevention. J Am Coll Nutr 22:18–35
Polidori M, Nelles G (2014) Antioxidant clinical trials in mild cognitive impairment and Alzheimer's disease - challenges and perspectives. Curr Pharm Des 20(18):3083–3092
Quinn J, Suh J, Moore M, Kaye J, Frei B (2003) Antioxidants in Alzheimer's disease-vitamin c delivery to a demanding brain. J Alzheimer's Dis 5:309–313
Ransom J, Morgan P, McCaffery P, Stoney P (2014) The rhythm of retinoids in the brain. J Neurochem 129:366–376
Reddy P, Beal M (2008) Amyloid beta, mitochondrial dysfunction and synaptic damage: implications for cognitive decline in aging and Alzheimer's disease. Trends Mol Med 14:45–53
Rinaldi P, Polidori M, Metastasio A, Mariani E, Mattioli P, Cherubini A, Catani M, Cecchetti R, Senin U, Mecocci P (2003) Plasma antioxidants are similarly depleted in mild cognitive impairment and in Alzheimer's disease. Neurobiol Aging 24:915–919
Robinson N, Grabowski P, Rehman I (2018) Alzheimer's disease pathogenesis: is there a role for folate? Mech Ageing Dev 174:86–94
Rota C, Rimbach G, Minihane A, Stoecklin E, Barella L (2005) Dietary vitamin E modulates differential gene expression in the rat hippocampus: potential implications for its neuroprotective properties. Nutr Neurosci 8(1):21–29
Sano M, Ernesto C, Thomas R, Klauber M, Schafer K, Grundman M, Woodbury P, Growdon J, Cotman C, Pfeiffer E, Schneider L, Thal L (1997) A controlled trial of selegiline, alpha-tocopherol, or both as treatment for Alzheimer's disease. The Alzheimer's disease cooperative study. N Engl J Med 336(17):1216–1222

Selhub J, Jacques P, Wilson P, Rush D, Rosenberg I (1993) Vitamin status and intake as primary determinants of homocysteinemia in an elderly population. JAMA 270:2693–2698
Shahar S, Lee L, Rajab N, Lim C, Harun N, Noh M, Then S, Jamal R (2013) Association between vitamin A, vitamin E and apolipoprotein E status with mild cognitive impairment among elderly people in low-cost residential areas. Nutr Neurosci 16:6–12
Shearer K, Stoney P, Morgan P, McCaffery P (2012) A vitamin for the brain. Trends Neurosci 35:733–741
Shen L, Ji H (2015) Associations between homocysteine, folic acid, vitamin B12 and Alzheimer's disease: insights from meta-analyses. J Alzheimer's Dis 46:777–790
Śliwińska S, Jeziorek M (2021) The role of nutrition in Alzheimer's disease. Rocz Panstw Zakl Hig 72(1):29–39
Sodhi R, Singh N (2014) Retinoids as potential targets for Alzheimer's disease. Pharmacol Biochem Behav 120:117–123
Soumare A, Elbaz A, Ducros V, Tavernier B, Alpérovitch A, Tzourio C (2006) Cross-sectional association between homocysteine and motor function in the elderly. Neurology 67:985–990
Stoney PN, McCaffery P (2016) A vitamin on the mind: new discoveries on control of the brain by vitamin A. World Rev Nutr Diet 115:98–108
Succari M, Garric B, Ponteziere C, Miocque M, Cals MJ (1991) Influence of sex and age on vitamin a and E status. Age Ageing 20:413–416
Sun H, Kawaguchi R (2011) The membrane receptor for plasma retinol-binding protein, a new type of cell-surface receptor. Int Rev Cell Mol Biol 288:1–41
Tan H, Narasimhalu K, Ting S, Hameed S, Chang H, De Silva D, Chen C, Tan E (2023) B-vitamin supplementation on mitigating post-stroke cognition and neuropsychiatric sequelae: a randomized controlled trial. Int J Stroke 18(2):163–172
Tarawneh R, Holtzman DM (2012) The clinical problem of symptomatic Alzheimer disease and mild cognitive impairment. Cold Spring Harb Perspect Med 2:a006148
Tardy A, Pouteau E, Marquez D, Yilmaz C, Scholey A (2020) Vitamins and minerals for energy, fatigue and cognition: a narrative review of the biochemical and clinical evidence. Nutrients 12:228
Tippmann F, Hundt J, Schneider A, Endres K, Fahrenholz F (2009) Up-regulation of the alpha-secretase ADAM10 by retinoic acid receptors and acitretin. FASEB J 23:1643–1654
Tousi B (2015) The emerging role of bexarotene in the treatment of Alzheimer's disease: current evidence. Neuropsychiatr Dis Treat 11:311–315
Touyarot K, Bonhomme D, Roux P, Alfos S, Lafenêtre P, Richard E, Higueret P, Pallet V (2013) A mid-life vitamin a supplementation prevents age-related spatial memory deficits and hippocampal neurogenesis alterations through CRABP-I. PLoS One 8:e72101
Travica N, Ried K, Sali A, Scholey A, Hudson I, Pipingas A (2017) Vitamin C status and cognitive function: a systematic review. Nutrients 9(9):960
Tucker K, Qiao N, Scott T, Rosenberg I, Spiro A 3rd (2005) High homocysteine and low B vitamins predict cognitive decline in aging men: the veterans affairs normative aging study. Am J Clin Nutr 82:627–635
Van Dam F, Van Gool W (2009) Hyperhomocysteinemia and Alzheimer's disease: a systematic review. Arch Gerontol Geriatr 48:425–430
Wang Q, Zhao J, Chang H, Liu X, Zhu R (2021) Homocysteine and folic acid: risk factors for Alzheimer's disease-an updated meta-analysis. Front Aging Neurosci 13:665114
Wang Z, Zhu W, Xing Y, Jia J, Tang Y (2022) B vitamins and prevention of cognitive decline and incident dementia: a systematic review and meta-analysis. Nutr Rev 80:931–949
Winterbourn CC (2014) The challenges of using fluorescent probes to detect and quantify specific reactive oxygen species in living cells. Biochim Biophys Acta 1840(2):730–738
Yao Y, Chinnici C, Tang H, Trojanowski J, Lee V, Pratico D (2004) Brain inflammation and oxidative stress in a transgenic mouse model of Alzheimer-like brain amyloidosis. J Neuroinflammation 1:21

Zandi P, Anthony J, Khachaturian A, Stone S, Gustafson D, Tschanz J, Norton M, Welsh-Bohmer K, Breitner J (2004) Cache County study group. Reduced risk of Alzheimer disease in users of antioxidant vitamin supplements: the Cache County study. Arch Neurol 61(1):82–88

Zeng J, Chen L, Wang Z, Chen Q, Fan Z, Jiang H, Wu Y, Ren L, Chen J, Li T, Song W (2017) Marginal vitamin a deficiency facilitates Alzheimer's pathogenesis. Acta Neuropathol 133:967–982

Zhou J, Sun Y, Ji M, Li X, Wang Z (2022) Association of Vitamin B status with risk of dementia in cohort studies: a systematic review and meta-analysis. J Am Med Dir Assoc 23(11):1826.e21–1826.e35

Zielonka J, Hardy M, Michalski R et al (2017) Recent developments in the probes and assays for measurement of the activity of NADPH oxidases. Cell Biochem Biophys 75(3–4):335–349

Dietary Polyphenols and the Aging Brain: Mechanistic Insights into Alzheimer's Disease Prevention

20

Prachi Patel, Ritu Khosla, Meenakshi Garg, Poonam Khare, and Navkiran Kaur

Abstract

Alzheimer's disease (AD) is more than just memory loss: It is a complex and progressive brain disorder that touches millions of lives worldwide. Recent research has revealed how nature's own compounds polyphenols found in foods like berries, tea, and red wine may help protect the brain and slow the progress of this devastating condition. These powerful plant-derived molecules can cross the blood-brain barrier and work in multiple ways: they neutralize harmful free radicals, reduce inflammation, and even prevent the buildup of toxic proteins like amyloid-beta and tau that damage brain cells in AD. This chapter unpacks the incredible journey of polyphenols inside the body and how they target key disease mechanisms by promoting healthier protein processing, improving mitochondrial function, and protecting the brain's communication pathways. From the antioxidant strength of quercetin to the neuro-healing touch of curcumin and resveratrol, polyphenols offer a multitargeted, natural approach to support cogni-

P. Patel · N. Kaur (✉)
Amity Institute of Biotechnology, Amity University, Noida, India
e-mail: pjpatel@amity.edu; navkirank@amity.edu

R. Khosla
Department of Biomedical Science, Acharya Narendra Dev College, University of Delhi, New Delhi, Delhi, India
e-mail: ritukhosla@andc.du.ac.in

M. Garg
Department of Food Technology, Bhaskaracharya College of Applied Science, New Delhi, Delhi, India
e-mail: Meenakshi.garg@bcas.du.ac.in

P. Khare (✉)
Department of Allied Health Sciences, Evergreen Valley College, San Jose, CA, USA
e-mail: Poonam.khare@evc.edu

J. Malik et al. (eds.), *Diet and Alzheimer's Disease: Let Food be Our Medicine*, Nutritional Neurosciences, https://doi.org/10.1007/978-981-95-2736-6_20

tive function. Beyond Alzheimer's, we also explore their broader therapeutic impact in conditions like heart disease, diabetes, and cancer. In a world searching for safer, more holistic treatments, polyphenols shine as nature's multitasking healers offering hope not just for managing disease but also for improving overall brain health and quality of life.

Keywords

Natural brain protectors · Plant-based antioxidants · Polyphenols and memory health · Neuroprotective nutrition · Alzheimer's · Diet connection

20.1 Introduction

Polyphenols are a class of naturally occurring organic compounds marked by the presence of at least one aromatic benzoind ring also known as the phenyl ring (C_6H_6) and one or more hydroxyl groups (-OH). Apart from the hydroxyl groups and aromatic rings, they often contain ether ester linkages and carboxylic acid groups. Basis of classification of polyphenols generally depends on the number of phenolic rings present, position of different functional groups, and the overall chemical structure. They are moderately soluble in water and have molecular weight in the range of 500–4000 daltons.

Polyphenols are generally secondary metabolites of plants and are particularly prevalent in fruits like watermelon, apples, berries, grapes, and vegetables like onion, soybean, and cereals. They are also present in coffee, red wine, and juices. Polyphenols hold significant synthetic, medicinal, and industrial value due to their unique properties. They are well known antioxidants, preventing free radical formation by reacting with one-electron oxidant species. In fact, this single-electron oxidation process is a key mechanism behind their therapeutic effects. They can protect the human body from harmful free radicals by three ways: hydrogen atom transfer (HAT), single electron transfer (SET), and transition metal chelation (TMC). In case of HAT, polyphenols neutralize the free radicals by breaking their own O–H bonds to donate the hydrogen atoms. During SET, they act as electron donors so as to stabilize the free radicals. Finally, in TMC mechanism, polyphenols form stable complexes upon binding with metal ions, which in turn prevents harmful oxidative reactions. Together, these processes enable polyphenols to play a vital role in safeguarding the body against oxidative stress and damage (Lv et al. 2021; Lang et al. 2024).

Owing to their antioxidant and therapeutic properties, polyphenols serve as potential drug candidates for diseases such as AIDS, heart ailments, ulcers, bacterial infections, mutagenesis, and neural disorders. Additionally, they are involved in modulating gene expression, enzymatic activity, and many intracellular signaling pathways. Furthermore, polyphenolic derivatives demonstrate properties like metal chelation, autophagy induction, and neuroprotection, making them crucial for maintaining cellular homeostasis and overall human health (Albadrani et al. 2024).

Recently, it has been documented that supplementation of dietary polyphenols can prevent cognitive decline (Caruso et al. 2022b). In conjunction, they exert neuroprotective effects due to their ability to cross the blood-brain barrier. Their potent role in treating Parkinson and Alzheimer's is now well established. In this chapter, we will be discussing about the role of polyphenols in managing various disorders with particular emphasis on their mechanism of action in controlling Alzheimer's disease (AD) (Bertelli et al. 2021).

20.2 Classification of Polyphenols

Till date, around 8000 different types of polyphenols have been identified and documented in literature. These compounds can be broadly categorized into phenolic acids, flavonoids, stilbenes (like resveratrol and piceatannol), and lignans (such as sesamol, pinoresinol, sinol, and enterodiol). Other notable polyphenols include tannins (which can be hydrolysable, nonhydrolysable, or condensed), lignins, xanthones, chromones, and anthraquinones. Among these, dietary polyphenols, mainly derived from phenolic acids, catechins, flavones, and isoflavones, have been particularly reported to have various health benefits (Prabhu et al. 2021).

20.2.1 Phenolic Acids

Phenolic acids are plant-based compounds consisting of an aromatic ring with a carboxylic acid group (-COOH). They are known for their strong antioxidant properties, since its aromatic ring donates an electron, generating a hydrogen free radical that functions as a reducing agent. This process neutralizes harmful free radicals, ultimately playing a protective role. These compounds are primarily found in seeds, fruit skins, leafy vegetables, cereals, and their derivatives like beer and also in olive oil, tea, and coffee. They have applications in medicine as well as cosmetic and food industry. Phenolic acids can be mainly classified into hydroxybenzoic acids (C_6-C_1) and hydroxycinnamic acids (C_6-C_3), as shown in Fig. 20.1. Hydroxybenzoic acids, which are derived from benzoic acid, include salicylic acid, vanillic acid, protocatechuic acid, gallic acid, and ellagic acid. They usually exist in bound form and are present in olive products with antioxidant and cardioprotective effects. Hydroxycinnamic acids, on the other hand, are derivatives of cinnamic acid and essentially include ferulic acid, caffeic acid, coumaric acid, sinapinic acid, and cinnamic acid. Easily digestible and rich in phytochemicals, phenolic acids provide numerous health benefits, including anti-inflammatory, neuroprotective, anticancer, antidiabetic, and food-preserving properties (Sehrawat et al. 2022).

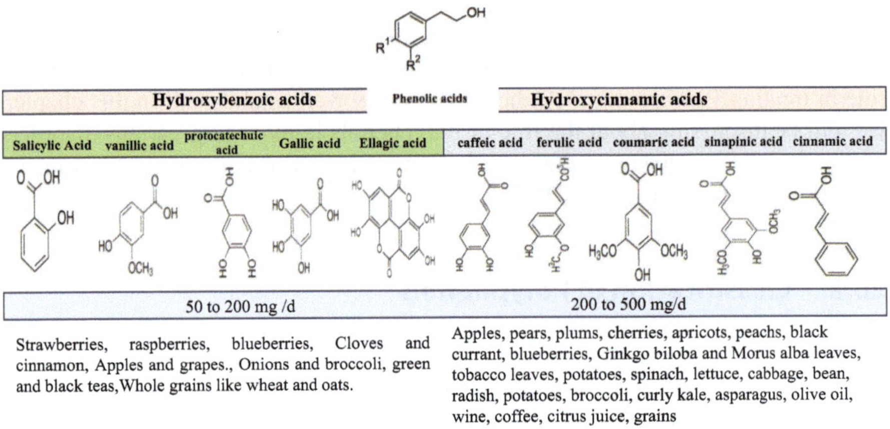

Fig. 20.1 Classification and dietary sources of phenolic acids

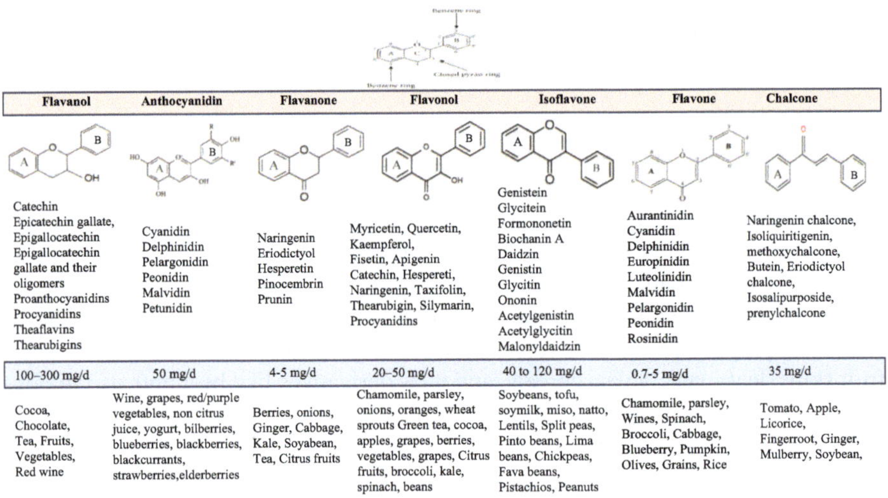

Fig. 20.2 Structural classification of flavonoids and their food sources

20.2.2 Flavonoids

Flavonoids, a major class of plant-based polyphenolic compounds, consist of 15 carbon atoms arranged in two aromatic rings linked through a three-carbon chain. Certain flavonoids help preserve vitamin C, which is why they are sometimes referred to as vitamin C2. These compounds are synthesized through the phenylpropanoid pathway, a process that not only produces flavonoids but also generates a diverse range of other phenolic compounds, as illustrated in Fig. 20.2. They are categorized based on the attachment and oxidation of their rings into flavones (e.g., chrysin, apigenin), flavonols (e.g., quercetin, kaempferol), isoflavones (e.g., daidzein, glycitein), flavan-3-ols (e.g., catechin, epicatechin), flavanones (e.g., hesperetin, naringenin), and anthocyanidins (e.g., delphinidin, cyanidin). Flavonoids are commonly found in foods such as berries, onions, tea, grapes, apples, and cocoa. Flavonoids play several crucial roles in plants, including protecting them from harmful UV radiation, preventing parasite attacks, regulating certain enzymatic activities, and acting as signaling molecules (Das et al. 2019). Apart from their role in plants, flavonoids offer substantial health benefits to humans including antioxidant, neuroprotective, antithrombogenic, and cell-signaling properties, making them valuable for our overall well-being (Busch et al. 2015; Panche et al. 2016).

20.2.3 Other Polyphenols

Other than flavonoids, other polyphenols like stilbenes, lignans, tannins, and lignins also offer various therapeutic and industrial applications as shown in Fig. 20.3 (Ciupei et al. 2024). Lignins are complex organic compounds found in plant cell walls, providing structural strength and rigidity. Common dietary sources include flaxseeds, tomatoes, peaches, apples, and certain berries. Silymarin, a flavonolignan extracted from milk thistle seeds, is known for its antioxidant properties and has been used traditionally to treat gallbladder- and liver-related disorders. Some reports also suggest that they may help in managing type 2 diabetes and cholesterol levels, but potential side effects like nausea and bloating might warrant medical supervision. Stilbenes, a group of phenol-derived metabolites, are known for their biological activity, but their exact health benefits and bioavailability are still under study. Resveratrol, a well-known stilbene, is found in red wine, grapes, and berries. Though its absorption in the body is limited, it has shown potential in protecting against cardiovascular diseases, lowering blood pressure, and preventing skin

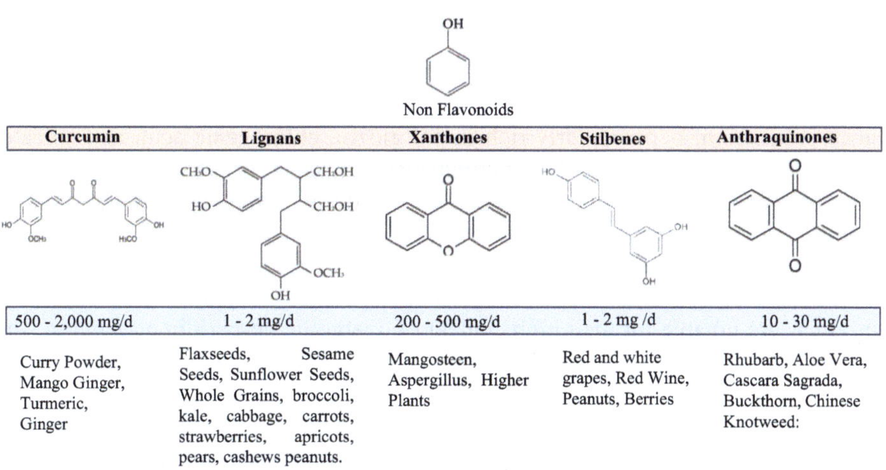

Fig. 20.3 Nonflavonoid polyphenols and their common dietary sources

cancer. Most studies suggest that resveratrol supplements may be more effective than obtaining it through diet, but further research is needed.

20.3 Sources of Polyphenols

Polyphenols are ubiquitously distributed across a wide array of botanical sources, encompassing fruits, vegetables, seeds, tea, and red wine, where they significantly contribute to the organoleptic (sensory) and nutritional attributes of plant-derived foods, such as astringency, pigmentation, and aroma (Bertelli et al. 2021). Certain food matrices exhibit a predominance of specific polyphenol subclasses. For instance, sesame oil is particularly enriched with lignans, predominantly sesamin and sesamolin, which constitute the principal polyphenolic constituents in sesame seeds (Reshma et al. 2010). In contrast, other food sources demonstrate a more diverse polyphenolic profile. Blueberries, for example, are notable for their high concentrations of phenolic acids, anthocyanins, and flavonoids, while red wine presents a complex polyphenolic matrix, containing stilbenes (notably resveratrol), phenolic acids, and flavonoids (Xu et al. 2021; Garcia-Viñola et al. 2024). Similarly, whole-grain cereals are characterized by the presence of various phenolic acids, including p-hydroxybenzoic, protocatechuic, vanillic, gallic, syringic, caffeic, p-coumaric, ferulic, and sinapic acids (Drawbridge et al. 2021).

The polyphenol concentration in plant-derived foods is highly variable, influenced by a multitude of agronomic and environmental factors such as cultivar genotype, climatic conditions, maturity at harvest, and postharvest processing techniques. For instance, the lignan content in sesame oil can vary significantly, typically accounting for approximately 1–2% of the oil's total composition, while cranberry fruits exhibit considerably higher phenolic diversity, containing 392.37 mg/100 g of total phenolics and 3.60 mg/100 g of total anthocyanins (Reshma et al. 2010; Pap et al. 2021).

By-products of fruits and vegetables also represent an abundant, yet underutilized source of polyphenols, offering substantial potential for valorization in the context of sustainable food processing. For instance, apple pomace, a by-product of juice production, is particularly rich in phloridzin, chlorogenic acid, and epicatechin, compounds known for their antidiabetic, anticancer, neuroprotective, and cardioprotective effects (Szabo et al. 2022). Similarly, cereal by-products contain significant concentrations of flavonoids (e.g., cyanidin 3-glucoside) and phenolic acids (e.g., ferulic acid), which have demonstrated protective effects against noncommunicable diseases (Szabo et al. 2022). In addition tomato and citrus by-products are notable for their high content of quercetin, hesperidin naringenin, and rutin (Drawbridge et al. 2021; Mahato et al. 2021). These bioactive compounds can be efficiently extracted from processing residues and incorporated into processed foods, thereby enhancing their nutritional and functional properties. For instance, the inclusion of pomegranate seed powder and grape seed proanthocyanidins in bakery products has been shown to significantly increase the punicic acid and proanthocyanidin content, respectively, thereby enriching their antioxidant capacity and health-promoting potential (Girard et al. 2019; Pamisetty et al. 2020). Additionally, extracts from *Caralluma fimbriata*, which are particularly rich in quercetin, have been incorporated into potato French fries to enhance their nutritional profile while also reducing lipid oxidation and acrylamide formation. This approach not only improves the nutritional value of processed foods but also enhances their textural and sensorial properties, including color stability, textural integrity, and oxidative stability, thereby contributing to a superior overall sensory experience (Kalita et al. 2013; Pamisetty et al. 2020).

While the incorporation of polyphenol-rich extracts offers substantial health benefits, it is important to consider that conventional food processing methods often lead to significant polyphenol losses due to thermal degradation, prolonged processing times, and mechanical stress. For example, thermal denaturation during pasteurization can cleave glycosidic bonds or induce polymerization of flavonoids, thereby reducing their antioxidant capacity (Yuste et al. 2020). Likewise, anthocyanins are particularly susceptible to degradation under acidic or thermally dynamic conditions due to their labile flavylium cationic structure, which readily degrades or polymerizes under such stresses (Oancea 2021).

In contrast, certain processing techniques such as fermentation have demonstrated the potential to enhance polyphenol content by liberating bound polyphenols and transforming their chemical structures, thus increasing their bioactivity and antioxidant potential. For instance, fresh cruciferous vegetables typically contain polyphenols in a glycosylated or acylated form, which can limit their bioavailability. Fermentation not only releases these bound forms but also modifies their structure, resulting in a significant increase in total polyphenol content up to 83% in the case of fermented broccoli puree compared to raw broccoli (Cai et al. 2019). This transformation is primarily driven by the enzymatic hydrolysis of complex polyphenol conjugates and microbial metabolism, which releases bioactive aglycones from their glycosylated precursors.

Additionally, the encapsulation of polyphenols in biopolymeric carriers has emerged as a promising strategy to mitigate the degradation typically associated with thermal and oxidative stress during food processing. For example, yogurt fortified with essential oils from lavender, basil, mint, and fennel, encapsulated in sodium alginate, has been shown to preserve the functional properties of these sensitive bioactives during processing and storage, thereby maintaining their antioxidant potential (Tiţa et al. 2020).

20.4 Role of Polyphenols in Disease Management

Polyphenols have been extensively exploited in the management of several chronic disorders: refer to Fig. 20.4.

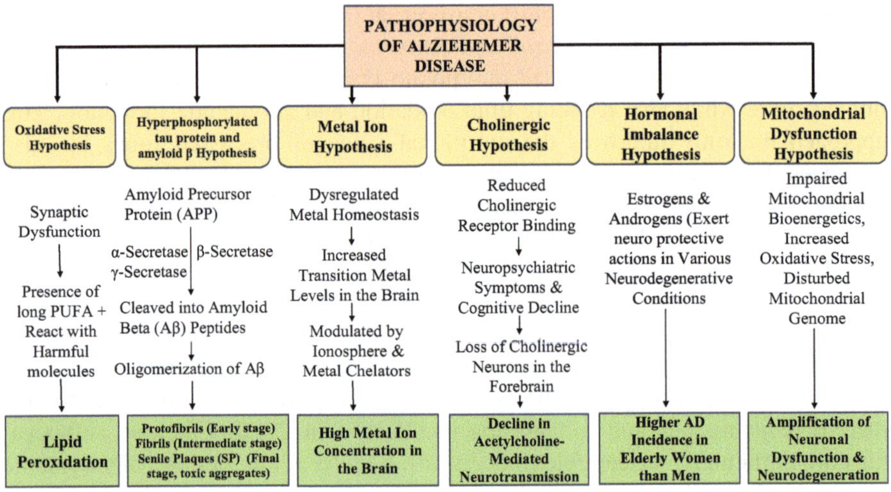

Fig. 20.4 Polyphenol actions in disease management

20.4.1 Polyphenols in Cardiovascular Diseases

Polyphenols have shown cardioprotective effects by improving endothelial function, reducing blood pressure, and preventing the formation of atherosclerotic plaques (Cheng et al. 2017; Yamagata 2019). Catechins from green tea and resveratrol from red wine help in vasodilation and improved blood flow through enhanced nitric oxide (NO) production. Polyphenols also inhibit the oxidation of low-density lipoproteins (LDLs), a key factor low-density lipoprotein (LDL) oxidation, which is a key factor in the development of atherosclerosis.

Polyphenols reduce platelet aggregation and prevent foam cell formation, thereby reducing the risk of plaque rupture and cardiovascular events (Santhakumar et al. 2014). Flavonoids such as quercetin and epicatechin have been shown to improve cardiac function and reduce myocardial damage following ischemic injury. Their anti-inflammatory properties further contribute to cardiovascular protection by reducing the levels of potent pro-inflammatory cytokines like tumor necrosis factor-alpha and interleukin-6 (IL-6) pro-inflammatory cytokines.

20.4.2 Polyphenols in Diabetes

Polyphenols improve glucose metabolism and insulin sensitivity through multiple mechanisms. They activate AMP-activated kinase (AMPK), which elevates the uptake of glucose and increases oxidation of fatty acids in muscle cells. Flavonoids such as quercetin and epicatechin inhibit alpha-glucosidase and alpha-amylase enzymes, thereby reducing postprandial hyperglycemia.

Polyphenols also improve pancreatic beta-cell function and increase the secretion of insulin. Resveratrol has been shown to enhance glucose transporter type 4

(GLUT4) translocation, promoting glucose uptake in muscle and adipose tissue (Yonamine et al. 2017). The antioxidant properties of polyphenols protect pancreatic beta cells from oxidative stress and improve overall glycemic control.

20.4.3 Polyphenols in Cancer

Polyphenols exhibit anticancer effects in general through the inhibition of cell proliferation, induction of apoptosis and reduction of angiogenesis (Abbaszadeh et al. 2019). For instance, epigallocatechin gallate (EGCG) present in green tea and resveratrol obtained from grapes inhibit tumor growth by downregulating cyclin D1 and activating tumor suppressor genes such as p53. Additionally, resveratrol and curcumin reduce the expression of vascular endothelial growth factor (VEGF), thereby restraining tumor angiogenesis. Further, polyphenols promote apoptosis through the activation of caspases and inhibition of antiapoptotic proteins such as Bcl-2. They also inhibit matrix metalloproteinases (MMPs), which are involved in cancer metastasis.

20.4.4 Polyphenols in Inflammatory and Autoimmune Diseases

Polyphenols exert anti-inflammatory effects by inhibiting the nuclear factor-kappa B (NF-κB) signaling pathway, which regulates the production of pro-inflammatory cytokines such as TNF-α, IL-6, and IL-1β (Yahfoufi et al. 2018). Flavonoids such as kaempferol and luteolin inhibit cyclooxygenase-2 (COX-2) and inducible nitric oxide synthase (iNOS) expression, reducing the production of inflammatory mediators. Polyphenols have shown benefits in managing inflammatory bowel disease (IBD) as well as autoimmune diseases like rheumatoid arthritis by modulating immune cell function and reducing oxidative stress. Their ability to regulate inflammatory responses and enhance tissue repair makes them effective in the treatment of chronic inflammatory conditions.

20.4.5 Polyphenols in Neurodegenerative Diseases

Polyphenols have shown neuroprotective effects by reducing oxidative stress, inflammation, and amyloid-beta (Aβ) aggregation in neurodegenerative diseases such as Alzheimer's and Parkinson's disease. Curcumin, catechins, and resveratrol improve synaptic plasticity and reduce tau hyperphosphorylation, which are key factors in cognitive decline (Freyssin et al. 2018). Polyphenols also enhance the expression of brain-derived neurotrophic factor (BDNF), which promotes neurogenesis and protects against neuronal damage. Their ability to modulate multiple signaling pathways involved in neuroinflammation, oxidative stress, and protein aggregation makes them promising therapeutic agents for neurodegenerative diseases.

20.5 What Is Alzheimer's Disease?

Alzheimer's disease (AD) is a progressive neurodegenerative disorder often marked by the presence of memory loss, behavioral changes, and cognitive decline (Sohrabi and Weinborn 2019; Zvěřová 2019). AD was first observed by Alois Alzheimer in 1906, who described it at a conference in Tübingen, Germany, as a "peculiar severe disease process of the cerebral cortex." It is one of the most prevalent causes of dementia, accounting for almost 70% cases worldwide. Dementia is a generic term describing a group of conditions that cause a decline in cognitive abilities, making execution of even simple daily routine tasks difficult to accomplish (Ayodele et al. 2021).

20.5.1 Pathophysiology of Alzheimer's

One of the major contributors of AD pathology is oxidative stress since excessive reactive oxygen species overwhelm antioxidant defence responses. Concomitantly, neuroinflammation also plays a critical role, with activated microglia releasing pro-inflammatory cytokines aggravating various AD associated pathologies as shown in Fig. 20.5. Peripheral immune cells further amplify the inflammatory response, exacerbating the process of neurodegeneration.

20.5.1.1 Accumulation of Amyloid-Beta (Aβ) Plaques and Neurofibrillary Tangles

The most striking feature of AD at the molecular level is the abnormal build-up of amyloid-beta (Aβ) plaques and neurofibrillary tangles between the neurons, which result due to disrupted protein processing and posttranslational modifications

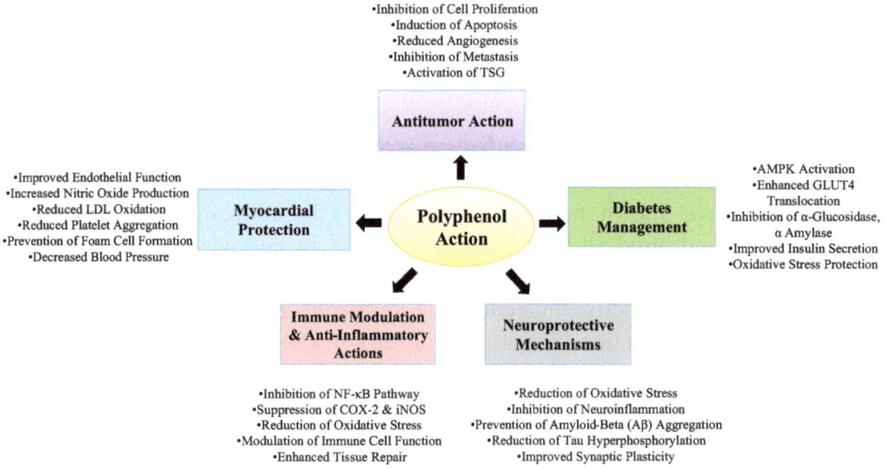

Fig. 20.5 Pathophysiology of Alzheimer's

(Ramesh et al. 2020). Under normal physiological conditions, the nonamyloidogenic pathway is initiated in the brain wherein α-secretase cleaves within the Aβ domain at the Lys16-Leu17 peptide bond. As a result, soluble amyloid precursor protein alpha (sAPPα) and an 83 amino acid long C-terminal fragment (C83) is formed. The C83 fragment is subsequently processed by γ-secretase, releasing a 24 or 26 residue long peptide called p3 into the extracellular region and the amyloid precursor protein intracellular domain (AICD) into the cytoplasm, thereby precluding Aβ formation (Crump et al. 2013).

However, during abnormal conditions the amyloidogenic pathway is activated wherein, Aβ peptides are formed from APP through the sequential two step cleavage by the enzymes β-secretase (BACE1) and γ-secretase. β-secretase cleaves the extracellular domain of APP forming soluble APPβ and a membrane-bound C-terminal fragment (C99), which is further processed by γ-secretase to release Aβ peptides of varying lengths, primarily Aβ40 and Aβ42. Aβ42, due to its high hydrophobicity and tendency to aggregate, forms oligomers and fibrils that accumulate into extracellular plaques, disrupting synaptic communication and inducing neuronal toxicity through the activation of microglial cells and astrocytes.

20.5.1.2 Abnormal Phosphorylation of Tau Protein

Tau proteins are basically microtubule associated proteins mainly found in neuronal cells. These proteins which normally stabilizes microtubules within neurons, becomes hyperphosphorylated due to dysregulation of certain kinases and phosphatases. Essentially, Tau functions as a microtubule-stabilizing protein and undergoes various posttranslational modifications, including phosphorylation for its activity. Several kinases including glycogen synthase kinase 3β (GSK-3β), protein kinase A, p38 MAPK, Erk1/2, and JNK1/3, are involved in regulating Tau phosphorylation. However, in AD, a disbalance between the activity of kinases and phosphatases leads to excessive phosphorylation of Tau (Rawat et al. 2022). Hyperphosphorylation of Tau detaches it from microtubules and causes it to aggregate within neurofibrillary tangles (NFTs). This detachment disrupts axonal transport, weakens the cytoskeleton, impairs mitochondrial function, induces oxidative stress, and ultimately leads to synaptic loss. Like Aβ, aggregated Tau proteins for β-sheet-rich amyloid fibrils, also referred to as paired helical filaments. A striking characteristic of Tau is its ability to spread between interconnected neurons, resembling the prion-like propagation seen in various proteinopathies (Zhang et al. 2021). The combination of synaptic dysfunction, chronic neuroinflammation, oxidative stress, and impaired intracellular transport culminates in progressive neuronal loss and cognitive decline.

Additionally, Aβ deposition can activate caspase-3 and calpain-1 enzymes that cleave at the C-terminal of Tau proteins, producing fragments that promote degeneration of neurites and death of neurons. The JNK pathway further exacerbates this process by activating caspases, which in turn enhance Tau cleavage.

While Tau toxicity is widely recognized as a major driver of AD-related neurodegeneration, the precise molecular mechanisms underlying Tau-induced neuroinflammation remain unclear. Recent studies, however, suggest that Tau can trigger a pro-inflammatory response in microglia by activation of p38 MAPK pathway.

Interestingly, blocking this pathway has been found to enhance microglial phagocytic activity and mitigate Tau toxicity (Piccialli et al. 2022).

Emerging research highlights additional underlying mechanisms in AD, including dysfunction of mitochondria, epigenetic modifications, vascular abnormalities and disturbances in the brain-gut axis. Mitochondrial dysfunction compromises the energy production and increases the level of oxidative stress, while vascular abnormalities cause abridged cerebral blood flow and metabolic disturbances. Epigenetic modifications affect gene expression patterns while altered gut-brain axis can lead to dysbiosis thereby increasing intestinal permeability and the consequent impairment of blood-brain barrier (Kamatham et al. 2024).

20.5.1.3 Inflammation, Oxidative Stress, and Mitochondrial Dysfunction

Aβ42 accumulation triggers a cascade of inflammatory responses, releasing cytokines and ROS, contributing to oxidative damage and mitochondrial dysfunction (Chatterjee 2016). Aβ peptides themselves contribute to oxidative damage by disrupting lipid membranes and increasing intracellular calcium, which further deteriorates mitochondrial function (Piccialli et al. 2022). ROS and reactive nitrogen species (RNS) both play a vital role in cellular functions, but their excessive accumulation leads to neuronal damage. Mitochondria, the primary source of ROS, become dysfunctional due to oxidative damage, impairing ATP production and further exacerbating ROS generation. Antioxidant enzymes like catalase, superoxide dismutase (SOD), and glutathione peroxidase (GPx) typically counteract ROS, but their activity declines in AD, leading to increased oxidative stress. Neurons, particularly in the hippocampus and amygdala, are highly susceptible to oxidative damage, resulting in protein degradation, lipid peroxidation, and DNA damage. Mitochondrial dysfunction in AD is well-documented, with evidence of increased mitochondrial DNA mutations and impaired DNA repair mechanisms. ROS-induced permeability transition pore (PTP) opening further amplifies oxidative stress, ultimately leading to apoptosis.

20.5.1.4 Cholinergic Injury

The cholinergic system of the brain consists of two main neuronal networks: the pons complex and the basal forebrain. Additionally, it includes motor nuclei that extend to the peripheral nervous system and the interneurons, that are primarily found in the striatum. The cholinergic injury hypothesis is a well-established mechanism underlying AD, highlighting the significant loss of cholinergic neurons of the basal forebrain as an early pathological feature. The progressive atrophy of the cholinergic basal forebrain occurs alongside normal age-related decline and is particularly evident in the early stages of AD. When the defects extend toward the central cholinergic system controlling learning and memory, they contribute to significant cognitive decline in AD patients (Christensen et al. 2010).

Another major factor behind AD is the abnormal increase in acetylcholinesterase (AChE) activity, which accelerates the breakdown of acetylcholine, a key neurotransmitter in cognitive processes. To counteract this, AchE inhibitors like donepezil, tacrine and galantamine have been developed to prevent acetylcholine hydrolysis, thereby boosting its levels in the brain and improving cholinergic transmission. On the other hand, many studies have also reported significant reductions in choline acetyltransferase (ChAT) activity (the enzyme responsible for synthesis of acetylcholine), in the hippocampus and neocortex. Together, the abnormal activity of both the enzymes leads to reduced levels of acetylcholine and severe cholinergic dysfunction (Pinto et al. 2011; Chen et al. 2022).

20.5.2 Risk Factors for AD

AD is a complex neurodegenerative disorder with both genetic and nongenetic contributing factors (Table). A small percentage of cases, primarily early-onset AD, result from inherited mutations in three genes—APP, presenilin-1 (PSEN1) and presenilin-2 (PSEN2)—linked to autosomal dominant familial AD. These mutations enhance APP cleavage by β- or γ-secretase, leading to the production and accumulation of Aβ42. However, these mutations are rare, occurring in less than 1% of the AD cases. Majority of AD cases fall under the category of late-onset AD (LOAD), which is sporadic and influenced by multiple risk factors. Among the genetic contributors, apolipoprotein E (APOE) allele on chromosome 19 is the most significant, with a dose-dependent impact on disease risk and age of onset. One or two copies of APOE-4, increase the likelihood of developing LOAD by 2.2–4.4 times and 5.1–17.9 times, respectively. APOE-4 appears to contribute by enhancing Aβ aggregation or impairing its clearance rather than increasing its production. Conversely, the APOE-2 isoform has been shown to have a protective effect with decreased risk of AD (Sims et al. 2020).

Additional to the genetic factors, lifestyle factors such as smoking, poor diet, sedentary lifestyle, diabetes, and hypertension also contribute to the risk of developing AD. Inflammation, driven by gut microbiota (GM) alterations, has also emerged as a significant factor in AD pathogenesis. Dysbiosis that can be caused due to excessive intake of processed food, elevated stress and anxiety levels or even long-term consumption of antibiotics, cause oxidative stress and systemic inflammation, activating microglia and compromising the blood-brain barrier (BBB). Metabolites like trimethylamine N-oxide (TMAO), produced by gut bacteria, have been linked to increased cytokine expression and neuronal damage. Interestingly, studies on transgenic AD mice have shown that fecal microbiota transplantation can reduce Aβ accumulation, tau pathology, and cognitive impairment, highlighting the gut-brain connection in AD (Shabbir et al. 2021).

Table: Genetic and lifestyle risk factors for Alzheimer's disease		
Risk factor type	Specific factors	Contribution to Alzheimer's
Genetic factors	Mutations in APP, PSEN1, PSEN2; APOE ε4 allele	Increased aβ aggregation, impaired protein clearance
Lifestyle factors	Smoking, poor diet, sedentary lifestyle	Increased oxidative stress and inflammation
Metabolic factors	Diabetes, hypertension	Disrupted vascular function and increased neuroinflammation
Age	Advanced age	Progressive synaptic loss and neuronal death
Environmental factors	Pollution, head trauma	Increased oxidative stress and inflammation

20.6 Role of Polyphenols in Management of AD

Polyphenols exert their neuroprotective effects in Alzheimer's disease through multiple interconnected mechanisms. As illustrated in Fig. 20.6, these natural compounds help regulate key pathological features of the disease, including oxidative stress, amyloid-beta toxicity, tau hyperphosphorylation, neuroinflammation, and cholinergic dysfunction.

20.6.1 Regulatory Role of Polyphenols in Modulating Aβ Production

Polyphenols can influence Aβ production by promoting nonamyloidogenic APP processing or inhibiting amyloidogenic processing. This modulation occurs through two primary mechanisms:

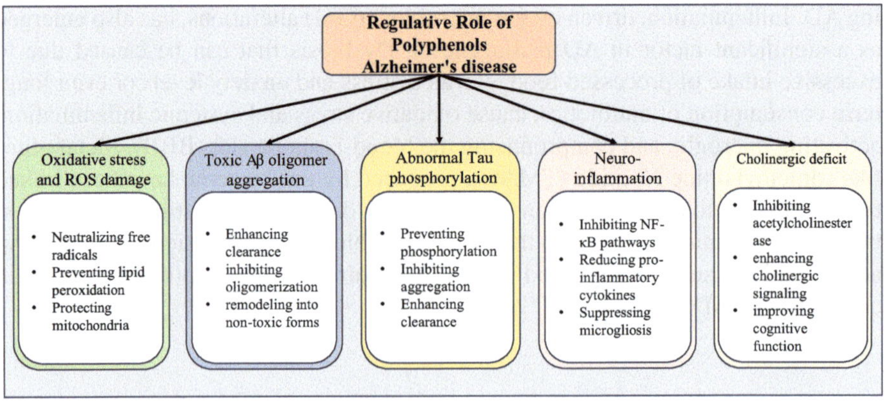

Fig. 20.6 Regulative role of polyphenols in Alzheimer's disease

20.6.1.1 Enhancement of α-Secretase Activity

Polyphenols enhance α-secretase activity and promote the nonamyloidogenic pathway, reducing Aβ production. Epicatechin, epigallocatechin, and epigallocatechin-3-gallate (EGCG) are flavonoids that possess antioxidant, anti-inflammatory, anticarcinogenic, and neuroprotective effects. They facilitate α-secretase-mediated cleavage, thereby favoring nonamyloidogenic processing and lowering Aβ levels in neuronal cell models. Similarly, curcumin boosts α-secretase activity, leading to increased sAPPα production and decreased amyloidosis. Flavones, including apigenin, nobiletin, and baicalein, have been shown to significantly reduce soluble and insoluble Aβ deposits in Alzheimer's disease mouse models. Quercetin, a dietary flavonol widely found in plants, fruits,and vegetables, also effectively modulates Aβ content in the brain, further supporting the protective role of polyphenols (Kostomoiri et al. 2013).

20.6.1.2 Inhibition of β-Secretase Activity

The β-secretase enzyme (BACE), also known as β-site amyloid precursor protein-cleaving enzyme, plays a crucial role in regulating Aβ production, making it a key target for Alzheimer's disease (AD) treatment. In addition to promoting α-secretase activity, polyphenols inhibit BACE activity, thereby preventing amyloidogenic processing. Both curcumin and EGCG have been identified as BACE inhibitors. Ferulic acid (FA), a BACE modulator, enhances the inhibitory effects of EGCG in APP/PS1 AD mice, leading to decreased amyloidosis and improved cognitive function. Natural compounds, such as curcuminoids found in turmeric, have been shown to inhibit BACE activity as well (Zheng et al. 2019).

20.6.2 Regulatory Role of Polyphenols on Toxic Aβ Oligomerization

Another key mechanism of action of polyphenols is targeting the toxic aggregation of amyloid-beta (Aβ) peptides, which play a central role in AD pathogenesis. Three major strategies have been identified through which polyphenols exert their beneficial effects in this context: (1) enhancing Aβ monomer clearance, (2) inhibiting Aβ oligomerization, and (3) remodeling toxic Aβ oligomers into nontoxic forms (El Gaamouch et al. 2022).

20.6.2.1 Enhancement of Aβ Monomer Clearance

Resveratrol (RES), a well-known polyphenol found in grapes and red wine, has been demonstrated to facilitate the clearance of Aβ monomers in neuronal cells through multiple proposed mechanisms. One hypothesis suggests that RES activates AMP-activated protein kinase (AMPK), which in turn inhibits the mammalian target of rapamycin (mTOR). This inhibition promotes autophagy, a cellular degradation pathway that leads to lysosomal clearance of Aβ. Additionally, RES has been shown to enhance proteasomal degradation of Aβ40 and Aβ42, further reducing their toxic accumulation. Another key mechanism involves the upregulation of insulin-degrading enzyme (IDE) in the hippocampus, which plays a crucial role in breaking down Aβ peptides. More recent findings indicate that RES sustains IDE activity by increasing the number of initial cleavage sites in Aβ42 monomers. This prevents the formation of disease-relevant Aβ fibrils and facilitates the degradation of Aβ fragments into smaller, less toxic pieces, thereby reducing the likelihood of aggregation. Collectively, these findings suggest that RES may serve as a potential therapeutic candidate for AD by promoting Aβ clearance and preventing its accumulation (Zheng et al. 2019).

20.6.2.2 Inhibition of Aβ Oligomerization

Curcumin (CUR) and rosmarinic acid (RA) have been identified as potent inhibitors of Aβ oligomerization. RA, when administered orally, has been shown to prevent Aβ-related neuropathology in transgenic AD mouse models. Biophysical studies, including atomic force microscopy, circular dichroism, nuclear magnetic resonance (NMR), and photo-induced cross-linking, reveal that RA directly inhibits the oligomerization of Aβ40 and Aβ42. Since Aβ oligomers are known to impair synaptic plasticity by inhibiting long-term potentiation (LTP) and enhancing long-term depression (LTD), RA's ability to block Aβ oligomerization has significant neuroprotective implications.

Curcumin, another polyphenol with strong antiaggregation properties, has been shown to inhibit Aβ42 oligomerization in a dose-dependent manner. Structural analysis suggests that CUR's inhibition of Aβ aggregation depends on its chemical structure, particularly the presence of two aromatic groups connected by a rigid linker. These structural features allow CUR to interfere with Aβ monomer interactions, preventing the formation of toxic oligomers. Interestingly, RA shares similar structural characteristics with CUR, supporting the hypothesis that specific molecular configurations contribute to the ability of polyphenols to prevent Aβ aggregation. By inhibiting the early stages of Aβ oligomerization, CUR and RA effectively reduce the formation of toxic Aβ species, thereby mitigating their neurotoxic effects (Elawad et al. 2024).

20.6.2.3 Remodeling of Toxic Aβ Oligomers into Nontoxic Forms

While some polyphenols inhibit Aβ oligomerization, others, such as RES and epigallocatechin gallate (EGCG), have been shown to remodel preformed toxic Aβ oligomers into structurally distinct, nontoxic assemblies. EGCG, a major polyphenol found in green tea, does not prevent Aβ oligomer formation but rather redirects the process toward the production of nontoxic, off-pathway oligomers that are unable to seed amyloid fibril formation. Biochemical and cellular studies have demonstrated that EGCG binds to large Aβ oligomers and fibrils, remodeling them into less toxic conformations. NMR studies suggest that EGCG interferes with the aromatic core region of Aβ40, altering its structural properties and reducing its capacity to form toxic aggregates. Similarly, RES has been found to attenuate the cytotoxicity of Aβ42 oligomers by remodeling them into nontoxic forms. Structural analyses indicate that RES disrupts the aggregation pathway, leading to the formation of amorphous, less harmful aggregates that do not elicit toxic effects on neuronal cells.

These findings suggest that while RES does not inhibit Aβ oligomerization directly, it plays a crucial role in detoxifying Aβ assemblies and preventing their neurotoxic impact (Rajasekhar et al. 2015; Gonçalves et al. 2024).

20.6.3 Regulatory Role of Polyphenols in Abnormal Tau Protein Phosphorylation

Polyphenols, particularly curcumin, EGCG, and other plant-derived compounds, play a significant role in mitigating Tau pathology in Alzheimer's disease (AD). EGCG, a green tea polyphenol, binds to specific phosphorylation sites on Tau, modifying its tridimensional structure, preventing aggregation, and enhancing its clearance by upregulating connexin expression in neurons. It interacts with full-length Tau at multiple residues, destabilizing its structure and dissolving Tau fibrils and oligomers, thereby maintaining Tau in an unfolded monomeric state. Similarly, rosmarinic acid prevents β-sheet assembly, and grape seed-derived polyphenol extract (GSPE) has been shown to interfere with Tau aggregation while significantly reducing AD-like Tau neuropathology in mouse models through ERK1/2 pathway modulation. Curcumin, widely studied for its neuroprotective effects, has demonstrated the ability to reduce soluble Tau dimers in human Tau (hTau) transgenic mice, restore the expression of heat-shock proteins (HSP90, HSP70, and HSP72), and improve spatial learning and memory. It inhibits Tau oligomerization by preventing misfolding, a critical step in aggregation, and decomposes already formed Tau filaments in vitro. Resveratrol enhances PP2A activity, promoting Tau dephosphorylation and inhibiting hyperphosphorylation at specific sites, including Ser396. Myricetin, rosmarinic acid, and epicatechin-3-gallate further contribute to neuroprotection by preventing the self-assembly of Tau into β-sheet-containing filaments. These findings highlight the potential of dietary polyphenols in modulating Tau pathology through multiple mechanisms, including aggregation inhibition, enhanced degradation, microtubule stabilization, and improved neuronal function, making them promising candidates for therapeutic intervention in AD (Freyssin et al. 2018).

20.6.4 Polyphenols Modulate Oxidative Stress, Mitochondrial Dysfunction, and Inflammation in Alzheimer's Disease

Polyphenols, known for their potent antioxidant properties, have shown significant therapeutic potential in mitigating oxidative damage. Flavonoids, such as quercetin, are highly effective in neutralizing free radicals, preventing lipid peroxidation, and chelating metal ions like iron, which otherwise contribute to neuronal toxicity. Resveratrol enhances superoxide dismutase activity, reducing ROS formation in vivo and alleviating oxidative stress-related damage. Curcumin stabilizes ROS by interacting with mitochondrial membranes, preventing excessive free radical production and lipid peroxidation. Additionally, nobiletin, administered chronically in AD mouse models, significantly reduces ROS levels and oxidative markers, offering neuroprotective benefits.

Polyphenol-rich foods, such as blueberries, have demonstrated protective effects against cognitive decline in individuals at high risk for dementia. Blueberry extracts enhance glutathione levels, thereby improving the brain's redox balance and protecting against amyloid-induced oxidative damage. Furthermore, catechins and epicatechins from green tea prevent ROS formation and lipid peroxidation in AD models. A single oral dose of epicatechin has been shown to effectively prevent Aβ-mediated oxidative stress in the hippocampus, reinforcing its neuroprotective potential (Vicente-Zurdo et al. 2024).

Polyphenols, particularly curcumin and resveratrol, have shown promise in protecting mitochondrial function. Curcumin enhances mitochondrial stability by preventing membrane depolarization and preserving oxidative phosphorylation, thereby reducing ROS formation. Similarly, resveratrol has been found to enhance mitochondrial biogenesis by activating SIRT1, a crucial regulator of cellular metabolism and neuroprotection (Caruso et al. 2022a).

In AD models, apigenin, a flavonoid, significantly reduces oxidative stress by enhancing the activities of antioxidant enzymes such as glutathione peroxidase and superoxide dismutase. These effects support neuronal survival by maintaining mitochondrial integrity and reducing oxidative damage. Curcumin, when administered intravenously to rodents, has been shown to prevent free radical propagation, further emphasizing its mitochondrial protective role in AD management (Yadav et al. 2022).

Several polyphenols have been observed to exhibit potent anti-inflammatory properties in both in vitro and in vivo studies. Resveratrol, one of the most well-researched polyphenols, attenuates Aβ-induced microglial activation by inhibiting key inflammatory signaling pathways such as TLR4/NF-κB and NLRP3/STAT. Additionally, resveratrol activates SIRT1, an epigenetic regulator that modulates inflammation and enhances neuronal survival. Resveratrol treatments have been associated with increased levels of neurotrophic factors, reduced neuroinflammation, and improved spatial memory in AD mouse models. Clinical studies in AD patients suggest that synthetic resveratrol reduces MMP9 levels in cerebrospinal fluid (CSF), suggesting improved blood-brain barrier integrity and reduced infiltration of inflammatory agents into the brain (Karunaweera et al. 2015).

Apart from resveratrol, curcumin effectively blocks NF-κB activity and its associated inflammatory cascades. It also inhibits Aβ-induced release of pro-inflammatory cytokines and chemokines, thereby reducing neuroinflammation. Other polyphenols, such as luteolin, quercetin, and fisetin, have demonstrated anti-inflammatory properties in AD models by suppressing microgliosis and astrogliosis, leading to reduced inflammation and neuronal damage (William Raja et al. 2023).

20.6.5 Regulatory Role of Polyphenols in the Context of Cholinergic Injury

Polyphenols are promising natural compounds for counteracting the cholinergic deficits seen in AD. These compounds work by inhibiting acetylcholinesterase (AchE), the enzyme responsible for breaking down ACh, thereby restoring cholinergic function and improving cognitive abilities. Curcumin, for example, has demonstrated the ability to regulate cholinergic signaling and enhance neuroprotection. Studies have shown that curcumin administration in cadmium-exposed neurotoxic rats significantly reduced AchE activity and gene expression in the cerebral cortex, confirming its inhibitory effects on AchE. Similarly, curcumin treatment has been effective in lowering AchE activity in the hippocampus and cortex of dementia-induced rats.

Apart from curcumin, other polyphenols have shown remarkable potential in mitigating cognitive decline in AD models. Apigenin, another polyphenol, was shown to activate ERK/CREB signaling, leading to enhanced learning and memory in. Additionally, nobiletin treatment improved cognitive function, while fisetin enhanced synaptic protein expression and activated ERK/MAPK signaling, resulting in cognitive benefits. Anthocyanins, which are abundant in berries and grapes, have been reported to reduce synaptic protein loss and enhance memory function. Grape-derived anthocyanins, in particular, have been found to counteract Aβ-induced deficits in long-term potentiation (LTP), a key mechanism of learning and memory. These findings suggest that incorporating polyphenol-rich foods into the diet may offer neuroprotective benefits and serve as a complementary approach to conventional AD treatments. Further research is needed to explore their full therapeutic potential and develop targeted interventions for neurodegenerative diseases (Aleksandrova et al. 2023).

20.7 Discussion

Polyphenols are widely distributed in the plant kingdom, occurring in various fruits, vegetables, seeds, cereals, and even their processing by-products. Key dietary sources include fruits like berries, grapes, and apples, which are rich in anthocyanins, flavonoids, and phenolic acids, while vegetables such as broccoli, onions, and spinach contain significant amounts of flavonols and phenolic acids. Additionally, beverages like green tea, red wine, and coffee are notable for their high polyphenol content. Valorizing fruit and vegetable by-products as polyphenol sources present a promising strategy for enhancing the nutritional and functional quality of processed foods while reducing waste. Techniques like fermentation and encapsulation can significantly improve polyphenol retention and bioactivity and have the ability to shield them from thermal and oxidative degradation, preserving their functional properties during processing and storage. Future research should focus on optimizing these approaches for scalability and cost-effectiveness to fully leverage the health benefits of these underutilized by-products.

References

Abbaszadeh H et al (2019) A review of molecular mechanisms involved in anticancer and antiangiogenic effects of natural polyphenolic compounds. Phytother Res 33(8):2002–2014

Albadrani HM et al (2024) Mechanistic insights into the potential role of dietary polyphenols and their nanoformulation in the management of Alzheimer's disease. Biomed Pharmacother 174:116376

Aleksandrova S et al (2023) Preventive and therapeutic effects of Punica granatum L. polyphenols in neurological conditions. Int J Mol Sci 24(3):1856
Ayodele T et al (2021) Early-onset Alzheimer's disease: what is missing in research? Curr Neurol Neurosci Rep 21:1–10
Bertelli A et al (2021) Polyphenols: from theory to practice. Foods 10:2595. s note: MDPI stays neutral with regard to jurisdictional claims in published …
Busch C et al (2015) Epigenetic activities of flavonoids in the prevention and treatment of cancer. Clin Epigenetics 7:1–18
Cai YX et al (2019) Fermentation for enhancing the bioconversion of glucoraphanin into sulforaphane and improve the functional attributes of broccoli puree. J Funct Foods 61:103461
Caruso G et al (2022a) Phenolic acids and prevention of cognitive decline: polyphenols with a neuroprotective role in cognitive disorders and Alzheimer's disease. Nutrients 14(4):819
Caruso G et al (2022b) Polyphenols and neuroprotection: therapeutic implications for cognitive decline. Pharmacol Ther 232:108013
Chatterjee S (2016) Oxidative stress, inflammation, and disease. In: Oxidative stress and biomaterials. Elsevier, pp 35–58
Chen Z-R et al (2022) Role of cholinergic signaling in Alzheimer's disease. Molecules 27(6):1816
Cheng Y-C et al (2017) Polyphenols and oxidative stress in atherosclerosis-related ischemic heart disease and stroke. Oxidative Med Cell Longev 2017(1):8526438
Christensen DZ et al (2010) Intracellular aβ triggers neuron loss in the cholinergic system of the APP/PS1KI mouse model of Alzheimer's disease. Neurobiol Aging 31(7):1153–1163
Ciupei D et al (2024) Polyphenols: from classification to therapeutic potential and bioavailability. Foods 13(24):4131
Crump CJ et al (2013) Development and mechanism of γ-secretase modulators for Alzheimer's disease. Biochemistry 52(19):3197–3216
Das AB et al (2019) Phenolic compounds as functional ingredients in beverages. In: Value-added ingredients and enrichments of beverages. Elsevier, pp 285–323
Drawbridge PC et al (2021) Bioaccessibility of phenolic acids in Canadian hulless barley varieties. Food Chem 358:129905
El Gaamouch F et al (2022) Benefits of dietary polyphenols in Alzheimer's disease. Front Aging Neurosci 14:1019942
Elawad MA et al (2024) Polyphenols and their biogenic Nano-formulations targeting BACE1 as anti-amyloid therapies; meeting the challenges of bioavailability, safety, and specificity for the treatment of Alzheimer's disease. Mol Nutr Food Res 68(24):2400525
Freyssin A et al (2018) Natural polyphenols effects on protein aggregates in Alzheimer's and Parkinson's prion-like diseases. Neural Regen Res 13(6):955–961
Garcia-Viñola V et al (2024) Simultaneous analysis of organic acids, glycerol and phenolic acids in wines using gas chromatography-mass spectrometry. Foods 13(2):186
Girard AL et al (2019) Effects of condensed vs hydrolysable tannins on gluten film strength and stability. Food Hydrocoll 89:36–43
Gonçalves PB et al (2024) Natural products targeting amyloid-β oligomer neurotoxicity in Alzheimer's disease. Eur J Med Chem 276:116684
Kalita D et al (2013) Role of polyphenols in acrylamide formation in the fried products of potato tubers with colored flesh. Food Res Int 54(1):753–759
Kamatham PT et al (2024) Pathogenesis, diagnostics, and therapeutics for Alzheimer's disease: breaking the memory barrier. Ageing Res Rev 101:102481
Karunaweera N et al (2015) Plant polyphenols as inhibitors of NF-κB induced cytokine production—a potential anti-inflammatory treatment for Alzheimer's disease? Front Mol Neurosci 1(8):16–24
Kostomoiri M et al (2013) Oleuropein, an anti-oxidant polyphenol constituent of olive promotes α-secretase cleavage of the amyloid precursor protein (AβPP). Cell Mol Neurobiol 33:147–154
Lang Y et al (2024) Classification and antioxidant assays of polyphenols: a review. J Future Foods 4(3):193–204

Lv Q-z et al (2021) Current state of knowledge on the antioxidant effects and mechanisms of action of polyphenolic compounds. Nat Prod Commun 16(7):1934578X211027745

Mahato N et al (2021) Biotransformation of citrus waste-I: production of biofuel and valuable compounds by fermentation. PRO 9(2):220

Oancea S (2021) A review of the current knowledge of thermal stability of anthocyanins and approaches to their stabilization to heat. Antioxidants 10(9):1337

Pamisetty A et al (2020) Rheological, physico-sensory and antioxidant properties of punicic acid rich wheat bread. J Food Sci Technol 57:253–262

Panche AN et al (2016) Flavonoids: an overview. J. Nutr. Sci. 5:e47

Pap N et al (2021) Berry polyphenols and human health: evidence of antioxidant, anti-inflammatory, microbiota modulation, and cell-protecting effects. Curr Opin Food Sci 42:167–186

Piccialli I et al (2022) Exploring the therapeutic potential of phytochemicals in Alzheimer's disease: focus on polyphenols and monoterpenes. Front Pharmacol 13:876614

Pinto T et al (2011) Revisiting the cholinergic hypothesis of behavioral and psychological symptoms in dementia of the Alzheimer's type. Ageing Res Rev 10(4):404–412

Prabhu S et al (2021) Classifications of polyphenols and their potential application in human health and diseases. Int J Physiol Nutr Phys Educ 6(1):293–301

Rajasekhar K et al (2015) Function and toxicity of amyloid beta and recent therapeutic interventions targeting amyloid beta in Alzheimer's disease. Chem Commun 51(70):13434–13450

Ramesh M et al (2020) Role of post-translational modifications in Alzheimer's disease. Chembiochem 21(8):1052–1079

Rawat P et al (2022) Phosphorylated tau in Alzheimer's disease and other tauopathies. Int J Mol Sci 23(21):12841

Reshma M et al (2010) Extraction, separation and characterisation of sesame oil lignan for nutraceutical applications. Food Chem 120(4):1041–1046

Santhakumar AB et al (2014) A review of the mechanisms and effectiveness of dietary polyphenols in reducing oxidative stress and thrombotic risk. J Hum Nutr Diet 27(1):1–21

Sehrawat R et al (2022) Phenolic acids-versatile natural moiety with numerous biological applications. Curr Top Med Chem 22(18):1472–1484

Shabbir U et al (2021) The potential role of polyphenols in oxidative stress and inflammation induced by gut microbiota in Alzheimer's disease. Antioxidants 10(9):1370

Sims R et al (2020) The multiplex model of the genetics of Alzheimer's disease. Nat Neurosci 23(3):311–322

Sohrabi HR, Weinborn M (2019) Cognitive impairments in Alzheimer's disease and other neurodegenerative diseases. In: Neurodegeneration and Alzheimer's disease: the role of diabetes, genetics, hormones, and lifestyle, pp 267–290

Szabo K et al (2022) Natural polyphenol recovery from apple-, cereal-, and tomato-processing by-products and related health-promoting properties. Molecules 27(22):7977

Tiţa O et al (2020) Use of yoghurt enhanced with volatile plant oils encapsulated in sodium alginate to increase the human body's immunity in the present fight against stress. Int J Environ Res Public Health 17(20):7588

Vicente-Zurdo D et al (2024) A comprehensive analytical review of polyphenols: evaluating neuroprotection in Alzheimer's disease. Int J Mol Sci 25(11):5906

William Raja TR et al (2023) Role of polyphenols in alleviating Alzheimer's disease: a review. Curr Med Chem 30(35):4032–4047

Xu L et al (2021) Anthocyanins, anthocyanin-rich berries, and cardiovascular risks: systematic review and meta-analysis of 44 randomized controlled trials and 15 prospective cohort studies. Front Nutr 8:747884

Yadav E et al (2022) Resveratrol: a potential therapeutic natural polyphenol for neurodegenerative diseases associated with mitochondrial dysfunction. Front Pharmacol 13:922232

Yahfoufi N et al (2018) The immunomodulatory and anti-inflammatory role of polyphenols. Nutrients 10(11):1618

Yamagata K (2019) Polyphenols regulate endothelial functions and reduce the risk of cardiovascular disease. Curr Pharm Des 25(22):2443–2458

Yonamine CY et al (2017) Resveratrol improves glycemic control in type 2 diabetic obese mice by regulating glucose transporter expression in skeletal muscle and liver. Molecules 22(7):1180

Yuste S et al (2020) Thermal and non-thermal processing of red-fleshed apple: how are (poly) phenol composition and bioavailability affected? Food Funct 11(12):10436–10447

Zhang H et al (2021) Interaction between aβ and tau in the pathogenesis of Alzheimer's disease. Int J Biol Sci 17(9):2181

Zheng Q et al (2019) Inhibition of the self-assembly of aβ and of tau by polyphenols: mechanistic studies. Molecules 24(12):2316

Zvěřová M (2019) Clinical aspects of Alzheimer's disease. Clin Biochem 72:3–6

MIX
Papier aus verantwortungsvollen Quellen
Paper from responsible sources
FSC® C105338

If you have any concerns about our products,
you can contact us on
ProductSafety@springernature.com

In case Publisher is established outside the EU,
the EU authorized representative is:
Springer Nature Customer Service Center GmbH
Europaplatz 3, 69115 Heidelberg, Germany

Printed by Libri Plureos GmbH
in Hamburg, Germany